WITHDRAWN

CONTEMPORARY HYDROGEOLOGY

THE GEORGE BURKE MAXEY MEMORIAL VOLUME

DEVELOPMENTS IN WATER SCIENCE, 12

advisory editor
VEN TE CHOW

Professor of Civil and Hydrosystems Engineering
Hydrosystems Laboratory
University of Illinois
Urbana, Ill., U.S.A.

OTHER TITLES IN THIS SERIES

1 G. BUGLIARELLO AND F. GUNTER
COMPUTER SYSTEMS AND WATER RESOURCES

2 H.L. GOLTERMAN
PHYSIOLOGICAL LIMNOLOGY

3 Y.Y. HAIMES, W.A. HALL AND H.T. FREEDMAN
MULTIOBJECTIVE OPTIMIZATION IN WATER RESOURCES SYSTEMS:
THE SURROGATE WORTH TRADE-OFF METHOD

4 J.J. FRIED
GROUNDWATER POLLUTION

5 N. RAJARATNAM
TURBULENT JETS

6 D. STEPHENSON
PIPELINE DESIGN FOR WATER ENGINEERS

7 V. HALEK AND J. SVEC
GROUNDWATER HYDRAULICS

8 J. BALEK
HYDROLOGY AND WATER RESOURCES IN TROPICAL AFRICA

9 T.A. McMAHON AND R.G. MEIN
RESERVOIR CAPACITY AND YIELD

10 K.W. HIPEL AND A.I. McLEOD
APPLIED BOX-JENKINS MODELLING FOR WATER RESOURCES ENGINEERS

11 W.H. GRAF AND C.H. MORTIMER (EDITORS)
HYDRODYNAMICS OF LAKES

CONTEMPORARY HYDROGEOLOGY

THE GEORGE BURKE MAXEY MEMORIAL VOLUME

EDITED BY

WILLIAM BACK

U.S. Geological Survey, Water Resources Division, National Center, Reston, VA 22092, U.S.A.

and

D.A. STEPHENSON

Woodward-Clyde Consultants, San Francisco, CA 94111, U.S.A.

Reprinted from *Journal of Hydrology,* Volume 43 (1979)

ELSEVIER SCIENTIFIC PUBLISHING COMPANY
Amsterdam–Oxford–New York   1979

ELSEVIER SCIENTIFIC PUBLISHING COMPANY
Jan van Galenstraat 335
P.O. Box 211, 1000 AE Amsterdam, The Netherlands

Distributors for the United States and Canada:

ELSEVIER/NORTH-HOLLAND INC.
52, Vanderbilt Avenue
New York, N.Y. 10017

ISBN 0-444-41848-2 (Vol. 12)
ISBN 0-444-41669-2 (Series)

© Elsevier Scientific Publishing Company, 1979

All rights reserved. No part of this publication may be reproduced, stored in a retrieval system or transmitted in any form or by any means, electronic, mechanical, photocopying, recording or otherwise, without the prior written permission of the publisher, Elsevier Scientific Publishing Company, P.O. Box 330, 1000 AH Amsterdam, The Netherlands

Printed in The Netherlands

Dedication

GEORGE BURKE MAXEY

(1917–1977)

This book is dedicated to the memory of George Burke Maxey and to Jane Maxey who remains a close friend of all the contributors.

Prefaces

Few people are fortunate enough to influence an aspect of science as much as George Burke Maxey influenced hydrogeology. The stimulation he provided through his teaching, writing, and lecturing was brought about by a combination of persistent encouragement, honest criticism, and friendly motivation. This book, originated with the Editors of the *Journal of Hydrology*, especially Ven Te Chow and George H. Davis, where Burke Maxey also served as Editor for several years. It is a collection of articles, many prepared especially for this volume, that represent current activities of some of his many close associates. We hope that these papers on a variety of hydrogeologic topics will serve as a tribute and demonstrate his widespread influence and inspiration. We thank each of the contributors who so willingly cooperated in this mutual effort and the many reviewers who provided insight and guidance.

August 1, 1979

WILLIAM BACK (Reston, Va.) and
D.A. STEPHENSON (San Francisco, Calif.)

It is a privilege to support the publication of this volume honoring George Burke Maxey. Burke was one of the founders of the Hydrogeology Division of the Geological Society of America, and served as our first chairman from 1959 through 1961. He was a strong and active supporter of the Division and of the science of hydrogeology. He influenced our science greatly, an influence felt by all hydrogeologists; those of us who knew and worked with Burke have been very fortunate.

KEROS CARTWRIGHT, Chairman
Hydrogeology Division
The Geological Society of America

Foreword

This volume, dedicated to the memory of George Burke Maxey as a scientific leader, has many points in common with the canonization of a religious leader. The volume includes chiefly the works of men who were Dr. Maxey's students and followers; thus it corresponds to the Acts of the Apostles, in an earlier testamentary volume that was first published for English readers in A.D. 1611. Professor Ven Te Chow was a contemporary of Dr. Maxey throughout most of the latter's teaching career which began in his early 30's. And Ven covers this academic activity more clearly and concisely than would have been possible if a four-man committee had been appointed, and each wrote a separate gospel.

For every leader there are formative years — the years before he becomes great or even shows promise of greatness — and these years are commonly slighted or even lost in history. Of all writers in this memorial volume, only we, the undersigned, knew him in these early, struggling years. We knew him as Burke, before he had bowed to the conformity implicit in government payroll designation as "George B." We knew him as a budding junior geologist and blooming assistant geologist. As to the science of hydrology, we baptized him in the faith.

The first fruit of Burke's B.S. in geology was a field assistantship for Smithsonian Institution in the Canadian Rockies, where he acquired a gourmet taste for Cambrian trilobites on the halfshell. Early in 1942 he was introduced to groundwater by the U.S. Geological Survey in Utah, under the tutelage of P. Eldon Dennis. In his first field study — in the aptly-named Flowell district of central Utah — he frequently looked beyond the limits of the groundwater reservoir and toward the headwaters of tributary streams, high in the Pavant Range. His disciples would doubtless take this as early recognition by Burke of the interrelations of snow on the mountains, the runoff therefrom, and the artesian water in the lowlands. But there were also trilobites in them thar mountains. His continuing taste for them is well confirmed in his Ph.D. thesis published in 1957. After about a year of groundwater apprenticeship in Utah, Burke's services were urgently required in Louisville, Kentucky, where he acquired a deep and abiding appreciation of arid spaces and their hydrology.

In 1944 George B. Maxey, assistant geologist of USGS, was transferred to Las Vegas, Nevada, to commence a study of the Las Vegas and Pahrump Ground Water Basins, a cooperative study between the State of Nevada and USGS. Following completion of these studies in southern Nevada he joined with Thomas Eakin of USGS to study groundwater conditions in several of the groundwater basins and valleys of eastern Nevada. In 1948 he left Nevada for graduate study at Princeton University. He spent the next 13 years teaching in the East (University of Connecticut) and Middle West (University of Illinois), with a two-year interim of groundwater studies in the Middle East (Libya).

In October 1961 while at the University of Illinois, Dr. Maxey was invited to present a paper at the Annual Nevada Water Conference. Professor Wendall Mordy, Director of the newly formed Desert Research Institute was favorably impressed and invited George to join the institute staff; and in 1962 he became a member of the institute and research professor of hydrology and geology at the University of Nevada. In mid-1967 he was appointed Head of the Center for Water Resources of the Desert Research Institute, a position he held until his death in February 1977.

HAROLD E. THOMAS (Woodside, Calif.) and
HUGH A. SHAMBERGER (Carson City, Nev.)

Foreword

BURKE'S ACADEMIC ACTIVITIES

In 1955 Dr. George Burke Maxey came to Urbana, Illinois, to assume the academic position of Professor of Geology in our University of Illinois and also the Head of Division of Ground-Water Geology and Geophysical Exploration of our Illinois Geological Survey, one of three scientific surveys housed on the campus and administered by the Department of Registration and Education, State of Illinois. However, it was not till 1957 at the General Assembly of the International Union of Geophysics and Geodesy in Toronto, Canada, that he became one of my most respected colleagues and beloved friends, known simply as "Burke" for the next two decades. As the University of Illinois has a large campus, where it is not easy to know all my colleagues on a personal basis, I regret that I could not have had three more years earlier of enjoying his friendship and learning his wisdom.

In 1960 I was developing an interdisciplinary course on groundwater in the Department of Civil Engineering, to be taught by volunteer faculty members from various departments on the campus. Burke was one of the most enthusiastic participants. He gave a series of three-week intensive lectures on hydrogeology which benefited not only the students but also other participating faculty members. The material used in his lectures was refreshing and challenging, since it came directly from his research at that time. A large portion of the material was later condensed in the *Handbook of Applied Hydrology* which I edited and was published by the McGraw-Hill Book Company in 1964.

In the beginning of scientific hydrology, the fields of groundwater and surface water were traditionally developed almost as separate professions. Almost exclusively groundwater hydrology was explored by geologists whereas surface water was explored by engineers. Burke was one of a few geologists who first recognized the importance of geology in surface-water hydrology and began to develop a dialogue with engineers on surface-water hydrology. On the basis of this insight, he wrote the following in the *Handbook of Applied Hydrology*:

> *"Geologic factors that act as controls on surface-water phenomena may be classified broadly, as (1) lithologic, depending primarily on the composition, texture, and sequence of rock types, and (2) structural, including chiefly faults and folds that interrupt the continuity or uniformity of occurrence of a rock type or sequence of rock types. Often structures such as beds and joints also materially influence movement of water and development of drainage patterns. These factors in combination with hydroclimatic processes control the development of soils and topography, which in turn profoundly affect distribution and movement of water.*

> *Together the geologic, pedologic, and topographic features influence regimes, stability, form and distribution of streams and other surface-water bodies. The magnitude of effects resulting from geologic controls alone or in combination with other controls ranges from very broad, such as the outlines, relief, and distribution of continents and ocean basins, to very small, affecting the character of the smallest creek."*

Burke was also one of a handful of hydrogeologists who first broadened their thoughts on hydrology and geohydrology in terms of units and systems, which inevitably enlightened and benefited many applied hydrogeologists as well as hydraulic and civil engineers. This can also be seen from his following writing in the *Handbook of Applied Hydrology*:

> *"Occurrence, movement, and storage of groundwater are influenced by the sequence, lithology, thickness, and structure of rock formations. Movement and storage capacity are chiefly controlled by permeability and porosity. An aquifer in a lithologic unit (or combination of such units) which has an appreciably greater transmissibility than adjacent units and which stores and transmits water commonly recoverable in economically usable quantities.*
>
> *The lithologic units of low permeability which bound the aquifer are commonly called confining beds, or aquicludes. Appreciable quantities of water may move through confining beds, and when this occurs, these beds have been referred to as aquitards. An aquifer or a combination of aquifers and confining beds that comprise a framework for a reasonably distinct hydraulic system may be considered as a geohydrologic unit."*

Burke left Illinois in 1962 to become Research Professor of Hydrology and Geology at the University of Nevada System, Reno. He also joined the then newly formed Desert Research Institute staff in Reno and was instrumental in leading that organization to its eminent position today. He eventually became the Director of the Nevada Center for Water Resources Research in 1967.

In the ensuing years, we had many occasions to meet and discuss academic as well as professional subjects, mainly in connection with the U.S. National Committee for the International Hydrological Decade, the Universities Council on Water Resources, the International Water Resources Association, and the Editorial Board of the *Journal of Hydrology*. In fact, the last time we met was in July 1977, in Honolulu, Hawaii, only about half a year before his death, where he was a featured speaker at the International Water Resources Educational Workshop sponsored by the Universities Council on Water Resources, The East—West Center, The University of Hawaii Water Resource Research Center, American Society for Engineering Education, and International Water Resources Association.

Before coming to Illinois, Burke had held many important positions including Geologist in the U.S. Geological Survey, serving in Utah, Kentucky and Nevada, Instructor of Geology at Princeton University and University of Connecticut, and Acting Chief of National Resources Division of Point 4 in Tripoli, Libya, for U.S. Technical Cooperation Administration (now U.S. Agency for International Development).

He was also appointed Distinguished Lecturer for the American Association of Petroleum Geologists; 1959—1960, Visiting Geoscientist for American Geological Institute, 1962, 1963 and 1972; Visiting Professor of Hydrogeology, Indiana University, 1965; and Visiting Scientist in Geophysics for the American Geophysical Union from 1965 to 1972.

According to some of his close friends and former students,

> "Burke was a man who had several families and in each he played numerous roles and for each he owned a different hat."

One of his families was his students who are now spread across the nations and the world.

> "To them he represented teacher, advisor, counselor, confidant, and well spring of inspiration and innovation. From each of us he expected and demanded the highest standards. But Professor Maxey expected nothing less than he gave! Example is the finest tool."

Burke was

> "a good man, highly intelligent, broad in knowledge and interest, a teacher in the highest sense, a geopolitician and a master scientist. But none of these attributes set him apart from many men".

> "What was the quality that made him so special: a great man who leaves a lasting contribution to so many."

> "He was a terrific teacher, he pushed all of his students to do their intellectual best as geologists and especially as groundwater hydrologists. He inspired us; supported us; pulled us through when the going got tough. It was an exciting atmosphere. There were always ideas, stimulating arguments, and meaningful projects to be worked on. Burke was there providing the leadership."

Burke's academic activities were many and had lasting significance. He was author and coauthor of more than 50 important publications and member of many National Academy of Sciences committees and other academic societies. His academic career was highlighted in 1972 when the Geological Society of America presented him with the Meinzer Award for his paper on *Hydrogeology of Desert Basins*.

I do not claim to know all about Burke's academic activities, as many of his other colleagues may know more, but I do sense that above all, Burke had a deep and abiding respect for knowledge and scholarship. Because of this respect for knowledge, one of the most important things in Burke's life was his students. Indeed his most lasting contribution in the academic field may well be the legacy of leaving hundreds of his former students in positions as water resources specialists throughout the U.S.A. and several foreign countries, in universities, state and federal agencies and in private enterprises.

Burke was one of the founding fathers of the *Journal of Hydrology* to which he had devoted much of his ideas and energy that had helped the Journal to become one of the most world-wide known publications on hydrology today. In order to acknowledge Burke's many accomplishments and contributions, the editors of the Journal have decided to sponsor this Memorial Volume in dedication to him. It was most gratifying that many of Burke's friends and colleagues have enthusiastically responded to this important project. To them and to the Coordinators Bill Back and Dave Stephenson in particular, the editors and publishers of the Journal wish to express deepest gratitude for they have now made this project a reality and have paid a truly meaningful homage to a great man in our scientific profession.

VEN TE CHOW (Urbana, Ill.)

Contents

Dedication VII

Prefaces

William Back (Reston, Va., U.S.A.) and D.A. Stephenson (Madison, Wis., U.S.A.) . . . VIII
Keros Cartwright, Chairman, Hydrogeology Division, Geological Society of America
    (Urbana, Ill., U.S.A.) VIII

Forewords

Harold E. Thomas (Woodside, Calif., U.S.A.) and Hugh A. Shamberger (Carson City,
    Nev., U.S.A.) IX
Burke's academic activities
    Ven Te Chow (Urbana, Ill., U.S.A.) XI

The Meinzer era of U.S. hydrogeology, 1910—1940
    George B. Maxey (†, Reno, Nev., U.S.A.) 1

Hydrologic Modeling and Groundwater Flow Systems

Hydrogeology of glacial-terrain lakes, with management and planning applications
    S.M. Born (Madison, Wis., U.S.A.), S.A. Smith (Tempe, Ariz., U.S.A.) and
    D.A. Stephenson (Madison, Wis., U.S.A.) 7
The role of groundwater in storm runoff
    M.G. Sklash (Windsor, Ont., Canada) and R.N. Farvolden (Waterloo, Ont.,
    Canada) 45
Hydraulic potential in Lake Michigan bottom sediments
    K. Cartwright, C.S. Hunt, G.M. Hughes and R.D. Brower (Urbana, Ill., U.S.A.) . . . 67
Unsteady streamflow modeling guidelines
    V.L. Gupta (Reno, Nev., U.S.A.), S.M. Afaq (Toronto, Ont., Canada),
    J.W. Fordham and J.M. Federici (Reno, Nev., U.S.A.) 79
Pore size distribution, suction and hysteresis in unsaturated groundwater flow
    C.M. Case and A. Welch (Reno, Nev., U.S.A.) 99
Contribution of groundwater modeling to planning
    J.E. Moore (Reston, Va., U.S.A.) 121
Application and analysis of a coupled surface and groundwater model
    A.B. Cunningham (Bozeman, Mont., U.S.A.) and P.J. Sinclair (Reno, Nev.,
    U.S.A.) 129
Progress in analytical groundwater modeling
    W.C. Walton (Twin Cities, Minn., U.S.A.) 149
Consideration of total energy loss in theory of flow to wells
    R.L. Cooley and A.B. Cunningham (Reno, Nev., U.S.A.) 161
Measurement of fluid velocity using temperature profiles: experimental verification
    K. Cartwright (Urbana, Ill., U.S.A.) 185
Utility of a computerized data base for hydrogeologic investigations, Las Vegas
Valley, Nevada
    R.F. Kaufmann and H.N. Friesen (Las Vegas, Nev., U.S.A.) 195
Regional carbonate flow systems in Nevada
    M.D. Mifflin and J.W. Hess (Las Vegas, Nev., U.S.A.) 217
Groundwater flow systems in the western phosphate field in Idaho
    D.R. Ralston and R.E. Williams (Moscow, Idaho, U.S.A.) 239

Geological Aspects of Hydrogeology

Cooling mechanisms and effects on mantle convection beneath Antarctica
    L.D. McGinnis (De Kalb, Ill., U.S.A.) 265
Major geochemical processes in the evolution of carbonate—aquifer systems
    B.B. Hanshaw and W. Back (Reston, Va., U.S.A.) 287
Effects of karst and geological structure on the circulation of water and permeability in carbonate aquifers
    V.T. Stringfield, J.R. Rapp and R.B. Anders (Reston, Va., U.S.A.) 313
Evaluation techniques of fractured-rock hydrology
    H. LeGrand (Raleigh, N.C., U.S.A.). 333
Secondary permeability as a possible factor in the origin of debris avalanches associated with heavy rainfall
    A.G. Everett (Rockville, Md., U.S.A.) 347

Hydrogeochemistry

Seasonal chemical and isotopic variations of soil CO_2 at Trout Creek, Ontario
    E.J. Reardon (Waterloo, Ont., Canada), G.B. Allison (Glen Osmond, S.A., Australia) and P. Fritz (Waterloo, Ont., Canada) 355
Arsenic species as an indicator of redox conditions in groundwater
    J.A. Cherry (Waterloo, Ont., Canada), A.U. Shaikh, D.E. Tallman (Fargo, N. Dak., U.S.A.) and R.V. Nicholson (Waterloo, Ont., Canada) 373
Modern marine sediments as a natural analog to the chemically stressed environment of a landfill
    M.J. Baedecker and W. Back (Reston, Va., U.S.A.). 393
Time-dependent sorption on geological materials
    P.R. Fenske (Reno, Nev., U.S.A.). 415
The volume-averaged mass-transport equation for chemical diagenetic models
    P.A. Domenico and V.V. Palciauskas (Urbana, Ill., U.S.A.) 427

Economic Hydrogeology

Problems of large-scale groundwater development
    S. Mandel (Jerusalem, Israel) 439
The impacts of coal strip mining on the hydrogeologic system of the Northern Great Plains: case study of potential impacts on the Northern Cheyenne Reservation
    W.W. Woessner (Las Vegas, Nev., U.S.A.), C.B. Andrews and T.J. Osborne (Lame Deer, Mont., U.S.A.). 445
Connector wells, a mechanism for water management in the Central Florida Phosphate District
    P.E. LaMoreaux (Tuscaloosa, Ala., U.S.A.). 469
Simulated changes in potentiometric levels resulting from groundwater development for phosphate mines, west-central Florida
    W.E. Wilson and J.M. Gerhart (Tampa, Fla., U.S.A.). 491
Depressurization of a multi-layered artesian system for water and grout control during deep mine-shaft development
    W.M. Greenslade (Phoenix, Ariz., U.S.A.) and G.W. Condrat (Salt Lake City, Utah, U.S.A.). 517
Geothermal well testing
    T.N. Narasimhan and P.A. Witherspoon (Berkeley, Calif., U.S.A.). 536

Epilogue

Groundwater: new directions — Where we've been and where we're going
    R.G. Kazmann (Baton Rouge, La., U.S.A.). 555

[2]

THE MEINZER ERA OF U.S. HYDROGEOLOGY, 1910—1940

GEORGE B. MAXEY (†)

Water Resources Center Desert Research Institute, University of Nevada System, Reno, Nev. (U.S.A.)

(Accepted for publication April 27, 1979)

ABSTRACT

Maxey, G.B., 1979. The Meinzer era of U.S. hydrogeology, 1910—1940. In: W. Back and D.A. Stephenson (Guest-Editors), Contemporary Hydrogeology — The George Burke Maxey Memorial Volume. J. Hydrol., 43: 1—6.

Acceleration of agricultural, industrial, and municipal development following the American Civil War resulted in unprecedented demands for knowledge of movement, occurrence, quality and availability of groundwater. By 1910, response to this demand resulted, in government, in establishment of specialized agencies. The agency most affecting development of hydrogeology was the Ground Water Division of the U.S. Geological Survey, especially after appointment of O.E. Meinzer as Chief in 1912. The Division established a system utilizing geologists and engineers (later other scientists) working as teams to assess groundwater resources. The U.S. Geological Survey developed cooperative programs allowing State Engineers and geologists to participate in resource studies, an effective way of encouraging interest in hydrogeological problems as well as establishing a strong funding base. By 1940, resource evaluation studies numbered into the hundreds and, more important, a sound scientific and engineering basis for hydrogeology was established.

Contributions were made by scientists in the petroleum industry chiefly in movement of fluids through porous media (works of M. Muskat and M.K. Hubbert and associates). Also C.F. Tolman and associates made contributions in special areas (water supply, saltwater intrusion, land subsidence).

By 1940, qualitative hydrology was well developed and documented and a firm base for quantitative work was eestablished. Contributions were made to specialized studies including budget methods, subsidence, saltwater intrusion, pumping effects, model development, and water quality. Printed summations of the "state of the art" in 1940 include O.E. Meinzer in *Physics of the Earth* (1942) and Tolman's *Groundwater* of 1937.

INTRODUCTION

Acceleration of agricultural, industrial and municipal development following the American Civil War resulted in unprecedented demands for knowledge in all areas of water-resources development. These demands were, in part, answered by the government in the organization and execution of the early geographical surveys sponsored by the U.S. War Department and by

Congress in the late 1860's and the early 1870's. These surveys produced much substantive data but, more important, resulted in the establishment of the United States Geological Survey (USGS) in 1879 which, in turn, cleared the way for systematic studies involving sources of water supply including groundwater investigations. By 1903, the USGS had established a Division of Hydrology which by 1908 had evolved into the Division of Ground Water. By this time, a Division of Surface Water and the roots of a later developed Division of Quality of Water were firmly established. O.E. Meinzer succeeded W.C. Mendenhall as Chief of the Division of Ground Water in 1912 and retained that position until his retirement in 1946, thus his period of service roughly spanned the period described in this paper. By 1910 the USGS had initiated and conducted many areal groundwater studies throughout the U.S.A. as well as many specialized studies involving investigation of the flow equations, geologic controls, artesian flow and tracer studies.

By 1910 a substantial scientific background had been developed both independently and based upon European studies. The basic equations of flow derived from Darcy's work and extended by A. Dupuit, A. and G. Thiem, and P. Forchheimer formed a rational foundation for the permeability investigations of F.H. King, Allen Hazen and others. C.S. Slichter also had developed theory and methodology in applications of the flow equations and in tracer studies. T.C. Chamberlain, A.J. Ellis, N.H. Darton, M.L. Fuller and Mendenhall developed hydrogeological aspects of far-reaching significance. Most of this work was independently developed and was not systematically put together, thus many experiments were duplicated or went unnoticed for many years. Early attempts to bring all of this work together include a textbook by Turneaure and Russell (1901, first edition) whose second edition in 1909 might be regarded as an incomplete "state of the art" work at that time, especially when combined with Fuller's publications, although no really complete descriptions of hydrogeological knowledge appeared until the middle 1920's when the works of Meinzer appeared.

INSTITUTIONAL AND ORGANIZATIONAL DEVELOPMENTS, 1910—1940

In 1911, a year before Meinzer became Geologist in Charge of the Division of Ground Water, the staff consisted essentially of four men: Meinzer, Ellis, Everett Carpenter, and, a year later, Kirk Bryan. Most of the older staff had transferred or resigned. However, Meinzer's earlier work (he joined the USGS as an aide in 1906) and his training under Mendenhall, his predecessor as Chief, had instilled several principles in his mind that later became guiding policies in the development of the Division of Ground Water. First of all, he learned from field experience that resource evaluation studies of immediately applicable value were sorely needed throughout the country and that these studies must be quantitative as data and knowledge would permit. Secondly, he recognized that knowledge of the science was incomplete and scattered and there was a real need to bring together existing knowledge and methodology and to develop new information in both categories. Thirdly, he recognized groundwater geology as a multidisciplinary field and that not only geologists but physicists, chemists and engineers were needed to develop basic principles and to apply those principles to immediate needs. Fourth, he recognized that a systematic approach to problem solution should be made by an organized team of experts. Lastly, he recognized that the public should be served and that data and information should be promptly and accurately released. All these principles were employed as Meinzer slowly developed the Division of Ground Water. By 1917 nine staff members were available. Unfortunately, World War I caused a break in the development of the Division and during and after the war the staff was much reduced. By 1929 there were only ten geologists and engineers on the staff. The great drought and depression resulted in strong demand for groundwater studies and by 1941, there were more than eighty members on the staff.

By 1929 one of the USGS staff's ambitions was fulfilled by the passage of Congressional legislation which established in principle the practice of 50:50 cooperation of the federal agencies with state and local governmental units. This made possible training of state and other scientists as well as a significant expansion in resource evaluation and other studies and was probably a chief factor in the accelerated growth of the Water Resources Branch in the decade following. In some state or local cooperative programs, part of the staff was employed directly by the state or local agency; in others only federal employees were involved. Also, some cooperating agencies preferred to publish all the results, thus enhancing the prestige of their publications and agencies. Other agencies saved publication costs by allowing all results to be published by the USGS.

By 1940, the Division of Ground Water and the closely coordinated Division of Quality of Water made up the largest and most able hydrogeological corps in the world. It contributed, by far, more to the literature than most of the other organizations then operating especially in the fields of areal studies, applications of the flow equations, and quality of water problems. It was an effective force in hydrogeologic training especially of practicing engineers and geologists in state, county and municipal service. Most of the hydrogeologists in America today have received training from the USGS, from ex-members of the Survey, or from students of ex-members.

During the latter part of the 19th century and onward, petroleum exploration and development took place at an explosive rate. Accompanying this growth problems regarding the flow of fluids through porous media developed which primarily individual scientists and engineers were invited to solve. Among these scientists the contributions of Muskat, Hubbert and their associates are notable.

The universities and some private consultants also made contributions. Meinzer's doctoral thesis at Chicago, Illinois, was concerned with the general field of hydrogeology and was published by the USGS in 1923 under the title, *Occurence of ground water in the United States with a discussion of the principles* (Meinzer, 1923a). C.F. Tolman and his associates at Stanford, California, made many contributions in special areas of water supply, saltwater intrusion, and land subsidence. Although degrees in hydrogeology were not offered courses were taught in many universities especially after about 1930. Tolman's textbook entitled, *Ground Water* was published in 1937.

HYDROLOGIC ADVANCES, 1910—1940

In hydrogeology, the trends set by the early surveys and the early work by the USGS continued until the late 1920's. The work consisted primarily of resource evaluation and areal studies. However, some research was carried on, as Meinzer said in one address given in the late 1930's, "almost surreptitiously" because of lack of funds and interest, and the collection, analysis, and consolidation of the Western world's fund of knowledge of hydrogeology was in large degree accomplished. Foremost among examples of works resulting from these activities must be mentioned *The occurrence of ground water in the United States, with a discussion of principles* (Meinzer, 1923a) and *Outline of ground-water hydrology, with definitions* (Meinzer, 1923b), literally became textbooks that were used internationally by nearly all workers in the field. They constituted a "state of the art" review of the groundwater field and gave clean and concise definitions, many of which persist to the present. Most of the experience and understanding of the USGS groundwater staff came from the background of the areal-resource evaluation studies which by 1930 numbered well over one hundred, conducted in all parts of the U.S.A. and dealing with most of the gamut of features of groundwater occurrence, availability, movement and quality. Concurrently geologic, hydraulic, hydrologic, geochemical and management and development disciplines also were developing, all of which formed a constantly-changing scientific and engineering background from which was drawn increasingly useful knowledge and methodology to be used in hydrogeology.

Accessory to geologic and engineering developments and of great utility in hydrogeologic practice were the development of drilling technology, pumping technology, various types of equipment development, and improved mapping techniques which peaked in the later 1930's with the introduction and widespread use of aerial photography and photogrammetry.

Geological progress during the 1910—1940 period that affected hydrogeologic advances is difficult to itemize in detail. In general, more precise and quantitative practice in sedimentology, increasingly quantiative understanding of principles and processes in stratigraphy, structural geology and geomorphology, and a growing interest in areal geology all contributed significantly.

In hydraulics and analytical activities, the development and wide use of steady-state flow equations, such as the Thiem method, dominated the early part of the period and paved the way for rapid acceptance of the Theis non-steady flow method (first introduced in 1935) and its expansion to more satisfactory solutions of many analytical problems. The Theis method and its derivatives resulted in the development of a highly quantitative technology that is still developing today especially in reservoir and regional analysis and in modeling methodology.

Non-analytical models based on flow-net approximations (finite-difference equations) came into use during the latter part of the period. Numerical analysis and applications of iteration and relaxation methods applied to groundwater flow problems began to appear, and many reports were published on these methods following World War II. Studies of problems in drainage, engineering geology and soil mechanics all contributed to our knowledge of flow through porous media during the 1920's and 1930's.

Hubbert's (1940) *The theory of ground-water motion* revived the early conformal mapping theories of Forchheimer and emphasized the necessity to understand and apply the principles of potential distribution in the reservoir or aquifer.

In general, hydrologic methods for determining and balancing the water budget were introduced resulting in several papers by Meinzer and others, most notably his and Stearns' paper on the Pomperaug Basin in Connecticut (Meinzer and Stearns, 1929). Later, in 1932, Meinzer published *Outline of methods for estimating ground-water supplies* (Meinzer, 1928, 1932—1933) which reasonably summarized knowledge of this area of study up to the time the Theis equation was introduced. Methods described by L.K. Wenzel, C.V. Theis, and later, C.E. Jacob, W.E. Guyton, and others provided additional means for determining the perennial yield of aquifers.

Early in the establishment of the Geological Survey, it was recognized that the chemical quality of water should be determined, especially in relation to its intended use. Thus, the early work of A.C. Lane, W. Lindgren, M.L. Fuller and associates, M.O. Leighton, R.B. Dole and others firmly established the practice of chemical analysis of waters in relation to use before 1910 and this practice continued throughout the period under discussion. Lane, Meinzer, V.C. Fishel, D.D. Jackson and others recognized the importance of utilizing water quality as a research tool, especially in relation to mineral deposits and to salt-water encroachment; thus, early in this period a foundation was established for low temperature aqueous geochemistry which became important in the latter part of the period and laid the foundation for work following 1940.

In groundwater development and management, it was abundantly clear by 1910 that aquifers and groundwater reservoirs were depletable. Much of the resource-evaluation work between 1910 and 1940 was directed, not only at depletion by water withdrawal, but also at quality effects. Studies in the 1930's by Tolman and his associates pointed up the effects of salt-water encroachment and subsidence resulting from withdrawal of underground fluids. Earlier the USGS had used the Ghyben—Herzberg formula in studies of encroachment along the Atlantic Coast (Brown, 1925). Later, concepts developed there were applied on the Gulf Coast and Hawaii. Thus, problems such as reservoir depletion, contamination, salt-water encroachment, and subsidence kept surfacing and most workers in the field were convinced that management and development should be planned carefully on the basis of properly collected and analyzed data. One large step toward this objective was the development of cooperative programs in which state, local, and private officials could participate, on the one hand sharing the knowledge of the experts, and on the other, pointing out vexing and often apparently unsolvable local problems.

Thus, by 1940, the science of hydrogeology had developed a strong theoretical foundation and many methodologies that were important to the science. This background was highly important in the solution of the many problems generated during World War II, and laid the basis for the rapid development of the science and hydrogeologic training which took place after World War II.

ACKNOWLEDGMENTS

Much of the material documenting this paper is from the unpublished files of the U.S. Geological Survey and the author wishes to thank the staff for their assistance. Much general reference information also came from the 50th Anniversary Volume (Part II) of the Society of Economic Geologists (1955), especially the article entitled, *The quantitative approach to ground-water investigations* by John G. Ferris and A. Nelson Sayre, for which the author is especially grateful.

REFERENCES

Brown, J.S., 1925. A study of coastal ground water, with special reference to Connecticut. U.S. Geol. Surv., Water-Supply Pap. 537, 101 pp.
Hubbert, M.K., 1940. The theory of ground-water motion. J. Geol., 48 (8) Part 1, 159 pp.
Meinzer, O.E., 1923a. Occurrence of ground water in the United States with a discussion of the principles. U.S. Geol. Surv., Water-Supply Pap. 489, 321 pp.
Meinzer, O.E., 1923b. Outline of ground-water hydrology, with definitions. U.S. Geol. Surv., Water-Supply Pap. 491, 71 pp.
Meinzer, O.E., 1928. Methods of estimating ground-water supplies. Part I. Outline of available methods. Soc. Econ. Geol., Tech. Sess. 1928, Lancaster, Pa., 25 pp.
Meinzer, O.E., 1937. The history and development of ground-water hydrology. Wash. Acad. Sci. J., 24: 6—32.
Meinzer, O.E. and Stearns, N.D., 1929. U.S. Geol. Surv. Water-Supply Pap., 597-b.
Tolman, C.F., 1937. Ground Water. McGraw-Hill, New York, N.Y., 593 pp.
Turneaure, F.E. and Russell, H.L., 1901. Public Water Supplies. Wiley, New York, N.Y., 808 pp. (2nd ed., 1909).
Turneaure, F.E. and Russell, M.L., 1932—1933. Methods for ground-water supplies. Water Water Eng., 34(411): 603—606; 35 (413): 25—28.

[2]

HYDROGEOLOGY OF GLACIAL-TERRAIN LAKES, WITH MANAGEMENT AND PLANNING APPLICATIONS

S.M. BORN[1], S.A. SMITH[2] and D.A. STEPHENSON[3,*]

[1] *Department of Urban and Regional Planning, University of Wisconsin, Madison, WI 53706 (U.S.A.)*
[2] *1423 South College Avenue, Tempe, AZ 85281 (U.S.A.)*
[3] *Water Resources Section, Wisconsin Geological and Natural History Survey and Department of Geology and Geophysics, University of Wisconsin, Madison, WI 53706 (U.S.A.)*

(Accepted for publication April 25, 1979)

ABSTRACT

Born, S.M., Smith, S.A. and Stephenson, D.A., 1979. Hydrogeology of glacial-terrain lakes, with management and planning applications. In: W. Back and D.A. Stephenson (Guest-Editors), Contemporary Hydrogeology — The George Burke Maxey Memorial Volume. J. Hydrol., 43: 7—43.

The subject of the relationship between groundwater and lakes is characterized by sparse information and, in general, has received limited attention by hydrologists. Nevertheless, the hydrogeologic regime of lakes must be adequately assessed in order to intelligently manage lakes and their related shorelands. This paper is a compilation of hydrogeologic data for numerous lakes in North America and presents a preliminary classification framework for lakes based on hydrogeologic considerations. The classification leads to systematic categorization of lake types for planning and management purposes.

The main hydrogeologic factors for assessing lake environments are: (1) *regime dominance*, the relative magnitude of groundwater in the total water budget of a lake; (2) *system efficiency*, a description of the rate aspects of surface and groundwater movement through a lake system; and (3) *position within a groundwater flow system*. We indicate the significance and difficulty of measuring these descriptive characteristics and provide examples of each category. Additionally, a variety of lake-related activities that illustrate the value of hydrogeologic information for planning and management purposes are presented.

PREFACE

As a consequence of extensive experience with lake resources management and applied research during the late 1960's and early 1970's, the authors became increasingly aware that the interrelationship between lakes and groundwater was not only poorly understood, but that what was known was often ignored. Although published literature was sparse, significant amounts of information seemed to exist in what might be termed "fugitive sources", i.e.,

---

Present address: Woodward-Clyde Consultants, Three Embarcadero Center, Suite 700, San Francisco, CA 94111, U.S.A.

unpublished project reports, governmental agency file documents and memoranda, and other such materials. To fill this void, we undertook a review paper which compiled and interpreted available information about the hydrogeology of lakes in glacial terrains of North America, and included planning and management implications of our analyses (Born et al., 1974). That report was aimed at applied researchers as well as those directly concerned with the planning and management of lakes and related resources. Unfortunately, that paper too was readily absorbed into the "fugitive literature"!

Since that time, there have been some meaningful advances — both scientifically and societally. Significant theoretical work has been done by Winter (1976, 1978) with regard to the interaction of lakes and groundwater. The United States Congress, in enacting Section 314 of the Federal Water Pollution Control Amendments of 1972, has launched the nation on a "Clean Lakes Program" that has stimulated a number of lake—groundwater studies to underpin lake restoration efforts. These and other recent achievements have persuaded us of the value of making our original work more widely available. We have made some modifications in the original paper but have not significantly updated our data base. Nevertheless, we believe that the information, concepts, and questions originally raised are equally valid today.

INTRODUCTION

Lakes are among our most valuable recreational, scenic, scientific and economic resources, and there are about 100,000 of them in the U.S.A. alone (U.S.H.R.-C.G.O., 1967). These resources, and their related shorelands, continue to be used more intensively every year. Of the conventional array of uses, recreational and residential development pressures are increasing most rapidly. These uses (sometimes misuses) are stressing lake environments, degrading and, in some cases, essentially destroying them. The resultant problems include lake water quality degradation (mainly eutrophication and/or sedimentation), contamination of drinking water supplies, waste disposal difficulties, loss or damage of shoreland values, and diminishment in the quality of recreational experiences, among others. These problems occur in man-made as well as natural lakes.

To intelligently protect, manage, or restore lakes and their shorelands requires substantial amounts of resource information. Most lake-related problems stem from (1) a lack of planning, or (2) insufficient data and understanding when planning for or managing the resource. Hydrogeologic information is one kind of environmental information that can be critical for sound planning and management of lake resources; unfortunately, all too often it is unavailable or unused, and its importance and value underestimated. Inferences made about the groundwater regime in the vicinity of lakes, if made at all, are commonly misleading and unsubstantiated, and critical planning and management decisions are frequently made on an assumed knowledge of the lake—groundwater relationship.

Born (1970), Maclay et al. (1972), and Winter (1976) note that there has been limited attention given to groundwater—surface water relationships pertaining to lakes; we are unaware of any summary or review papers that deal comprehensively with the theoretical aspects, field verification, and applications of the subject. Even in recent publications intended as basic guides to lake—environment planners and managers (e.g., Britton et al., 1975), little mention is made of the role of groundwater in lake management. Misconceptions exist. For example, Hutchinson's (1957) widely-referenced limnological text cites Broughton's (1941) evidence to support the conclusion that:

> "In most small lakes not in rock basins, the (lake) water is separated from the groundwater by a seal."

The shoreland management zoning regulations of some states (e.g., Wisconsin Statutes, Chapter 144) apply uniform zoning to the entire lake shoreline. This is done without considering the different hydrogeologic settings, i.e., lacking adequate background data the assumption is made that groundwater discharges to a lake around its entire perimeter (see Kusler, 1971).

We have focused on lakes in the glaciated temperate states of North America (see Frey, 1966) because: (1) this area includes a large proportion of the lakes in the U.S.A. (e.g., there are more than 30,000 lakes in Michigan, Wisconsin and Minnesota alone); (2) we wished to constrain the number of environmental (climatic and geologic) variables to consider in this preliminary summarizing effort; and (3) our own collective experience in dealing with lakes has largely been in glacial hydrogeologic settings.

The specific objectives of this paper are to:

(1) Compile and summarize information about lakes for which hydrogeologic data are available.

(2) Review and evaluate hydrogeologic aspects of lakes, with special emphasis on groundwater—surface water relationships.

(3) Specify critical hydrogeologic informational needs for evaluating lake environments, concluding with a tentative management-oriented hydrogeologic classification framework for lakes.

(4) Demonstrate the applicability and utility of hydrogeologic information for lake management and development activities.

We first present an inventory of glacial-province lakes that have been the subject of previous hydrogeologic investigations. The inventory was generated in part from an extensive literature search and from a direct mail survey. Questionnaires were sent to: (a) state and federal water-resource and geological agencies; (b) selected agency personnel; and (c) university hydrologists and hydrogeologists. To this data base we added results from our collective hydrogeologic experience on lakes in the Upper Midwest (see, e.g., Hackbarth, 1968; Stephenson, 1971; Smith et al., 1972; Born et al., 1973; Possin, 1973; Born and Stephenson, 1974; Hennings, 1978).

Second, we identify and review hydrogeologic factors or descriptors that are prerequisites for assessing lake environments. These factors include:
(1) regime dominance (relative magnitude of groundwater in a lake's water

budget); (2) system efficiency (the rate aspects of groundwater—lake interchange); and (3) position of a lake within a groundwater flow system. Each of these main factors is defined by criteria and quantified (using standards). Measurement techniques are outlined. Examples of lakes exhibiting the array of characteristics are described.

Third, we have analyzed lake population data, critical information needs, and information availability, in an attempt to suggest a systematic grouping or classification of lakes from a hydrogeologist's viewpoint in a user-oriented way.

Finally, we illustrate and discuss the application of specific hydrogeologic data in selected management and planning situations (including: lake management and rehabilitation, shoreland use and development, water supply, waste disposal and environmental impact assessment).

INVENTORY

The inventory for this study was developed in 1974 (S.M. Born, S.A. Smith and D.A. Stephenson). At that time, we identified 63 lakes for which groundwater information was available within twelve states and provinces of North America. These lake sites are primarily in the north-central U.S.A. and the central Canadian provinces of Ontario and Saskatchewan. The Appendix summarizes the lake population with name, location, size, and the published source of the data. Additional data, including water budget and geologic setting are given in Born et al. (1974, pp. 74—81).

By itself, the originally-described lake population represents an important addition to the hydrologic literature, but it is by no means a complete listing. Many additional lake—ground water studies have been made since 1974; e.g., there have been nearly 100 studies in Wisconsin alone (D.R. Knauer, pers. commun., 1978).

In addition, the original sample is a reflection of lakes that had been studied (those with special problems or those that could be studied given constraints of budgets and technology), rather than a random sample of the total population of all lakes in humid-zone glacial terrains. Additional limitations which should be considered in any attempt to extrapolate our results to other lakes arise from the great variability in the amount and quality of data available for lakes. For many of the lakes, the available information is rather rudimentary. Furthermore, many of our sample lakes are small (possibly related to the expense of intensive groundwater studies). We have not imposed any size limits, in that information from even the smallest lakes may have some transfer value to larger bodies. We have extended conventional definitions of lakes to include ponds, potholes, sloughs, swamps, and marshes — any closed topographic depression that contains standing water at least periodically.

HYDROGEOLOGIC CONSIDERATIONS IN THE ANALYSIS OF LAKE ENVIRONMENTS

Many factors ultimately affect lake—groundwater relationships and should be considered in any comprehensive lake management plan. We have concentrated on those hydrogeologic descriptors which have a direct bearing on the magnitude, rate and direction relationships of groundwater movement relative to a lake. The descriptors we chose can be used to define any lake with respect to its hydrogeologic setting. In conjunction with supporting data (e.g., water-quality records), or by themselves, these descriptors provide a valuable input to the planning and management process.

For each of three hydrogeologic descriptors, we defined a set of idealized criteria, standards and measurements. Because sophisticated hydrogeologic analyses are available for relatively few lakes, a set of alternate measures were used when the "ideal" could not be attained. A certain degree of accuracy is sacrificed through the use of less definitive substitute measures, but the approach allowed us to work with the available lake population.

Regime dominance

Regime dominance is a descriptive term pertaining to the relative importance of groundwater in a lake water budget. This criterion could be based on the ratio of groundwater inflow or outflow to total inflow or outflow, which in turn could be quantified to set standards for different classes of regime dominance: groundwater dominated lakes vs. surface water dominated lakes. We considered that groundwater is dominant if it represents a significant portion of the overall water budget from either a qualitative or quantitative viewpoint. Thus the standard for characterizing regime dominance becomes a subjective judgement as to whether groundwater is a significant or an insignificant hydrogeologic factor. Measurement of regime dominance is via water-budget analysis for the lake basin, accomplished either directly in the field, or with the aid of modeling techniques (see, e.g., Ferris et al., 1962; Pinder and Bredehoeft, 1968; Dettman and Huff, 1972).

Because there are only a few lakes in our study population for which adequate water-budget data were available, we adopted a proxy criterion to reflect regime dominance. It was based on surface-water characteristics: *seepage lakes* are groundwater dominated, and *drainage lakes* are surface water dominated. In order to include consideration of both influent and effluent groundwater, we have extended the original seepage lake definition of Birge and Juday (1934) to include all lakes which are not on a throughgoing drainage system. Seepage lakes may have no surface-water connection, or they may have an inlet *or* outlet, but not both. Drainage lakes have an inlet *and* an outlet.

Although the proxy criterion is straightforward, there are some limitations to its use. For some lakes which do not have inlets or outlets, precipitation, intermittent and non-channelized surface runoff, and evapotranspiration can be responsible for nearly all of the water inputs and outputs; yet according to our scheme, they would be classified as groundwater dominated lakes. Similarly, groundwater can constitute a significant fraction of the water budget of drainage lakes. However, lacking the needed data, we assumed that for lakes as a whole, groundwater is more important in seepage lakes than in drainage lakes, especially in humid areas.

Examples. Pickerel Lake, a 18-ha (45-acre) kettle-hole lake in the central Sand Plains of Wisconsin is an excellent example of a groundwater dominated lake. The hydrogeology of the lake was investigated by Hennings (1978) as part of a comprehensive lake study prior to a lake rehabilitation project (Knauer and Peterson, 1974). Soils (used here to mean unconsolidated surficial materials) in the Pickerel Lake basin directly overlie Precambrian crystalline bedrock, and consist of a 30-m (100-ft.) thick sequence of stratified sand outwash with thin discontinuous silt and clay stringers. No surface streams enter or leave the lake, and because of the highly permeable soils and the very small surface drainage basin, runoff accounts for less than 3% of the total annual water input. Groundwater makes up about 72% of the inflow on a

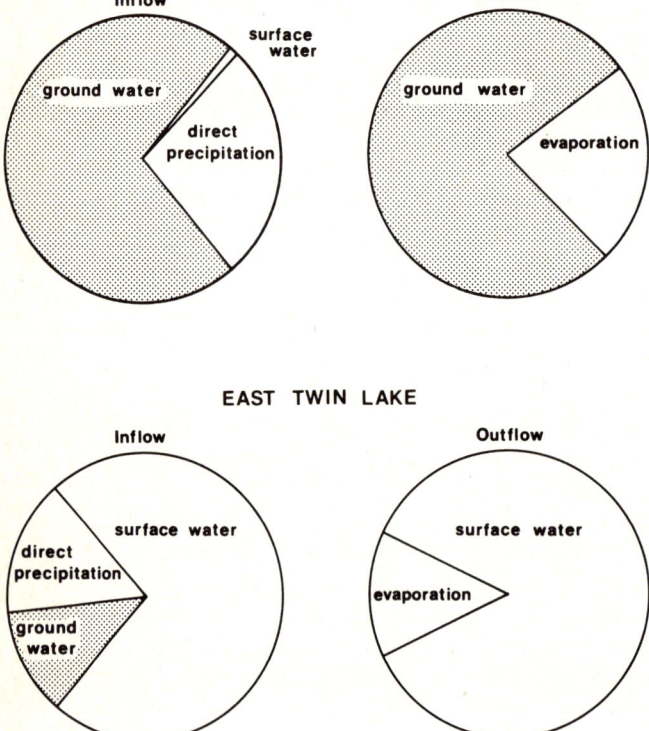

Fig. 1. Water-budget data for Pickerel and East Twin lakes (after Hennings, 1978; Cooke et al., 1973).

yearly basis, with direct precipitation on the lake surface making up the remainder. Water loss from the lake is entirely through groundwater sepage (77%) and evapotranspiration (23%).

East Twin Lake in northeastern Ohio provides an example of a surface water dominated lake. East Twin Lake is the second of a pair of lakes investigated by Cooke et al. (1973). The lake has a surface area of 27 ha (66 acre) and a maximum depth of 12 m (39 ft.). Soils in the 210-ha (520-acre) drainage basin are generally sand and gravel. However, a continuous silt and clay layer impedes groundwater movement into the lake.

The two surface streams which enter East Twin Lake contribute 72% of the annual water input. Direct precipitation and groundwater account for 15% and 12%, respectively. About 85% of the average annual water outflow is by surface drainage; evapotranspiration accounts for the remainder. There is no significant groundwater seepage from the lake.

The diagrams in Fig. 1 show the water budget data for Pickerel and East Twin lakes. Based on water-budget criteria, Pickerel Lake is clearly groundwater dominated and East Twin is surface water dominated. Groundwater is an insignificant part of the inflow to East Twin Lake, and there is no groundwater outflow. The proxy standards for regime dominance correspond well with the ideal standards. Pickerel is a seepage lake, with no surface water connection, and East Twin is a drainage lake, receiving input from two perennial streams and losing water via surface drainage.

System efficiency

System efficiency is a descriptor used to describe the rate of water movement through a lake. Although we are mainly concerned with groundwater dominated lakes, the system efficiency concept can be applied to all lakes, regardless of regime dominance, using lake flushing rate as a criterion.

Lake flushing rates are commonly expressed in terms of the time required to replace the equivalent of an entire lake volume. For example, a flushing rate of two years indicates that an amount of water equal to the lake volume enters and leaves the lake every two years. (If there is a change in storage, outflow only is usually the basis for the calculation, and if changes in water quality are of interest, evapotranspiration, which removes no dissolved chemical constituents, is neglected.)

Apparent lake flushing rates, which do not consider lake hydrodynamics, and which assume volume-for-volume displacement, are highly variable. They range from one lake volume every few days for man-made reservoirs on major river systems to one lake volume every few centuries for Lake Superior (see Rainey, 1967; Sonzogni and Lee, 1974). Actual flushing rates, however, are less rapid than apparent flushing rates and depend to a great degree on the mixing mechanics of the lake. Surface inflows to a stratified lake may replace the epilimnetic waters many times over, but the hypolimnetic waters will be affected mostly during the fall and spring mixing periods. Viessman (1971) presents a comprehensive discussion of flushing rate calculations and their use in lake water quality modeling.

Although flushing rate is a convenient criterion for defining system efficiency from the standpoint of the rate of water movement through a lake, it is only indirectly related to the rate of groundwater flux through a lake system. Thus, where groundwater is an important consideration in a resource decision (e.g., the rate at which a pollutant in the groundwater will move to a lake), flushing rate itself is an inadequate "system efficiency" criterion. Instead of flushing rate, therefore, we have used what is best termed the "Darcy characteristics" of the saturated zone as a criterion for determining the system efficiency of groundwater dominated lakes. "Darcy characteristics" include: (a) groundwater gradients; and (b) the permeabilities of lake sediments and contiguous materials expressed in conventional units such as gal. day^{-1} ft.$^{-2}$ or ft. day^{-1} (cm s^{-1}). Darcy characteristics embody the concept of the degree of hydraulic connection between the lake itself and the surrounding groundwater regime. Used in conjunction with the saturated thickness of the aquifer which is in communication with a lake, Darcy characteristics can be used to quantify the amount of groundwater flow across the lake—groundwater interface.

It is just as difficult to justify quantitative standards for system efficiency as it is for regime dominance; the final decision will depend to a great extent on the specific circumstances. We have used a subjective judgement, and the standard is simply the designation of Darcy characteristics as "low" or "high." Darcy characteristics are measured with observation wells and piezometers to determine groundwater gradients, and pump testing or laboratory analysis to determine permeability.

We have retained the notion of Darcy characteristics as a proxy criterion that can be used with the existing data, but have placed major emphasis on the permeability characteristics of the soils *surrounding* the lake. We thus give only limited consideration to: (a) thickness, distribution, and permeability of the lake sediments; and (b) values of groundwater gradients. Generally, if a lake is situated in coarse-grained glacial outwash or coarse till, we have classified it as having a high rate of hydraulic communication; the groundwater flow system is efficient in transmitting water to a lake. Lakes situated in fine-grained glaciolacustrine deposits or clayey till have a lower rate of hydraulic communication. Some indication of the nature of hydraulic communication can be obtained from correlating groundwater and lake-level fluctuations, where such information exists. This measure more-or-less integrates the many variables influencing the "Darcy characteristics". The correlation can be for a period of years, or for short-term periods during which the lag time between adjustments in lake level and the groundwater table can be evaluated (i.e., storms, lake-level manipulations).

There is some justification for neglecting the effect of lake-sediment permeability and distribution in classifying lakes according to rate of hydraulic communication, especially in the case of large lakes with wave-washed shorelines. McBride (1972) has shown theoretically, and Hackbarth (1968) has confirmed experimentally, that most groundwater moves into a lake through a relatively narrow band in the littoral zone, regardless of sediment distribution. This was more recently confirmed by McBride and Pfannkuch (1975). Of course, in some cases the use of the simplifying assumptions in applying the proxy criterion would be both naive and misleading. Therefore, appropriate adjustments are required in situations where idealized "Darcy characteristics" data are lacking and there is some evidence suggesting that lake sediments are a major factor affecting system efficiency.

Examples. Pickerel Lake, Wisconsin, situated in outwash sands, is representative of a lake having good hydraulic communication between the lake and groundwater (Hennings, 1978). Based on an average gradient of 0.0015 cm cm^{-1}, a permeability of 0.047 cm s^{-1} (1000 gal. day^{-1} ft.$^{-2}$), and an effective porosity of 0.20, the groundwater velocity is ~30 cm day^{-1} (1 ft. day^{-1}). Near the lake shore, groundwater gradients are as high as 0.003 cm cm^{-1}, and flow velocities are ~60 cm day^{-1} (2 ft. day^{-1}). Although organic lake sediments reach a thickness of more than 15 m (50 ft.) in the center of the lake, they do not substantially interfere with groundwater movement into the lake. Wind and waves maintain a 15--30 m (50—100 ft.) sandy strip around the lake shore. Further evidence for a high degree of hydraulic connection between Pickerel Lake and the groundwater is the high degree of correlation (99.99% confidence level) between continuous records of the water-table elevation and the lake level.

A willow-ring slough in southeastern Saskatchewan (Meyboom, 1966) is an example of a lake with a low rate of hydraulic communication. This is a shallow (1 m deep), semi-permanent 0.1-ha (0.25-acre) pond in sandy to clayey glacial till. A ring of willows and other phreatophytic vegetation surround it, greatly influencing the groundwater flow regime and causing seasonal reversals in the flow direction. Groundwater gradients vary from +0.062 to 0.147 cm cm^{-1} in the vicinity of the slough. (A positive gradient indicates flow toward the slough; a negative gradient indicates flow away from the slough.) Although the gradients are high, the permeability of the soils is very low, and averages $1.4 \cdot 10^{-5}$ cm s^{-1} (0.30 gal. day^{-1} ft.$^{-1}$). Assuming a minimum effective porosity for the till of 0.20, groundwater flow velocities probably do not exceed $1.0 \cdot 10^{-5}$ cm s^{-1} (0.03 ft. day^{-1}) in the vicinity of the slough, two orders of magnitude less than at Pickerel Lake.

Position of a lake in a groundwater flow system

The classification of lakes according to regime dominance and rate of hydraulic communication requires the same kind of essential data. The two classification factors, in fact, are closely related. It is not possible, for instance at current levels of technology, to develop an accurate water budget for a lake with any amount of groundwater inflow and outflow without first determining the velocity of groundwater movement in the saturated zone. The final step in the water-budget analysis requires delineation of the area through which groundwater communicates with a lake, plus the direction of movement.

Field identification of groundwater flow systems has been demonstrated by several researchers (e.g., Winograd, 1962; Back, 1966; Meyboom, 1966; Meyboom et al., 1966; Maxey, 1968; Mifflin, 1968; R.E. Williams, 1968). These in-field analyses are technically feasible, reasonably economical, and generally corroborative of theoretical analyses (e.g., Hubbert, 1940; Meyboom, 1962; Toth, 1962, 1963; Freeze and Witherspoon, 1966, 1967, 1968). However, there is a paucity of field documentation of groundwater flow systems around lakes.

Fig. 2 is an example of a generalized groundwater flow condition within which lakes in humid-zone glacial terranes occur. Lakes situated in groundwater recharge areas (recharge lakes) can contribute to the groundwater through the entire lake bottom. Lakes in groundwater discharge areas (discharge lakes) gain groundwater through the entire lake perimeter as well as partially through lake-bottom sediments. In areas of lateral groundwater flow, lakes lose to the groundwater on one side and gain groundwater on the other side (flow-through lakes). In addition to these three simple situations (Fig. 3) it is possible for lakes to intersect both a shallow and a deep flow system, with the result that groundwater movement near the lake margins may be opposite to the direction of movement near the lake center. Theoretically, all six of the configurations shown in Fig. 4 are possible; Meyboom (1967)

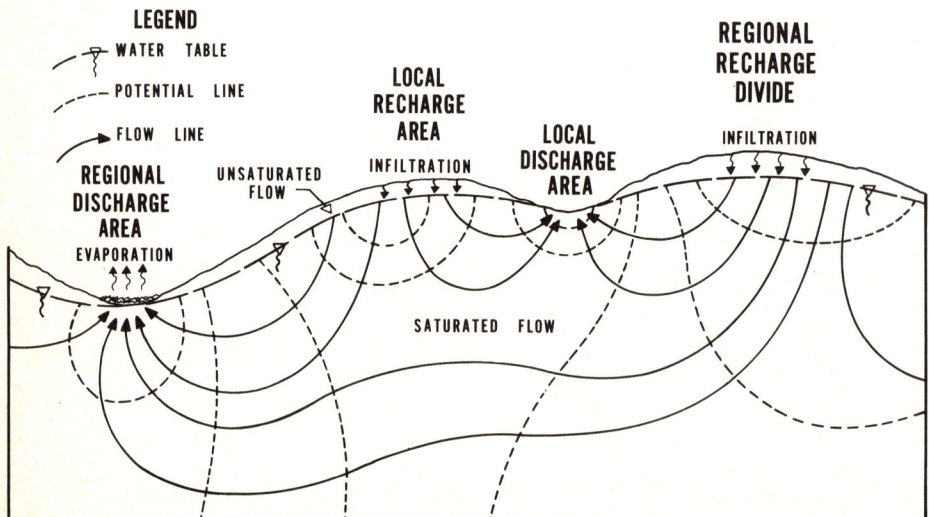

Fig. 2. Generalized groundwater flow-system configuration. The *dotted line* separates the areas of shallow and deep groundwater flow.

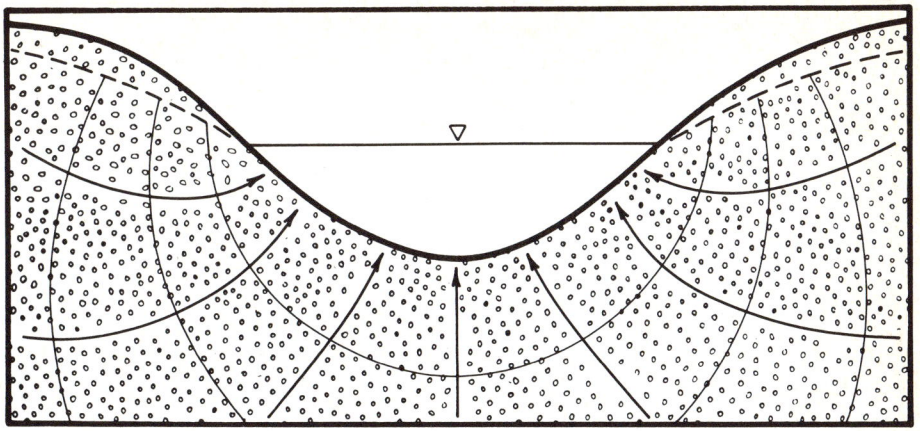
a) Discharge lake

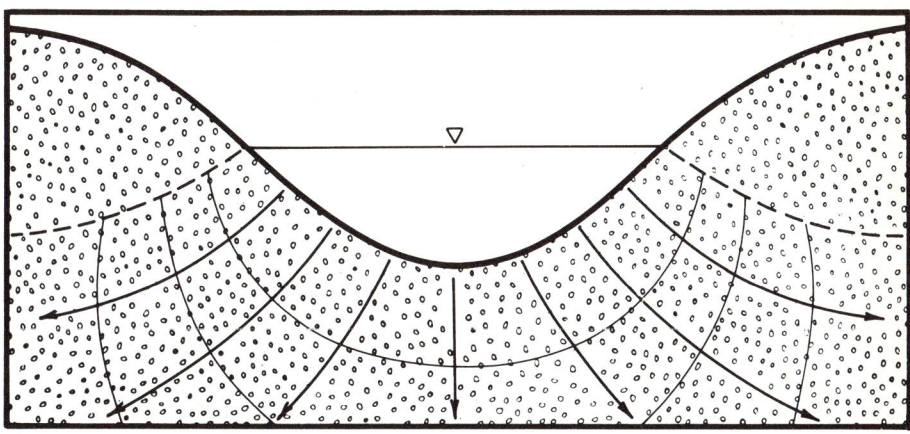
b) Recharge lake

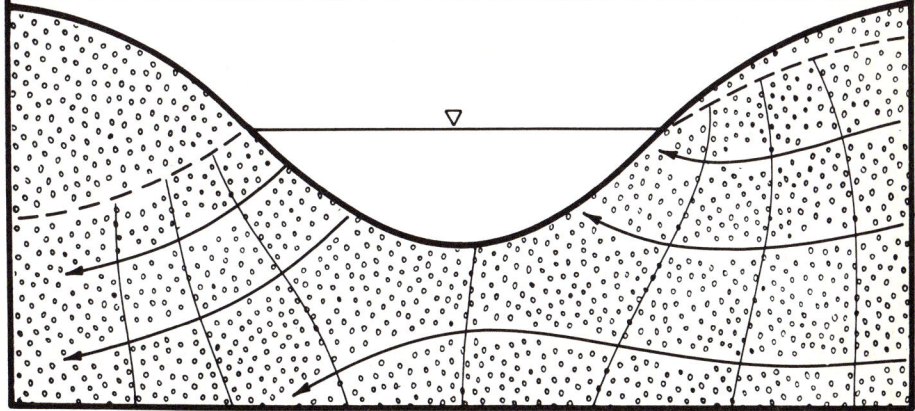
c) Flow-through lake

Fig. 3. Three possible configurations of groundwater flow systems around lakes.

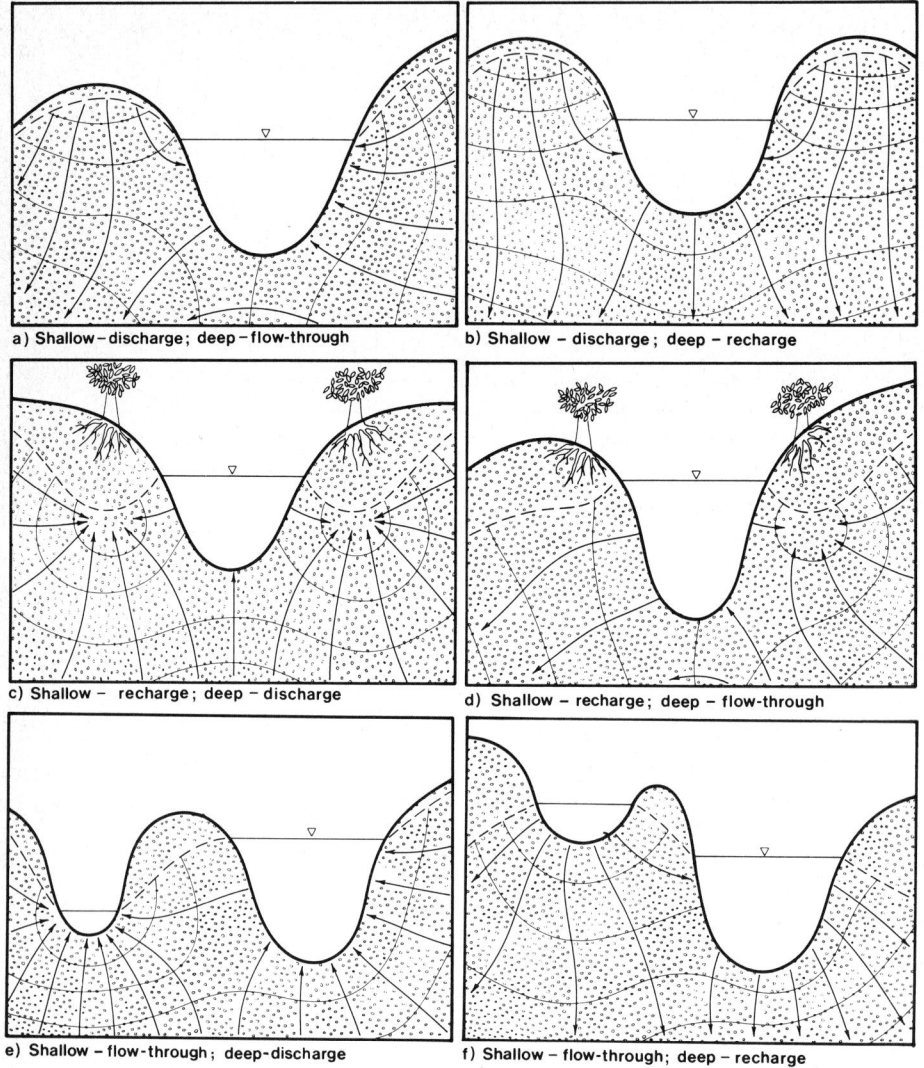

Fig. 4. Conceptual groundwater flow around lakes which intersect two flow systems.

has found evidence for a) and c), and Maclay et al. (1972) hypothesize the existence of b).

Fluctuations in the flow system around small lakes are not uncommon. During the spring, runoff can raise the lake level before there is an appreciable increase in the elevation of the water table, producing a condition of groundwater recharge. During the summer, however, evapotranspiration may lower the lake level more than the water table, resulting in groundwater discharge. Meyboom (1967) has shown that the deterioration of local groundwater mounds between adjacent lakes can result in a transition from a discharge to a flow-through situation.

In addition to lakes which are part of a dynamic groundwater flow system, there are probably some lakes which are truly isolated from groundwater, i.e. perched. This is an extremely difficult situation to document because of the necessity of installing properly-cased piezometers through a lake bottom to confirm or deny the existence of an unsaturated zone. Stearns (1942) presents evidence for a perched condition at Mud Lake, Idaho. Wells near one end of Mud Lake indicate that there are two water tables; the shallow water table coincides approximately with lake level and is perched above the regional water table, which is in the under-lying basalt. The existence of truly perched lakes in glacial terrains remains questionable. Schwartz and Manson (1964) and Manson et al. (1968) concluded from their study of more than 40 lakes and ponds in Minnesota that perched ponds were a common situation. However, with perhaps a single exception, none of their cross-sections indicate a water table that could not be interpreted as continuous with lake level.

The ideal criterion for determining the position of a lake in a groundwater flow system is the groundwater potential distribution beneath the lake bottom, and the vertical gradient is the ideal standard. Measurement of this factor is done on-site with piezometers. Such information can be expensive and difficult to obtain.

Fortunately, there are numerous proxy criteria which can be used to classify a lake according to its position within the groundwater flow system. The groundwater potential distribution around the lake shore gives relevant information which can be more readily determined than information from beneath the lake. Temperature anomalies, water chemistries, certain ecological attributes, soil characteristics, and visible evidence of groundwater discharge such as springs and seeps, all can help establish the flow-system configuration. None, however, are as definitive as water-level measurements via piezometers.

Examples. The flow-system configuration around Pickerel Lake in Wisconsin was confirmed by numerous water-table observation wells and piezometers around the lake shore, two piezometers driven through the lake sediments, and a shoreline temperature survey. The water-table configuration and the groundwater potential distribution beneath the lake are shown in Fig. 5. Groundwater moves from west to east through the lake. The influent and effluent portions of the shore line are fairly well defined on the basis of both water-table elevations and the temperature distributions in the shallow sands along the shore. A more visible measure of the flow system is the open-water condition on the west side of the lake during winter when the rest of the lake is frozen.

Jyme Lake in north-central Wisconsin, provides an example of a lake in a groundwater recharge area. The hydrogeology of this small bog lake [0.4-ha (1-acre) surface area; 3-m (10-ft.) maximum depth] was studied by Smith et al. (1972). The lake and the associated sphagnum bog are located on a

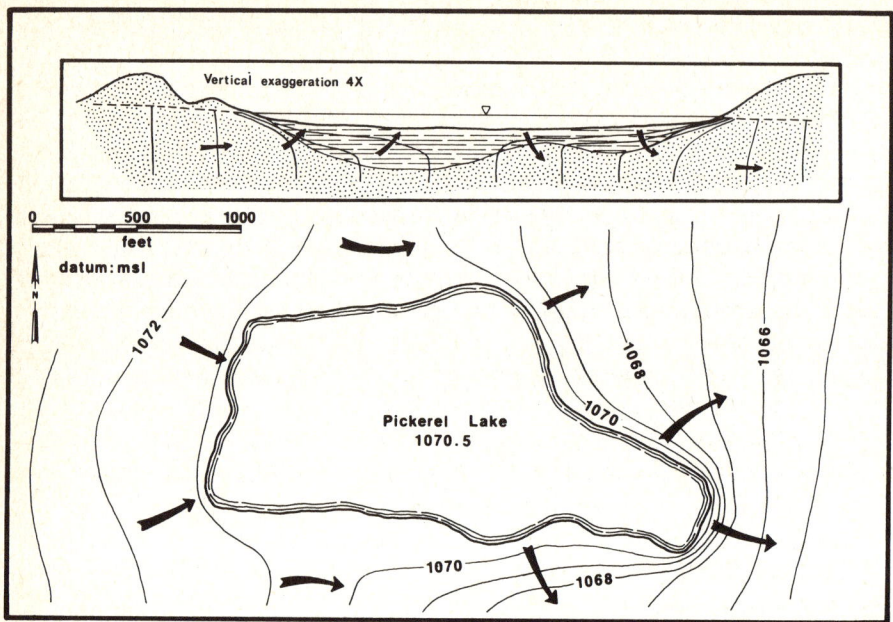

Fig. 5. Cross-section and plan view of groundwater flow around Pickerel Lake, Wisconsin (after Hennings, 1978).

peninsula jutting out into a large 1500-ha (3600-acre) lake which is 2.6 m (8.5 ft.) lower than Jyme Lake, and at the closest point, only 150 m (500 ft.) away. Low hills of poorly stratified sand and gravel outwash surround Jyme Lake, but up to 6 m (20 ft.) of organic lake sediments and peat are present beneath the lake and bog. The groundwater flow system at Jyme Lake was monitored with twelve piezometers and observation wells, most of which were driven through the lake and bog sediments into the underlying sand and gravel. The water table in the bog is at the ground surface, nearly coinciding with lake level, but at the outside border of the bog, the water table slopes steeply downward. The water level in all of the piezometers driven through the lake and bog sediments is considerably lower than the lake level. Groundwater moves nearly vertically downward through the peat and lake sediments, and except for perhaps a small amount of shallow flow near the bog surface, there is no groundwater input to Jyme Lake.

Lake Sallie, Minnesota is not a groundwater dominated lake (groundwater accounts for only 14% of the total inflow and none of the outflow), but a detailed hydrogeologic investigation by McBride (1972) has shown that the groundwater flow regime is representative of a discharge lake. Lake Sallie has a surface area of 490 ha (1200 acre) and a maximum depth of 17 m (55 ft.). It lies in a pitted glacial outwash plain, and sand and gravel thicknesses as great as 34 m (110 ft.) have been recorded near the lake shore. A thick till layer underlies the outwash, and many of the hills along the lake shore are till. A thin discontinuous till unit mantles the sand and gravel in some parts of the basin.

More than thirty water-table observation wells and piezometers were installed in the Lake Sallie area and, groundwater levels along with soil thickness and permeability measurements were used in a digital computer program to determine the flow system configuration. Fig. 6 shows a water-table map of the Lake Sallie area. Groundwater flows toward the lake around nearly the entire perimeter and upward through the lake bottom. Because of the very low permeability of the till (less than one-thousandth of the permeability of the outwash), groundwater movement through the till-cored hills near the lake is insignificant.

The movement of groundwater around Lake Sallie points out a minor discrepancy in our flow-system categories. In humid areas, precipitation exceeds evaporation, and a "pure" groundwater discharge lake cannot exist throughout the year. Water must leave the lake either by surface or ground-

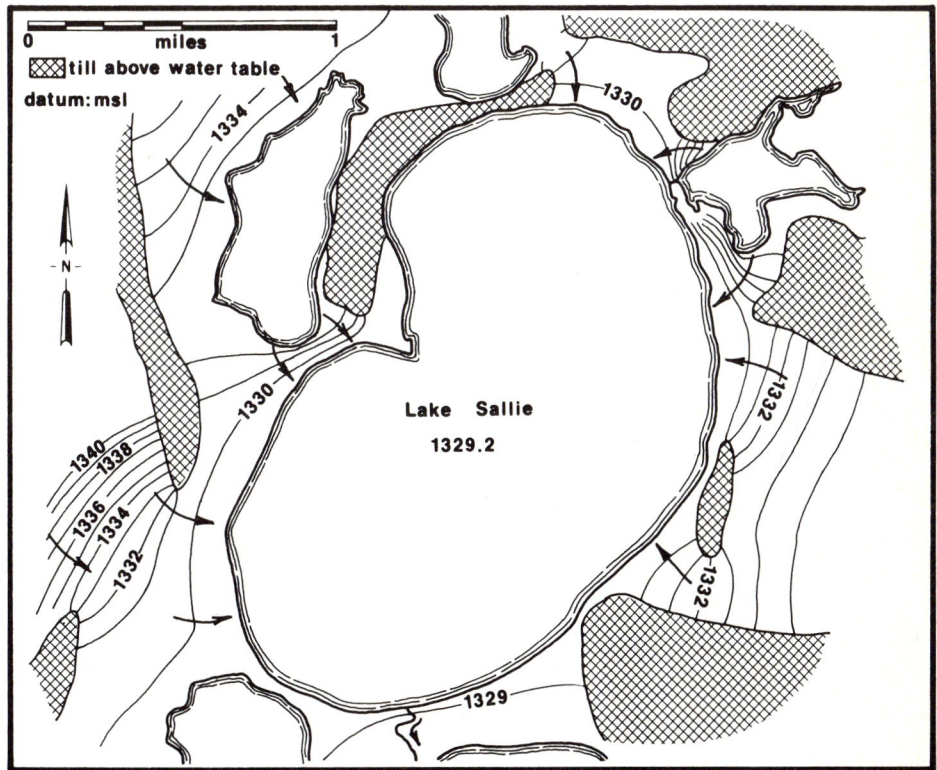

Fig. 6. Map of the Lake Sallie, Minnesota, area showing water-table configuration and the direction of groundwater movement. The quantity of flow through the till-cored hills near the lake is insignificant compared to flow through the sand and gravel outwash (after McBride, 1972).

water flow, and if there is a surface outlet, there is undoubtedly some groundwater underflow through the fluvial sediments. This is the situation at Lake Sallie, and near the outlet, groundwater is moving from the lake. However, it seems reasonable to classify lakes as groundwater discharge lakes even if there is some local outlet-associated groundwater movement away from the lake.

There are very few lakes in the study population for which vertical groundwater gradients have been determined, either by piezometers through the lake bottom or near the lake shore. Consequently, we have had to rely on proxy criteria to determine the groundwater flow system for most of the lakes in the study lake population. In most cases, this has meant relying heavily on the configuration of the water table around the lake. Usually, the water table slopes toward discharge lakes, away from recharge lakes, and toward one end and away from the other end of flow-through lakes (Fig. 3). This is not a completely definitive standard, however. The water-table configuration gives only the direction of the horizontal component of groundwater flow at the lake margin and tells nothing about vertical gradients or the distribution of flow beneath the lake. As shown in Fig. 4, for example, a lake with groundwater moving toward the lake at the edges may be losing to the groundwater reservoir at the center. Other shoreline observations which help define the

TABLE I

Terminology used in categorizing lakes

Descriptor[1]	Criterion[2]	Standard[3]	Measurement[4]
Regime dominance	relative magnitude of groundwater in a lake's water budget (the ratio of groundwater to total input or output)	high or low (subjective)	water-budget analysis
System efficiency	*surface water:* lake flushing rate *groundwater:* "Darcy characteristics" (gradient, permeability)	high or low (subjective)	*surface water:* water budget and lake volume *groundwater:* field and laboratory testing
Position in flow system	groundwater potential distribution beneath lake	vertical gradients positive, negative, or zero	piezometers

[1] A descriptive term used to summarize important hydrogeologic factors; each descriptor is defined in the text.
[2] Qualitative definition of the descriptors.
[3] Quantification of criteria.
[4] How the criterion is measured or monitored.

position of a lake within the groundwater flow system are subject to similar misinterpretations, but frequently comprise the best available data.

Summary

The foregoing discussion of hydrogeologic considerations in analyzing lake environments is summarized in Table I. The organization is intended to provide a frame of reference for describing the hydrogeology of lakes. Any important hydrogeologic parameters not emphasized in Table I, for example, water quality (nutrient budgets, contaminant movement, etc.), can be readily evaluated in terms of the hydrogeologic characterization of the entire system.

A SUGGESTED LAKE CLASSIFICATION SCHEME BASED ON HYDROGEOLOGIC FACTORS

The objectives of a hydrogeologic lake classification scheme should not be to merely impose artificial order on nature, but rather to systematize common elements and distinctive differences in complex natural (lake) systems. This allows us to more fully examine, and hopefully understand, these systems, keeping in mind that the scientist making the classification may not be the person who must ultimately utilize the results. Thus, classification attempts might lead to systematic categorization of lake types with management and planning implications (e.g., all lakes in a given class might be undevelopable for septic-tank installation; another lake type might be classified as difficult to manage by drawdown or dilution; characteristics of a third lake class might influence a decision on whether and where to discharge urban storm waters, etc.). In addition, a viable classification would allow one to infer how a lake—groundwater system will behave without having to first identify all the critical hydrogeologic information needs described earlier. This, in turn, permits data needs for decision-making to be weighed against data-collection capabilities and costs, thereby reducing the total cost of data acquisition for sound decision-making. Of course the cost of "misinterpreting", and the consequences of doing so, is important in this decision.

One possible classification scheme growing out of our hydrogeologic discussion is presented in Table II.

Applying this classification scheme to our study of lake population is illuminating.

Of the 63 lakes in our inventory, 45 are seepage lakes (by our definition) and are therefore groundwater dominated, according to the proxy standard for regime dominance. For those seepage lakes with water-budget data, groundwater makes from 21 up to 98% of total inflow and from zero to about 77% of total outflow. For drainage lakes having water budget data, groundwater makes up no more than 44% of total inflow and less than 2% of total outflow. Although this limited water-budget information does suggest

TABLE II

Possible classes of lakes based on hydrogeologic considerations*[1]

Regime dominance	System efficiency	Position within flow system
Ground water dominated	low*[2]	discharge recharge flow-through perched
	high*[2]	discharge recharge flow-through
Surface water dominated	low*[3]	negligible importance
	high*[3]	

*[1] For example, a lake might be classed as simply groundwater dominated; or more thoroughly as a groundwater dominated, high system efficiency, flow-through lake.
*[2] Based on "Darcy characteristics".
*[3] Based on flushing rate.

fairly good agreement between the ideal (water-budget analysis) and proxy (surface-water characteristics) criterion for determining regime dominance, the correlation is less than perfect. For example, the groundwater input to Muskellunge Lake, northern Wisconsin, (44% of the total) and Shadow Lake, central Wisconsin, (41% of the total) would probably be regarded as significant by any definition, yet both lakes are surface water dominated according to the proxy criterion (Born et al., 1974, appendix 1).

We did not have sufficient data to apply the ideal criterion for determining rate of hydraulic communication (Darcy characteristics of the saturated zone) to more than a few lakes. Consequently, we used the proxy criterion, which is largely based on the hydrogeologic properties of soils in the lake basin. Of the lakes for which data are available, about two-thirds have a high system efficiency. Most of these are situated in glacial outwash deposits. The remining lakes, with a low system efficiency, are situated in glacial tills or fine-grained glaciolacustrine deposits.

The position of lakes within groundwater flow systems probably represents the most accurate hydrogeologic data for the study population. About 40% of the lakes are discharge lakes, and about 40% are flow-through lakes. The remainder are recharge lakes. There is no obvious correlation between this classification factor and the other two, except that no surface water dominated lakes in the study population are recharge lakes. This is reasonable, in view of our proxy standard for regime-dominance classification. Surface water dominated lakes are on through-going drainage systems and therefore occupy a regional topographic low — an unlikely area for groundwater recharge in most humid zones.

We concluded that a perfectly workable classification could not be adopted at the time of our original work. Subsequent testing suggests its utility, but the requisite hydrogeologic data base may still be too sparse. The effort is valuable, however, in identifying knowledge and information gaps that may help guide research efforts. For example, about 80% of the lakes in the study population are classified as either flow-through or discharge lakes, and are in a position to intercept groundwater. This obviously has major implications from the standpoint of water quality.

Furthermore, we want to stress that there is no way to do an accurate water-budget analysis on flow-through lakes by treating groundwater as a residual. Problems arise when the amount of groundwater inflow approximates groundwater outflow. If at Pickerel Lake, for instance, the groundwater component of the water budget had been calculated by difference, the rate of water turnover in the lake would have been underestimated by a factor of 4.

APPLICATION OF LAKE—GROUNDWATER INFORMATION IN PLANNING AND MANAGEMENT

Hydrogeologic information is an important and sometimes essential consideration in planning for or managing lake environments. Two representative cases are presented to concretely illustrate the utility of hydrogeologic information in planning-management decisions involving lakes. One example involves a lake management (rehabilitation) project in northern Wisconsin and the other involves the planning of a large lake-oriented recreational new town in northwestern Wisconsin. They are typical cases, and are illustrative of our theme.

The Snake Lake Rehabilitation Project

Snake Lake, a 5.0-ha (12.3-acre) soft-water lake in northern Wisconsin, received the direct discharge of municipal wastes for more than 20 years. Consequent accelerated eutrophication destroyed the lake's recreational and aesthetic values. Annual dissolved oxygen depletion, resulting in major fish-kills and nuisance aquatic plant growth, converted a natural resource and asset into a community liability. A lake restoration project that employed a lake flushing scheme using dilutional pumping was undertaken at Snake Lake (Born et al., 1973; Jaquet, 1976).

Within the study area, approximately 60 m (200 ft.) of Wisconsin Stage glacial-drift deposits overlie Precambrian crystalline bedrock; Snake Lake occupies a kettle hole in a relatively low-relief pitted outwash plain. Extensive hydrogeologic and geologic background studies were conducted as part of the multi-disciplinary scientific investigations necessary for establishing baseline pre-treatment data and for defining the problem. A piezometer network was installed around the lake and water-level measurements were taken monthly. The groundwater flow system is fairly simple; water enters the lake from the adjacent aquifer on the east side, and is discharged to the groundwater

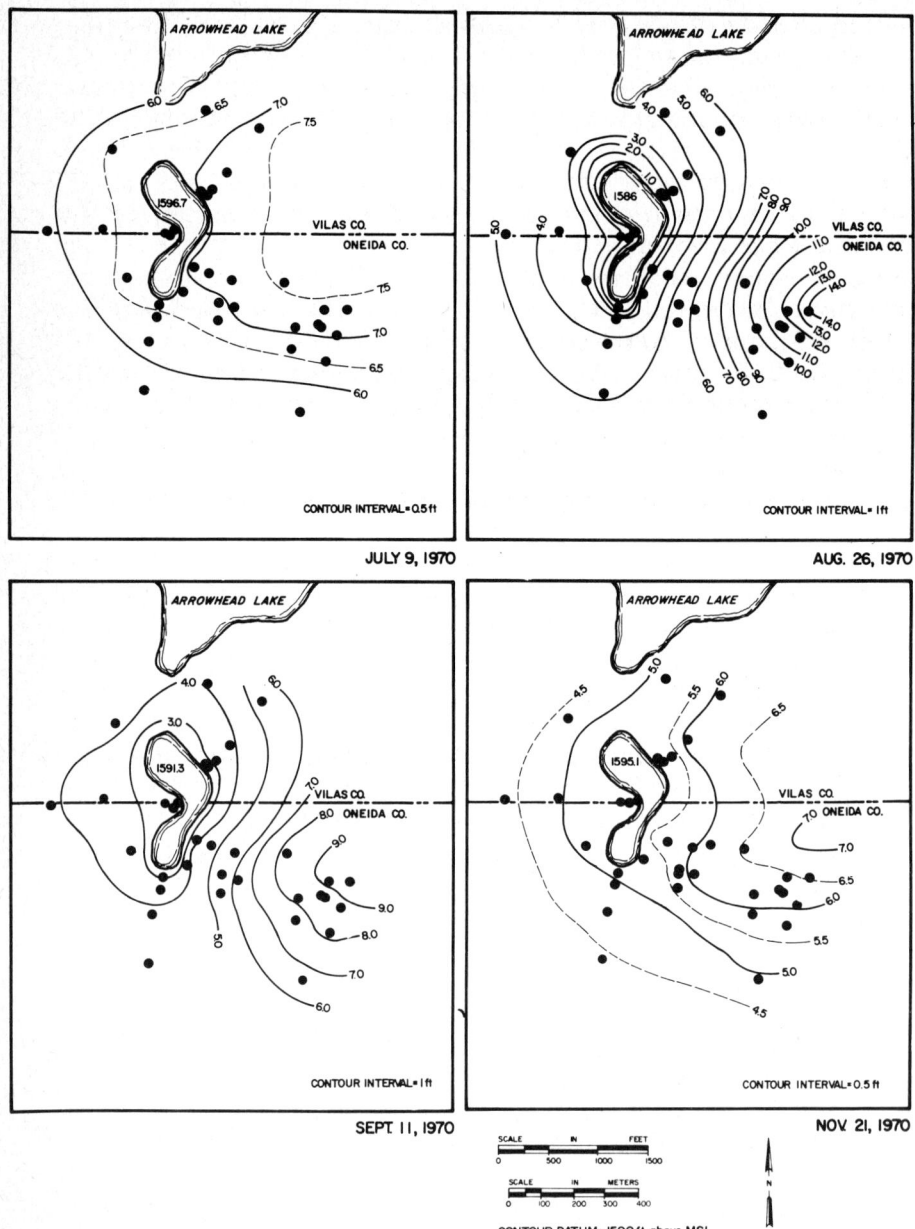

Fig. 7. Sequential water-table maps of Snake Lake, Wisconsin. Maps show water-table configuration prior to start of pumping (July 9), near end of pumping, two weeks after cessation of pumping, and twelve weeks after cessation of pumping (Nov. 21) (after Born et al., 1973).

reservoir along the western, and to some extent northern and southern perimeters of the lake. The water table is relatively flat; water moves under the influence of a gradient of about 0.6—1.2 m km^{-1} (3—6 ft. mi.$^{-1}$). In comparison to Snake Lake, groundwater quality is high, particularly upgradient of the lake.

The local geology and hydrology suggested an innovative flushing method in which nutrient-rich lake water was pumped to a nearby land-disposal area, allowing dilution by precipitation and influent groundwaters. Past lake flushing-dilution schemes applied elsewhere have used either municipal water supplies or diversions from nearby rivers (Dunst et al., 1974). The response of the hydrogeologic system to the stress of pumping is shown in Fig. 7. As a result of the project, the concentrations of phosphorus (a critical plant nutrient) in the lake were reduced by about one half, and have remained substantially lower than pre-treatment values in succeeding years; nuisance aquatic vegetation has decreased (Born et al., 1973).

Hydrogeologic information was critical for identifying and designing a rehabilitation approach, and for monitoring pumping operations. Groundwater flow-system data and water-quality information suggested the methodology. More detailed information about the lake flushing rate, involving water-budget and hydraulic-connection studies, permitted evaluation of the efficacy of the dilutional pumping scheme. Water-table condition and infiltration studies governed the selection and design of the pumped-water disposal area. Monitoring of the groundwater flow system during treatment influenced pumping rates and disposal area operations; furthermore, it insured protection of nearby individual well supplies and soil absorption fields. Lake sediment studies were invaluable in interpreting the results of treatment. The Snake Lake restoration project represents an excellent example of how hydrogeologic information can be used in the management of lakes; we suggest other lake management applications in a later section.

The Voyager Village Recreational Development Project

Voyager Village is a lake-oriented recreational land development in northwestern Wisconsin, typical of recent large-scale land developments in the nation. It is an innovative project and has been designed and planned with attention to environmental concerns. The shoreland area is being left undeveloped; none of the 3000 planned 0.2-ha (0.5-acre) lots, located on 1600 ha (4000 acre) of land, front directly on a water body. About 45% of the land is subdivided and extensive recreational facilities are planned on the remaining 55%.

The Voyager Village Project is located in a geologic environment of glacial-drift deposits. These deposits are predominantly sandy out-wash soils, averaging 46 m (150 ft.) thick and overlying Precambrian crystalline bedrock. The land surface is rolling and typical of "knob and kettle" topography. Extensive pre-development hydrogeologic and water-chemical investigations were conducted on-site. These investigations addressed aspects of: (1) site developability; (2) water supply; (3) waste disposal; (4) water quality available for recreational and consumptive purposes; (5) environmental impact analysis; and (6) the feasibility of creating two man-made lakes (originally planned to take anticipated user pressure off the eight natural lakes, and later eliminated from the development plan when hydrogeologic investigations illustrated the infeasibility of attaining desired lake levels because of calculated excessive seepage losses). Hydrogeologic studies included: water-level measurements from specially installed observations wells and surface-water control points; monthly cumulative precipitation measurements; geologic logs of observation wells and borings; evaluation of topographic and soil conditions; stream gauging; ground- and surface-water quality investigations; and water-budget analyses for several possible reservoir configurations for the man-made lakes.

Particular attention was given to identification of the shallow groundwater flow system and its relation to surface waters and to potential land-use patterns. At Voyager Village, the water table occurs in glacial drift, has a low gradient between lakes, and is almost flat beneath uplands. Despite the fact that the project area is in a humid region, high soil permeability was found

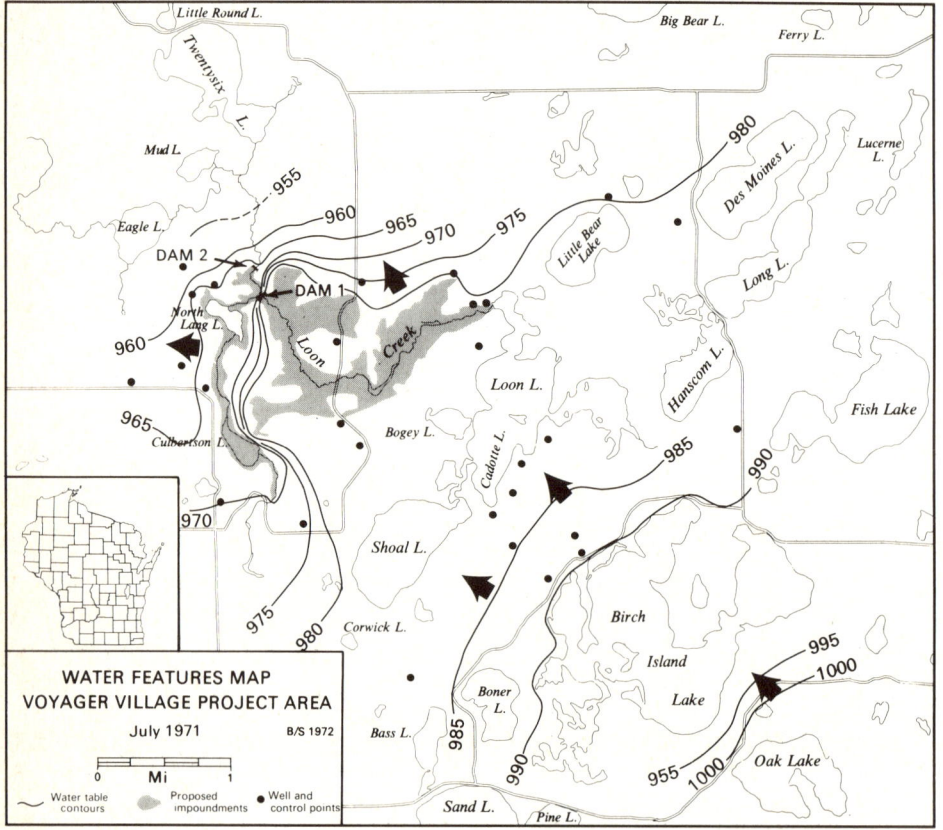

Fig. 8. Map of the Voyager Village, Wisconsin, development area showing water-table configuration and direction of groundwater movement (after Born and Stephenson, 1974).

to be a more important factor than topography in controlling water-table configuration. The groundwater flow system is unidirectional across the project site — a "flow-through". system (Fig. 8).

Hydrogeologic information was of prime interest to the developer for determination of septic-tank effluent migration direction and rates, and in evaluating the lake—groundwater relationships. The developer acquired information to help plan development patterns that would minimize both on-site and off-site environmental impacts. He also received data on water-table fluctuations (which were useful in deciding depth of water-supply wells) and background water quality and trophic state of the lakes prior to development (useful as lake management guidelines).

Table III summarizes the relationships between the type of hydrogeologic data collected and their application in planning and managing a large water-oriented recreational development.

TABLE III

Major impacts of water-regime variables on developers' concerns (after Born and Stephenson, 1974)

Hydrologic Factors	Developer Concerns	Site Developability				Water Supply	Waste Disposal			Surface Water Quality & Quantity	Environmental Impacts
		Building Sites	Roads	Utilities	Natural Hazards		Collection & Treatment	On-Site	Landfill		
GROUND-WATER REGIME	Water-table Position & Fluctuation	**X**	**X**	**X**	**X**		X	**X**	**X**	X	**X** (Wetlands)
	Aquifer Characteristics					**X**	X	X	X	**X**	
	Flow Systems	X			X		X	**X**	X	**X**	
	Ground-Water Quality					**X**	X	**X**	X	**X**	
SURFACE-WATER REGIME	Runoff	**X**	**X**	**X**	**X**	X	X			**X**	**X**
	Lakes & Wetlands	X	X	X		**X**	X	X	X	**X**	**X**
	Surface-Water Quality					**X**	X	X	X	**X**	**X**
GROUND WATER-SURFACE WATER RELATIONSHIPS						X	X	X	X	**X**	X
OTHER HYDROLOGIC BUDGET ASPECTS										**X**	

X — important relationship   X — consideration

GENERAL DISCUSSION

Lake management

We noted earlier that lakes are valued resources and that the public is greatly concerned about their protection and management. This concern has heightened as lake degradation, resulting from misuse and mismanagement, has become a major environmental problem (U.S.H.R.-C.G.O., 1967; Ketelle and Uttormark, 1971; Björk et al., 1972; Crossland and McCaull, 1972). Although lake quality has been impaired by the addition of heat and toxic substances, the most prevalent problems result from lake aging, i.e. eutrophication and sedimentation (Stewart and Rohlich, 1967; Vollenweider, 1968; N.A.S., 1969; G.F. Lee, 1970a; Born and Yanggen, 1972). Eutrophication and sedimentation are closely related, and when accelerated, commonly result in one or more of the following familiar lake problems: (1) algae blooms, with attendant odor problems; (2) nuisance rooted aquatic plant growth; (3) sediment infilling, diminished usable water surface, and related water-quality problems; and (4) declining fisheries and fishkills, attributable to dissolved oxygen depletion. In some situations, the problem is related to water quantity (lake-level fluctuations), rather than water quality. This array of problems has led federal, state and local governments, and private interests to greatly increase their management activities with regard to these vital resources (Kusler, 1971). The goal of these management efforts is generally to protect or maintain lake resources; however, in some cases it is too late for protection or maintenance, and rehabilitation must be the objective (Kinney et al., 1973; Dunst et al., 1974; Born, 1979).

The Snake Lake case history is only one of a very large number of lake-management examples (see Dunst et al., 1974). Many kinds of lake management efforts require hydrogeologic information (Born, 1970; Stephenson, 1971). Clearly, physical or water-quantity problems (declining or fluctuating water levels) demand hydrologic budget analyses that include both ground- and surface-water dimensions. Outstanding examples of the use of groundwater information in dealing with such problems are the Crystal Lake, Illinois, (Sasman, 1957) and Kingsbury Pond, Massachusets, (J.R. Williams, 1967) studies.

Many techniques have been suggested to deal with lake quality problems (Table IV). These techniques include two general approaches: (1) limiting fertility — either by restricting nutrient inputs or by direct in-lake schemes aimed at reducing nutrient recycling or accelerating nutrient outflows; and (2) treating or managing the symptoms of the water-quality problem. Hydrogeologic data are an important part of the background information needed to assess the nature and cause of a particular lake-quality problem as well as to define a management strategy. Regime dominance and system efficiency, along with groundwater quality are principal components in calculating a lake's flushing rate and nutrient budget, which in turn are primary considerations in water-quality management in lakes.

Any lake management or rehabilitation scheme that involves limiting nutrient inputs must be cognizant of the levels of natural nutrient loading from uncontrollable sources, including groundwater (G.F. Lee, 1970a). Major financial investments in advanced waste treatment for nutrient removal or waterworks for diverting nutrient-rich inflows may be wasted if background nutrient levels exceed critical values. Uttormark et al. (1974) provide a thorough discussion of nutrient loading from various point and diffuse sources, and document the substantial nutrient contributions that can be made by groundwater. At Mirror Lake, Wisconsin [4.5-ha (11-acre) surface

TABLE IV

Lake management and renewal techniques (after Dunst et al., 1974)

(I) Limiting fertility and controlling sedimentation

    (A) Curbing nutrient influx:
        (1) Wastewater treatment
        (2) Diversion
        (3) Land-use practices
        (4) Treatment of inflow
        (5) Product modification

    (B) In-lake schemes to accelerate nutrient outflow or prevent recycling:
        (1) Dredging for nutrient control
        (2) Nutrient inactivation/precipitation
        (3) Dilution/flushing
        (4) Biotic harvesting
        (5) Selective discharge
        (6) Sediment exposure and desiccation
        (7) Lake-bottom sealing

(II) Managing the consequences of lake aging

    (A) Aeration and/or circulation

    (B) Lake deepening:
        (1) Dredging
        (2) Drawdown and sediment consolidation

    (C) Other physical controls:
        (1) Harvesting
        (2) Water-level fluctuation
        (3) Habitat manipulation

    (D) Chemical controls:
        (1) Algicides
        (2) Herbicides
        (3) Pesticides

    (E) Biological controls:
        (1) Predator—prey relationships
        (2) Intra- and interspecific manipulation
        (3) Pathological reactions

area; 14-m (45-ft. maximum depth] for example, groundwater is responsible for an estimated 25% of incoming phosphorus and 34% of nitrogen (Possin, 1973).

Even where nutrient inputs have been curbed, nutrient transfer from lake sediments can maintain overfertile conditions for years and retard recovery of water quality (G.F. Lee, 1970b). Lake Trummen, a 1-km^2 (250-acre) lake in southern Sweden, was a recipient of waste waters from domestic and industrial sources for many years, and suffered from severe eutrophic conditions. Inflows were halted in 1959, but no improvement in water quality was observed during the next ten years. Investigations of sediment and interstitial water chemistry indicated that the nutrient-rich sediments in Lake Trummen were continuing to serve as a nutrient source (Björk et al., 1972). Such information, essentially hydrogeologic, dictated a lake rehabilitation technique involving removal of these sediments. This in-lake rehabilitation scheme — hydraulic dredging of the nutrient-rich sediments — resulted in the restoration of Lake Trummen.

Lake management schemes involving dredging raise a number of questions of a hydrogeologic nature. What are the hydraulic properties of the sediment to be removed, and what will be the effect of dredging with regard to hydraulic connection (lake sealing)? How will changes in the hydrogeologic regime affect the lake flushing rate? What changes can be anticipated in terms of nutrient recycling? Another important factor in lake dredging is disposal of the dredged spoils (Pierce, 1970). Dredged materials must be disposed of in a hydrogeologically innocuous location to prevent nutrient-rich leachate from re-entering the lake.

Some lake management schemes (see Table IV: lake-bottom exposure; lake-bottom manipulation; lake deepening via consolidation; water-level fluctuation; rough fish concentration and removal) require drawdown of lake levels; and the feasibility of drawdown may be governed by the hydrogeologic regime. Hydrogeology of a lake may be a comparatively minor concern where management schemes such as nutrient inactivation (see Peterson et al., 1973; Dunst et al., 1974) are employed. Even in these situations, however, hydrogeologic studies are needed to determine critical pre-treatment information with regard to flushing rates and nutrient budgets. In short, intelligent management of lakes generally requires some hydrogeologic understanding of the lake environs.

Development around lakes

In many situations, developmental activities on lake shorelands have detrimentally altered lakes. Hydrogeologic information about the shoreland area can be valuable for planning and managing shorelands. Attention should be given to those activities that might adversely affect: (1) water quantity available for residential supply or maintenance of lake levels (e.g., ascertain effects of pumping for water supply, of changed groundwater flow regime around lakes, or of alterations in water-table position and fluctuations); and

(2) water quality (e.g., what will be the impacts on both lake and adjacent groundwater from septic systems, landfills and altered surface-water runoff patterns).

One cause of changed water-quantity condition is related to development of domestic water supplies in the shoreland area. Low population densities coupled with high costs of treating surface water commonly leave groundwater development as the only viable option; then, the decision is whether individual wells or a central supply can be utilized. In either case, depth to water table and its seasonal fluctuations plus the groundwater reservoir characteristics governing withdrawal rates are critical factors. One lake-related supply concern is the potential for lake-level drawdown induced by pumpage.

Kingsbury Pond in Norfolk, Massachusetts, illustrates the adverse impact on lake level from development of a municipal water supply well. There, reversal of the water-table gradient between Kingsbury Pond and the pumping well caused a 4-m (13-ft.) decline in pond level. As a result, the pond's surface area was reduced from 3.6 to 1.1 ha (26 to 9 acre), leaving many residents without shore frontage and/or without use of docks and other shore facilities. Most existing private wells had to be deepened. Property values declined, and the market for homes was eroded (J.R. Williams, 1967).

The creation of Legend Lake, a recreational lake complex in northern Wisconsin, and the adjustment of the groundwater flow regime to impoundment is another example of adverse impact due to development (Hoffman and Meland, 1971; Born and Stephenson, 1974). During filling of this manmade lake, levels in natural lakes nearby were raised several feet. A groundwater divide that originally existed between the developing lakes and the natural lakes shifted, reversing groundwater flow gradients. The Legend Lake basin, once a discharge area, became a recharge area. Seepage across the land divide resulted in localized tree kills and other shorezone flooding adjacent to the "down-gradient" lakes. Such adverse impact is predictable with a minimum of hydrogeologic analysis in the pre-development stage. (In this case, lacking such investigations, the developer found it necessary to attempt remedial measures. Seepage water was pumped back to Legend Lake, and a clay seal was applied to the lake bottom, which decreased seepage losses but did not halt them.)

Development can alter the water budget by reducing the amount of water reaching the groundwater reservoirs. However, the greatest impacts of development and use of lake shorelands relate to potential water-quality degradation. Pollutants from businesses, streets, parking lots, yards and septic systems impair water quality in lakes. Over-fertilization is the primary problem, but in some lakes a health hazard may be posed where water supplies or beaches are contaminated.

Establishing the trophic status of lakes (Vollenweider, 1968; Shannon and Brezonik, 1972) is an important step in assessing their vulnerability to development. Although many lakes are naturally eutrophic, others are oligotrophic (infertile), and proper shore-zone development is necessary to minimize the potential for eutrophication and sedimentation. Developing a nutrient budget for a lake in order to utilize the resource wisely requires information about the groundwater flow system as well as the surface-water regime.

Solid and liquid waste disposal operations in the vicinity of lakes can adversely alter both ground- and surface-water quality. As with water supply, an initial low population density and a slow rate of growth in lake developments generally limit the options available for waste disposal. Centralized sewage treatment facilities may be too costly, and septic systems are widely utilized. Leachates from disposal operations (especially bacteria, nitrates, and some phosphates from septic systems; and chlorides, sulfates, iron, and manganese from landfills) can impair lake water quality as well as potability of groundwater supplies. Whereas groundwater rarely contains significant phosphorus concentrations (phosphorus is sorbed on most aquifer materials), nitrate-nitrogen is very poorly sorbed and thus travels with the groundwater flow system (Corey et al., 1967; Scalf et al., 1968; G.F. Lee, 1970a; Dudley and Stephenson, 1973).

Efforts to control or reduce nutrient contributions to lakes from groundwaters may include proper positioning of disposal systems around lakes in terms of location relative to the groundwater flow system (see Kusler, 1971). With the advent of clustered development, the aggregate downgradient impact of effluents is becoming a major concern for state agency personnel charged with issuing permits for septic-tank systems. The impact of a single, properly installed system is not as crucial as the collective impact where human activity and wastes have been concentrated.

Additional potential impacts of lake-oriented developments having hydrogeologic implications include: surface flooding upgradient from newly-constructed reservoirs because of altered water-table configurations, destruction of wetlands and springs, local changes in climate and reservoir bank erosion, among others. The ability to predict such hydrogeologic events prior to development is especially significant in this day of the environmental-impact statement.

CONCLUSIONS

The hydrogeologic regime of lakes must be adequately assessed in order to intelligently manage lakes and their related shorelands. The subject of groundwater—surface water relationship around lakes is characterized by sparse information and, in general, has received very limited attention. This paper represents a compilation and interpretation of the available information about the hydrogeology of lakes in glacial terrains. In that respect, this is intended to be a review paper.

We have inventoried and summarized information about lakes for which hydrogeologic data exist. The main hydrogeologic factors which are prerequisites for assessing lake environments are: (1) *regime dominance* (the relative magnitude of groundwater in the water budget of a lake); (2) *system efficiency* (a term describing the rate aspects of surface and groundwater movement through a lake system); and (3) *position within a groundwater flow system*. We have indicated the significance of measuring these descriptive characteristics, and have provided "type" examples of each hydrogeologic category. A preliminary classification framework for differentiating among lakes, based on hydrogeologic considerations, is presented; unfortunately, the available data base is too meager to thoroughly assess its applicability. However, as data accrue, our classification scheme can be subjected to further testing. We also believe that the classification approach is a useful conceptual tool for those individuals responsible for planning and managing lakes and related resources. To that end, we have attempted to show the value and use of hydrogeologic information in a variety of lake-related planning and management activities.

RECENT ADVANCES

Since this paper was first written in 1974 (Born et al., 1974), relatively little quantitative work has been done with regard to lake classification using hydrogeologic characteristics. The primary exception is the work of Winter (1976, 1977, 1978), described below.

Much of the new work has been qualitative and has involved the use of seepage meters designed by John and Lock (1977) and by D.R. Lee (1977). One reason for the surge in applications of this new tool may be that non-hydrologists seek out the most economical method of measuring hydrologic parameters, including groundwater seepage into lakes. Seepage meters are a comparatively low-cost method for acquiring groundwater data. But they are now being used so extensively in North America, and elsewhere, that it is time to take a rigorous look at these devices in order to evaluate the reproducibility of data obtained through their use.

The first application of our classification scheme was by Tolman (1975). Using our criteria he determined that White Clay Lake, in northern Wisconsin, is surface water dominated with a high system efficiency. However, where we have indicated that, under these conditions, position within a groundwater flow system would be of minor importance, Tolman has shown that flow-system position is significant enough to be considered in lake-management decisions.

McBride and Pfannkuch (1975) developed a two-dimensional digital model to examine the distribution of seepage below and through lake-bottom sediments. They tested results of their model on Lake Sallie, Minnesota, and found nearly all seepage flowing into a lake is concentrated in a narrow zone near shore; this near-shore seepage occurs independently of the presence of fine-grained low-permeability bottom sediments. There is an exponential decrease in seepage away from the shore.

The quantity of lake water contributed from adjacent groundwater systems has always been a difficult and expensive parameter to measure in the field.

Jaquet (1976) approached this problem by using numerous observation wells and chloride tracer analyses. He was able to identify the effective aquifer thickness in communication with Snake Lake, in northern Wisconsin.

Downing and Peterka (1978) found a correlation between rainfall and groundwater inflow rates to a lake in northern North Dakota using seepage meters described and evaluated by D.R. Lee (1977).

The most definitive work in North America in recent years on the interaction of lakes and groundwater is that of Winter (1976, 1978). He has shown through numerical simulation of groundwater flow that the existence, position and head value in the stagnation zone of the groundwater flow field relative to the head represented by the lake level is the key to understanding this interaction. He describes the position of a stagnation zone relative to a lake shore and is able to identify the area of a lake bed through which seepage occurs. Winter argues against using just a few observation wells around a lake, unless they are properly placed, and measuring only horizontal components of groundwater gradients relative to a lake.

Winter (1977) also attempted to develop a classification of the hydrologic setting of selected lakes in the north-central U.S.A. He classified 150 lakes according to their interchange with atmospheric water, surface water and groundwater. He concludes that the lake—groundwater relationship remains as the least known aspect of lake hydrology. Of the groundwater characteristics relative to a given lake, local relief and regional slope of the land surface are more important than is regional position. Texture of the soil or bedrock, which is related to hydraulic conductivity, is also very important.

Thus, it can be said that assorted scientists, managers, and other interested groups are all still "feeling their way" with regard to: (1) how to incorporate hydrogeologic data in lake-environment evaluations; (2) what data to incorporate; and (3) what is an acceptable efficacy—cost trade-off in a qualitative vs. a quantitative in-field investigation. Significant new information and the methodology to collect it does appear to be available on the subject of lake hydrology. However, we do not yet know the best way to economically evaluate the lake—groundwater relationship.

ACKNOWLEDGEMENTS

The investigations upon which this paper are based were supported by numerous agencies over a period of many years. Foremost of these support agencies is the Upper Great Lakes Regional Commission, an agency to which we are indebted. We gratefully acknowledge the following individuals for their helpful reviews at different stages in the writing of this paper: Messrs. C.W. Fetter, P.L. Hilpman, R.F. Kaufmann, R.S. McLeod, R.P. Novitzki, P.G. Olcott, J.O. Peterson, P.D. Uttormark and T.C. Winter. We also thank those staff of the Wisconsin Geological and Natural History Survey who lent assistance in preparation of the figures and who typed and proofed the manuscript. The authors share equal responsibility for this paper.

APPENDIX — LAKE INVENTORY

Lake name	Location	Area*1 (acre)	Maximum depth*2 (ft.)	Reference

(A) Lakes which have received the most intensive hydrogeologic analysis [this group includes: (a) lakes for which the groundwater flow system has been accurately defined, and/or (b) lakes for which the groundwater component of the water budget has been accurately determined]:

Lake name	Location	Area (acre)	Max depth (ft.)	Reference
Basin A	northeast Ill.	1	1	R.E. Williams (1968)
Basin B	northeast Ill.	8	1	R.E. Williams (1968)
Basin E	northeast Ill.	5	1	R.E. Williams (1968)
Deep Lake	Grand Coulee, Wash.	120	67 (mean depth)	Friedman and Redfield (1971)
East Twin Lake	Portage Co., Ohio	66	40	Cooke et al. (1973)
Jyme Lake	Oneida Co., Wis.	1	10	Smith et al. (1972)
Lenore, Lake	Grand Coulee, Wash.	1,100	27 (mean depth)	Friedman and Redfield (1971)
Little St. Germain Lake	Vilas Co., Wis.	950	16	Hackbarth (1968)
Mirror Lake	Waupaca Co., Wis.	11	45	Possin (1973)
Muskellunge Lake	Vilas Co., Wis.	275	18	Hackbarth (1968)
Pickerel Lake	Portage Co., Wis.	40	15	Hennings (1978)
Sallie, Lake	Becker Co., Minn.	1,200	55	Mann and McBride (1972); McBride (1972)
Shadow Lake	Waupaca Co., Wis.	40	38	Possin (1973)
Slough No. 3	Moose Mt., Sask.	0.7		Meyboom (1967)
Slough No. 15	Moose Mt., Sask.	4		Meyboom (1967)
Slough No. 16	Moose Mt., Sask.	1		Meyboom (1967)
Snake Lake	Vilas and Oneida Cos., Wis.	14	17	Born et al. (1973)
Soap Lake	Grand Coulee, Wash.	830	36 (mean depth)	Friedman and Redfield (1971)
West Twin Lake	Portage Co., Ohio	71	38	Cooke et al. (1973)
Willow Ring Slough	Allan Hills, Sask.	0.3	2	Meyboom (1966)

(B) Lakes which have received less intensive hydrogeologic analysis [this group includes: (a) lakes for which the water-table configuration has been determined, and/or (b) lakes for which the groundwater component of the water budget has been evaluated as a residual]:

Lake name	Location	Area (acre)	Max depth (ft.)	Reference
Booster Club Pond	Anoka Co., Minn.	7.5	3—6	Allred et al. (1971)
Browns Lake	Racine Co., Wis.	396	44	Cotter et al. (1969); S.E.W.R.P.C. (1969)
Butternut Lake	Forest Co., Wis.	1,292	45	Oakes et al. (1970)
Cadotte Lake	Burnett Co., Wis.	127	18	D.D. Huff and D.A. Stephenson, unpublished data (1970s)
Crystal Lake	Dane Co., Wis.	571	9	Deer, upublished data (1960s)
Crystal Lake	McHenry Co., Ill.	234	42	Sasman (1957)
Culbertson Lake	Burnett Co., Wis.	28	34	D.D. Huff and D.A. Stephenson, unpublished data (1970s)
Fish Lake	Dane Co., Wis.	252	62	Deer, unpublished data (1960s)

APPENDIX (continued)

Lake name	Location	Area[*1] (acre)	Maximum depth[*2] (ft.)	Reference
Franklin Lake	Forest Co., Wis.	881	53	Oakes et al. (1970)
Good Spirit Lake	Saskatchewan	~10,000	~15	Freeze (1968)
Kampeska, Lake	Codington Co., S. Dak.	4,800	14	Barari (1971b)
Kangaroo Lake	Door Co., Wis.	1,019	13	W.R.M. Workshop (1973)
Kingsbury Pond	Norfolk, Mass.	26	~30	J.R. Williams (1967)
Legend Lakes	Menominee Co., Wis.	1,270		Hoffman and Meland (1971); Born and Stephenson (1974)
Lily Lake	Forest Co., Wis.	206	20	Sharkawy and Graaskamp (1971)
Madison Lakes	Dane Co., Wis.	16,000	84	Cline (1965)
Morgan Lake	Florence Co., Wis.	43	10	Oakes et al. (1970)
Mud Lake	Sibley Co., Minn.	33	5—7	Allred et al. (1971)
Ontario, Lake (Canadian side)	Ontario	7,500 (mi.²)	800	Haefeli (1972)
Oyster Pond	near Woods Hole, Mass.	62	20	Emery (1969)
Pitcher Lake	near St. Paul, Minn.	8	9	Allred et al. (1971)
Poinsett, Lake	Hamlin Co., S. Dak.	9,800 (includes smaller nearby lake)	15	Barari (1971a)
Pothole C-1	near Buchanan, N. Dak.	~40		Sloan (1972)
Ria Lake	near Newport,	11	7.5	Allred et al. (1971);
Section 21 Potholes	Stutsman Co., N. Dak.	<2	0—2?	Sloan (1972)
Sherwood and Camelot Lakes	Adams Co., Wis.	500 (together)	30	B.N. Possin, unpublished data (1970s)
Shoal Lake	Burnett Co., Wis.	247	4	D.D. Huff and D.A. Stephenson, unpublished data (1970s)
Volo Bog Pond	Lake Co., Ill.	~1	~5	McComas et al. (1972)
Weber Lake	Vilas Co., Wis.	11	8	Juday and McLoche (1943); Fries (1938)

(C) Lakes which have received the least intensive hydrogeologic analysis [this group inclu includes: (a) lakes for which the water-table configuration has been approximated, and/ or (b) lakes for which the groundwater component of the water budget has been estimated]:

Duck Lake	Oakland Co., Mich.	253	10	Twenter and Knutilla (1972)
Elizabeth Lake	Oakland Co., Mich.	363	80	Twenter and Knutilla (1972)
Fresh Pond	Middlesex Co., Mass.	~125		Hallberg and Roberts (1949)
Houghton Lake	Roscommon Co., Mich.	19,600	20	Ellis and Childs (1973)

APPENDIX (continued)

Lake name	Location	Area*1 (acre)	Maximum depth*2 (ft.)	Reference
Judah Lake	Oakland Co., Mich.	115	10	Twenter and Knutilla (1972)
Krause Springs	Langlade Co., Wis.	0.9	12	Carline (1973)
Medicine Lake	Codington Co., S. Dak.	400	32.5	Tipton and Stockdale (1971)
Orchard Lake	Oakland Co., Mich.	788	60	Twenter and Knutilla (1972)
Quinsigamond, Lake	Worcester Co., Mass.	722	84	Salo and Cooperman (1972)
Silver Lake	Barron Co., Wis.	~200		Young and Hindall (1972)
Spiritwood Lake	Stutsman Co., N. Dak.	700	48	Schulte (1972)
Sunshine Springs	Langlade Co., Wis.	1	12	Carline (1973)
Tipisco Lake	Oakland Co., Mich.	301	25	Twenter and Knutilla (1972)
White Lake	Oakland Co., Mich.	540	25	Twenter and Knutilla (1972)

[*1] 1 acre = 0.40468 ha; 1 mi.^{2} = 2.5899 km^2.
[*2] 1 ft. = 0.3048 m.

REFERENCES

Allred, E.R., Manson, P.W., Schwartz, G.M., Golany, P. and Reinke, J.W., 1971. Continuation of studies on the hydrology pf ponds and small lakes. Univ. Minn. Agric. Exp. Stn., Tech. Bull. 274, 62 pp.

Back, W., 1966. Hydrochemical facies and groundwater flow patterns in northern part of Atlantic Coastal Plain. U.S. Geol. Surv., Prof. Pap. 498-A, 42 pp.

Barari, A., 1971a. Hydrology of Lake Kampeska. S.D., Geol. Surv., Rep. Invest. 103, 84 pp.

Barari, A., 1971b. Hydrology of Lake Poinsett. S.D., Geol. Surv., Rep. Invest. 102, 69 pp.

Birge, E.A. and Juday, C., 1934. Particulate and dissolved organic matter in inland lakes. Ecol. Monogr., 4: 440—474.

Björk, S., Bengtsson, L., Berggren, H., Cronberg, G., Digerfeldt, G., Fleischer, S., Gelin, C., Lindmark, G., Malmer, N., Plejmark, F., Ripl, W. and Swanberg, P.O., 1972. Ecosystem studies in connection with the restoration of lakes. Verh. Int. Ver. Limnol., 18: 379—387.

Born, S.M., 1970. The role of hydrogeology in lake management. Annu. Meet. Geol. Soc. Am., Abstr. Programs, 2(7): 499.

Born, S.M., 1974. Inland Lake Demonstration Project final report (1968—1974). Univ. Wis. Ext.—Wis. Dep. Nat. Resour., Upper Great Lakes Reg. Comm., 31 pp.

Born, S.M., 1979. Lake rehabilitation: a status report. Environ. Manage., 3(2): 27—35.

Born, S.M. and Stephenson, D.A., 1974. Environmental geologic aspects of planning, constructing, and regulating recreational land developments. Univ. Wis. Ext.—Wis. Dep. Nat. Resour. Inland Lake Demonstr. Proj., Rep., Upper Great Lakes Reg. Comm., 40 pp.

Born, S.M. and Yanggen, D.A., 1972. A resource-protective approach to shoreland development, including innovative private controls for recreational land development. Proc. Conf. on Inland Lake Renewal and Shoreland Management. Univ. Minn. Water Resour. Res. Cent., Bull. 53, pp. 17—21.

Born, S.M., Wirth, T.L., Peterson, J.O., Wall, J.P. and Stephenson, D.A., 1973. Dilutional pumping at Snake Lake, Wisconsin — a potential renewal technique for small eutrophic lakes. Wis. Dep. Nat. Resour., Tech. Bull. 66, 32 pp.

Born, S.M., Smith, S.A. and Stephenson, D.A., 1974. The hydrogeologic regime of glacial-terrain lakes, with management and planning applications. Univ. Wis. Ext.—Wis. Dep. Nat. Resour., Inland Lake Demonstr. Proj., Rep., Upper Great Lakes Reg. Comm., 81 pp.

Britton L.J., Averett, R.C. and Ferreira, R.F., 1975. An introduction to the processes, problems, and management of urban lakes. U.S. Geol. Surv., Circ. 601-K, 22 pp.

Broughton, W.A., 1941. The geology, ground water and lake basin seal of the region south of the Muskellunge Moraine, Vilas County, Wisconsin. Wis. Acad. Sci. Arts Lett. Trans., 33: 5—20.

Carline, R.F., 1973. Spring pond research — a progress report. Wis. Dep. Nat. Resour., 18 pp.

Cline, D.R., 1965. Geology and ground-water resources of Dane County, Wisconsin. U.S. Geol. Surv., Water-Supply Pap. 1779-U, 64 pp.

Cooke, G.D., McComas, M., Bhargava, T.N. and Heath, R., 1973. Monitoring and nutrient inactivation on two glacial lakes (Ohio) before and after nutrient diversion. K. State Univ., Cent. Urban Reg., Interim Res. Rep., 92 pp.

Corey, R.D., Hasler, A.D., Schraufnagel, F.H. and Wirth, T.L., 1967. Excessive water fertilization. Madison, Wis., Rep. to Water Subcomm., Nat. Resour. Counc. State Agencies (unpublished).

Cotter, R.D., Hutchinson, R.D., Skinner, E.L. and Wentz, D.A., 1969. Water resources of Wisconsin Rock-Fox River Basin. U.S. Geol. Surv., Hydrol. Invest. Atlas HA-360 (4 sheets).

Crossland, J. and McCaull, J., 1972. Overfed. Environment, 14: 30—37.

Dettman, E.H. and Huff, D.D., 1972. A lake water balance model. At. Energy Comm.—Oak Ridge Lab., U.S. Int. Biol. Program Eastern Deciduous For. Biom. Rep. 72-126, 22 pp.

Downing, J.A. and Peterka, J.J., 1978. Relationship of rainfall and lake groundwater seepage. Limnol. Oceanogr., 23(4): 821—825.

Dudley, J.G. and Stephenson, D.A., 1973. Nutrient enrichment of ground water from septic tank disposal systems. Univ. Wis.—Wis. Dep. Nat. Resour., Inland Lake Demonstr. Proj. Rep., Upper Great Lakes Reg. Comm., 131 pp.

Dunst, R.D., Born, S.M., Uttormark, P.D., Smith, S.A., Nichols, S.A., Peterson, J.O., Knauer, D.R., Serns, S.L., Winter, D.R. and Wirth, T.L., 1974. Survey of lake rehabilitation techniques and experiences. Wis. Dep. Nat. Resour., Tech. Bull. 75, 179 pp.

Ellis, B. and Childs, K.E., 1973. Nutrient movement from septic tanks and lawn fertilization—Water Quality Protection Project, Houghton Lake, Michigan. Mich. Dep. Nat. Resour., Tech. Bull. 73-5, 69 pp.

Emery, K.O., 1969. A Coastal Pond Studied by Oceanographic Methods. American Elsevier, New York, N.Y., 80 pp.

Ferris, J.G., Knowles, D.B., Brown, R.H. and Stallman, R.W., 1962. Theory of aquifer tests. U.S. Geol. Surv., Water-Supply Pap. 1536-E, 174 pp.

Freeze, R.A., 1968. Quantitative interpretation of regional ground water flow patterns as an aid to water balance studies. In: Groundwater. Proc. Gen. Assem., Int. Assoc. Sci. Hydrol. Publ. No. 77, pp. 154—173.

Freeze, R.A. and Witherspoon, P.A., 1966. Theoretical analysis of regional ground-water flow, 1. Analytical and numerical solutions to the mathematical model. Water Resour. Res., 2: 641—656.

Freeze, R.A. and Witherspoon, P.A., 1967. Theoretical analysis of regional ground-water flow, 2. The effect of water-table configuration and subsurface permeability variation. Water Resour. Res., 3: 623—634.

Freeze, R.A. and Witherspoon, P.A., 1968. Theoretical analysis of regional ground-water flow, 3. Quantitative interpretations. Water Resour. Res., 4: 581—590.

Frey, D.G. (Editor), 1966. Limnology in North America. University of Wisconsin Press Madison, Wis., 734 pp.

Friedman, I. and Redfield, A.C., 1971. A model of the hydrology of the lower Grand Coulee, Washington. Water Resour. Res., 7: 874—898.

Fries, Jr., C., 1938. Geology and ground water of the Trout Lake region, Vilas County, Wisconsin. Wis. Acad. Sci. Arts Lett. Trans., 31: 305—322.

Hackbarth, D.A., 1968. Hydrogeology of the Little St. Germain Lake Basin, Vilas County, Wisconsin. M.S. Thesis, University of Wisconsin, Madison, Wis., 76 pp.

Haefeli, C.J., 1972. Groundwater inflow into Lake Ontario from the Canadian side. Can. Dep. Environ., Inland Waters Branch Sci. Ser., No. 9, 101 pp.

Halberg, H.N. and Roberts, C.M., 1949. Recovery of ground water supplies by pumping from water-table ponds. Am. Geophys. Union Trans., 30: 283—292.

Hennings, R.G., 1978. The hydrogeology of a sand plain seepage lake, Portage County, Wisconsin. M.S. Thesis, University of Wisconsin, Madison, Wis., 69 pp.

Hoffman, J.I. and Meland, N., 1971. The effect of an artificial lake development complex on the groundwater system. Univ. Wis.—Oshkosh, Dep. Geol. Geogr., 10 pp.

Hubbert, M.K., 1940. The theory of ground-water motion. J. Geol., 48: 785—944.

Hutchinson, G.E., 1957. A Treatise on Limnology, Vol. 1. Wiley, New York, N.Y., 1015 pp.

Jaquet, N.G., 1976. Ground-water and surface-water relationships in the glacial province of northern Wisconsin — Snake Lake. Ground Water, 14(4): 194—199.

John, P.H. and Lock, M.A., 1977. The spacial distribution of groundwater discharge into the littoral zone of a New Zealand lake. J. Hydrol., 33: 391—395.

Juday, C. and Meloche, V.W., 1943. Physical and chemical evidence relating to the lake basin seal in certain areas of the Trout Lake region of Wisconsin. Wis. Acad. Sci. Arts Lett. Trans., 35: 157—174.

Ketelle, M.J. and Uttormark, P.D., 1971. Problem lakes in the United States. Univ. Wis. Water Resour. Cent., Tech. Rep. 16010 EHR, 282 pp.

Kinney, W.L., Lewis, J.I., Guarrziz, L.J., Czting, J.P., Boynton, D.K., Wilkes, F.G., Dyer, R.S., Maloney, T.E., Malueg, K.W., Shults, D.W., Powers, C.F., Peterson, S.A., Szuville, W.D. and Stzy, Jr., F.S., 1973. Measures for the restoration and enhancement of quality of freshwater lakes. U.S. Environ. Prot. Agency, Rep. EPA-430/9-73-005, 238 pp.

Knauer, D.R. and Peterson, J.O., 1974. Nutrient inactivation as a lake management technique — a summary report. Wis. Dep. Nat. Resour. Tech. Bull.

Kusler, J.A., 1971. Regulations for disposal of rural domestic liquid wastes in Wisconsin: a review. Univ. Wis. Ext. Inland Lake Demonstr., Proj. Rep., Upper Great Lakes Reg. Comm., 94 pp.

Lee, D.R., 1977. A device for measuring seepage flux in lakes and estuaries. Limnol. Oceanogr., 22(1): 140—147.

Lee, G.F., 1970a. Eutrophication. Univ. Wis. Water Resour. Cent., Eutrophication Inf. Program, Occas. Pap. 2, 39 pp.

Lee, G.F., 1970b. Factors affecting the transfer of materials between water and sediments. Univ. Wis. Water Resour. Cent., Eutrophication Inf. Program, Lit. Rev. 1, 50 pp.

Maclay, R.W., Winter, T.C. and Bidwell, L.E., 1972. Water resources of the Red River of the North drainage basin in Minnesota. U.S. Geol. Surv., Water Resour. Invest. 1-72, 129 pp.

Mann IV, W.B. and McBride, M.S., 1972. The hydrologic balance of Lake Sallie, Becker County, Minnesota. U.S. Geol. Surv., Prof. Pap., 800-D: D189—D191.

Manson, P.W., Schwartz, G.M. and Allred, E.R., 1968. Some aspects of the hydrology of ponds and small lakes. Univ. Minn. Agric. Exp. Stn., Tech. Bull. 257, 88 pp.

Maxey, G.B., 1968. Hydrogeology of desert basins. Ground Water, 6: 10—22.

McBride, M.S., 1972. Hydrology of Lake Sallie, northwestern Minnesota, with special attention to ground water—surface water interactions. M.S. Thesis, University of Minnesota, St. Paul, Minn., 62 pp.

McBride, M.S. and Pfannkuch, H.O., 1975. The distribution of seepage within lakes. U.S. Geol. Surv., J. Res., 3(5): 505—512.

McComas, M.R., Kempton, J.P. and Hinckley, K.C., 1972. Geology, soils, and hydrogeology of Volo Bog and vicinity, Lake County, Illinois. Ill. State Gol. Surv., Environ. Geol. Note 57, 27 pp.

Meyboom, P., 1962. Patterns of groundwater flow in the prairie profile. Proc. 3rd Can. Hydrol. Symp., Natl. Res. Coun. Can., pp. 5—33.

Meyboom, P., 1966. Unsteady groundwater flow near a willow ring in hummocky moraine. J. Hydrol., 4: 38—62.

Meyboom, P., 1967. Mass transfer studies to determine the groundwater regime of permanent lakes in hummocky moraine of western Canada. J. Hydrol., 5: 117—142.

Meyboom, P., VanEverdingen, R.D. and Freeze, R.A., 1966. Patterns of ground-water flow in seven discharge areas in Saskatchewan and Manitoba. Geol. Surv. Can. Bull. 147, 48 pp.

Mifflin, M.D., 1968. Delineation of groundwater flow systems in Nevada. Univ. Nev. Desert Res. Inst., Tech. Rep. Ser. HW, No. 4, 54 pp.

N.A.S. (National Academy of Sciences), 1969. Eutrophication: Causes, Consequences, and Correctives. Government Printing Office, Washington, D.C., 661 pp.

Oakes, E., Field, S.J., Seeger, L.P., 1970. The Pine-Popple River basin — hydrology of a wild river area, northeastern Wisconsin. U.S. Geol. Surv., Water-Supply Pap. 2006.

Peterson, J.O., Wall, J.P., Wirth, T.L. and Born, S.M., 1973. Eutrophication control: nutrient inactivation by chemical precipitation at Horseshoe Lake, Wisconsin. Wis. Dep. Nat. Resour., Tech. Bull. 62, 20 pp.

Pierce, N.D., 1970. Inland lake dredging evaluation. Wis. Dep. Nat. Resour., Tech. Bull. 46, 68 pp.

Pinder, G.F. and Bredehoeft, J.D., 1968. Application of digital computer for aquifer evaluation. Water Resour. Res., 4: 1069—1093.

Possin, B.N., 1973. Hydrogeology of Mirror and Shadow Lakes in Waupaca, Wisconsin. M.S. Thesis, University of Wisconsin, Madison, Wis., 85 pp.

Rainey, R.H., 1967. Natural displacement of pollution from the Great Lakes. Science, 155: 1242—1243.

Salo, J.E. and Cooperman, A.N., 1972. Lake Quinsigamond water quality study — 1971. Mass. Water Resour. Comm., Div. Water Pollut. Control, Publ. 6306, 78 pp.

Sasman, R.T., 1957. The water level problem at Crystal Lake, McHenry County, Illinois. Ill. State Water Surv., Rep. Invest., 32, 27 pp.

Scalf, M.R., Hauser, V.L., McMillion, L.G., Dunlap, W.J. and Keeley, J.W., 1968. Fate of DDT and nitrate in ground water. U.S. Dep. Inter.—U.S. Dep. Agric., Agric. Res. Serv. Texas.

Schulte, F.J., 1972. Groundwater of the Spiritwood Lake area, Stutsman County, North Dakota. Ph.D. Thesis, University of North Dakota, Grand Forks, N. Dak., 329 pp.

Schwartz, G.M. and Manson, P.W., 1964. Pothole drainage and deep ground water supplies. Minn. Farm Home Sci., 21: 13—15.

S.E.W.R.P.C. (Southeastern Wisconsin Regional Planning Commission), 1969. A Comprehensive Plan for the Fox River Watershed. Waukesha, Wis.

Shannon, E.E. and Brezonik, P.L., 1972. Relationships between lake trophic state and nitrogen and phosphorus loading rates. Environ. Sci. Tech., 6: 719—725.

Sharkawy, M.A. and Graaskamp, J.A., 1971. Inland lakes renewal and management demonstration — Lily Lake forest recreational environ. Univ. Wis. Environ. Awareness Cent., 154 pp.

Sloan, C.E., 1972. Ground-water hydrology of prairie potholes in North Dakota. U.S. Geol. Surv., Prof. Pap. 585-C, 28 pp.

Smith, S.A., Peterson, J.O., Nichols, S.A. and Born, S.M., 1972. Lake deepening by sediment consolidation — Jyme Lake. Univ. Wis. Ext.—Wis. Dep. Nat. Resour. Inland Lake Demonstr. Proj. Rep., Upper Great Lakes Reg. Comm., 36 pp.

Sonzogni, W.C. and Lee, G.F., 1974. Diversion of wastewaters from Madison lakes. Am. Soc. Civil Engrs. Trans., Jour. Environmental Engineering Div., 100: 153—170.

Stearns, H.T., 1942. Hydrology of volcanic terranes. In: O.E. Meinzer (Editor), Hydrology. McGraw-Hill, Toronto, Ont., pp. 678—703.

Stephenson, D.A., 1971. Groundwater flow system analysis in lake environments, with management and planning implications. Water Resour. Bull., 7: 1038—1047.

Stewart, K. and Rohlich, G.A., 1967. Eutrophication: a review. Calif. State Water Qual. Control Board, Publ. 34, 188 pp.

Tipton, M.J. and Stockdale, R.G., 1971. A geologic study of the chemical quality of Medicine Lake. S. Dak. Univ. Water Resour. Res. Inst., 50 pp.

Tolman, A.L., 1975. The hydrogeology of the White Clay Lake area, Shawano County, Wisconsin. M.S. Thesis, University of Wisconsin, Madison, Wis., 55 pp.

Toth, J., 1962. A theory of groundwater motion in small drainage basins in central Alberta, Canada. J. Geophys. Res., 67: 4375—4387.

Toth, J., 1963. A theoretical analysis of ground water flow in small drainage basins. J. Geophys. Res., 68: 4795—4812.

Twenter, F.R. and Knutilla, R.L., 1972. Water for a rapidly growing urban community — Oakland County, Michigan. U.S. Geol. Surv. Water-Supply Pap. 2000, 150 pp.

U.S.H.R.-C.G.O. (U.S. House of Representatives, Committee on Government Operations), 1967. To save America's small lakes (water pollution control and abatement). U.S. House Reprent., Comm. Gov. Oper., House Rep. 594, 20 pp.

Uttormark, P.D., Rohlich, G.A., Chapin, J.D. and Green, K.M., 1974. Estimating nutrient loadings of lakes. U.S. Environ. Prot. Agency Rep.

Viessman, W., 1971. Estimation of lake flushing rates for water quality control planning and management. Proc. Workshop-Conf. on Reclamation of Maine's Dying Lakes, March 24—25, 1971. Univ. Maine, Bangor, Maine, Water Resour. Cent., pp. 50—66.

Vollenweider, R.A., 1968. The scientific basis of lake and stream eutrophication, with particular reference to phosphorus and nitrogen as eutrophication factors. Organ. Econ. Coop. Dev., Tech. Rep. DAS/C51/68, No. 27, 182 pp.

Williams, J.R., 1967. Drastic lowering of Kingsbury Pond, Norfolk, Massachusetts. U.S. Geol. Surv., Open-File Rep., 33 pp.

Williams, R.E., 1968. Flow of ground water adjacent to small, closed basins in glacial till. Water Resour. Res., 4: 777—784.

Winograd, I.J., 1962. Interbasin movement of ground water at the Nevada test site, Nevada. U.S. Geol. Surv. Prof. Pap. 450-C: C108—C111.

Winter, T.C., 1976. Numerical simulation analysis of the interaction of lakes and ground water. U.S. Geol. Surv., Prof. Pap. 1001, 45 pp.

Winter, T.C., 1977. Classification of the hydrologic settings of lakes in the north-central United States. Water Resour. Res., 13(4): 753—767.

Winter, T.C., 1978. Numerical simulation of steady state threedimensional groundwater flow near lakes. Water Resour. Res., 14(2): 245—254.

W.R.M. (Water Resources Management) Workshop, 1973. Groundwater quality — Door County, Wisconsin. Univ. Wis. Inst. Environ. Stud., Rep. 7, 202 pp.

Young, H.L. and Hindall, S.M., 1972. Water resources of Wisconsin — Chippewa River basin. U.S. Geol. Surv., Hydrol. Invest. Atlas HA-386 (4 sheets).

[2]

THE ROLE OF GROUNDWATER IN STORM RUNOFF

MICHAEL G. SKLASH and ROBERT N. FARVOLDEN

Department of Geology, University of Windsor, Windsor, Ont. N9B 3P4 (Canada)
Department of Earth Sciences, University of Waterloo, Waterloo, Ont. N2L 3G1 (Canada)

(Accepted for publication April 25, 1979)

ABSTRACT

Sklash, M.G. and Farvolden, R.N., 1979. The role of groundwater in storm runoff. In: W. Back and D.A. Stephenson (Guest-Editors), Contemporary Hydrogeology — The George Burke Maxey Memorial Volume. J. Hydrol., 43: 45—65.

Groundwater plays a much more active, responsive and significant role in the generation, of storm and snow-melt runoff in streams than the recent literature on the subject suggests. Basin-wide tracer experiments using environmental isotopes (^{18}O, deuterium, tritium) and hydrometric studies carried out in hydrogeologically diverse watersheds, indicate that for all except the most intense rain storms and the most prolific melting days, groundwater dominates the runoff hydrographs in the study basins. The increased groundwater discharge during runoff events is apparently related to a rapid rise in hydraulic head along the perimeter of transient and perennial discharge areas. This groundwater ridging phenomenon probably arises from the almost instantaneous conversion of the near-surface tension-saturated capillary fringe into phreatic water. The ridging precedes, and is apparently independent of the response of the rest of the basin. In addition to its compatibility with several of the field observations commonly associated with contemporary concepts of runoff generation, the groundwater discharge theory explains some of the temporal variations in stream water chemistry which are not adequately accounted for by other theories.

INTRODUCTION

Most of the recent literature on storm runoff generation has overlooked true groundwater flow as a significant and active factor in the storm and snow-melt runoff process. Freeze (1974) summarized the hydrologic thought on the subject as:

"True groundwater flow is seldom the cause of the major runoff during storms. Its primary role is in sustaining streams during low-flow periods between rainfall and snow-melt events . . ."

Indeed, it is difficult to conceptualize how slow-moving groundwater can respond rapidly enough to contribute to a storm or snow-melt runoff peak. In the past decade, however, basin-wide tracer experiments using environmental isotope techniques have demonstrated that groundwater often dominates snow-melt runoff in humid to sub-humid regions (e.g., Dincer et al., 1970; Martinec, 1975). Fritz et al. (1976) and Sklash (1978) have reported the occurrence of large groundwater components in both storm and snow-melt runoff in their environmental isotope studies of eight hydrogeologically diverse watersheds in Canada. The exact mechanism which enables groundwater to appear so quickly and in such large quantities in the stream during high runoff events, however, has yet to be established.

The purpose of this paper is two-fold. It presents further documentation for the active, responsive and significant role of groundwater in storm runoff. Secondly, it introduces a theory for storm (and snow-melt) runoff generation which, in addition to explaining how large quantities of groundwater can appear in the peak storm or snow-melt runoff, is compatible with some field phenomena which one might ascribe to other theories for runoff generation.

CONCEPTUAL MODELS FOR STORM RUNOFF GENERATION

The study of storm (and snow-melt) runoff generation can be approached in a variety of ways. For example, one may consider only what portion of the runoff water existed in the basin prior to the runoff event and what portion was added by the runoff-inducing event. Alternatively, one may choose only to examine the manner in which the runoff water travels over the last several tens of metres to the stream. A third approach could be concerned only with the history of the water from its arrival in the basin to its ultimate delivery to the stream. These three approaches, termed: time aspects, ultimate delivery mechanism aspects, and historical aspects, respectively, are outlined in Fig. 1.

Currently, the most widely accepted theories for storm (and snow-melt) runoff generation stem from the ultimate delivery mechanism approach to the subject. Each concept attempts to account for both the rapid response of the stream to runoff-inducing events and the observed increase in stream discharge. Freeze (1974) has summarized the most popular of these theories as: (1) partial area—overland flow; (2) variable source area—overland flow; and (3) variable source area—subsurface storm flow. Brief explanations of these mechanisms follow, however, the reader is referred to Freeze (1974) which provides an excellent review and reference list for each mechanism.

The partial area—overland flow concept suggests that runoff water is produced mainly from certain fixed portions of the watershed (usually controlled by soil characteristics) where the soil becomes saturated from above by the infiltrating water. After surface detention requirements have been satisfied, the excess water runs off rapidly to the stream as overland flow.

Perhaps the most widely accepted runoff generation mechanism concept is the variable source area—overland flow theory. In essence, runoff is generated from watershed areas (usually controlled by topography, geology and soil type) which have become saturated from below by a rising water table. The source areas, generally located near the stream, may expand and

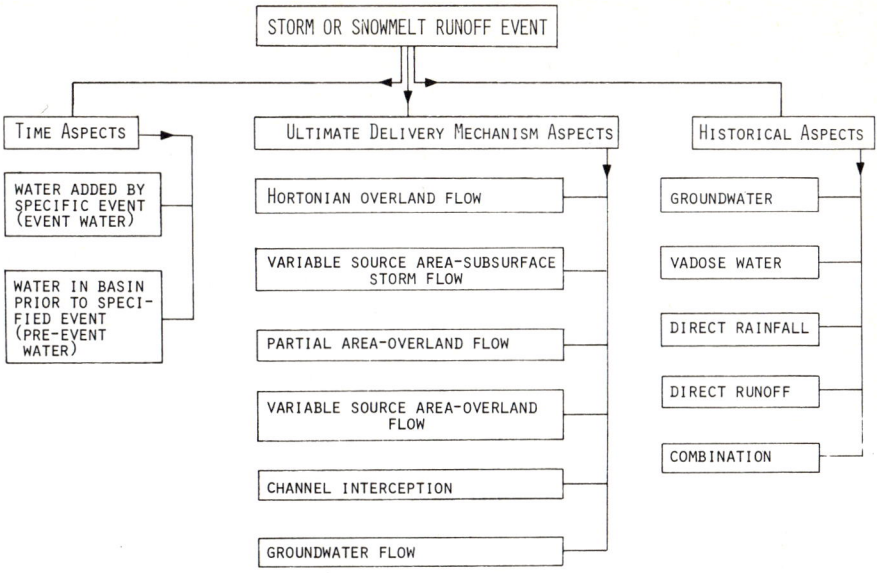

Fig.1. Classification chart for runoff generation terminology.

contract in response to climatic factors. The runoff from the variable source area, consisting of both rain and "return flow", runs off to the stream rapidly as overland flow.

The variable source area—subsurface storm flow concept states that the areas contributing to storm runoff expand and contract in response to climatic factors. Unlike the variable source area—overland flow theory, the transfer of water from the hillside to the stream is accomplished through subsurface routes. An expanding channel network and translatory flow (displacement or bumping of subsurface water toward the stream) allows the runoff water to reach the stream quickly. The subsurface storm flow may be either saturated—unsaturated Darcian flow through the porous soil matrix or turbulent flow through root channels, animal burrows and soilcracks (Cheng et al., 1975; Beasley, 1976).

Most of the research indicating groundwater as a significant factor in the storm (and snow-melt) runoff process can be classified under the heading of time aspect studies. Under this classification, the water which existed in the basin prior to a specified event is considered pre-event water and consists of groundwater, vadose water and surface storage. Water added to the basin by the specified runoff-inducing event, either a rain or snow-melt event, is called event water. Basin-wide tracer experiments involving such parameters as specific conductance, environmental isotopes, and conservative ions, have been used to distinguish pre-event water from event water (e.g., Pinder and Jones, 1969; Martinec, 1975). Groundwater contributions to peak storm and snow-melt runoff have been reported to exceed 50% in some instances.

The time aspect studies have all been stream-oriented; that is, their conclusions are based mainly on the examination of temporal variations of selected tracers in the stream. The ultimate delivery mechanism studies have all been slope-oriented; that is, their conclusions are based mainly on the examination of processes occurring somewhere above the stream. The conclusions from these two approaches have generally been divergent. Historical aspect studies attempt to identify changes in the runoff water chemistry while enroute to the stream in order to reconcile the slope- and stream-oriented results. Kennedy (1977) suggests that rain may infiltrate into microridges (furrows), dissolve soluble materials, discharge from the furrows, and then run off to the stream as overland flow. In this manner, the chemistry of the runoff water may approach that of the groundwater and thereby confound any tracing attempts.

ENVIRONMENTAL ISOTOPE AND SPECIFIC CONDUCTANCE TECHNIQUES

Dincer et al. (1970), Martinec et al. (1974), Fritz et al. (1976), and others have demonstrated that environmental isotopes such as: ^{18}O, deuterium (D) and tritium (T), can be used to distinguish pre-event water from event water in the stream during runoff events. The storm (or snow-melt) runoff hydrograph can then be separated into its simple time components (pre-event and event water) by the simultaneous solution of the steady-state mass balance equations describing the fluxes of water and the tracer isotope in the stream. These equations are of the form:

$$Q_t = Q_p + Q_e \quad \text{and} \quad C_t Q_t = C_p Q_p + C_e Q_e \qquad (1),(2)$$

where Q is the discharge (L^3/T), C represents the tracer concentration, and subscripts t, p and e refer to the total stream discharge, pre-event component, and event component, respectively. Q_p and Q_e are unknowns which can be determined readily subject to the following criteria:

(1) The isotopic content (^{18}O, D, or T) of the event component is significantly different from that of the pre-event component.

(2) The event component maintains a constant isotopic content.

(3) The groundwater and vadose water are isotopically equivalent or vadose water contributions to runoff are negligible due to hydrogeologic constraints.

(4) Surface storage contributes minimally to the runoff event.

The environmental isotope technique for hydrograph separation is the dominant tool in this study. Isotopic, chemical and specific-conductance analyses of the various runoff components enhance the hydrograph interpretations.

HYDROMETRIC TECHNIQUES

The main hydrometric technique in this study involves the establishment of groundwater stage—groundwater discharge rating curves for hydrograph separation. According to this method, curves relating groundwater stage to stream discharge during baseflow periods (when stream discharge is assumed to be only groundwater) are assumed to hold true during high runoff episodes. By monitoring groundwater stage during runoff events, groundwater discharge is estimated by referring to the groundwater stage—groundwater discharge rating curve.

Unlike many of the previous studies (Schicht and Walton, 1961; Stevenson, 1967; Visocky, 1970) which used average groundwater stages for a number of index wells throughout the watershed, small-diameter recording wells installed in the stream bed and a few metres from the stream provide the groundwater stage data for one of the basins in this study. Although the individual wells may not represent the average response of the basin, they are indicative of the nature of the near-stream groundwater response.

Water level responses of wells and piezometers located farther from the stream offer some indication of the groundwater response of the remainder of the basin to runoff-inducing events.

WATERSHED MODELLING

The response of small hypothetical watersheds to runoff-inducing events is examined with the aid of a two-dimensional saturated—unsaturated transient finite-element flow model developed by Segol (1976). Observations are focussed on the groundwater response in the near-stream area when near-stream watershed relief and basin width are varied.

THE STUDY AREAS

Although several small watersheds have been examined during this study, only two, Ruisseau des Eaux Volées and Hillman Creek, are discussed here. The results from the other study basins, dealing mainly with snow-melt runoff, corroborate the findings reported in this paper.

The Ruisseau des Eaux Volées experimental watershed, situated at latitude 47°16′N and longitude 71°09′W, lies some 80 km north of Quebec City, Quebec, Canada. The basin is typical of many of the hanging valley tributaries found in the Laurentian Uplands of the Canadian Shield. Two subbasins, the west branch of Ruisseau des Aulnaies (subbasin 7A) and the upper portion of Ruisseau des Eaux Volées (subbasin 6), were monitored during the study (Fig.2). Subbasin 7A occupies some 1.2 km² of heavily forested land between the 880- and 760-m elevation contours. Subbasin 6 covers 3.9 km² of heavily forested and experimentally logged land between 780 and 720 m in elevation. The climate of the area is considered moist cold temperate (Plamondon and Naud, 1975).

Surficial deposits, mainly of glacial origin, cover 80% of the basin to depths of 1—20 m. Jointed charnockitic gneiss, underlying the entire basin, outcrops only in the high areas (Rochette, 1971).

The groundwater system is considered unconfined with the bedrock groundwater demonstrating hydraulic connection to the surficial materials (Rochette, 1971). The water table, even in high areas, is close to ground surface.

Groundwater in the surficial materials is characterized by $\delta^{18}O$ and T values of approximately $-12.8^0/_{00}$ and 75 TU, respectively (all ^{18}O analyses are given relative to the SMOW standard). Bedrock groundwater has ^{18}O and T values of about $-11.9^0/_{00}$ and 175 TU, respectively. Baseflow ^{18}O values at the stream gauging stations for both subbasins range from $-12.6^0/_{00}$ in the winter to $-11.9^0/_{00}$ in the summer.

Suction lysimeters (soil moisture samplers) installed between depths of 0.26 and 0.47 m below ground surface were used to extract samples of vadose water from several sites (Fig.2). The ^{18}O analyses of the vadose water indicate some influence by the infiltrating rain but the values are consistently heavier than the groundwater by as much as $6^0/_{00}$.

The Hillman Creek study area, situated at latitude 42°04'N and longitude 82°39'W, is approximately 50 km SE of Windsor, Ontario, Canada. The drainage area above the stream gauging station is in the order of 1 km² which corresponds to a channel length of nearly 1.1 km (Fig.3). Relief over the basin is low with maximum and minimum elevations of approximately 212 and 202 m, respectively. Intensive agriculture in the area leaves only minimal tree cover and exposed soil conditions. An average of 740 mm of precipitation is distributed evenly throughout the year.

The uppermost unit in the basin, from 1 to 5 m of fine to coarse sand with a hydraulic conductivity in the order of 10^{-2} to 10^{-3} cm/s (Gillham et al.,

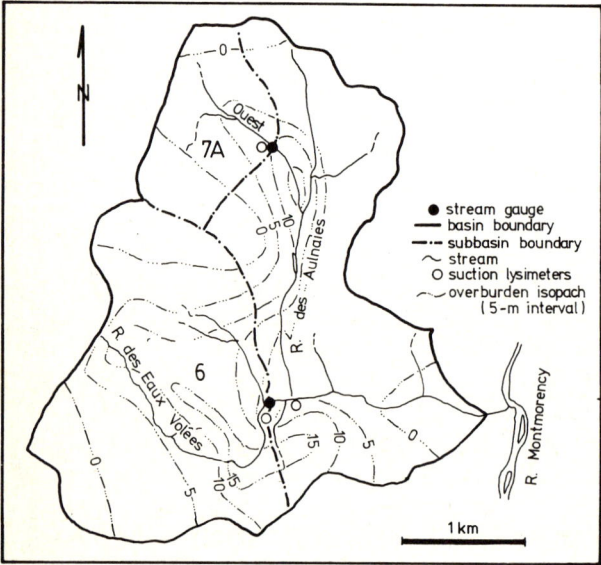

Fig.2. Instrumentation and depth of overburden in Ruisseau des Eaux Volées (after Rochette, 1971).

1978), overlies an impervious silty-clay unit. The water table, with a gradual slope from west to east, lies 1—4 m below ground surface. From water table maps (Gillham et al., 1978) and seepage meter data (Lee et al., 1977), it appears that groundwater discharge is low over the upper 600 m of the stream. From that point downstream beyond the gauging station, groundwater discharge is significant

The groundwater originates both from gravel pits along the western margin of the watershed and from local infiltration. The $\delta^{18}O$ values of the gravel pit groundwater are somewhat heavier than those of the locally infiltrated groundwater. Baseflow $\delta^{18}O$ values at the gauging station, ranging from −6.3 to −8.4°/$_{oo}$, indicate a mixture of the two groundwaters. The specific conductance of the baseflow is approximately 750 μS.

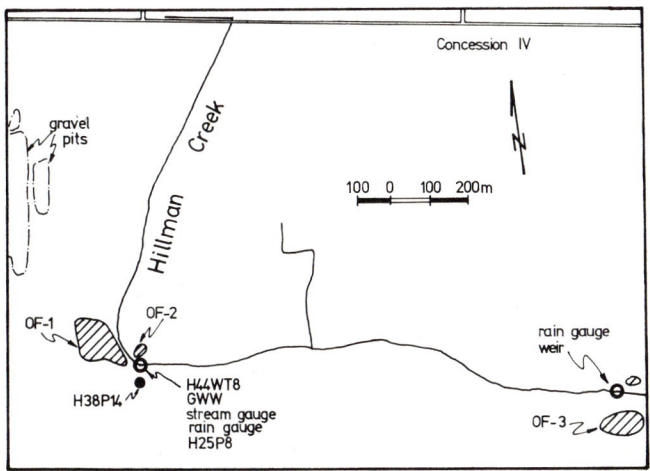

Fig.3. Instrumentation in the Hillman Creek study area.

RESULTS FROM RUISSEAU DES EAUX VOLÉES

Five rain events between August 5 and 25, 1976, resulted in stream discharge increases ranging from 45 to 820% in the two subbasins. Equipment malfunctions and storms unsuitable for the environmental isotope technique reduced the number of usable events which could be analyzed.

On August 5, a 32-mm rain event resulted in a discharge increase from subbasin 7A of approximately 450% over its pre-storm baseflow discharge (Fig.4). The runoff hydrograph accounts for only 12% of the rainfall. Prior to the storm, baseflow $\delta^{18}O$ values were approximately −12.0°/$_{oo}$. Although the $\delta^{18}O$ value of the rain was −8.3°/$_{oo}$, the peak discharge $\delta^{18}O$ value of the stream only reached −10.8°/$_{oo}$ (Fig.4). Using eq. 1 and 2 and assuming that the vadose water contributions were negligible, it appears that groundwater contributed more than 65% of the peak discharge in the stream.

The validity of the assumption that vadose water contributions are negligible can be evaluated by examining a plot of stream $\delta^{18}O$ vs. stream discharge (Fig.5A). A simple two-component groundwater—rain mixture in the stream should result in essentially collinear $\delta^{18}O$-discharge data points for the rising and falling limbs of the hydrograph. As discussed previously, the vadose water in the basin is isotopically heavier than the groundwater.

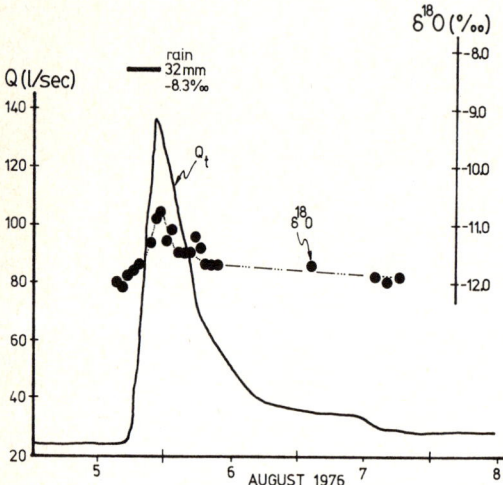

Fig.4. Storm runoff hydrograph and stream $\delta^{18}O$, subbasin 7A, Ruisseau des Eaux Volées, August 5—8, 1976.

Significant vadose water contributions to storm runoff should cause a shift in the recession limb data to the isotopically heavier side of the groundwater —rain mixing line (which is defined by the data from the earlier part of the rising limb). If the shift exceeds the analytical error for ^{18}O analysis ($\pm 0.2^\circ/_{oo}$), the assumption is questionable. Also, the magnitude of the shift should be directly related to the magnitude of the vadose water contribution.

Fig. 5B is a plot of the stream $\delta^{18}O$ vs. stream discharge for the August 5 storm in subbasin 7A. The small shift between the rising and falling limb values of $\delta^{18}O$ supports the simple two-component groundwater—rain mixing assumption.

On August 12, a 35-mm rain event in subbasin 6 instigated a discharge increase of approximately 340% over the pre-storm baseflow discharge (Fig.6). The runoff hydrograph accounts for nearly 20% of the rainfall. Prior to the event, the baseflow $\delta^{18}O$ value was $-11.9^\circ/_{oo}$. Although the rain $\delta^{18}O$ value was $-5.7^\circ/_{oo}$, the stream attained a $\delta^{18}O$ value of only $-10.1^\circ/_{oo}$ at peak discharge (Fig.6).

A plot of stream $\delta^{18}O$ vs. stream discharge for this event (Fig.7) suggests a significant vadose water contribution. Although a simple two-component hydrograph separation cannot be made in this case, it is apparent from the isotopic character of the various runoff components that pre-event water (groundwater and/or vadose water) dominated the storm runoff.

On August 22, a very brief but intense storm dropped 6 mm of rain on the basin in less than 15 min. Stream discharges from subbasins 6 and 7A increased by 55 and 45%, respectively (Fig. 8A and B). Storm yields for both basins were about 6%.

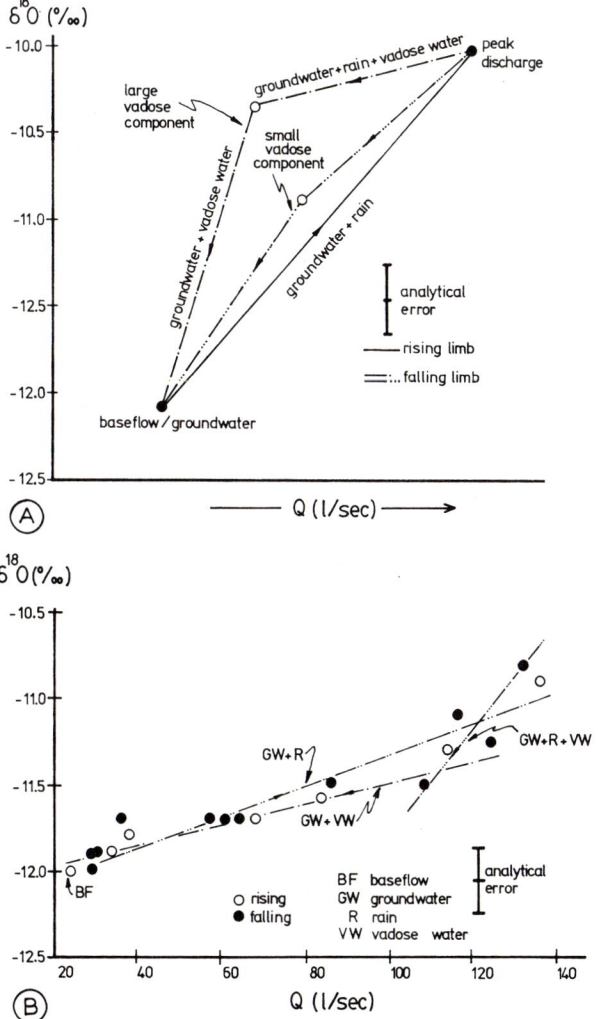

Fig.5. A. Model of $\delta^{18}O$ variations during storm runoff for the Ruisseau des Eaux Volées watershed.
B. $\delta^{18}O$ vs. discharge during storm runoff, subbasin 7A, Ruisseau des Eaux Volées watershed, August 5—8, 1976.

The rain event was ideal for hydrograph separation by the isotope technique. The short duration and high intensity of the rain meant that the event water was added "instantaneously" to the watershed. Baseflow $\delta^{18}O$ prior to the event was $-12.1°/_{oo}$ in subbasin 6 and $-11.9°/_{oo}$ in subbasin 7A. Although the rain $\delta^{18}O$ value was $-6.6°/_{oo}$, peak discharge attained $\delta^{18}O$

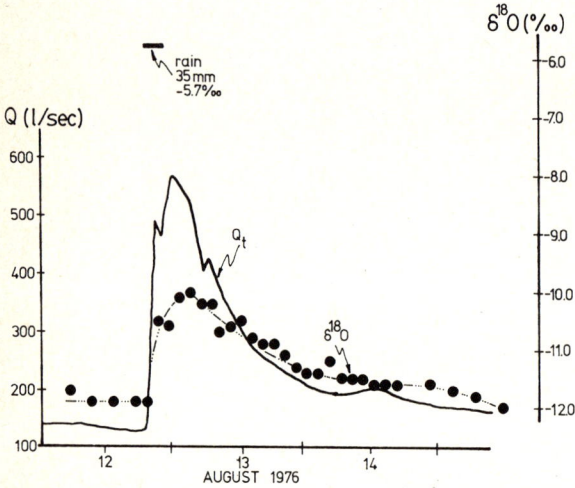

Fig.6. Storm runoff hydrograph and stream $\delta^{18}O$, subbasin 6, Ruisseau des Eaux Volées watershed, August 12–15, 1976.

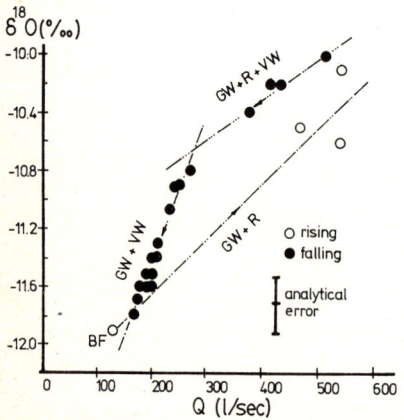

Fig.7. $\delta^{18}O$ vs. discharge during storm runoff event subbasin 6, Ruisseau des Eaux Volées watershed, August 12–15, 1976.

values of $-11.4^0/_{00}$ and $-11.5^0/_{00}$ for subbasins 6 and 7A, respectively (Fig.8). These correspond to groundwater components of more than 80% in peak discharge if vadose water contributions were negligible. A plot of stream $\delta^{18}O$ vs. stream discharge would support that assumption. Deuterium analyses for subbasin 6 gave similar groundwater contributions.

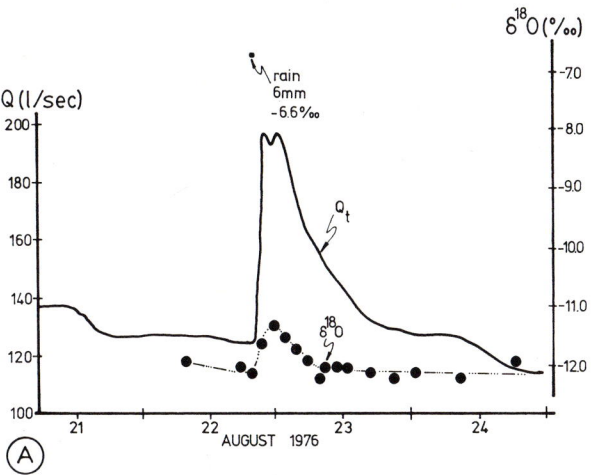

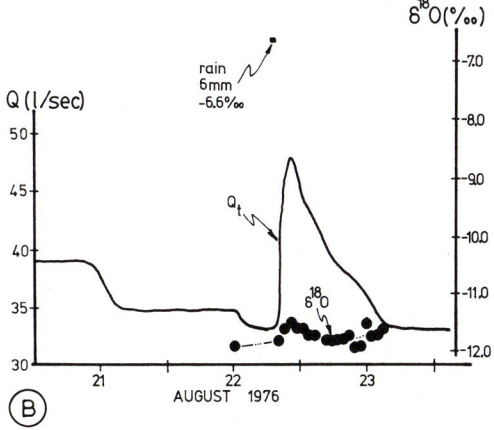

Fig.8.A. Storm runoff hydrograph and stream $\delta^{18}O$, subbasin 6, Ruisseau des Eaux Volées watershed, August 21–24, 1976.
B. Storm runoff hydrograph and stream $\delta^{18}O$, subbasin 7A, Ruisseau des Eaux Volées watershed, August 21–24, 1976.

RESULTS FROM HILLMAN CREEK

Groundwater stage—groundwater discharge rating curves for two near-stream observation wells were determined. H44WT8 is a 10-cm diameter, 2.44 m deep, recording water table well located approximately 3 m from the stream bank at the stream gauging site. GWW (Fig.9) is a 9-cm diameter recording piezometer open between 80 and 90 cm below the stream bed and adjacent to H44WT8. Vertical and horizontal hydraulic gradients during base-flow periods were approximately 0.12 and 0.03, respectively, between H44WT8 and GWW. These gradients are an order of magnitude greater than those given for the watershed average by Gillham et al. (1978).

Hydrogeologic constraints in the watershed; namely, low topographic relief and high hydraulic conductivities of the sand, preclude significant

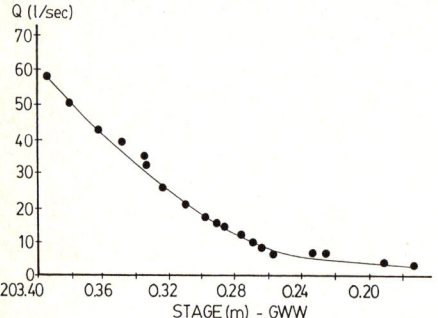

Fig.9. Groundwater stage—groundwater discharge rating curve for GWW, Hillman Creek study area.

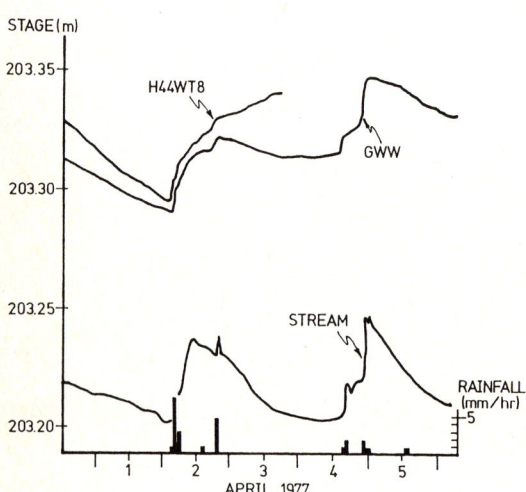

Fig.10. Stream and groundwater hydrographs for Hillman Creek study area, April 1—5, 1977.

vadose water contributions to storm runoff. Groundwater, therefore, is synonymous with pre-event water in this watershed.

Rain events of 19 mm on April 2 and 6 mm on April 4, 1977, caused two distinct runoff events at the Hillman Creek gauging site (Fig.10). Stream discharge increased from 15 l/s prior to the storm to 25 and 30 l/s at the first and second discharge peaks, respectively. The most noteworthy feature of these events is the rapid response of the near-stream groundwater exhibited by H44WT8 and GWW. The stream remained effluent throughout the events and although the hydraulic gradient to the stream decreased slightly during the first event, the gradient in the second event exceeded the pre-storm baseflow gradient. According to the groundwater stage—groundwater discharge rating curve technique, groundwater contributed more than 60 and 80% of the peak discharge on April 2 and 4, respectively. No isotope data are available for these events.

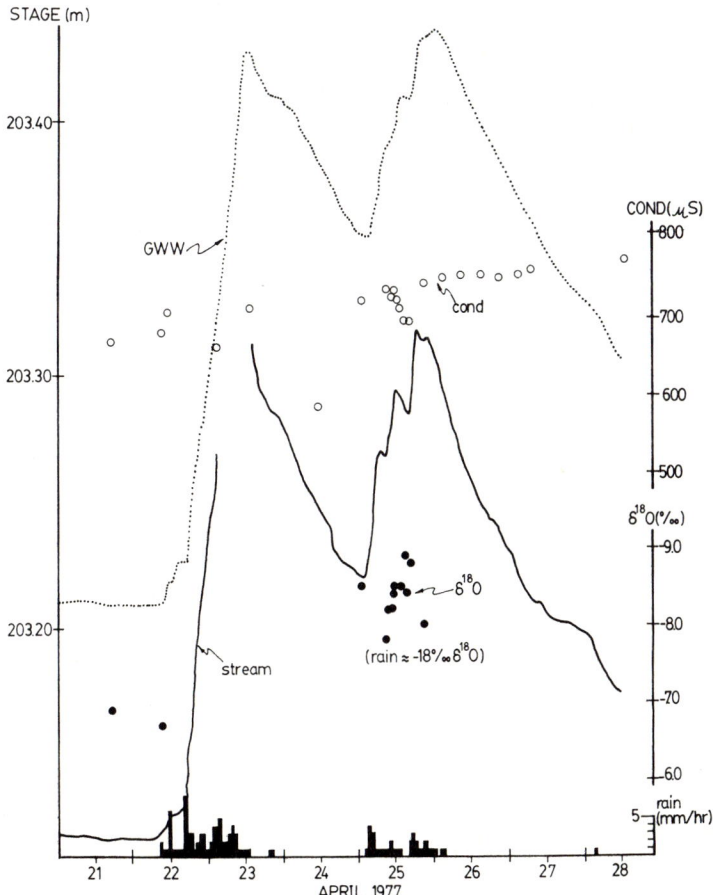

Fig.11. Stream and groundwater hydrographs, and stream $\delta^{18}O$ and specific conductance, Hillman Creek study area, April 21—28, 1977.

Rain events of 38, 35 and 31 mm occurring on April 22, 23 and 25, 1978, respectively, produced dramatic responses in both stream discharge and groundwater stage (Fig.11). Stream discharge increased from about 6 l/s on April 22 to peak discharge of at least 85 and 90 l/s on April 23 and 25, respectively. The length of the storms and instrument malfunctions curtailed some of the isotopic and specific conductance studies, however, many significant observations were made.

The stream remained effluent throughout the events even though stream discharge increased by more than an order of magnitude over the pre-storm discharge. The near-stream groundwater responded quickly to the rain events; however, the groundwater responsiveness apparently decreased away from the stream. Table I shows that over the duration of the study period, the stream discharge was more closely related to the near-stream groundwater stage than to the more remote groundwater. Although the near-stream hydraulic gradient toward the stream decreased slightly at the first discharge peak, the gradient at the second discharge peak was greater than the pre-storm base-flow gradient.

TABLE I

Relationship between stream discharge and depth to groundwater near H44WT8, Hillman Creek Watershed

Well or piezometer	Approximate lateral distance to stream (m)	Linear correlation coefficient between depth to water and stream stage
H38P14	65	−0.175
H25P8	5	−0.869
H44WT8	3	−0.954
GWW	0	−0.925

Visual observations were made and water samples were taken only on April 25. One area west of the stream gauging station (OF-1 on Fig.3) produced sufficient overland flow to erode a channel several centimetres deep to the stream. Two other overland flow producing areas, OF-2 and OF-3, produced much less overland flow.

The $\delta^{18}O$ values of the rain on April 25 ranged from −16.5 to −20.0°/$_{00}$ with a weighted average of −18.1°/$_{00}$. Since the baseflow prior to April 22 had $\delta^{18}O$ values of about −6.9°/$_{00}$ and since the groundwater had $\delta^{18}O$ values of less than −10.0°/$_{00}$ (Gillham et al., 1978), a very good distinction can be drawn between the groundwater/base flow (pre-event) and rain (event) components. Also, the specific conductance of the rain was less than 100 μS, whereas baseflow specific conductance is about 750 μS.

Fig.12A and B illustrates the temporal variations in $\delta^{18}O$ and specific conductance for the rain, overland flow and stream samples on April 25. The most prolific source of overland flow, OF-1, had $\delta^{18}O$ and specific conductance values indicative of large groundwater contributions. The two other observed sources, OF-2 and OF-3, had rain-like $\delta^{18}O$ and specific conductance. The stream $\delta^{18}O$ and specific conductance strongly suggest a large groundwater component. The groundwater stage—groundwater discharge rating curve technique supports these findings, with both methods giving more than 80% of the discharge as groundwater on April 25. Observed increases in the stream nitrate concentration on April 25 are apparently related to the high nitrate concentrations in the groundwater-dominant overland flow issuing from OF-1.

On May 4, 1977 a 17-mm rain event produced another storm runoff event in which the near-stream groundwater responded rapidly (Fig.13). Stream

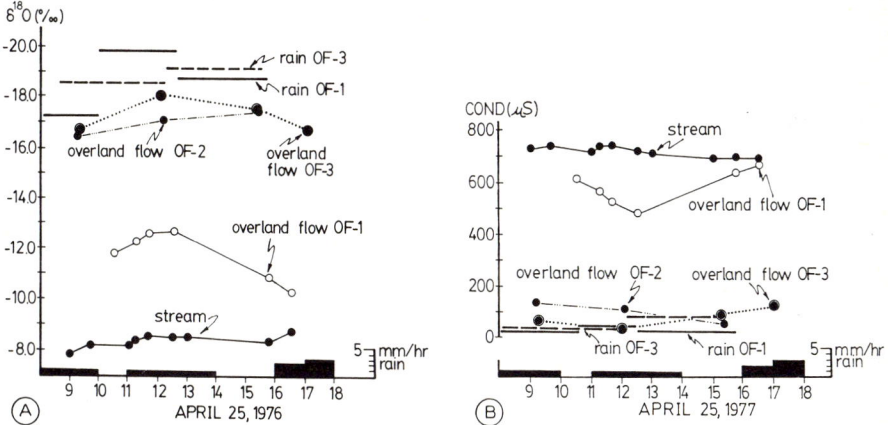

Fig.12. A. Temporal variations in $\delta^{18}O$ of overland flow, rain and the stream, Hillman Creek study area, April 25, 1977.
B. Temporal variations in specific conductance of overland flow, rain, and the stream, Hillman Creek study area, April 25, 1977.

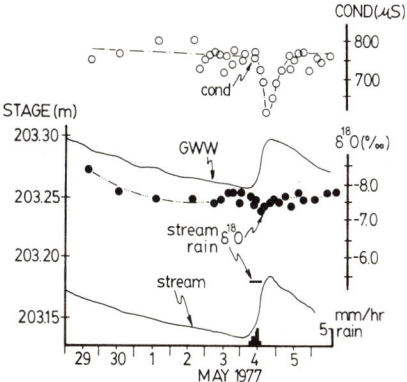

Fig.13. Stream and groundwater hydrograph, and stream $\delta^{18}O$ and specific conductance, Hillman Creek study area, April 29—May 5, 1977.

discharge increased from 9 to 22 l/s at peak discharge, yet the stream remained effluent. No overland flow was observed and baseflow $\delta^{18}O$ and specific conductance were diluted only slightly during the event. Both the groundwater stage—groundwater discharge rating curve and isotope techniques suggest that groundwater contributed over 80% of the peak discharge.

On June 6, 1977, following a month with only 20 mm of rain, a 36-mm rain event produced very prominent hydrographs in the stream and near-stream groundwater (Fig.14). Stream discharge increased from less than 5 l/s before the storm to over 200 l/s at peak discharge. A brief reversal of hy-

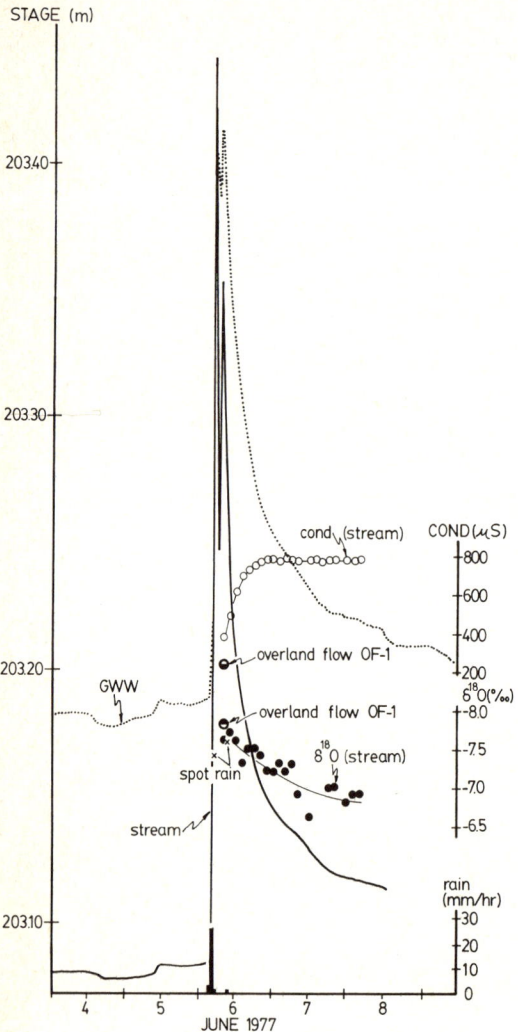

Fig.14. Stream and groundwater hydrographs, and stream $\delta^{18}O$ and specific conductance, Hillman Creek study area, June 4—8, 1977.

draulic gradient caused the stream to become influent for the only time during the study. The near-stream groundwater, however, did respond to the storm very rapidly. Significant amounts of overland flow emanated from OF-1, OF-2 and OF-3.

Fig.14 shows the temporal variations in stream $\delta^{18}O$ and specific conductance. Although the $\delta^{18}O$ value of the rain ($-7.8^0/_{00}$) was too close to the base-flow value ($-6.8^0/_{00}$) for very accurate separations, it is obvious that the stream runoff was dominated by rain runoff. The overland flow from OF-1, which was mostly groundwater during the April 25 storm, had rain-like $\delta^{18}O$ ($-7.8^0/_{00}$), and specific conductance (234 μS). The overland flow from OF-1 also had nitrate concentrations (4.5 mg/l) similar to those of peak discharge in the stream (4.2 mg/l). The stream nitrate concentrations increased to between 8 and 10 mg/l when the groundwater reasserted itself in the stream.

RESULTS FROM COMPUTER SIMULATIONS

Four small hypothetical watershed configurations (Fig.15) are examined to determine how near-stream watershed relief and basin width affect near-stream groundwater response. Watersheds *1*, *2* and *3* have comparatively low, medium and high near-stream topographic relief. Each configuration is 9 m from mid-stream to divide and varies in thickness above an impermeable base from 1 m at mid-stream to 2.9 m at the divide. The finite-element grid for each contains 464 nodes and 420 quadrilateral elements with the grids varying only in their near-stream topographic configuration. Watershed *4* is identical to watershed *2* in element configuration except that it is truncated 5 m from the stream leaving fewer nodes (400) and elements (360). The uppermost elements in the grids represent real thicknesses of no more than 3 cm.

The watersheds are homogeneous and isotropic. The saturated—unsaturated characteristics of the porous media are those of the Botany Sand as described by Watson (1967). The sand has a saturated hydraulic conductivity of 0.0186 cm/s, a porosity of 35%, and a capillary fringe of 39 cm. The initial conditions are static and the stream level is held constant throughout the simulation period. Seepage faces are allowed to form.

A rainfall rate equal to one-tenth of the saturated hydraulic conductivity (0.00186 cm/s) was applied to each watershed for 2.3 h after which drainage was allowed for at least 4.6 h. Computer output, consisting of total and pressure heads, for specified times during the event, are available for all the nodes.

Fig.16 compares the responses in total head for a common near-stream point *a* (Fig.15). Watershed *1*, with the lowest near-stream relief, exhibited the most rapid response to the storm. Watersheds *2* and *3* responded less rapidly but had higher peak stages. A comparison of the early responses of long watershed *2* and short watershed *4* reveals that the upland area has no effect on the early near-stream groundwater response.

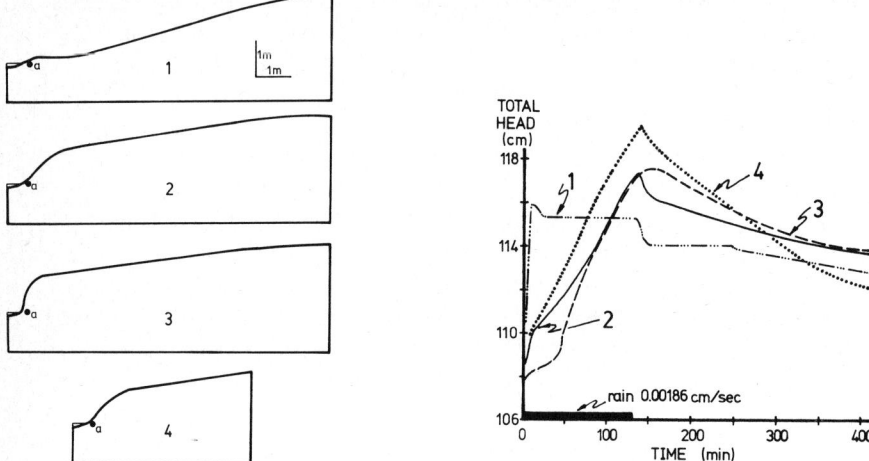

Fig.15. Watershed configurations used in mathematical simulations.

Fig.16. The effects of the near-stream unsaturated zone thickness and upland areas on early near-stream groundwater response to rain.

Fig.17 illustrates the temporal variations in the water table configuration for the rain event on watershed 2. An apparent groundwater ridge, similar to the one observed by Ragan (1968), forms in the near-stream area prior to the response of the water table in the upland area. A seepage face forms in this case as it does in all the others. The discharges of the various runoff components in a 1-cm slice of watershed 2 are compared in Fig.18. Groundwater dominates the total discharge from the watershed.

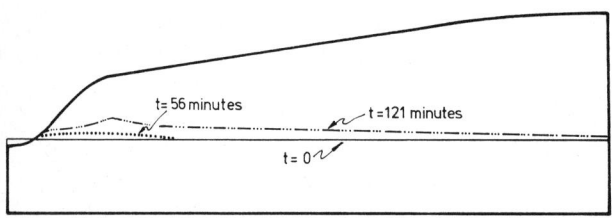

Fig.17. Formation of a near-stream "groundwater ridge" in response to a rain event.

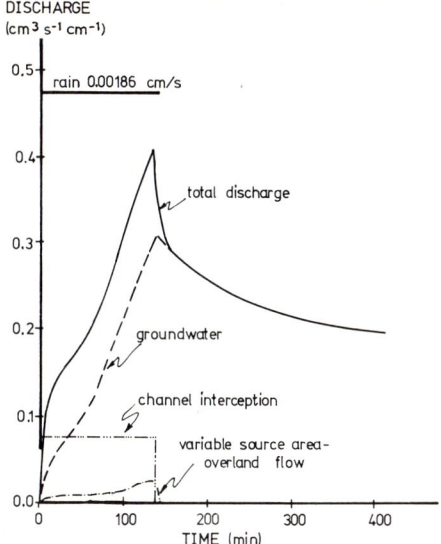

Fig.18. Discharge hydrograph from watershed 2.

CONCLUSIONS

Groundwater was found to be a major component in virtually all of the runoff events documented in this study. Pinder and Jones (1969), Dincer et al. (1970), Martinec et al. (1974), Fritz et al. (1976), and others have arrived at the same conclusion, that is, groundwater is a significant and active factor in runoff generation. Martinec (1975), however, summarized the dilemma of how the groundwater appears in the stream so quickly as:

"The exact nature of the propagation is not known".

Several clues to the mechanism can be gathered from this study. Both field observations and computer simulations suggest that very large and rapid increases in the hydraulic head in the near-stream groundwater occur soon after the onset of rain. These responses precede, and are apparently independent of, the upland area response. The computer simulations also reveal the formation of a groundwater ridge adjacent to the stream in response to a rain event with the lag time brief and inversely related to the near-stream unsaturated zone thickness. Field observations indicate, however, that the near-stream hydraulic gradient may decrease slightly from its pre-storm value.

Isotopic and specific conductance data for overland flow in the Hillman Creek watershed imply that for one very prolific site (OF-1), overland flow is generated in more than one way. During a moderate intensity storm on a very wet basin, both the overland flow and stream flow were dominated by groundwater. In response to a very intense storm on a much drier basin, both the overland flow and stream were dominated by rain water.

The following theory outlines a physical mechanism which could explain the responsive, active and significant role of groundwater in storm and snow-melt runoff generation. Along the perimeter of transient and perennial discharge areas, the water table and its associated capillary fringe lie very close to the ground surface. Soon after a rain or snow-melt event begins, infiltrating water readily converts the near-surface tension-saturated capillary fringe into a pressure-saturated zone or groundwater ridge (Ragan, 1968). This groundwater ridge not only provides the early increased impetus for the displacement of the groundwater already in a discharge position, but it also results in an increase in the size of the groundwater discharge area which is essential in producing large groundwater contributions to the stream. The response of the upland area groundwater may become important at later times in the runoff event but has little influence in the early part of the runoff event.

The groundwater may discharge directly into the stream through the stream bed or it may issue from the growing near-stream and/or remote seep areas and flow as overland flow to the stream (as in the variable source area— overland flow theory). Following periods of drought during which the water table has fallen far below ground surface, intense storms may result in surface saturation from above and rain-like overland flow (partial area—overland flow) before the water table can emerge.

The recognition of groundwater as an important factor in storm and snow-melt runoff in humid to subhumid areas raises many questions regarding runoff events. What control does groundwater have on surface water quality during high runoff events? How do the increased heads in the near-stream groundwater affect the erodibility of soils? How can we improve predictive hydrologic models?

ACKNOWLEDGEMENTS

John Cherry, Emil Frind, Peter Fritz and Bob Gillham of the University of Waterloo provided many helpful comments during this study. Funds for this study were provided through National Research Council of Canada grants A7712, A6112 and A8368, and from Agriculture Canada and the Ontario Ministry of Agriculture and Food. The senior author wishes to acknowledge the NRCC for the support of his postgraduate education.

REFERENCES

Beasley, R.S., 1976. Contribution of subsurface flow from the upper slopes of forested watersheds to channel flow. Soil Sci. Soc. Am. J., 40: 955—957.
Dincer, T., Payne, B.R., Florkowski, T., Martinec, J. and Tongiorgi, E., 1970. Snow-melt runoff from measurements of tritium and oxygen-18. Water Resour. Res., 6: 110—124.

Cheng, J.D., Black, T.A. and Wellington, R.P., 1975. The generation of stormflow from small forested watersheds in the Coast Mountains of Southwestern British Columbia. Proc. Can. Hydrol. Symp., Winnipeg, Man., pp. 542—549.

Freeze, R.A., 1974. Streamflow generation. Rev. Geophys. Space Phys. 12: 627—647.

Fritz, P., Cherry, J.A., Weyer, K.V. and Sklash, M.G., 1976. Runoff analyses using environmental isotope and major ions. In Interpretation of Environmental Isotope and Hydrochemical Data in Groundwater Hydrology. I.A.E.A., Vienna, pp. 111—130.

Gillham, R.W., Hendry, M.J., Cherry, J.A., Frind, E.O. and Pucovsky, G.M., 1978. Studies of agricultural contribution to nitrate enrichment of groundwater and the subsequent nitrate loading to surface waters. Part I. Field investigation of the processes controlling the transport of nitrate in groundwater. Waterloo Res., Inst., Waterloo, Ont., 203 pp.

Kennedy, V.C., 1977. Hypothesis to explain the rapid release of solutes from soil during storm runoff. EOS (Trans. Am. Geophys. Union) 58: 386 (abstract).

Lee, D.R., Fritz, P. and Hynes, H.B.N., 1977. Groundwater flow into lakes and streams mass movement, transport of nutrients and organic matter. Waterloo Res. Inst., Waterloo, Ont., 57 pp.

Martinec, J., 1975. Subsurface flow from snow-melt traced by tritium. Water Resour. Res., 11: 496—497.

Martinec, J., Siegenthaler, H., Oescheger, H. and Tongiorgi, E., 1974. New insight into the runoff mechanism by environmental isotopes. Proc. Symp. on Isotope Techniques in Groundwater Hydrology. I.A.E.A., Vienna, 1: 129—143.

Pinder, G.F. and Jones, J.F., 1969. Determination of the groundwater component of peak discharge from the chemistry of total runoff. Water Resour. Res., 5: 438—445.

Plamondon, A.P. and Naud, R.C., 1975. The watershed management program at the Montmorency Experimental Forest. Symp. on Montmorency Experimental Forest, 27 pp.

Ragan, R.M., 1968. An experimental investigation of partial area contribution. Int. Assoc. Sci. Hydrol., Publ., 76: 241—249.

Rochette, F.J., 1971. Hydrogeological study of "Ruisseau des Eaux Volées" Experimental Basin. M.Sc. Thesis, University Western Ontario, London, Ont. (unpublished).

Schicht, R.J. and Walton, W.C., 1961. Hydrologic budgets for three small watersheds in Illinois. Ill., State Water Surv., Rep. Invest. No. 40, 40 pp.

Segol, G., 1976. A three-dimensional Galerkin finite element model for the analysis of contaminant transport in variably saturated porous media. Univ. Waterloo, Waterloo, Ont., Rep. (unpublished).

Sklash, M.G., 1978. The role of groundwater in storm and snowmelt runoff generation. Ph.D. Thesis, University of Waterloo, Waterloo, Ont. (unpublished).

Stevenson, D.R., 1967. Geological and groundwater investigations in the Marmot Creek Experimental Basin of Southwestern Alberta, Canada. M.Sc. Thesis, University of Alberta, Edmonton, Alta. (unpublished).

Visocky, A.P., 1970. Estimating the groundwater contribution to storm runoff by the electrical conductance method. Groundwater 8: 5—10.

Watson, K.K., 1967. Experimental and numerical study of column drainage. J. Hydraul. Div., Proc. Am. Soc. Civ. Eng., 93(HY2): 1—15.

[2]

HYDRAULIC POTENTIAL IN LAKE MICHIGAN BOTTOM SEDIMENTS

KEROS CARTWRIGHT, CATHY S. HUNT, GEORGE M. HUGHES and ROSS D. BROWER

Illinois State Geological Survey, Urbana, IL 61801 (U.S.A.)

(Accepted for publication April 25, 1979)

ABSTRACT

Cartwright, K., Hunt, C.S., Hughes, G.M. and Brower, R.D., 1979. Hydraulic potential in Lake Michigan bottom sediments. In: W. Back and D.A. Stephenson (Guest-Editors), Contemporary Hydrogeology — The George Burke Maxey Memorial Volume. J. Hydrol., 43: 67—78.

The magnitude and direction of groundwater flux in the bottom sediments of Lake Michigan were deduced from measurements made during three shipboard cruises between 1973 and 1975. These factors affect the geochemical environment of the sediments and therefore the distribution of trace elements reported to be present. The near-shore, sandy-bottom and fine-grained, soft, deep-lake sediments were investigated; areas of hard till or bedrock were not included in the study.

Thirty-three piezometers were placed in near-shore sands in waters 5—15 m deep. The piezometers were placed an average of 3 m into the bottom sediment. Water levels from the piezometers averaged 0.6 cm above the lake level, equivalent to an upward hydraulic gradient of about 0.002 cm/cm. Water samples taken from the piezometers have a distinctly different chemical composition from that of the lake water. The total dissolved mineral content and hardness of the groundwater are about twice those of the lake water.

Twenty-two hydraulic gradient measurements were made in the fine-grained soft deep-lake sediments in waters 48—140 m deep by using a differential-pressure transducer dropped into the sediments. These measurements show an upward gradient averaging 0.2 cm/cm. No chemical data were obtained for the groundwater in the deep-lake sediments.

The results of this study indicate that the groundwater flux is upward through the bottom sediments into Lake Michigan and that there is a chemical change in the water near the water—sediment contact.

INTRODUCTION

This study was initiated in 1972 as a result of a discussion of the geochemistry of trace elements in the bottom sediments of Lake Michigan. During the study of the geochemistry of these sediments by geologists and chemists at the Illinois State Geological Survey (Ruch et al., 1970; Shimp et al., 1970, 1971; Kennedy et al., 1971; Lineback and Gross, 1972) a question arose concerning the reason for the distribution of the trace elements in the bottom sediments. In general, the concentration of certain trace elements (e.g., lead, zinc and mercury) decreases downward from the sediment—lake water interface. While there is no question that man's activities are the primary source of these elements, there is the question as to whether this distribution is related solely to deposition of sediment containing trace elements introduced to the lake by man, or whether it is also affected by groundwater movement into the lake. If, as we will argue, there is movement of water from the groundwater reservoir into the lake, and if the interstitial water of the bottom sediments has different chemical characteristics from the lake water, we must determine whether the observed distribution could also result from upward mobilization and precipitation of the ions in the sediment and, lastly, whether some of the ions present in the upper sediment could actually have been in solution in the groundwater and precipitated out of solution at the groundwater—lake water boundary?

A great deal more work must be undertaken on the geochemistry of the groundwater—sediment—lake water system to satisfactorily answer these questions. This study was undertaken to determine whether groundwater is moving into the lake through the bottom. Incidental to this, volumetric estimates of the groundwater contribution to the lake may be made from groundwater flux.

This study was conducted over three field seasons. The first year's program, during the summer of 1973, concentrated on a study of a near-shore sandy-bottom area. The program was carried out aboard the R.V. "Inland Seas" operated by the University of Michigan Great Lakes Research Division under sponsorship of the National Science Foundation. During the summers of 1974 and 1975, studies of the deep-lake sediments were conducted. In 1974, the program was carried out on the R.V. "Laurentian" operated by the University of Michigan, and in 1975, on the C.S.S. "Limnos" operated by the Canada Centre for Inland Waters. The 1974 program concentrated on testing the instruments designed to measure differential pore pressures in soft mud. Most of the usable data for the deep-water program were obtained during the summer of 1975. We are grateful to the captains and crews of these vessels for their assistance and patience and their skills in holding station for long periods.

NEAR-SHORE RESULTS

Thirty-three piezometers were placed in a near-shore sand body which extends from Waukegan, Illinois, north to Kenosha, Wisconsin (Fraser and Hester, 1974). The piezometers were set utilizing a 1-in. pipe and water from the ship's water system to jet the piezometer tips into the sand. The piezometer (Fig. 1) consisted of flexible tubing 10 mm in diameter with a screen covering the end. It was taped to a pipe coupling which was loosely placed on the jetting pipe. Wings were welded to the coupling to facilitate unscrewing the pipe. Once the desired depth had been reached (approximately 3 m) the jetting pipe was unscrewed from the coupling and withdrawn, and

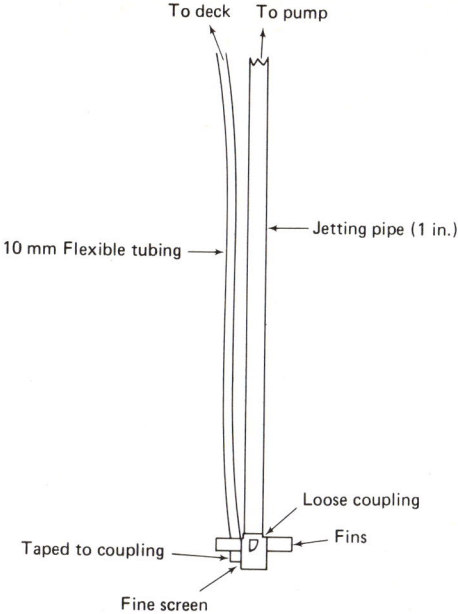

Fig.1. Construction details of the piezometer jetting pipe.

the sediment allowed to fall back. It is not known how large a cavity, if any, was left on the lake floor, but all the piezometers did seal, as they offered resistance when pulled from the bottom sediment. It is not known how this affected the measured heads, but we presume it to be minimal.

During jetting, water was pumped down the flexible tubing using a small peristaltic pump to prevent sand from clogging the screen. Once set, the pump was turned off and water in the tubing above deck was drained. The time necessary for the stabilization of the water levels in the piezometers ranged from only a few minutes to nearly an hour. Stabilization was rechecked by causing the water level in the tubing to rise and/or fall and then restabilize. Some piezometers became plugged during insertion or in the stabilization process and gave poor readings.

Hydraulic gradients at each station (Fig.2) were determined by comparing the water levels in a reference tube suspended from the boat to a point 1—2 m above the lake bottom. The difference in water levels between the piezometer and reference tube divided by the depth of burial of the piezometer in bottom sediments is the vertical hydraulic gradient. The water levels in the piezometers ranged from approximately —3.8 to +45 cm. However, those piezometers which showed no evidence of being plugged and in which readings were taken under ideal atmospheric conditions had water levels between +0.5 and +1.0 cm above lake levels. This is an upward gradient of about +0.002 cm/cm. The values of hydraulic gradient on the map (Fig.2) that vary significantly from the mean are believed to result from plugged piezometers and sudden lake-level fluctuations.

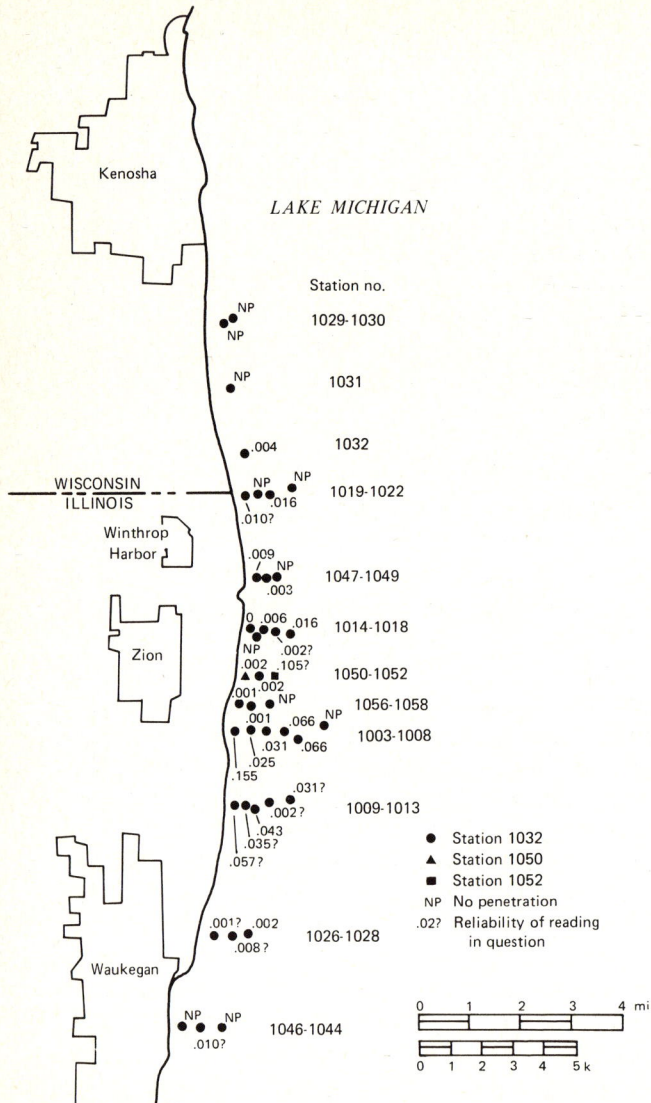

Fig.2. Location of near-shore piezometers and the hydraulic gradients measured at each station.

Some problems were encountered in using the lake level as a reference elevation. Onshore wind piles up water against the shore, locally raising the lake level while offshore winds have the opposite effect. During the 1973 cruise, although the winds were generally calm, several wind events did occur. One particularly strong squall with offshore winds caused sudden fluctuations in water levels at station *1052* (third station to the east of Zion). The maximum water level of +45 cm was recorded at this time. This station was reoccupied when the wind calmed, a reading of +0.6 cm was obtained. These lake fluctuations seem to explain both the very high positive gradients and the negative gradients.

To confirm that we had sealed the piezometer in the bottom sediment and that we were dealing with a groundwater flow system, water samples were obtained from the 13 piezometers that could be pumped successfully. Samples of lake water were also taken using the reference tube suspended in the lake at all sample points to compare with the samples from the piezometers. The samples of water from the piezometers and from the lake were sent to the Illinois State Water Survey for analysis; in addition, analyses for manganese, mercury and tritium were made by the Illinois State Geological Survey (ISGS); and analyses for several common pesticides were made by the Illinois Natural History Survey. The water samples from the piezometers showed varying mixtures of groundwater and lake water resulting from the use of lake water for jetting.

The groundwater is thought to be best represented by the sample at station *1032* (Table I) and illustrates the different chemical character of the two waters. Iron, sodium, potassium, zinc, silica and boron are consistently found in higher concentrations in the groundwater. Manganese (not analyzed for in sample from station *1032*) is present in all groundwater samples analyzed (0.02—0.1 mg/l) and was below detectable levels in lake water. On the other hand, chloride and sulfate occur in higher concentrations in the lake water than in the groundwater. The groundwater is harder, more alkaline, and higher in total dissolved solids (TDS) than the lake water. Tritium content of the lake water is about 100 TU and the tritium from the groundwater is about 40 TU. In water samples tested by the Illinois Natural History Survey, seven selected pesticides (ppDDT, ppDDD, opDDT, Dieldrin®, ppDDE, opDDE and HEO) were consistently found at low concentrations in the lake water whereas only opDDT and opDDE were above detectable limits in the groundwater. The presence of these pesticides in the groundwater suggests some contamination by lake water during jetting. The chemistry of the lake water and the chemistry of the groundwater are distinctly different. The upward hydraulic gradient and difference in water chemistry show that there is positive groundwater flux into the lake.

RESULTS FROM DEEP LAKE

Twenty-two measurements of hydraulic gradient were made in the fine-grained soft lake sediments in waters 48—140 m deep. The measurements were made utilizing a differential-pressure probe built in the instrument shop of the ISGS (Fig.3). This instrument utilizes a 0.5-lb. in^{-2} differential-pressure transducer with an accuracy of ±0.5% and precision of 0.1%. The tilt of the probe is determined by the pressure a mercury column exerts on a second transducer; the accuracy of the tilt measurements varies from ±2° at near vertical angles to ±1° at tilt angles greater than 20°. The electronic output is read in the laboratory of the ship by means of a strip-chart recorder

TABLE I

Representative chemical analyses of lake water and groundwater illustrating the differences in quality (given in milligrams per liter unless otherwise indicated)

	TDS	Hardness (as $CaCO_3$)	Alkalinity (as $CaCO_3$)	Fe	Mn	Na+K	Zn	Silica	B	Cl	SO_4 calculated	Tritium
Groundwater	442	380	398	14.0	57±2*	16.7	0.02	36.2	0.2	4	11	40*
Lake water	170	136	112	2.8	n.d.*	6.2	0.00	1.8	0.0	8	24	100*

n.d. = not detectable.
*Water sample from station other than *1032*.

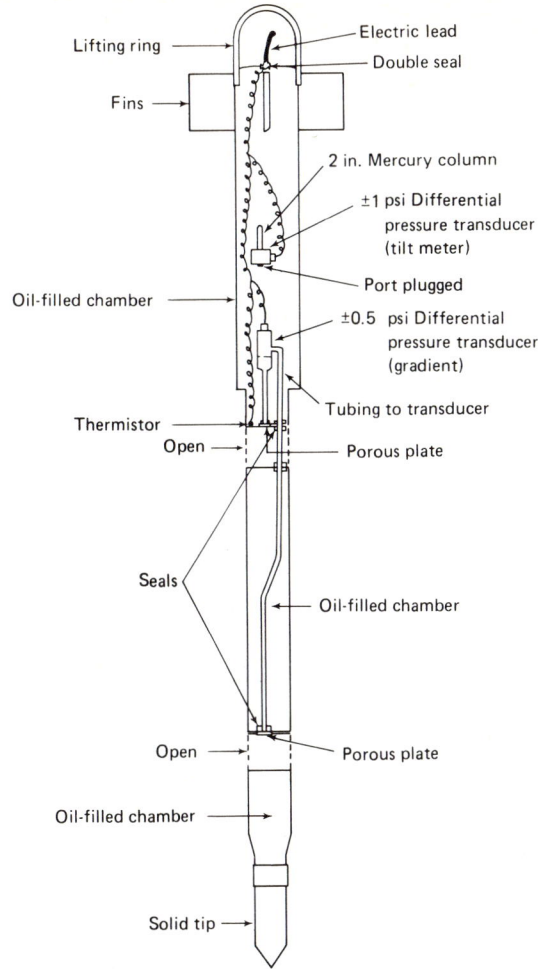

Fig.3. Construction details of the differential pore pressure probe.

for the pressure response and a null meter for the tilt.

Differences in hydrostatic pressure between the two porous plates are balanced by the water in the internal tubing. Thus, the pressure difference measured by the differential-pressure transducer is the actual pressure available to cause groundwater flow and represents the potential gradient. The probe is lowered into the lake (Fig.4) and the hydrostatic balance is checked. The probe is then allowed to freefall the final 8—10 m and penetrate the bottom.

When penetration occurs, it appears on the strip chart in the laboratory as a sharp deflection. Fig.5 shows a typical response curve which has been smoothed. It generally takes 30—45 min. for the shock of insertion to dissipate and a stable reading to be obtained. After a stable reading is obtained, the probe is pulled. The presence of mud found in the upper opening and on

Fig.4. Photograph of the differential pore pressure probe being lowered into the lake. The opening for pore pressure parts, lead weight and fins are visible.

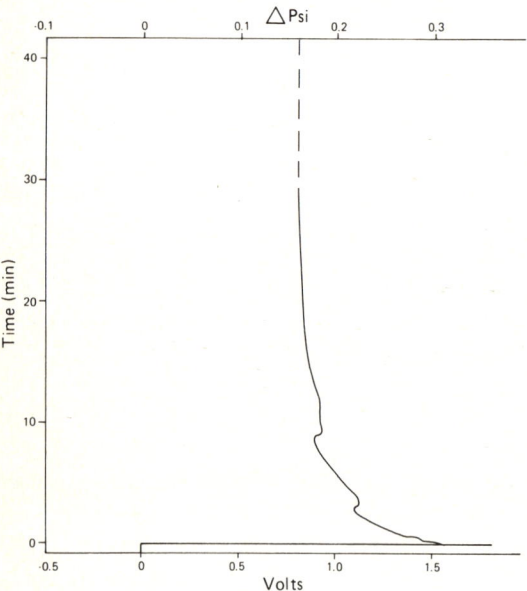

Fig.5. Typical pressure response curve of the differential pore pressure probe after insertion into soft bottom sediments.

the fins of the probe clearly indicates that penetration has occurred beyond both openings. However, the total depth of penetration generally cannot be determined.

The instrument is monitored for tilt angle and for any possible pulling on the cable by the ship moving off station. Tilt of the probe was less than 5° at all stations and presented no problem. At greater tilt angles, the probe would slowly fall to its side. In such cases, the probe was lifted and reset.

The results of the deep-lake probes (Fig.6) show the hydraulic gradient varies from a low of 0.180 cm/cm to a maximum of 0.350 cm/cm. The hydraulic gradient values are relatively consistent over much of the lake basin. The small variations, however, may be quite significant. They seem to be related to two factors: the bedrock geology and the thickness of unconsolidated sediments. The hydraulic gradients in the western half of the lake are somewhat higher than those on the eastern half. The exact boundary between the Silurian (limestones and dolomites which form a major aquifer on the western shore) and the Devonian (shales) under Lake Michigan is not known, but if the boundary is projected out into the lake basin, it divides the western area of higher hydraulic gradients (over more permeable Silurian bedrock) from the eastern area of lower hydraulic gradients (over less permeable Devonian bedrock). The one higher hydraulic gradient near South Haven, Michigan, may be related to a change to Mississippian bedrock (sandstones and limestones which are aquifers on the eastern shore).

The profiles run for this study closely (but not precisely) followed previous seismic and geologic profiles (Lineback et al., 1972). The relationship between hydraulic gradient and thickness of unconsolidated sediment thickness appears to be very strong. The gradients tend to decrease with increasing sediment thickness.

DISCUSSION

During this three-year program, we demonstrated that there are distinct differences in dissolved mineral contents of the groundwater and lake water. In addition, the chemical data show the possibility that some of the ions, most notably zinc and boron which were present only in the groundwater, may be carried by the groundwater to the lake sediment in addition to being deposited as part of the sediment load of the lake. The primary source of trace elements in the bottom sediments is undoubtedly man's activities; however, the vertical distributions noted may not be solely the result of increasing activities and contamination.

We also demonstrated that there is an upward hydraulic gradient in the bottom sediments of Lake Michigan, at least over the southern portion of the lake involved in this study. Using the gradients we obtained, groundwater flux in the near-shore and deep-lake sediments can be calculated if the hydraulic conductivities are known. Approximate hydraulic conductivities for the sediments were calculated using the methods of Rose and Smith (1957)

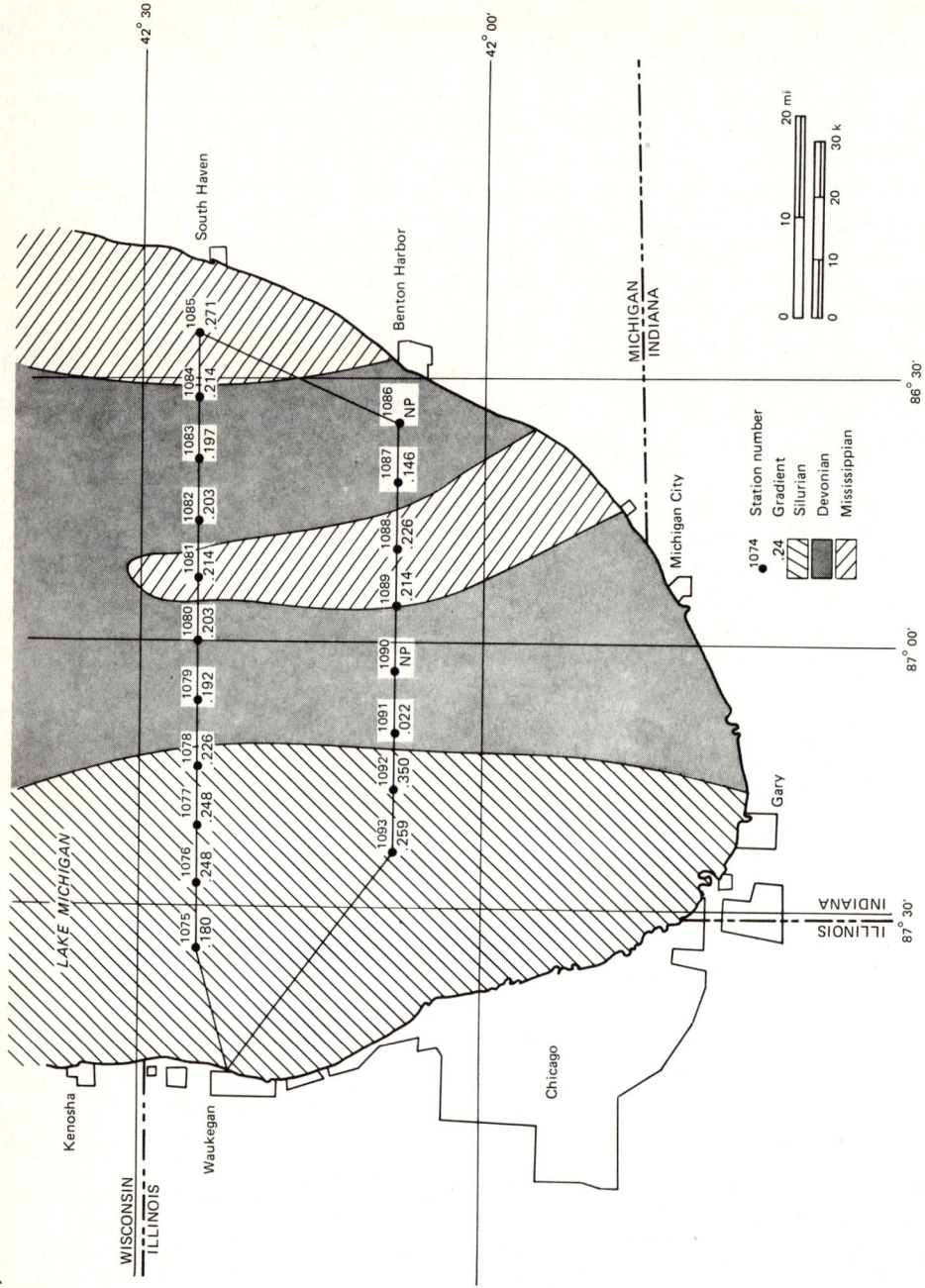

Fig. 6. Location of station where hydraulic gradients were measured in deep-lake sediments. Geologic boundaries projected from the U.S. Geological Survey geologic map of the U.S.A.

from the grain-size distribution for the near-shore sediments (Frazer and Hester, 1974) (mean value of 98 calculations; $k = 3 \cdot 10^{-2}$ cm/s) and determined in the laboratory by falling-head tests for the deep-lake bottom sediments (mean value of 11 measurements; $k = 5 \cdot 10^{-8}$ cm/s). Although hydraulic conductivities calculated by these methods are not very accurate, we calculated the flux to be in excess of 19 m/yr. in the near-shore sandy sediments and about 0.5 cm/yr. in the deep-lake clays. These fluxes are much greater than the rate of sediment accumulation, and therefore should be sufficient to prevent the lake water from entering the interstitial pore spaces of the sediments.

From such a limited data base, it is hazardous to estimate the groundwater contribution to Lake Michigan. However, some observations seem necessary and appropriate at this time. In the past, the groundwater contribution has generally been neglected as insignificant, or has been assumed not to exist at all.

Using a water budget, Bergstrom and Hansen (1962) estimated the total ground-water contribution or "shore discharge" to Lake Michigan to be $7.2 \cdot 10^8$ m³/yr. (800 ft.³/s); assuming a 58,000-km² lake area (Hough, 1958), this equals $1.2 \cdot 10^4$ m³ yr.⁻¹ km⁻². Based on the groundwater flux determined in this study, discharges of $1.9 \cdot 10^7$ m³ yr.⁻¹ km⁻² from sandy bottom areas and $5.0 \cdot 10^3$ m³ yr.⁻¹ km⁻² from fine-grained areas are calculated. In a 58,000-km² lake, discharge through the fine-grained lake sediments amounts to about $2.9 \cdot 10^8$ m³/yr., similar to that calculated by Bergstrom and Hansen without considering areas of sand with a flux similar to that calculated for this one sand body. Our best estimate of total near-shore sand area is 300 km² based on unpublished maps of Lake Michigan (D.L. Gross, pers. commun., 1976); this calculates to be a discharge of $5.7 \cdot 10^9$ m³/yr. Combining the discharges from the near-shore sands and the fine-grained lake sediments, we calculate the total groundwater discharge to the lake to be $6.0 \cdot 10^9$ m³/yr. (6700 ft.³/s) or $1.0 \cdot 10^5$ m³ yr.⁻¹ km⁻². This suggests that the total discharge to the lake is directly related to the area underlain by the near-shore sand bodies and may be much greater than that previously calculated.

REFERENCES

Bergstrom, R.E. and Hanson, G.F., 1962. Ground-water supplies in Wisconsin and Illinois adjacent to Lake Michigan. In: H.J. Pincus (Editor), Symposium on Great Lake Basin. American Association for the Advancement of Science, 18 pp.

Fraser, G.S. and Hester, N.C., 1974. Sediment distribution in a beach ridge complex and its application to artificial beach replenishment. Ill. State Geol. Surv., Environ. Geol. Note 67, 26 pp.

Hough, J.L., 1958. Geology of the Great Lakes. University of Illinois Press, Urbana, Ill.

Kennedy, E.J., Ruch, R.R. and Shimp, N.F., 1971. Studies of Lake Michigan bottom sediments, 7. Distribution of mercury in unconsolidated sediments from southern Lake Michigan. Ill. State Geol. Surv., Environ. Geol. Note 44, 18 pp.

Lineback, J.A. and Gross, D.L., 1972. Studies of Lake Michigan bottom sediments, 10. Depositional patterns, facies, and trace element accumulation in the Waukegan Member of the Late Pleistocene Lake Michigan Formation in southern Lake Michigan. Ill. State Geol. Surv. Environ. Geol. Note 58, 25 pp.

Lineback, J.A., Gross, D.L. and Meyer, R.P., 1972. Geologic cross sections derived from seismic profiles and sediment cores from southern Lake Michigan. Ill. State Geol. Surv., Environ. Geol. Note 54, 43 pp.

Rose, H.G. and Smith, H.F., 1957. A method for determining permeability and specific capacity from effective grain size. Ill. State Water Surv., Circ. No. 59.

Ruch, R.R., Kennedy, E.J. and Shimp, N.F., 1970. Studies of Lake Michigan bottom sediments, 4. Distribution of arsenic in unconsolidated sediments from southern Lake Michigan. Ill. State Geol. Surv. Environ. Geol. Note 37, 16 pp.

Schleicher, J.A. and Kuhn, J.K., 1970. Studies of Lake Michigan bottom sediments, 5. Phosphorous content in unconsolidated sediments from southern Lake Michigan. Ill. State Geol. Surv., Environ. Geol. Note 39, 15 pp.

Shimp, N.F., Leland, H.U. and White, W.A., 1970. Studies of Lake Michigan bottom sediments, 2. Distribution of major, minor, and trace constituents in unconsolidated sediments from southern Lake Michigan. Ill. State Geol. Surv., Environ. Geol. Note 32, 19 pp.

Shimp, N.F., Schleicher, J.A., Ruch, R.R., Heck, D.B. and Leland, H.V., 1971. Studies of Lake Michigan bottom sediments, 6. Trace element and organic carbon accumulation in the most recent sediments of southern Lake Michigan. Ill. State Geol. Surv., Environ. Geol. Note 41, 25 pp.

[2]

UNSTEADY STREAMFLOW MODELING GUIDELINES

VULLI L. GUPTA[1], SYED M. AFAQ[2], JOHN W. FORDHAM[1] and JAMES M. FEDERICI[3]

[1] *Water Resources Center, Desert Research Institute, and Civil Engineering Department, University of Nevada System, Reno, NV 89506 (U.S.A.)*
[2] *Department of Environment, Toronto, Ont. (Canada)*
[3] *Water Resources Center, Desert Research Institute, University of Nevada System, Reno, NV 89506 (U.S.A.)*

(Accepted for publication March 6, 1979)

ABSTRACT

Gupta, V.L., Afaq, S.M., Fordham, J.W. and Federici, J.M., 1979. Unsteady streamflow modeling guidelines. In: W. Back and D.A. Stephenson (Guest-Editors), Contemporary Hydrogeology — The George Burke Maxey Memorial Volume. J. Hydrol., 43: 79—97.

Presented herein is a case study of one-dimensional modeling of spatially varied unsteady flow regime in a river, utilizing the continuity and momentum formulations of Saint-Venant equations. The emphasis is on sensitivity analyses of the modeling framework for obtaining guidelines for modeling exercises. Finite-difference schemes were compared relative to their efficacy in simulating flow regimes. Model response was examined with respect to changes in six parameters, namely: (1) grid size; (2) lateral inflow or outflow rates; (3) discrete and composite description of hydraulic elements; (4) discrete and composite description of bed slopes; (5) roughness coefficient of stream bed; and (6) magnitude of weighting parameter.

INTRODUCTION

Application of successive approximation and finite-difference methods to the study of spatially varied and unsteady flow in channels has a relatively long history. In previous studies such as estuary modeling (Hann and Young, 1972), hydraulic surge propagation (Chaudry and Contractor, 1973), unsteady flow in streams and aquifers (Isaacson et al., 1956; Amein and Fang, 1969; Garrison et al., 1969; Strelkoff, 1970; Contractor and Wiggert, 1972; McDowell and Prandle, 1972; Quinn and Wylie, 1972), simulation efforts were based on simplifications of river geometry and its spatial variability, channel roughness and other hydraulic properties. Very few studies are reported in the literature where steep slopes and streams of rapidly varying geometry were involved. Furthermore, numerical simulation models hitherto were generally structured with very little input of hydraulic and physiographic information obtainable from field data.

This paper describes some of the principal findings related to the modeling of the unsteady flow regime of Truckee River, California—Nevada. Explicit and implicit finite-difference techniques and their modifications were utilized for solving the pair of partial differential equations describing the principles of continuity and momentum.

SCOPE AND OBJECTIVES

The scope of the investigation reported herein is limited to the main stem of Truckee River from Tahoe City to Nixon. For modeling purposes, the stretch of the entire river was considered to be made up of five reaches. River reaches were further represented by a total of 24 subreaches based on field survey with regard to cross-sections and other hydraulic elements. The principal objectives of the study efforts were as follows.

(1) To develop and test a numerical simulation model for the unsteady flow regime of Truckee River from Lake Tahoe to Nixon, 172 km in length, using a variety of finite-difference techniques.

(2) To compare the relative effectiveness of the finite-difference techniques in terms of: (a) reproducibility of observed flow events during floods and low flow regime; (b) propagation of a hypothetical hydraulic surge; and (c) routing the "Intermediate Regional" and "Standard Project" flood waves of the U.S. Army Corps of Engineers.

(3) To conduct sensitivity studies to investigate the model behavior due to the influence of: (a) varying lateral inflow and outflow; (b) variation of weighting parameter; (c) time—step changes; (d) smoothing the river-bed roughness coefficient; and (e) relative comparison using discrete and composite descriptions of channel geometry.

STUDY AREA

The study area encompasses the Truckee River basin which is situated on the eastern slopes of the Sierra Nevada Mountains and has a drainage area of 7925 km^2. Drainage area with runoff potential, however, is 2764 km^2 out of which 1310 km^2 directly drain into Lake Tahoe. The river originates from Tahoe City located on the northwestern rim of the lake where a dam regulates the outflow from the lake. The flow is further regulated by a number of reservoirs. The setting of the study area is illustrated in Fig. 1. The drainage network of Truckee River system is characterized by two types of stream channels. The first type occurs in the upper reaches of Truckee River and tributary stream draining into Lake Tahoe with bed slopes of the order of 95 m/km, whereas the second type occurs in middle and lower reaches below Reno where the bed slopes average about 2 m/km. The general layout of Truckee River system, indicating the gage locations and regulation aspects, is illustrated in Fig. 1.

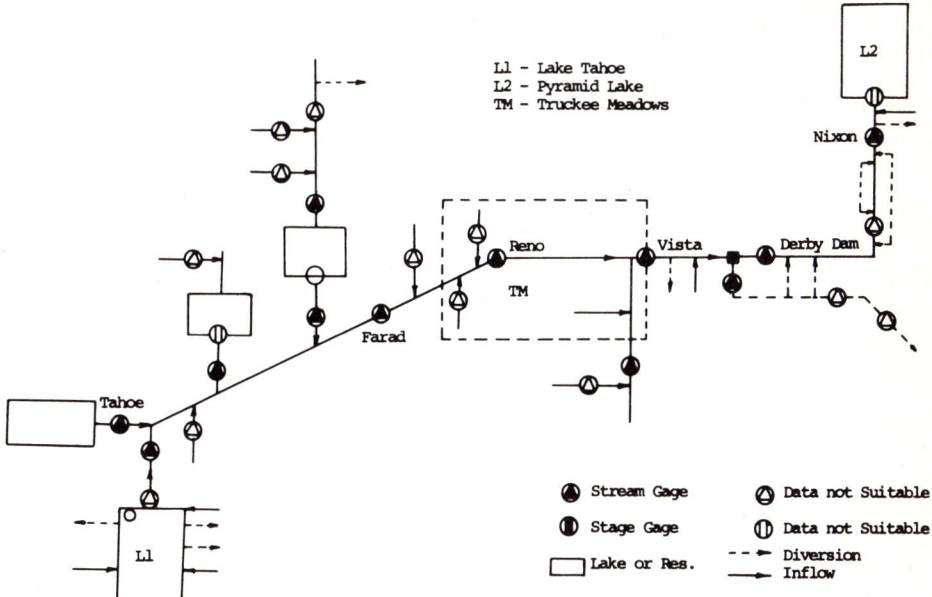

Fig. 1. Truckee River system.

METHODOLOGY

The two basic laws of fluid mechanics generally found to be applicable to the description of unsteady flow regime in rivers are: (a) continuity of unsteady flow or conservation of mass; and (b) dynamics of unsteady flow or conservation of momentum. Expressed in mathematical terms, these two principles lead to Saint-Venant's equations (Chow, 1959; Strelkoff, 1970), stated as:

$$\partial y/\partial t + V(\partial y/\partial x) + D(\partial v/\partial x) = q/T \tag{1}$$

and

$$\frac{1}{g}\frac{\partial v}{\partial t} + \frac{v}{g}\frac{\partial v}{\partial x} + \frac{\partial y}{\partial x} = (S_0 - S_f) - \frac{q(V - u_1)}{Ag} \tag{2}$$

Multiplying eq. 1 by T and substituting $T = \partial A/\partial y$ and $V = Q/A$, the resulting expression becomes:

$$T(\partial y/\partial t) + \partial Q/\partial x = q \tag{3}$$

Substituting $V = Q/A$ and simplifying, eq. 2 becomes:

$$\frac{1}{Ag}\frac{\partial Q}{\partial t} - \frac{Q}{2}\frac{\partial A}{\partial t} + \frac{Q}{2}\frac{\partial Q}{\partial x} - \frac{Q^2}{A^2 g}\frac{\partial Q}{\partial x} + \frac{\partial y}{\partial x} = S_0 - S_f - \frac{(V - u_1)q}{Ag} \tag{4}$$

Replacing $\partial A/\partial t$ by $T(\partial y/\partial t)$ and $\partial A/\partial x$ by $T(\partial y/\partial x)$, eq. 4 can be re-arranged as:

$$\frac{1}{Ag}\frac{\partial Q}{\partial t} - \frac{V}{Ag}\frac{\partial Q}{\partial x} - \frac{Q^2 T}{A^3 g}\frac{\partial y}{\partial x} + \frac{\partial y}{\partial x} = S_0 - S_f - \frac{(V-u_1)q}{Ag} \quad (5)$$

Substituting $q - \partial Q/\partial x$ for $T(\partial y/\partial t)$ and F^2 for $Q^2 T/A^3 g$, eq. 5 can be simplified as:

$$\frac{1}{Ag}\frac{\partial Q}{\partial t} + \frac{2V}{Ag}\frac{\partial Q}{\partial x} + (1-F^2)\frac{\partial y}{\partial x} = S_0 - S_f + \frac{u_1 q}{Ag} \quad (6)$$

In eqs. 1—6, A = waterway cross-sectional areas; D = hydraulic depth expressed as the ratio of waterway cross-sectional area to the width of free water surface; F = Froude number; g = acceleration due to gravity; Q = flow rate in the waterway; q = rate of lateral inflow or outflow; S_0 = bed slope of stream; S_f = slope of hydraulic gradient; T = width of free water surface; t = time; U_1 = component of lateral flow velocity in the direction of streamflow in the river.

Finite-difference formulations

In the most general form, equations of continuity and momentum are represented by the following weighted average approximations. Finite-difference formulation of eq. 3 can be stated as:

$$T_{i,j}\frac{Y_{i,j+1} - Y_{i,j}}{\Delta t} + \frac{1}{2\Delta x}\theta(Q_{i+1,j+1} - Q_{i-1,j+1}) + (1-\theta)(Q_{i+1,j} - Q_{i-1,j}) = Q_{IN} \quad (7)$$

Similarly, finite-difference formulation of eq. 6 can be expressed as:

$$\frac{1}{A_{i,j}g}\left[\frac{Q_{i,j+1} - Q_{i,j}}{\Delta t}\right] + \frac{2V_{i,j}}{A_{i,j}g\, 2\Delta x}\left[\theta(Q_{i+1,j+1} - Q_{i-1,j+1}) + \right.$$

$$\left. (1-\theta)(Q_{i+1,j} - Q_{i-1,j})\right] + \left[\frac{1-F^2_{i,j}}{2\Delta x}\right]\left[\theta(Y_{i+1,j+1} - Y_{i-1,j+1}) + \right.$$

$$\left. (1-\theta)(Y_{i+1,j} - Y_{i-1,j})\right] = S_0 - [\theta S_{f_{i,j+1}} + (1-\theta)S_{f_{i,j}}] \quad (8)$$

In eqs. 7 and 8, i and j denote the position of the variable with reference to space and time grids, respectively; θ is the weighting parameter; Q_{IN} is the rate of lateral inflow or outflow; and the remaining symbols are as defined before. The effect of lateral flow in the finite-difference formulation of momentum principles, eq. 8 has been ignored. Collection of field data, as discussed in the subsequent sections, did not include information related to the orientation of tributaries and other situations of lateral flow regime. The dependent variables in the foregoing formulations are discharge, Q, and depth, y. In eqs. 7 and 8, values of θ of 0, 0.5 and 1 lead to explicit, Crank—Nicholson's implicit, and fully implicit formulations.

Finite-difference formulations for the weighted average implicit schemes, $0 < \theta < 1$, were also structured.

A computer program was developed, in FORTRAN IV, with the capability of handling the foregoing finite-difference formulations. A program listing is presented by Moin (1974).

FIELD SURVEY

Efforts toward systematic fulfillment of the study objectives were based on two categories of data, namely, field data and historic hydrologic data. In summer 1973, field surveys in the study area were conducted to obtain information related to: (a) cross-sections at 25 locations on the main stem of river and 13 locations on the tributaries; (b) concurrent discharge measurements using current meter; and (c) river-bed level survey using information of bench-mark levels supplied by the U.S. Coast and Geodetic Survey. Field survey data were processed to yield the following:

(a) bed slopes for each of the subreaches and reaches.

(b) Manning's roughness coefficient values for each subreach calculated from the flow measurement bed slope and corresponding waterway area.

(c) graphical relationships among the hydraulic parameters: depth—area, depth—top width, depth—wetted perimeter, top width—area and depth—discharge.

(d) regression equations for each of the graphical relationships stated in (c).

Length and slope of the five main reaches of the river are presented in Table I.

TABLE I

Reaches of the main stem of the Truckee River

No.	Station	Elevation (km)	Length (km)	Slope
1	Tahoe City Gage	1.896	53.13	0.0061
2	Farad Gage	1.572	36.63	0.0059
3	Reno Gage	1.354	10.87	0.0020
4	Vista Gage	1.333	29.06	0.0018
5	Derby Dam	1.281	41.78	0.0019
6	Nixon Gage	1.202		

Field survey information was processed to yield empirical relationships relating pairs of hydraulic elements of major concern to modeling endeavors. For each of the 25 subreaches and five main reaches, regression equations were developed in the following form:

$$A = K_1 Y^{n_1}, \quad Q = K_2 Y^{n_2} \qquad (9), (10)$$

$$P = K_3 Y^{n_3}, \quad T = K_4 Y^{n_4} \qquad (11), (12)$$

where A = waterway area; K_1–K_4 = coefficients obtainable from regression

TABLE II

Regression equations of hydraulic elements — composite reaches

No.	Reach	Length (km)	Area–depth $A = K_1 Y^{n_1}$		Discharge–depth $Q = K_2 Y^{n_2}$		Wetted perimeter–depth $P = K_3 Y^{n_3}$		Top width–depth $T = K_4 Y^{n_4}$	
			K_1	n_1	K_2	n_2	K_3	n_3	K_4	n_4
1	Tahoe–Farad	53.13	21.12	1.68	21.80	2.62	39.56	0.69	38.63	0.62
2	Farad–Reno	36.63	37.48	1.72	25.94	2.65	63.59	0.66	64.79	0.65
3	Reno–Vista	10.87	38.24	1.54	11.25	2.96	56.00	0.63	57.16	0.62
4	Vista–Derby Dam	30.04	26.62	1.74	15.79	2.77	37.34	0.79	38.71	0.77
5	Derby Dam–Nixon	40.80	21.42	1.80	12.23	3.13	36.94	0.79	37.85	0.77

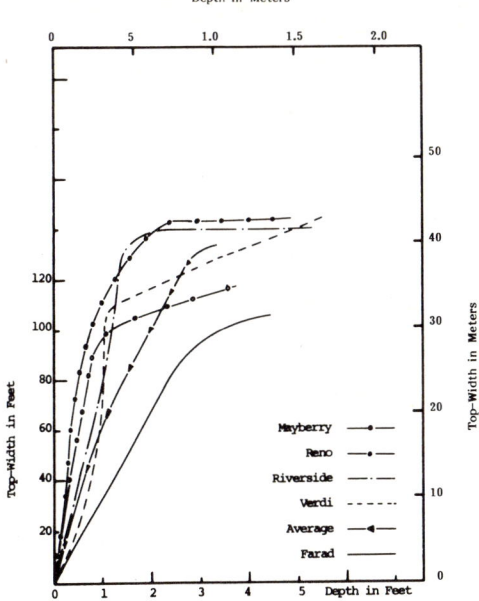

Fig. 2. Depth—top width relationship.

Fig. 3. Depth—area relationship.

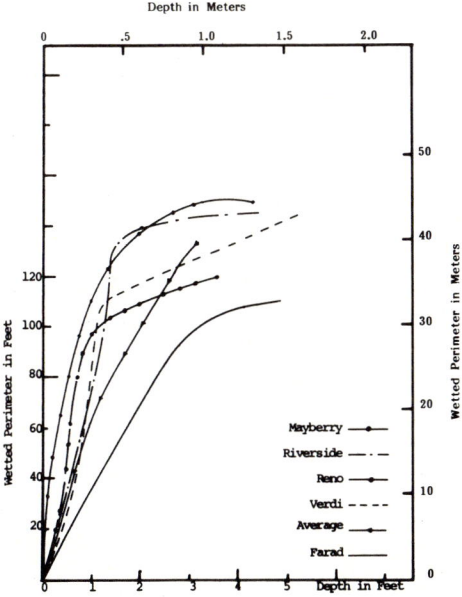

Fig. 4. Depth—wetted perimeter relationship.

analysis of logarithms of the respective pairs of hydraulic elements; n_1-n_4 = exponents obtainable from regression analysis of logarithms of the respective pairs of hydraulic elements; P = wetted perimeter; Q = discharge; and T = width of the free water surface. The resulting equations for the five main reaches are presented in Table II. A sampling of graphic relationships of hydraulic elements at selected subreaches of the river are illustrated in Figs. 2—4.

DATA ANALYSIS AND MODELING EXERCISE

Data analysis in terms of stressing the model consisted of the following tasks:

(1) Routing of the flood wave which occurred during January—February, 1963.

(2) Study of flow regime for the Farad—Vista reach of the river for a five-day duration, November 10—14, 1973.

(3) Routing of synthesized hydrographs for "intermediate regional" and "standard project" flood events estimated by the U.S. Army Corps of Engineers, Sacramento, California for the Truckee Meadows reach from Mayberry to Vista.

(4) Effects of surge propagation due to hypothetical flood event wherein the flow rate was increased from 14 to 280 m³/s in a time span of 20 min.

Identification of the specific reaches where each of the aforementioned modeling exercises were conducted is presented in Table III. Pertinent findings of the various modeling efforts are described below.

SIMULATION OF A HISTORIC FLOOD EVENT

Recorded streamflow data (U.S. Army Corps of Engineers, 1970; Young and Harris, 1966) relative to the 1963 flood occurrence in the Truckee River basin was utilized in investigating the modeling capabilities toward reproducing the observed flood flow. For the five main reaches of the river, the model was operated separately, and in each instance the observed hydrograph at the boundary was compared with the computed hydrograph. In the absence of gaged data for describing the lateral flow regime, several trials for various assumed values of lateral inflows were made. By trial-and-error approach, Q_{IN} was altered in a variety of ways such that a reasonable agreement between the observed and computed hydrographs was found. Findings of this phase of modeling exercises are presented below.

(1) *Tahoe—Farad reach.* For this 53.3 km long reach, smoothed or composite values of hydraulic elements were chosen. Even though the topography appears to warrant it, use of discrete sections within this reach would not have been advantageous because of hydrologic operation at Lake Tahoe in terms of storage in the lake and releases to the river. The lateral inflow regime was, after several trials, hypothesized to vary with time.

TABLE III

Modeling exercises

Task	Tahoe—Farad	Farad—Reno	Reno—Vista	Vista—Derby Dam	Derby Dam—Nixon
1963 flood simulation	×	×	×	×	×
1973 short-term flow simulation		×			
Fully implicit scheme	×	×	×	×	×
Weighted average scheme			×	×	
Composite-hydraulic element model	×	×	×	×	×
Discrete-hydraulic element model		×	×		
Intermediate regional flood-routing		×	×		
Standard Project Flood routing		×	×		
Surge propagation		×			

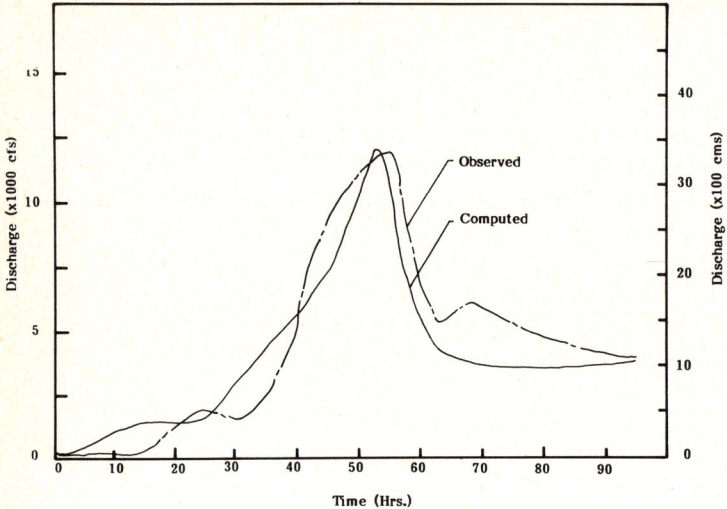

Fig. 5. Farad hydrograph, 1963 flood event.

A sample of trial exercises for Tahoe—Farad reach is illustrated in Fig. 5 wherein a reasonable agreement of observed and computed profiles is apparent with respect to peak time and crest segment. Recession segment registered a relatively poor agreement. Modeling exercises for this reach were found to be very cumbersome due to: (a) unavailability of tributary flow data; and (b) difficulties in describing the regulation at Lake Tahoe in mathematical terms.

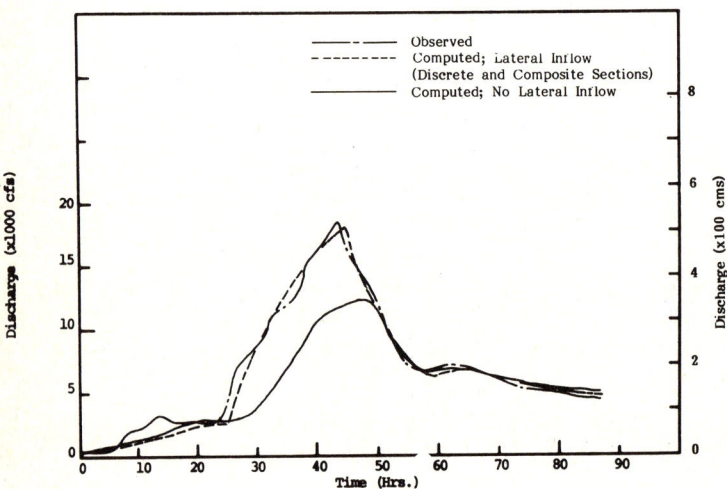

Fig. 6. Reno hydrograph, 1963 flood event, discrete and composite sections.

(2) *Farad—Reno reach*. Except for a small stretch in the vicinity of Farad, this reach of the river lies in Truckee Meadows, characterized by withdrawals and accretions at several locations. Initially, a no-lateral inflow situation was assumed and the resulting hydrograph is illustrated in Fig. 6. Significant disparity was evident between the computed and observed flow profiles during the rising limb and at the crest of the flood wave. This disparity is attributable to the total neglect of the lateral inflow. After several trial runs, an adequate lateral flow regime was derived and included to improve the agreement between the observed and the computed flood hydrographs as illustrated in Fig. 6. These runs were calculated using composite hydraulic elements. Similarly, Fig. 6 also shows the results based on discrete hydraulic elements and the associated properties. The phrase "composite" refers to consideration of the river in the entire reach to be prismatic. In contrast, the phrase "discrete" refers to the consideration of actual cross-sections of the subreaches based on the field survey. Comparison of results, as illustrated in Fig. 6, indicates that there is practically no gain in the modeling capability using discrete descriptions of the hydraulic elements instead of using composite or average properties.

(3) *Reno—Vista reach*. Alignment of the river in this reach is important from floodplain delineation considerations since it flows through the urbanized area of Reno—Sparks. This reach was modeled as being made up of fifteen subreaches, with composite hydraulic geometry, constant n of 0.05, and a uniform bed slope of 0.002. Steamboat Creek is the only major tributary to the river within the Reno—Vista reach and was superimposed on the Vista hydrograph in Fig. 7.

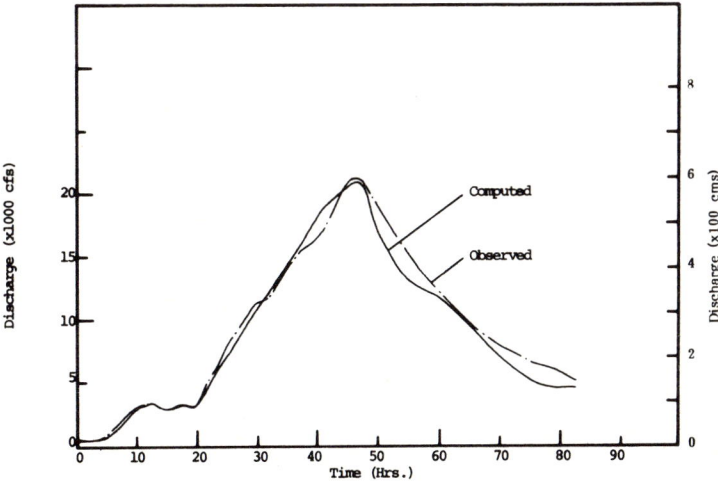

Fig. 7. Vista hydrograph, 1963 flood event.

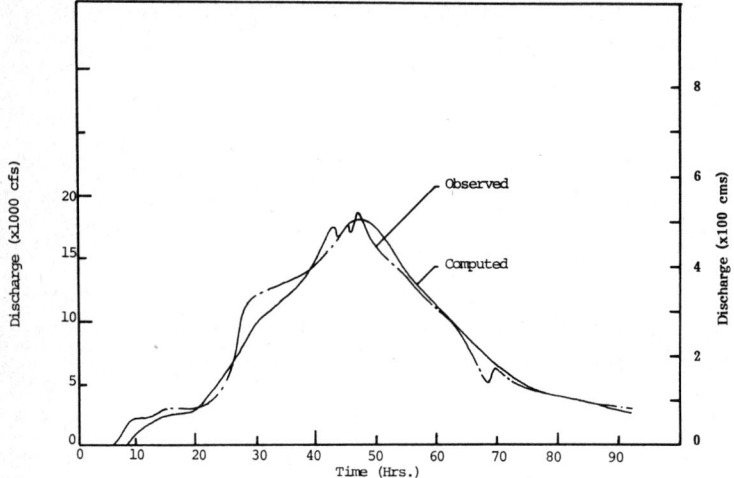

Fig. 8. Derby Dam hydrograph, 1963 flood event.

(4) *Vista—Derby Dam reach.* This reach of the river is characterized by rugged topography and alignment, no tributary flow inputs, and a major flow diversion via Truckee Canal at Derby Dam. There is no documented operational policy relative to flood control practice at Derby Dam. Consequently, several runs were made in modeling efforts with a variety of assumed storage policies at Derby Dam. Results of one of these policies were found to yield a close match between the computed and observed hydrographs and are illustrated in Fig. 8.

(5) *Derby Dam—Nixon reach.* This reach of the river is characterized by flat bed slopes, relatively wide floodplain, a number of flow diversions and accretions for which no systematic data are available. After several runs, lateral flow regime criteria were found to yield a reasonably close agreement between the observed and computed hydrograph profiles. The resulting hydrograph is compared against the historic hydrograph illustrated in Fig. 9.

Simulation efforts throughout the five reaches were characterized by trial-and-error approaches relative to the lateral flow regime. If information with regard to the linkages of surface water and subsurface water in the river basin were available, trial-and-error approaches might not be warranted.

Simulation of bi-hourly streamflow

Modeling exercises were conducted to investigate the simulation capabilities for the Farad—Reno reach of the river. Bi-hourly flows for the time span of November 10—14, 1973 were modeled using a fully implicit finite-difference scheme with discrete sections, discrete bed slopes, and discrete values of

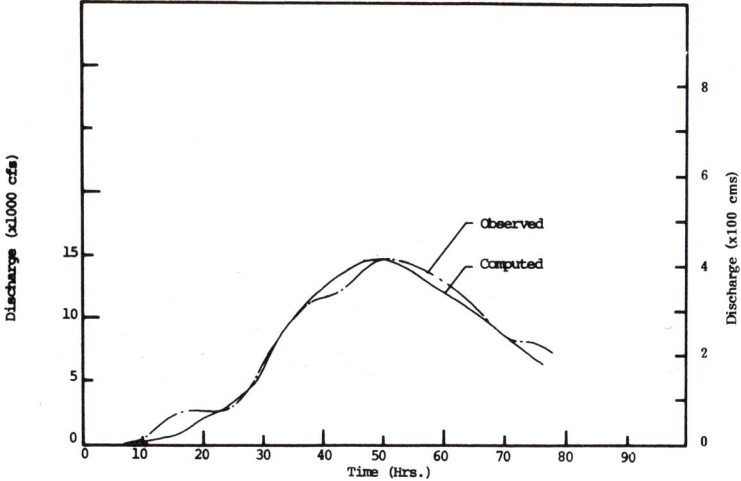

Fig. 9. Nixon hydrograph, 1963 flood event.

Manning's roughness coefficient. Field survey data were thus utilized to the fullest possible extent. In addition to lateral inflows, flow withdrawals from the river were also incorporated. The routed hydrograph for Reno is illustrated in Fig. 10. A comparison of observed and computed hydrographs reveals reasonable agreement relative to peaks and troughs in terms of timing and magnitude. The profiles of the rising and recession limbs, however, do not match satisfactorily.

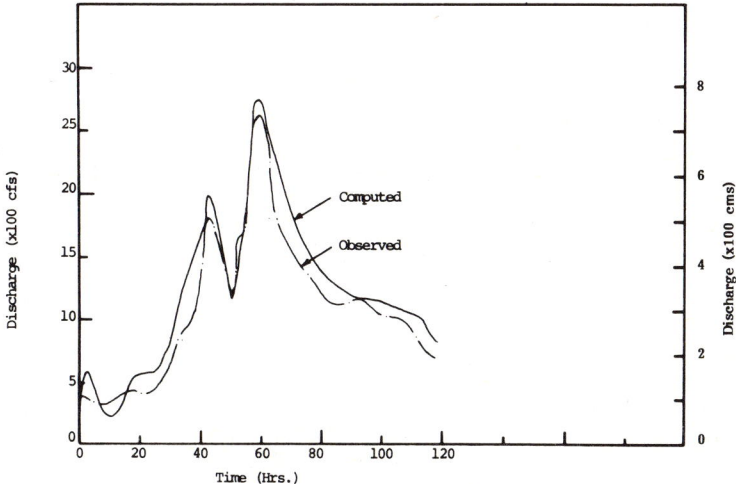

Fig. 10. Reno hydrograph, 1973 short-term flow study; time step = 2 hr.

Intermediate regional and standard project flood wave propagation

This aspect of modeling was conducted for Truckee Meadows reach of the river, 21.82 km long, stretching from Lawton near Mayberry bridge to Vista, Nevada. The study reach includes the urbanized area of Reno—Sparks. There have been several flood events in the past in this area. Consequently, floodplain studies for Truckee Meadows have received the attention of the U.S. Army Corps of Engineers, Sacramento District, California. Two measures of flood flow hazard were suggested in their studies (U.S. Army Corps of Engineers, 1970, 1971), Intermediate Regional Flood (IRF) and Standard Project Flood (SPF). IRF is associated with a recurrence interval of 100 years, and such estimates are usually based on frequency analysis of streamflow records. SPF, however, implies the consequence of a combination of the most severe meteorologic and hydrologic conditions.

In view of the significance of the urban area, a study was conducted with the objective of testing the model capabilities for routing the two flood waves separately. Fully implicit scheme, discrete sections, discrete values for bed slope and roughness coefficients were utilized. Flood routing exercises require the hydrograph description for the flood inflow. Available data (Young and Harris, 1966; U.S. Army Corps of Engineers, 1970) include peak rates of runoff and timing of the peak occurrences.

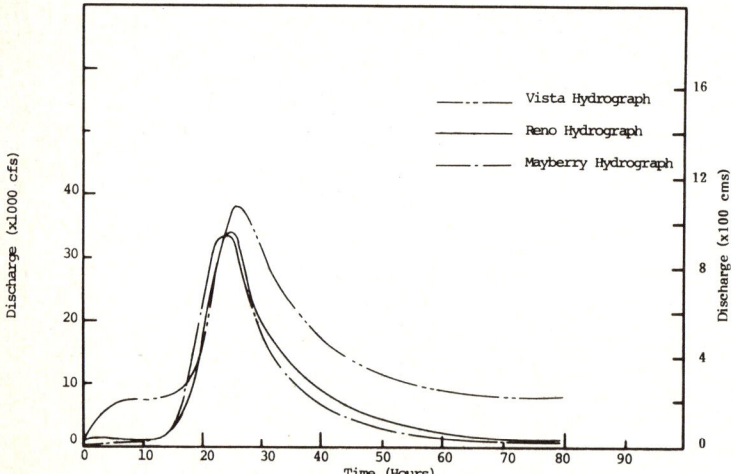

Fig. 11. Propagation of the Standard Project Flood wave, Truckee Meadows reach.

Utilizing the findings of Gupta and Moin (1974), profiles for inflow hydrographs were synthesized and applied at Lawton, one each for IRF routing and SPF routing. Results of the flood routing exercises are presented in Table IV. The computed hydrographs for the SPF are illustrated in Fig. 11.

TABLE IV

Comparison of estimated flow by U.S. Corps of Engineers and computed flows

Station	Intermediate Regional Flood (m³/s)			Standard Project Flood (m³/s)		
	estimated*	computed	% diff.	estimated*	computed	% diff.
Reno	512.5	532.1	3.8	982.6	963.9	−1.9
Vista	608.8	577.9	−5.1	1115.7	1083.1	−2.9

*U.S. Corps of Engineers (1970, 1971).

Hydraulic surge propagation

Modeling capabilities for routing a transient hydraulic phenomenon such as a severe surge wave were studied, on a limited scale, for the Farad—Reno reach. Fully implicit model with discrete cross-sections, discrete values of bed slopes and roughness coefficients, was used in modeling the hydraulic bore propagation.

A uniform flow rate of 14 m³/s was assumed at Farad, and this flow rate was arbitrarily assumed to increase from 14 to 269 m³/s at Farad in a time span of 20 min. The resulting hydrograph rise at Reno is illustrated in Fig. 12. Time delay in the surge development at Reno was found to be 4.4 hr. This exercise attests to the modeling capabilities in analyzing a transient hydraulic phenomenon such as a severe flow surge resulting from a possible failure of upstream dams of regulation structures.

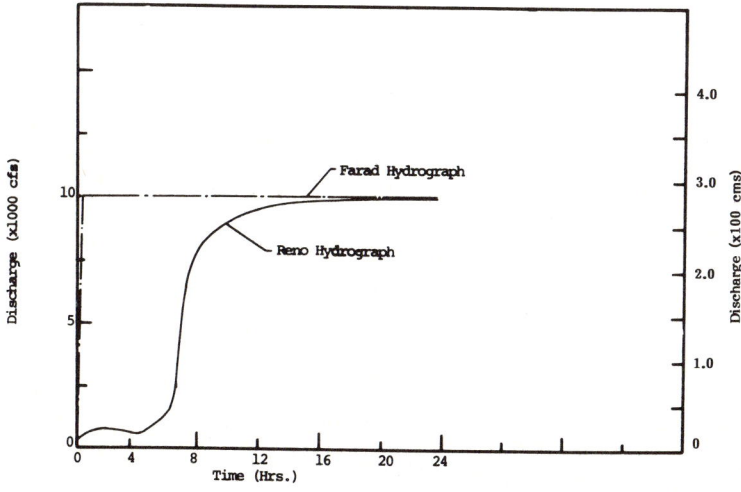

Fig. 12. Reno hydrograph, propagation of hydraulic surge.

SENSITIVITY STUDIES

Six parameters were considered throughout the modeling exercises. The influence of changing the parameters upon the computed hydrographs was investigated. The parameters are: (1) mesh size; (2) lateral inflow and/or outflow rates; (3) discrete and composite description of hydraulic elements; (4) discrete and composite description of bed slope; (5) discrete and composite description of Manning's roughness coefficient; and (6) weighting parameter, θ.

(1) *Mesh size.* Mesh size refers to the magnitude of time and space steps. Shorter distances and time increments tend to yield accurate results but may not necessarily be economical in the simulation exercises. Model behavior in simulating the bi-hourly flow within Farad—Reno reach during November 10— 14, 1973 was studied using three different values of time steps of the finite-difference scheme. These were 0.5, 1 and 2 hr. Results from the model are illustrated in Fig. 13, and it was observed that the model behavior is relatively independent of the time step used in the computations.

(2) *Lateral flow regime.* In all the modeling exercises, lateral flow regime was found to be the most influential parameter. Trial-and-error approaches by imposing several different rates of lateral flow were used to obtain a reasonable match between the computed and observed hydrographs. It cannot be inferred that the rates of lateral flow so used are unique values.

Extreme sensitivity of the model to the lateral flow regime can be readily inferred from examining Fig. 6, where the results of the 1963 flood event for Farad—Reno reach are illustrated. By ignoring lateral flow, the peak was found to be underestimated by about 32%. Computed peak rate was found to lag behind the observed peak by about 3.5 hr. Superimposition of lateral flow,

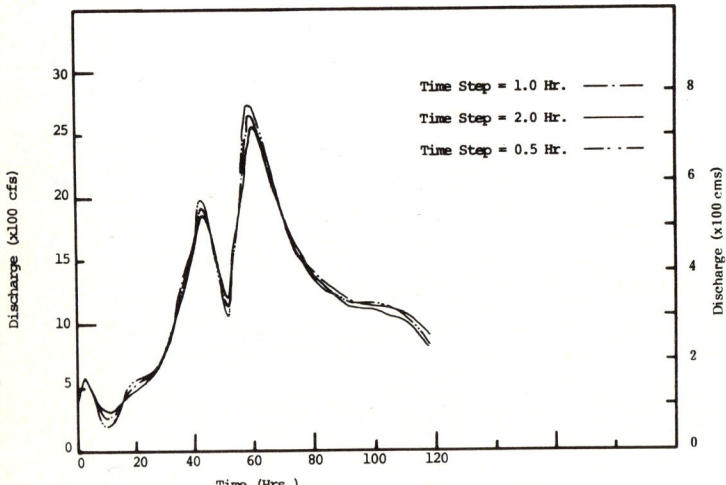

Fig. 13. Reno hydrograph, 1973 short-term flow study, effect of time step.

by trial and error, has yielded a much closer agreement, as can be seen in Fig. 6. Consideration of composite sections will tend to damp out the variations in depths. Flow modeling exercises in the Truckee Meadows reach and Farad—Reno reach were based on discrete descriptions, and in all other reaches composite descriptions were found to be sufficient.

(4) *Discrete and composite stream bed slopes.* Examination of Fig. 6 indicates that the model behavior is relatively the same whether varying or constant channel bed slope was used.

(5) *Discrete or composite Manning's roughness coefficient.* Examination of Fig. 6 also indicates that the model behavior is the same whether averaged and composite values for n or varying and discrete values are used. Discrete values were used in modeling Truckee Meadows reach, and in all other instances composite values were used.

(6) *Weighting parameter, θ.* Magnitude of weighting parameter θ was found to be influential in controlling the computed hydrograph profile. The 1963 flood wave propagation for Vista—Derby Dam reach of the river was investigated by setting $\theta = 0.5, 0.75$ and 1.0. The resulting hydrographs at Derby Dam are illustrated in Fig. 14. Model behavior was found to be relatively insensitive to the weighting parameter. It was found that the hydrograph corresponding to $\theta = 0.75$ lies between those for $\theta = 0.5$ and 1.0.

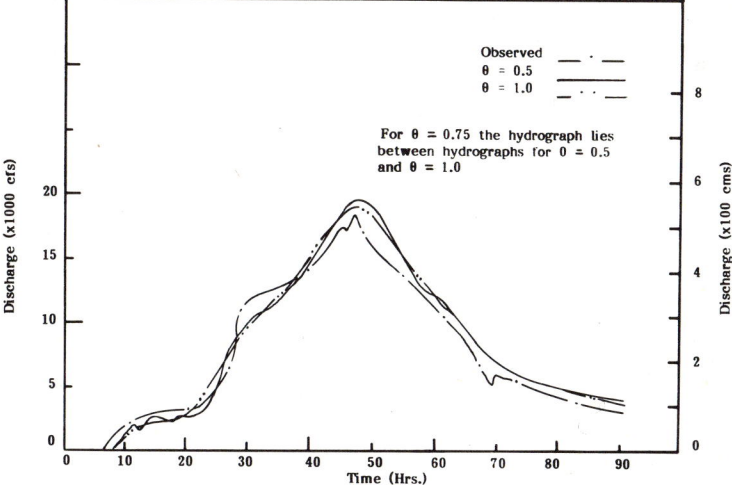

Fig. 14. Derby Dam hydrograph, 1963 flood event; effect of varying θ.

SUMMARY AND CONCLUSIONS

A simulation model has been developed and tested for its capability to solve the unsteady-flow equations using finite-difference schemes of analyses. The model was applied to simulate the 1963 flood wave movement in the Truckee River. The river was modeled as being made up of five main reaches, each of which has been further subdivided into a number of subreaches of equal length. The reaches are Tahoe—Farad, Farad—Reno, Reno—Vista, Vista—Derby Dam and Derby Dam—Nixon. The following list comprises the modeling exercises:

(1) Simulation of the 1963 flood hydrographs using the recorded hydrographs at the U.S. Geological Survey gaging stations as boundary conditions.

(2) Simulation of short-term flow for the period of November 10—14, 1973, in the Farad—Reno reach of the river.

(3) Routing the IRF and SPF floods in the Truckee Meadows reach from Mayberry Bridge to Vista using synthesized flood hydrographs.

(4) Evaluation of hydraulic bore propagation due to a hypothetical situation wherein a uniform flow of 14 m^3/s in Farad—Reno reach was increased to 269 m^3/s at Farad in a time span of 20 min.

(5) Sensitivity studies for monitoring the model behavior due to changes in six parameters, namely: (a) mesh size; (b) lateral inflow and/or outflow rates; (c) discrete and composite description of the hydraulic elements; (d) discrete and composite descriptions of bed slopes; (e) discrete and composite description of Manning's roughness coefficient; and (f) magnitude of the weighting parameter, θ.

The conclusions of the study, listed below, provide useful guidelines for modeling the unsteady flow regime of rivers:

(1) It was found that a computational time step of 1 hr. was best suited among the three time steps, 0.5, 1 and 2 hr. examined.

(2) The model was found to be capable of reproducing 1963 flood wave in the river from Lake Tahoe to Nixon.

(3) The model was found to be capable of simulating short-term flow hydrographs for the flow regime in Farad—Reno reach for November 10—14, 1973.

(4) Based on synthesized hydrograph descriptions of IRF and SPF flood events, it was concluded that the Truckee Meadows reach can be modeled with a high degree of reliability.

(5) The model is capable of handling a transient hydraulic phenomenon such as a surge propagation in the river.

(6) Sensitivity analyses have indicated that the most influential parameter controlling the model behavior is the lateral flow regime. Throughout the modeling exercises, it was found that lateral flow rates play an important role in the reproducibility of recorded hydrographs. Based on these findings, the most critical aspects in modeling include the explicit description of tributary flow inputs and rating curves for the river in the immediate vicinity of the points of confluence.

(7) It was found that for all values of θ between 0.5 and 1.0, the computed hydrograph is strikingly the same, with very little difference. This indicates that any value of θ between 0.5 and 1.0 is adequate for modeling purposes.

ACKNOWLEDGMENTS

The authors are grateful to Richard Cooley for his enthusiastic guidance throughout the study. Sincere thanks are extended to Thomas Gallagher for his timely cooperation in the acquisition of field data.

Work reported herein was supported in part by the United States Department of the Interior, Office of Water Resources Research, Project No. A-057-NEV and B-096-NEV, as authorized under the Water Resources Research Act of 1964, PL 88-379, and in part by funds provided by the Desert Research Institute, University of Nevada System, Reno, Nevada.

REFERENCES

Amein, M. and Fang, C.S., 1969. Streamflow routing (with applications to North Carolina River). Water Resour. Res. Inst., Univ. of North Carolina, Raleigh, N.C., Tech. Rep. No. 17.
Amein, M. and Fang, C.S., 1970. Implicit flood routing in natural channels. Am. Soc. Civ. Eng., J. Hydraul. Div., 96(HY12): 2548—2564.
Chaudry, Y.M. and Contractor, D.N., 1973. Application of implicit method to surges in open channels. Water Resour. Res., 9(6): 1605—1612.
Chow, V.T., 1959. Open Channel Hydraulics. McGraw-Hill, New York, N.Y., 680 pp.
Contractor, D.N. and Wiggert, J.M., 1972. Numerical studies of unsteady flow in the James River. Water Resour. Res. Cent., Va. Polytech. Inst.—State Univ., Blacksburg, Va., Bull. 51, 56 pp.
Garrison, J.M., Granju, J.P.P. and Price, T.T., 1969. Unsteady flow simulation in rivers and reservoirs. Am. Soc. Civ. Eng., J. Hydraul. Div., 95(HY5): 1559—1576.
Gupta, V.L. and Moin, S.M.A., 1974. Surface runoff hydrograph equation. Am. Soc. Civ. Eng., J. Hydraul. Div., 100(HY10): 1352—1368.
Hann, Jr., R.W. and Young, P.J., 1972. Mathematical models of water quality parameters for rivers and estuaries. Water Resour. Inst., Texas A & M Univ., College Station, Texas, Tech. Rep. 45, 424 pp.
Isaacson, E., Stoker, J.J. and Troesch, 1956. Numerical solution of flood prediction and river regulation problems, Inst. Math. Sci. New York Univ., New York, N.Y., Rep. No. IMM-235.
McDowell, D.M. and Prandle, 1972. Mathematical model of river Hooghly. Am. Soc. Civ. Eng., J. Waterw., Harb. Coastal Eng. Div., 98(WW2): 225—242.
Moin, S.M.A., 1974. Numerical simulation of unsteady flow hydraulics of Truckee River. M.S. Thesis, University of Nevada, Reno, Nev., 134 pp.
Quinn, F.H. and Wylie, E.B., 1972. Transient analysis of the Detroit River by the implicit method. Water Resour. Res., 8(6): 1461—1469.
Strelkoff, T., 1970. Numerical solution of Saint Venant equations. Am. Soc. Civ. Eng., J. Hydraul. Div., 96(HY1): 223—252.
U.S. Army Corps of Engineers, 1970. Flood plain Information — Truckee River Reno—Sparks—Truckee Meadows, Nevada. Department of Army, Sacramento District, Calif., 43 pp.
U.S. Army Corps of Engineers, 1971. Master Report on Reservoir Regulation for Flood Control — Truckee River Reservoirs, Nevada and California. Department of Army, Sacramento District, Calif., 43 pp.
Young, L.E. and Harris, E.E., 1966. Floods of January—February 1963 in California and Nevada. U.S. Geol. Surv., Water-Supply Pap. 1830-A, 472 pp.

[2]

PORE SIZE DISTRIBUTION, SUCTION AND HYSTERESIS IN UNSATURATED GROUNDWATER FLOW

C.M. CASE and A. WELCH*

Desert Research Institute, Water Resources Center, University of Nevada, Reno, NV 89506 (U.S.A.)

(Accepted for publication February 6, 1979)

ABSTRACT

Case, C.M. and Welch, A., 1979. Pore size distribution, suction and hysteresis in unsaturated groundwater flow. In: W. Back and D.A. Stephenson (Guest-Editors), Contemporary Hydrogeology — The George Burke Maxey Memorial Volume. J. Hydrol., 43: 99—120.

Formalism for calculating soil moisture suction via average relative humidity in the soil is reviewed and then extended via the concept of pore size distribution. Using an assumed form for pore size distribution and a form of the Kelvin equation for the vapor pressure of a fluid in a capillary allows direct calculation of the capillary contribution to suction under certain idealized conditions. The change in suction with changes in the concentration of total dissolved solids is then examined via the difference between the pressure of vapor over a pure solvent and that over a solution and changes in surface tension that occur when impurities are dissolved in water.

The pore size distribution and an assumed form of volume per pore/capillary are used to construct and calculate the degree of saturation $\theta_s(r)$ where the assumption is made that all pores/capillaries in the sample under consideration up to radius r are filled and those having radii larger than r are dry. Later in the development it is shown how this initial assumption can be refined. Negative suction and the degree of saturation θ_s are calculated as explicit functions of r and plotted as implicit functions of each other, thereby yielding one of the family of hysteresis loops appropriate to the current formalism. If r_0 is the value of r for which the pore size distribution curve has its peak, the value of r_0 for the wetting branch is larger than that for the drying branch, since the drying curve "sees" smaller pores/capillaries on the average than does the wetting curve. The area enclosed by the hysteresis loop is calculated and represents the energy per unit volume lost to the formation on adsorbing and then desorbing moisture.

INTRODUCTION

Much of the current work on the flow of water in the vadose zone either considers the equations of flow (e.g., Topp and Miller, 1966; Poulovassilis, 1970) and their solution in certain special cases, or focuses on hysteresis —

Present address: Carson City District, U.S. Geological Survey, 702 Plaza Street, Carson City, NV 89701, U.S.A.

either experimental measurements of saturation vs. suction (Liakopoulos, 1965; Vachaud and Thony, 1971) field measurements of suction (Reisenauer et al., 1975) or theoretical constructs to fit measured hysteresis curves. [There are many such constructs, most based in some way, however indirect, on geometrical assumptions about the porous medium. See, for example, Haines (1930), Everett and Whitton (1952), Everett (1954, 1955), Everett and Smith (1954), Enderby (1955, 1956), and Fatt (1956a, b).]

In this work, which is partly expository, the relations between pore size distribution and soil matric suction, and pore size distribution and degree of saturation, respectively, are considered. The relation between suction and the average vapor pressure of the water in the soil is derived and the required psychrometer (Brown and Van Havern, 1972) accuracy for meaningful measurements is considered. A form of the relation between the equilibrium vapor pressure above a capillary of radius r filled with water and the equilibrium vapor pressure above a flat water surface (the Kelvin relation) is then given (it is derived for completeness in Appendix A) and used, along with an assumed form of pore size distribution, to calculate the capillary contribution to the average vapor pressure measured by a psychrometer probe exposed to a partial or full pore size distribution of water-filled pores/capillaries in the soil.

The two branches of a hysteresis loop (degree of saturation vs. suction) reflect the fact that the suction apparatus used in experiments "sees" a different average, effective, pore size distribution, depending on whether the sample under study is being saturated or desaturated. Since the drying (desaturation) branch has higher moisture content for a given degree of suction than the wetting (saturation) branch it follows that the soil is more retentive for drying than for wetting and hence the average effective pore size distribution for drying is shifting toward smaller values of pore/capillary radius than the one for the wetting branch. In terms of the formalism of this paper, this is expressed by values of r for which the pore size distribution has its maximum, r_0, being smaller — and thus shifting the pore size distribution curve toward smaller values of r — for drying than for wetting. In symbols, $r_{od} < r_{ow}$ where the subscripts d and w refer to drying and wetting, respectively. Curves of suction vs. degree of saturation plotted for the same ranges of moisture content, but different values of r_0, form narrow hysteresis loops. The formalism given here, in which soil moisture suction is an explicit function of the vapor pressure of the water in the soil, lends itself naturally to incorporation of the lowering of the vapor pressure of a solution compared to that of a pure solvent (Huang, 1963, p. 50). To first approximation, this lowering of vapor pressure can be calculated using values for the amount of total dissolved solids without regard to their nature.

The area enclosed by the hysteresis loop can be calculated as pressure—volume work (Haines, 1930) and represents the energy per unit volume lost by the soil—water system, partly as a result of spatial redistribution of the water in the pore space of the sample, in the process of adsorbing and desorbing moisture.

MATRIC SUCTION AND THE VAPOR PRESSURE OF SOIL WATER

For water in equilibrium with its own vapor the chemical potentials of the two phases are equal. Thus:

$$\mu_l = \mu_g \tag{1}$$

where μ_l is the Gibbs free energy per unit mass of the liquid; and μ_g is the Gibbs free energy per unit mass of the gas (vapor) above it. Taking the total differential of eq. 1 with respect to pressure at constant temperature yields:

$$(\partial \mu_l / \partial P) dP_l = (\partial \mu_g / \partial P) dP_g \tag{2}$$

or, since the partial derivative of the Gibbs free energy with respect to pressure at constant temperature is volume (Edelfson and Anderson, 1943, p. 136):

$$v_l dP_l = v_g dP_g \tag{3}$$

where v_l and v_g are the volume per unit mass of the liquid and vapor, respectively. The volume per unit mass of the liquid, $1/\rho$, where ρ is the mass density of the liquid, is assumed to vary very slowly with pressure, and the volume per unit mass of the gas, assumed to be ideal, is, using the ideal gas law in the form:

$$P_g V_g = N_g k T: \tag{4}$$

$$v_g = 1/\rho_g = V_g / N_g m = kT / P_g m \tag{5}$$

where ρ_g is the mass density of the vapor in equilibrium with the liquid; k = Boltzmann constant, the universal gas constant per particle $\simeq 1.38 \cdot 10^{-16}$ erg K^{-1} P_g is the vapor pressure of the gas immediately above the liquid surface; m is the mass of a water molecule, $\simeq 3 \cdot 10^{-23}$ g; T is absolute temperature in kelvin; and N_g is the number of gas molecules in a volume V_g.

Combining eqs. 3 and 5 and integrating over pressure (assuming ρ is constant) yields:

$$\frac{1}{\rho} \int_{P_{ol}}^{P_l} dP' = \frac{kT}{m} \int_{P_o}^{\bar{P}} \frac{dP'}{P'} \tag{6}$$

or

$$P_l - P_{ol} = \frac{\rho k T}{m} \ln (\bar{P}/P_o) \tag{7}$$

where P_{ol} is the pressure of the water when the surface is flat; P_l is the pressure it has when it is in a porous medium; P_o is the pressure of the vapor above a flat water surface at temperature T, i.e., the usual vapor pressure; and $\bar{P}$ is the pressure of the vapor above the water lodged in the pores/capillaries of the porous medium. Identifying $P_l - P_{ol}$ with ψ, the negative pressure of water in the porous medium, we have:

$$\psi = \frac{\rho k T}{m} \ln(\bar{P}/P_0) = \psi_0 \ln(\bar{P}/P_0) \qquad (8)$$

where ψ_0 is plotted as a function of temperature in Fig. 1. Note that $\bar{P}/P_0$ is the quantity measured by thermocouple psychrometers. Also, since $\bar{P}$ is less than P_0, ψ is negative, and $-\psi$ is the usual convention for matric suction.

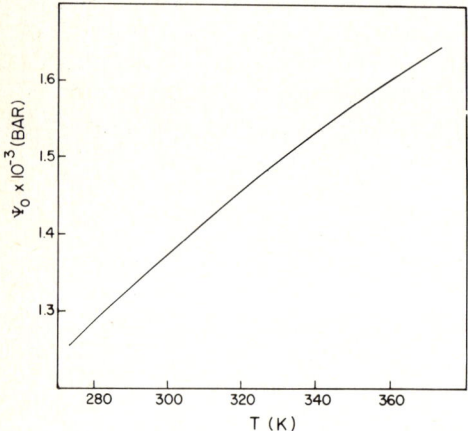

Fig. 1. The lumped parameter $\psi_0 = \rho k T/m$ as a function of temperature.

Gradients of ψ as a function of position give rise to flow in the unsaturated zone and experimental knowledge of $\bar{P}$ as a function of position, temperature, and saturation as well as hydraulic conductivity as a function of ψ are sufficient for input to a numerical model of unsaturated flow. Eq. 8 may be cast in more familiar form by noting that if n is the number of moles of liquid in a volume V and N is the number of molecules in the same volume, then:

$$nR = Nk \qquad (9a)$$

where R is the universal gas constant, and thus:

$$\rho = Nm/V = nRm/kV = Rm/kV_w \qquad (9b)$$

or

$$\rho k/m = R/V_w \qquad (10)$$

where $V_w = V/n$ is the volume of water per mol. Substituting eq. 10 into eq. 8 yields:

$$\psi = \frac{RT}{V_w} \ln(\bar{P}/P_0) \qquad (11)$$

which has been given previously by Brown and Van Havern (1972). Note that this result holds for the case where there are impurities dissolved in the water if the solution is assumed to be ideal and the solute non-volatile. In this case, P_0 is replaced by $P_0 - \Delta P$ where ΔP is the amount by which the vapor pressure over a flat surface of solvent is lowered when it contains a solute, and a corresponding change is made in $\bar{P}$. We will examine eq. 8 with a view toward determining the kind of accuracy required to measure differences in negative suction (ψ) in the −50 to −1 bar range. Fig. 2 shows the relation between the relative humidity ($\bar{P}/P_0$) and suction for a reasonable range of temperatures encountered in the field. Note that about a 3.5% change in relative humidity results in changes in $-\psi$ from ~50 to ~1 bar, which is the main range of interest in soil moisture suction measurements. We see from this that the accuracy of a psychrometer required for useful measurements of suction in this range is greater than 0.1% and the useful range of relative humidities is ~0.930—0.999.

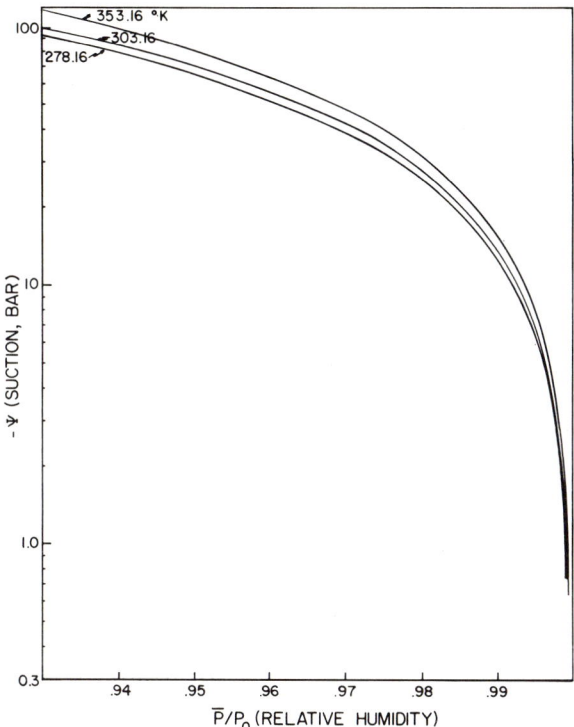

Fig. 2. Suction as a function of relative humidity with no regard to the physical situation giving rise to a given value of relative humidity.

PORE SIZE DISTRIBUTION

We now introduce the notion of a pore size distribution and show how it is related to porosity, ϵ, and the degree of saturation (the ratio of the volume of water in the pores of the sample to the total pore volume of the sample).

Let $N(r)dr$ be the number of pores/capillaries of radius r in a range of radii r to $r + dr$ in the sample per unit volume. If N is the total number of pores/capillaries in the sample per unit volume, then:

$$N = \int_0^\infty N(r')dr' \tag{12}$$

where r' is a dummy variable of integration and the range of integration is over all radii. Letting:

$$f(r)dr = N(r)dr/N$$

be the fraction of the total number of pores/capillaries in the sample that have radii in the range of r to $r + dr$, i.e., the usual pore size distribution, we have:

$$\int_0^\infty f(r')dr' = 1 \tag{13}$$

If $v(r)$ is the volume of a pore/capillary of average, effective radius r, then the volume of pore space per unit sample volume in a range of radii $0 \leq r' \leq r$ is:

$$\phi(r) = \int_0^r v(r')N(r')dr' \tag{14}$$

The usual porosity, ϵ (pore volume of the sample divided by the bulk sample volume), is:

$$\epsilon = \lim_{r \to \infty} \phi(r) = \int_0^\infty v(r')N(r')dr' \tag{15}$$

Differentiating eq. 14 using the Leibnitz rule (Abramowitz, 1965) yields:

$$\phi'(r) = v(r)N(r) = Nv(r)f(r) \tag{16}$$

Thus:

$$f(r) = \phi'(r)/Nv(r) \quad \text{and} \quad N(r) = \phi'(r)/v(r) \tag{17a}, (17b)$$

Thus, using eq. 12 and eqs. 17a and 17b, we have:

$$f(r) = \frac{\phi'(r)/v(r)}{\int_0^\infty [\phi'(r')/v(r')]dr'} \tag{18}$$

Note that the above expressions involving pore size distribution are independent of the specific forms of $f(r)$ and $v(r)$.

The degree of saturation, θ_s, can be expressed as the volume of water in the pores per unit sample volume divided by the pore volume per unit sample volume. Thus, if we assume that all pores/capillaries up to the radius r are filled with water and those having larger radii are empty, we have:

$$\theta_s(r) = \int_0^r v(r')N(r')dr' \bigg/ \int_0^\infty v(r')N(r')dr' =$$

$$\int_0^r v(r')f(r')dr' \bigg/ \int_0^\infty v(r')f(r')dr' \qquad (19)$$

Note that $\lim_{r \to 0} \theta_s(r) = 0$ and that $\lim_{r \to \infty} \theta_s(r) = 1$ as expected. Now if matric suction, $-\psi$, were known as a function of r, then via eq. 19, it would be known implicitly as a function of $\theta_s(r)$ and thus some aspects of hysteresis could be examined theoretically. We will now explore this point.

CAPILLARY CONTRIBUTION TO SUCTION

A relation between the vapor pressure above a pore/capillary of radius r, $P(r)$, and the vapor pressure over a flat (infinite radius of curvature) surface, P_0, of the same liquid, may be derived by considering the liquid mass near the top of the capillary (the derivation is given in Appendix A) and is for zero contact angle:

$$P(r) = P_0 \exp(-A_0/r) \qquad (20)$$

where

$A_0 = 4m\sigma/\rho kT$; σ is the surface tension of the liquid at temperature T; and the other quantities have been previously defined in connection with eq. 8. (A_0 as a function of temperature is shown in Fig. 3.) The capillary contribution to the average value of vapor pressure, $\bar{P}$, of water in the porous medium, which appears in eq. 8, can be calculated by averaging $P(r)$ over the pore size distribution $f(r)$. Thus:

$$\bar{P}(r) = P_0 \int_0^r f(r') \exp(-A_0/r')dr' \qquad (21)$$

where r is the largest filled pore radius. The limiting value of $\bar{P}(r)$ is:

$$\lim_{r \to \infty} \bar{P}(r) = \bar{P}_\infty = P_0 \int_0^\infty f(r') \exp(-A_0/r')dr' \qquad (22)$$

Substituting eq. 21 into eq. 8 yields:

$$\psi(r) = \frac{\rho kT}{m} \ln(\bar{P}(r)/P_0) \quad \text{and} \quad \lim_{r \to \infty} \psi(r) = \frac{\rho kT}{m} \ln(\bar{P}_\infty/P_0) \qquad (23a), (23b)$$

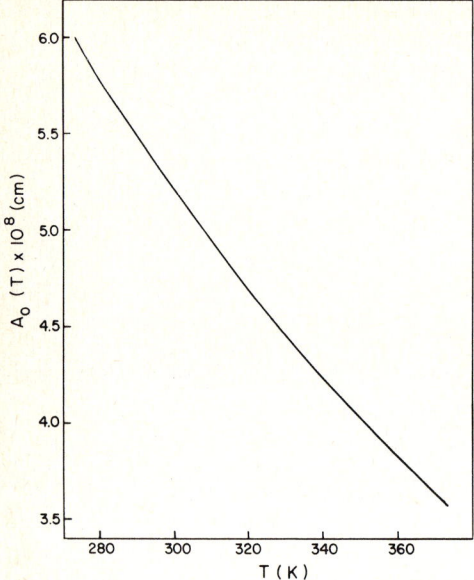

Fig. 3. The lumped parameter $A_0 = 4m\sigma/\rho kt$ as a function of temperature.

Note that since for every finite r, $P(r) < P$, and $\bar{P}/P < 1$, in an unconfined, unsaturated soil, ψ is always less than zero.

Thus: (1) both $\theta_s(r)$ and $\psi(r)$ depend on an integral over the pore size distribution, $f(r)$; and (2) since both $\theta_s(r)$ and $\psi(r)$ are parameterized on r, they are implicitly functionally related. Thus, curves of ψ vs. θ_s can be plotted directly from theoretical considerations using assumed forms of $f(r)$ and $v(r)$. Further, if differences in $f(r)$ for saturating and desaturating can be identified, hysteresis curves of θ_s vs. ψ can be plotted directly. This program will now be carried out by way of illustration for pure water.

ILLUSTRATION OF THE THEORY FOR PURE WATER

The following assumed form for pore size distribution (Case, 1977) will be used for the remainder of this work:

$$f(r) = r_0^{-2} r \exp(-r/r_0) \tag{24}$$

where r_0 is the value of r for which this distribution has its maximum and eq. 24 satisfies eq. 13. The mth moment of the distribution is:

$$M_m = \int_0^\infty r_0^{-2} r^{m+1} \exp(-r/r_0) dr = r_0^m \Gamma(m+2) \tag{25}$$

where $\Gamma(m+2)$ is the usual gamma function (Davis, 1965). A conventional and somewhat reasonable form for $v(r)$ is:

$$v(r) = \pi r_0^2 r \qquad (26)$$

This form has been used in similar calculations by Fatt (1956a, b). Substituting eqs. 24 and 26 into eq. 19 and performing the indicated integrations yields:

$$\theta_s(r) = \frac{\gamma(3, r/r_0)}{\Gamma(3)} = P\left(\frac{2r}{r_0} \middle| 6\right) \qquad (27)$$

where

$$\gamma(a, x) = \int_0^x t^{a-1} \exp(-t) \, dt$$

is the incomplete gamma function (Davis, 1965), and $P(\chi^2 | \nu)$ is the usual chi-square distribution (Zelen and Severo, 1965) with $\chi^2 = 2r/r_0$ and $\nu = 6$. Similarly, substituting eqs. 20 and 24 into eq. 21 yields (the development is given in Appendix B):

$$\frac{\bar{P}(r)}{P_0} = \sum_{l=0}^{\infty} \frac{1}{l!} (-r/r_0)^{l+2} E_{l+3}(A_0/r_0) \qquad (28a)$$

where (Gautschi and Cahill, 1965):

$$E_n(z) = \int_1^{\infty} [\exp(-zt)/t^n] \, dt$$

is the exponential integral of order n. Eq. 28a combined with eq. 23a yields the required function $\psi(r)$. The limiting case of $\bar{P}(r)/P_0$ as $r \to \infty$ is (Gradshteyn and Ryzhik, 1965, p. 340, No. 9)

$$\frac{\bar{P}_\infty}{P_0} = \frac{2A_0}{r_0} K_2[2(A_0/r_0)^{1/2}] \qquad (28b)$$

where $K_2[x]$ (Olver, 1965) is the modified Bessel function of the second kind of order two and eq. 28b is to be used with eq. 23b.

Numerical calculation of eqs. 27, 28a and 23a was performed, taking into account the temperature dependence of P_0, ρ and σ. Plotting $\theta_s(r)$ vs. $\psi(r)$ and using a smaller value of r_0, r_{od}, to represent the drying curve than the value of r_0 for the wetting curve, r_{ow}, since the moisture retention of the drying curve is greater than that of the wetting curve for a given value of suction, a family of theoretically-based hysteresis curves was generated. An example is shown in Fig. 4. The area enclosed by the hysteresis loop may be calculated approximately as follows. Fig. 5 shows a sketch of one branch of a hysteresis loop.

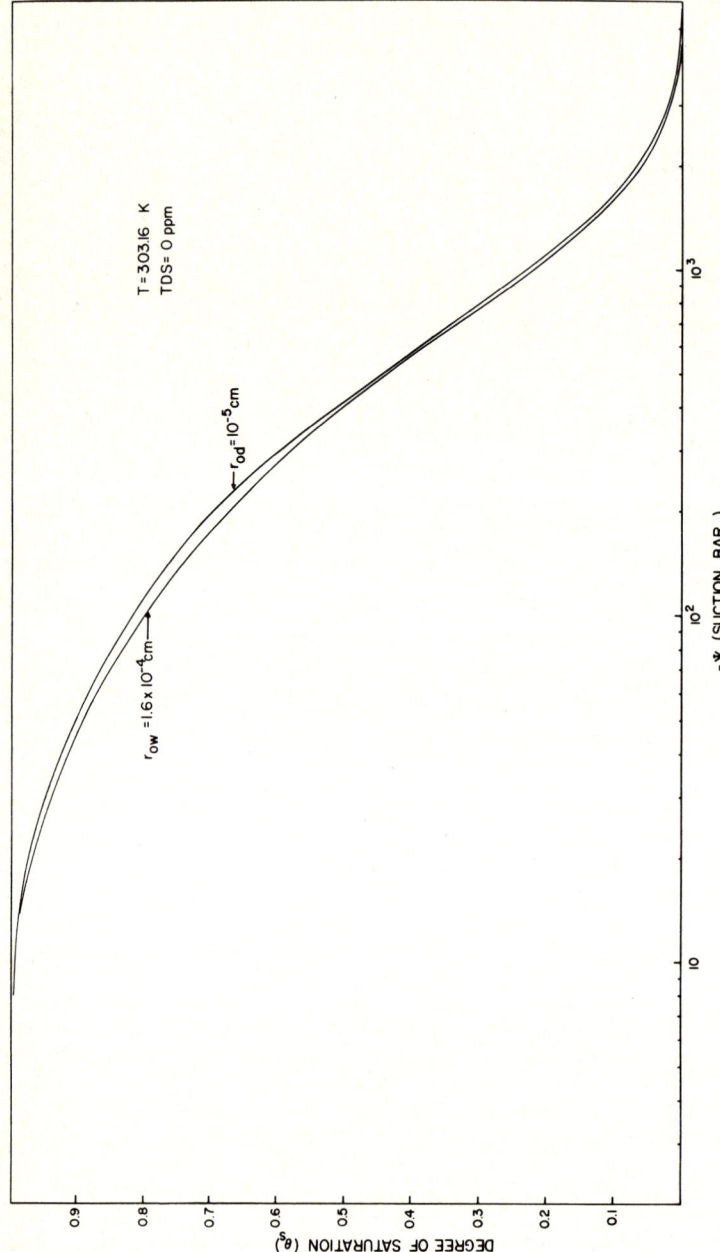

Fig. 4. Hysteresis loop for $T = 303.16$ K, $r_{ow} = 1.6 \cdot 10^{-5}$ cm, $r_{od} = 1.0 \cdot 10^{-5}$ cm and $\tau = 0$.

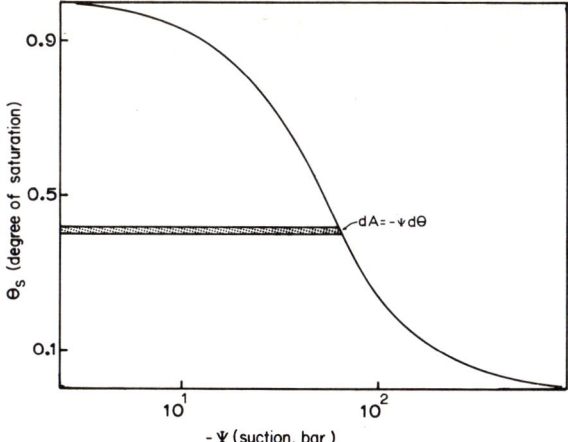

Fig. 5. Element of area used in computing the area between one branch of a hysteresis loop and the θ_s, $-\psi$ axes.

From this figure we see that the area between the axes and the curve is given by (since $-\psi$ is the quantity being plotted):

$$A = \int -\psi \, d\theta_s = -\int_0^r \psi(r')[\partial \theta_s(r')/\partial r'] \, dr' \tag{29}$$

Differentiating eq. 19 using the Leibnitz rule yields:

$$\partial \theta_s(r')/\partial r' = c^{-1} v(r') f(r') \tag{30a}$$

where

$$c = \int_0^\infty v(r'') f(r'') \, dr'' = 2\pi r_0^3 \tag{30b}$$

when eqs. 24 and 26 are substituted into eq. 30b and the integration is carried out. Also:

$$\psi(r') = \psi_0 \ln \left[\int_0^{r'} f(r'') \exp(-A_0/r'') \, dr'' \right] = \psi_0 \ln [\bar{P}(r')/P_0] \tag{31a}$$

where we recall that:

$$\psi_0 = \rho k T/m \tag{31b}$$

Substituting eqs. 30a and 31a into eq. 29 and using eqs. 24 and 26 yields, after some rearrangement:

$$A = \frac{\pi \psi_0}{c} \int_0^r \ln [\bar{P}(r')/P_0] r' \exp [\{r_0 \ln(r') - r'\}/r_0] \, dr' \tag{32}$$

Since the maximum of the argument of the exponential is at $r' = r_0$, we may, to first approximation, replace the lower limit by r_0. This casts the integral into the form:

$$f(x) = \int_\alpha^\beta g(t) \exp[xh(t)] \, dt \sim g(\alpha) \exp[xh(\alpha)] [-\pi/2xh''(\alpha)]^{1/2} \quad (33)$$

where $\sim$ means asymptotically equal to;

$$t = r', \quad \alpha = r_0, \quad \beta \to \infty \quad (34a, b, c)$$

$$g(r') = \frac{\pi \psi_0}{c} r' \ln[\bar{P}(r')/P_0], \quad h(r') = r'_0 \ln(r') - r'$$

and $\quad x = 1/r_0 \quad (34d, e, f)$

The right-hand side of eq. 33 is an asymptotic representation of the left-hand side due to Laplace (Erdélyi, 1956, p. 37) and is valid under the conditions that: (1) $h(t)$ has at most a finite number of maxima in the interval $\alpha \leq t \leq \beta$; (2) $h'(\alpha) = 0$; (3) $h''(\alpha) < 0$; (4) $x > 0$; and (5) certain conditions of regularity are satisfied by $g(t)$.

Using eqs. 33 and 34 to evaluate eq. 32 yields:

$$A(r_0) = -\frac{\pi \psi_0}{c} r_0 \ln[\bar{P}(r_0)/P_0] (\pi/2)^{1/2} r_0^3 e^{-1} \quad (35a)$$

where

$$\frac{\bar{P}(r_0)}{P_0} = \sum_{l=0}^{\infty} \frac{(-1)^l}{l!} E_{l+3}(A_0/r_0) \quad (35b)$$

the integral indicated in eq. 31a yielding eq. 35b having been calculated via the formalism of Appendix B.

The area enclosed by a hysteresis loop is the difference between two expressions of the form of eq. 35a with different values of r_0, one for saturation and one for desaturation. Using r_{ow} for the wetting curve and r_{od} for the drying curve we have for the required area:

$$W = [A(r_{od}) - A(r_{ow})] = \tfrac{1}{2}\psi_0 (\pi/2)^{1/2} [\ln\{\bar{P}(r_{ow})/P_0\} - \ln\{\bar{P}(r_{od})/P_0\}] \quad (36)$$

The area calculated has units of energy per unit specific area and represents, at least in part, the energy necessary to break the bonds between the water molecules and the soil surfaces to which they are attached after the wetting phase of the saturation—desaturation process has been completed. A plot of the area as a function of temperature is shown in Fig. 6. The values of r_{ow} and r_{od} are the same as those of the hysteresis loop of Fig. 4. From Fig. 6 we see that as temperature increases the area decreases, which means, at least in part, that there is less binding energy of adsorption to be overcome at higher

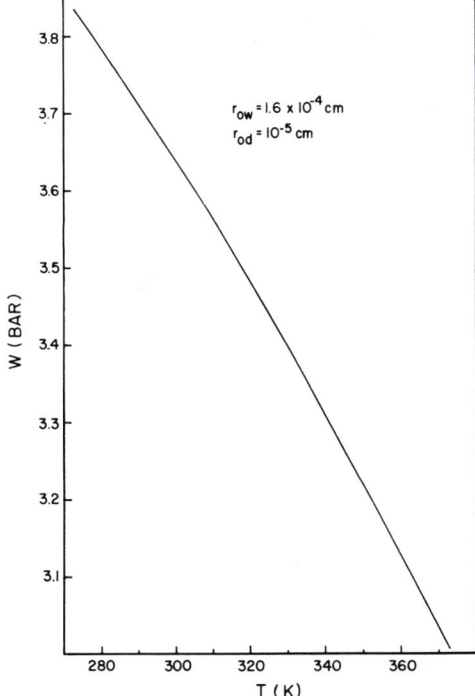

Fig. 6. Area, W, of a hysteresis loop having $r_{ow} = 1.6 \cdot 10^{-5}$ cm, $r_{od} = 1.0 \cdot 10^{-5}$ cm and $\tau = 0$ as a function of temperature.

temperatures since the water molecules have more thermal energy than at lower temperatures.

EFFECT OF DISSOLVED IMPURITIES

The effects of impurities dissolved in water enter the theory of soil suction via the changes in surface tension that occur and the reduction in the vapor pressure of a fluid that is observed when it contains a solute at low concentration. This reduction in vapor pressure, ΔP, over a flat water surface is:

$$\Delta P = \rho' n_1 RT / \rho_0 n_0 v_0 \qquad (37)$$

where ρ' is the density of the vapor above the solution; ρ_0 is the mass density of the solution; n_1 is the number of moles of solute in the solution; v_0 is the volume per mole of n_0 moles of solvent; R is the universal gas constant; and T is the absolute temperature of the solution—vapor system. The density, ρ', of the vapor may be calculated using the ideal gas law as follows:

$$\rho' = N_g m / V_g = m(P_0 - \Delta P)/kT \qquad (38)$$

The ratio of n_1 to n_0 is the amount of total dissolved solids, τ, for example, in parts per million, in the solution.

Combining eqs. 37, 38 and 20, and assuming τ is small enough so that $\rho_0 \cong \rho$, yields for the vapor pressure over a pore/capillary of radius r in the presence of impurities dissolved in the water, $P_1(r)$:

$$P_1(r) = (P_0 - \Delta P) \exp(-A_0/r) \tag{39a}$$

and for the relative humidity of the water in the soil in the presence of impurities:

$$P_1(r)/(P_0 - \Delta P) = \exp(-A_0/r) \tag{39b}$$

where the only difference between this and the result for the case in which there are no impurities in the water is in the surface tension dependence of A_0.

COMPARISON WITH EXPERIMENT

Comparison of Fig. 4 with experimentally derived hysteresis loops is complicated by the fact that hysteresis loops measured in the laboratory have suctions less than about four bars. See, for example, Youngs (1964), Topp and Miller (1966), Poulovassilis (1970) and Narasimhan and Witherspoon (1977). By contrast, field measurements of soil matrix suction vary over a wider range. For example, Brownell et al. (1975) observed variations of suction up to 22 bar near the soil surface on the Hanford Reservation. Also, suctions of more than 40 bar have been observed near Pullman, Washington, and psychrometers capable of measuring suctions up to 1000 bar are available (Campbell and Wilson, 1972). The fact that suctions of the size of those calculated from the assumed forms for $f(r)$ and $v(r)$ used here have not been observed in the field and that field and laboratory measurements do not entirely agree, may be partially resolved by noting that when a column is drained in the laboratory, suctions are sometimes measured as the height above a datum (Youngs, 1964), and so, comparable suctions could be obtained by columns of soil about 15 m high (or 760 m high to compare with some of the suctions calculated here). Therefore, the measurements are not really comparable as they stand, but higher suctions applied to laboratory samples might well show a comparable behavior.

There is another type of experiment, however, which does agree in a reasonable way with the results of the assumed model calculated here. This is an adsorption experiment in which the heat of adsorption of water, H, onto the pore/capillary surfaces of Millville loam is measured as a function of ambient vapor pressure (Cary et al., 1964). The adsorption data of Cary et al. are given in calories/gram of adsorbed water. This can be converted to ergs per unit sample volume as follows:

$$H \text{ (in ergs per unit sample volume)} = H \text{ (in calories per gram of water)} \times \rho \theta_s \epsilon c_1 \tag{40}$$

where $c_1 = 4.186 \cdot 10^7$ erg/cal. is a constant which converts from cal./g to erg/g; and ρ, θ_s and ϵ have been previously defined. Note that θ_s may be viewed as the fraction of pore volume that is filled with water. Assuming that the ambient values of relative humidity are close to those in the sample, we select for comparison the experimental values $\bar{P}/P_0 = 0.25$, $H = 543 \pm 50$ cal./g at $T \cong 298$ K, $\rho = 0.99708$, and $\theta_s = 0.0731$ which corresponds to a theoretical $\bar{P}/P_0$ of 0.25. Assuming a value of $\epsilon = 0.47$ for Millville loam, we apply eq. 40 and obtain $H = 7.89 \cdot 10^8$ erg/cm^3 for the energy of adsorption per unit volume of sample material. This is to be compared with the area between the wetting branch of the appropriate hysteresis curve and the θ_s and $-\psi$ axes. This can be done using the approximate formalism of eqs. 35a and 35b. The units of this area are energy per unit specific volume and so must be multiplied by ϵ to yield energy per unit sample volume. Using $T = 298$ K, the appropriate values of ρ and σ, and $r_0 = 5.12 \cdot 10^{-3}$ cm, yields $1.96 \cdot 10^8$ erg/cm^3, which, considering the nature of the approximations involved, is good order of magnitude agreement.

CONNECTION WITH EQUATIONS DESCRIBING UNSATURATED FLOW

Equations for the flow of water in the vadose zone have been considered by Reisenauer et al. (1975), Case (1977), and many other investigators.

The driving force for flow is proportional to the position gradient of suction head, which is $-\psi/\rho g$ where g is the acceleration due to gravity. Temperature gradients are also important in vadose zone flow but will not be considered here. The position dependence of suction head in terms of the formalism of this paper may be expressed for the case of impurities dissolved in the water as follows:

$$h(r) = h_0 \ln [\bar{P}(r,r)/(P_0 - \Delta P)] \tag{41}$$

where r is a position vector, and:

$$h_0 = -kT/mg \tag{42a}$$

$$\frac{\bar{P}(r,r)}{P_0 - \Delta P} = \int_0^r f(r',r) \exp [-A_0(r)/r'] dr' \tag{42b}$$

in which

$$f(r',r) = r_0(r)^{-2} r' \exp [-r'/r_0(r)] \tag{43a}$$

and

$$A_0(r) = [4m\sigma\{\tau(r)\}]/\rho(r)kT \tag{43b}$$

where $\rho[\tau(r)]$ expresses the dependence of ρ on τ and the dependence of τ on position. Taking the gradient of $h(r)$ yields:

$$\nabla h(r) = D_1 \nabla r_0(r) + D_2 \nabla \rho(r) - D_3 \nabla \tau(r) \tag{44}$$

where

$$D_1 = \left[h_0 \int_0^r r' K(r',r) dr'\right] \Big/ [r_0(r)^2 D_0] - 2h_0 \left[\int_0^r r'^{-1} K(r',r) dr'\right] \Big/ [r_0(r) D_0]$$

(45a)

$$D_2 = \left[h_0 A_0 \int_0^r r'^{-1} K(r',r) dr'\right] \Big/ [\rho(r) D_0]$$ (45b)

$$D_3 = \left[h_0 A_0(r) \int_0^r r'^{-1} K(r',r) dr'\right] \Big/ [\sigma(r) D_0] \cdot [\partial \sigma(\tau)/\partial \tau]$$ (45c)

$$D_0 = \int_0^r K(r',r) dr'$$ (45d)

and

$$K(r',r) = f(r',r) \exp[-A_0(r) r']$$ (45e)

We see that in this approximation, $\nabla h(r)$ is the sum of terms proportional to gradients of medium characteristics, water density and impurities.

SUMMARY AND CONCLUSIONS

The role of pore size distribution in calculating both suction and volumetric saturation has been clarified. The influence of dissolved solids on suction was computed via the changes in surface tension of the solution over that of the solvent and a start toward an irreversible thermodynamics formulation indicated. The experimentally observed fact of a temperature rise of soil samples when water vapor is adsorbed merits further detailed consideration relative to the area enclosed by a hysteresis loop. It is seen that the kinds of calculations presented here are valuable from the point of view of clarifying many aspects of suction as related to unsaturated flow, but that detailed field measurements are necessary for the implementation of a model of unsaturated flow.

ACKNOWLEDGEMENTS

The authors wish to thank Drs. M.E. Campana, C.L. Carnahan and P.R. Fenske for helpful discussions of this problem. The work reported here was initiated using U.S. Department of Energy funds and completed while the senior author was on sabbatical leave from the Nevada Desert Research Institute. In this latter connection, the senior author wishes to acknowledge the warm and gracious hospitality of Dr. P.A. Witherspoon and the personnel

APPENDIX A

In this appendix we will derive the relation between the vapor pressure over a flat surface of a liquid (infinite radius of curvature), and that over a liquid forming a meniscus in a pore or capillary of radius r.

For a system at constant temperature and pressure the state of equilibrium is represented by the minimum of the Gibbs free energy for the system. We write the Gibbs free energy for the water—vapor system as:

$$G(P,T) = M_1 \mu_1 + M_2 \mu_2 - \sigma A \qquad \text{(A-1)}$$

where P is pressure; T is temperature; M_1 is the mass of water being considered; M_2 is the mass of water vapor; μ_1 and μ_2 are the chemical potentials (Gibbs free energy per unit mass) of the water and vapor, respectively; σ is the surface tension of water; A is the area of the water surface being considered; and the term σA represents the energy "stored" in the water surface. The geometry of the system is shown in Fig. A-1. From the figure we see that, (Hodgman, 1961, p. 401):

$$A = 2\pi R h \quad \text{and} \quad h = 1 - \sin(\alpha) \qquad \text{(A-2a, b)}$$

and so:

$$G(P,T) = M_1 \mu_1 + M_2 \mu_2 - 2\pi \sigma R^2 [1 - \sin(\alpha)] \qquad \text{(A-3)}$$

where α is the angle of contact between the water and the capillary wall. Taking a small variation in G, δG, which at equilibrium is zero, we have:

$$\delta G = 0 = \mu_1 \delta M_1 + \mu_2 \delta M_2 - 2\pi\sigma [1 - \sin(\alpha)] 2R (\partial R/\partial M_1) \delta M_1 \qquad \text{(A-4)}$$

Since the variation in G results, by definition (Zemansky, 1957, p. 418) in this

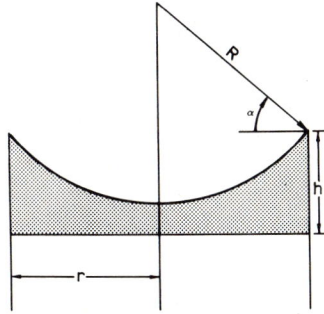

Fig. A-1. This figure shows the relation between the radius of a pore/capillary, the radius of curvature of the associated meniscus, and the contact angle between the shaded volume of water being considered and the pore/capillary wall.

case from an exchange of mass between the liquid and vapor, $\delta M_1 = \delta M_2$, and we have:

$$\mu_2 - \mu_1 = 4\pi\sigma R[1 - \sin(\alpha)](\partial R/\partial M_1) \tag{A-5}$$

From the figure we see that the mass of water being considered is given by:

$$M_1 = \rho v = \rho[\pi h r^2 - \frac{1}{3}\pi h^2(3R - h)] \tag{A-6}$$

where ρ is the density of water; and volume v is the shaded cross-sectional area shown in Fig. A-1.

Substituting eq. A-2b into eq. A-6 (after re-arrangement and simplification), results in:

$$M_1 = \frac{1}{3}\rho\pi R^3 f_0(\alpha) \tag{A-7a}$$

where

$$f_0(\alpha) = [1 - \sin(\alpha)]f_1(\alpha) \tag{A-7b}$$

and

$$f_1(\alpha) = 2\cos^2(\alpha) + \sin(\alpha) - 1 \tag{A-7c}$$

Thus, we have for $\partial R/\partial M_1$, which in this case is the reciprocal of $\partial M_1/\partial R$:

$$\partial R/\partial M_1 = 1/\pi\rho R^2 f_0(\alpha) \tag{A-8}$$

Substituting eq. A-8 into eq. A-5 yields:

$$\mu_2 - \mu_1 = -4\sigma/\rho R f_1(\alpha) \tag{A-9}$$

Differentiating eq. A-9 with respect to pressure at constant temperature, we have:

$$\left(\frac{\partial \mu_2}{\partial P}\right)_T - \left(\frac{\partial \mu_1}{\partial P}\right)_T = \frac{4\sigma}{f_1(\alpha)}\left[\frac{-1}{R\rho^2}\left(\frac{\partial \rho}{\partial P}\right)_T - \frac{1}{\rho R^2}\left(\frac{\partial R}{\partial P}\right)_T\right] \tag{A-10}$$

Recalling the standard thermodynamic relation:

$$V = [\partial G(P,T)/\partial P]_T \tag{A-11}$$

where V is volume, we see that:

$$(\partial \mu_2/\partial R)_T = 1/\rho_g \quad \text{and} \quad (\partial \mu_1/\partial P)_T = 1/\rho \tag{A-12a, b}$$

where ρ_g is the density of the water vapor.

Noting that $1/\rho_g \gg 1/\rho$, and also that $(\partial \rho/\partial P)_T$ is very small, eq. A-10 becomes:

$$\frac{1}{\rho_g} \cong \frac{4\sigma}{\rho R^2 f_1(\alpha)}\left(\frac{\partial R}{\partial P}\right)_T \tag{A-13}$$

Assuming the water vapor behaves like an ideal gas and invoking the ideal gas law in the form:

$$PV_g = N_g kT \qquad (A\text{-}14)$$

where the symbols have been previously defined, we have:

$$\rho_g = mN_g/V_g = P/kT \qquad (A\text{-}15)$$

Substituting eq. A-15 into A-13, re-arranging terms, separating variables, and indicating integrations over P and R, yields:

$$\int_{P(R)}^{P_0} dP'/P' = \frac{4m\sigma}{\rho k T f_1(\alpha)} \int_R^{\infty} dR'/(R')^2 \qquad (A\text{-}16)$$

where $P(R)$ is the vapor pressure above the meniscus having radius of curvature R, corresponding to the lower limit of the integral on the right-hand side of eq. A-16; P_0 is the vapor pressure over a flat surface at the same temperature, corresponding to an infinite radius of curvature — the upper limit on the right-hand side of eq. A-16; and P' and R' are dummy variables of integration. Carrying out the indicated integration, rearranging terms, and substituting $R = r/\cos(\alpha)$ yields:

$$P(r) = P_0 \exp\left[-4\sigma m \cos(\alpha)/\rho k T r f_1(\alpha)\right] \qquad (A\text{-}17)$$

which is the required relation.

Note that this derivation is unchanged for the case of impurities in the water if the chemical potentials are interpreted as the partial potentials of water and vapor respectively and the solution is assumed to be ideal.

APPENDIX B

We shall evaluate the integral:

$$I = \int_\alpha^\beta r_0^{-2} r \exp(-r/r_0 - A_0/r) dr \qquad (B\text{-}1)$$

Expanding $\exp(-r/r)$ in the usual Maclaurin series and breaking up the range of integration yields:

$$I = \sum_{l=0}^{\infty} (-r_0)^{-l-2} \frac{1}{l!} \left[\int_\alpha^0 r^{l+1} \exp(-A_0/r) dr + \int_0^\beta r^{l+1} \exp(-A_0/r) dr \right] \qquad (B\text{-}2)$$

Making the change of variable $\alpha/r = t$ in the first integral and $\beta/r = t$ in the second integral yields:

$$I = \sum_{l=0}^{\infty} (-r_0)^{-l-2} \frac{1}{l!} [\beta^{l+2} E_{l+3}(A_0/\beta) - \alpha^{l+2} E_{l+3}(A_0/\alpha)] \tag{B-3}$$

where

$$E_n(z) = \int_1^{\infty} [\exp(-zt)/t^n] \, dt \tag{B-4}$$

is the usual exponential integral of order n (Gautschi and Cahill, 1965). By inspection we see that if $\alpha = 0$, we have:

$$I = \sum_{l=0}^{\infty} (-\beta/r_0)^{l+2} \frac{1}{l!} E_{l+3}(A_0/\beta) \tag{B-5}$$

If $\alpha = 0, \beta \to \infty$, we have, using standard tables (Gradshteyn and Ryzhik, 1965):

$$I = \frac{2A_0}{r_0} K_2[2(A_0/r_0)^{1/2}] \tag{B-6}$$

where $K_2[x]$ is the modified Bessel function of the second kind of order two (Olver, 1965).

APPENDIX C

In this appendix we shall discuss some possible refinements of the theory presented thus far. The principal assumption that has been made is that all the pores/capillaries from radii 0 to r are filled and those having radii larger than r are empty. This formalism can be improved by introducing $\xi(r)$ which is the fraction of pores/capillaries of radius r that are filled. This yields for zero contact angle:

$$\psi(r) = \psi_0 \ln \left[\int_0^r \xi(r') f(r') \exp(-A_0/r') \, dr' \right] \tag{C-1}$$

and

$$\theta_s(r) = \int_0^r \xi(r') v(r') f(r') \, dr' \bigg/ \int_0^{\infty} \xi(r') v(r') f(r') \, dr' \tag{C-2}$$

In this formulation changes in $\xi(r)$ between wetting and drying, which are basically changes in the geometric distribution of water in the sample, are responsible for hysteresis and the average effective pore size distribution "seen" by the apparatus is $\xi(r)f(r)$ while the true pore size distribution, $f(r)$, remains the same, as it should. Note also, that there is probably some critical radius below which vapor transport and not liquid transport is important and so eqs. C-1 and C-2 would have the lower limits of the integrals replaced by this critical radius so as to exclude that range of radii from liquid transport consideration.

REFERENCES

Abramowitz, M., 1965. Elementary analytical methods. In: M. Abramowitz and I.A. Stegun, Editors, Handbook of Mathematical Functions, Natl. Bur. Stand., No. 55, U.S. Government Printing Office, Washington, D.C.

Brown, R.W. and Van Havern, B.P. (Editors), 1972..Psychrometry in Water Relations Research. Utah State University Press, Logan, Utah, 342 pp.

Brownell, L.E., Backer, J.G., Isaacson, R.E. and Brown, D.J., 1975. Soil moisture transport in arid sites — vadose zones. Atlantic Richfield Co., Richland, Wash., Tech. Rep. ARH-ST-123, 230 pp.

Campbell, G.S. and Wilson, A.M., 1972. Water potential measurements of soil samples. In: R.W. Brown and B.P. Van Havern (Editors), Psychrometry in Water Relations Research, Utah State University Press, Logan, Utah, pp. 142—149.

Carnahan, C.L., 1976. Non-equilibrium thermodynamics of groundwater flow systems: symmetry properties of phenomenological coefficients and considerations of hydrodynamics dispersion. J. Hydrol., 31: 125—150.

Cary, J.W. et al., 1964. Water adsorption by dry soil and its thermodynamic functions. Soil Sci. Soc. Proc., 26: 309—314.

Case, C.M., 1977. Constitutive relations for flow of water in the unsaturated zone. Water Resour. Cent. Desert Res. Inst., Univ. Nev. System, Reno, Nev., Tech. Rep. NVO-1253-11, 15 pp.

Davis, P.J., 1965. Gamma function and related functions. In: M. Abramowitz and I.A. Stegun (Editors), Handbook of Mathematical Functions. Natl. Bur. Stand., No. 55, U.S. Government Printing Office, Washington, D.C.

Edelfson, N.E. and Anderson, A.B.C., 1943. Thermodynamics of soil moisture. Hilgardia, 15(2): 31—298.

Enderby, J.A., 1955. The domain model of hysteresis, Part 1. Independent domains. Trans. Faraday Soc., 51: 835—848.

Enderby, J.A., 1956. The domain model of hysteresis, Part 2. Interesting domains. Trans. Faraday Soc., 52: 106—120.

Erdélyi, A., 1956. Asymptotic expansions. Dover, New York, N.Y., 108 pp.

Everett, D.H., 1954. A general approach to hysteresis, Part 3. A formal model of the independent domain model of hysteresis. Trans. Faraday Soc., 50: 1077—1096.

Everett, D.H., 1955. A general approach to hysteresis, Part 4. An alternative formulation of the domain model. Trans. Faraday Soc., 51: 1551—1557.

Everett, D.H. and Smith, F.W., 1954. A general approach to hysteresis, Part 2. Development of the domain theory. Trans. Faraday Soc., 50: 187—192.

Everett, D.H. and Whitton, W.I., 1952. A general approach to hysteresis. Trans. Faraday Soc., 48: 749—757.

Fatt, I., 1956a. The network model of porous media, I. Capillary pressure characteristics. Trans. Am. Inst. Min. Metall. Pet. Eng., 207: 144—159.

Fatt, I., 1956b. The network model of porous media, II. Dynamic properties of a single size tube network. Trans. Am. Inst. Min. Metall. Pet. Eng., 207: 160—178.

Gautschi, W. and Cahill, W.F., 1965. Exponential integral and related functions. In: M. Abramowitz and I.A. Stegun (Editors), Handbook of Mathematical Functions, Natl. Bur. Stand., No. 55, U.S. Government Printing Office, Washington, D.C.

Gradshteyn, I.S. and Ryzhik, I.M., 1965. Table of Integrals, Series and Products. Academic Press, New York, N.Y., 4th ed., 1086 pp.

Haines, W.B., 1930. Studies in the physical properties of soil, 5. The hysteresis effect in capillary properties, and the models of moisture distribution associated therewith. J. Agric. Sci., 20: 97—116.

Hodgman, C.D., 1961. CRC Standard Mathematical Tables. Chemical Rubber Publishing Co., New York, N.Y., 12th ed., 525 pp.

Letey, J., Kemper, W.D. and Noonon, L., 1969. The effect of osmotic pressure gradients on water movement in unsaturated soil. Soil Sci. Soc. Am. Proc., 33: 15—18.

Liakopoulos, A.C., 1965. Transient flow through unsaturated porous media. D. Eng. Dissertation, University of California, Berkeley, Calif.

Narasimhan, T.N. and Witherspoon, P.A., 1977. Numerical model for saturated—unsaturated flow in deformable porous media, 1. Theory. Water Resour. Res., 13(3): 657—664.

Olver, F.W.J., 1965. Bessel functions of integer order. In: M. Abramowitz and I.A. Stegun (Editors). Handbook of Mathematical Functions. Natl. Bur. Stand., No. 55, U.S. Government Printing Office, Washington, D.C.

Poulovassilis, A., 1970. Hysteresis of pore water in granular porous bodies. Soil Sci., 109(1): 5—12.

Reisenauer, A.E., Cearlock, D.B. and Bryan, C.A., 1975. Partially saturated transient groundwater flow model — Theory and numerical implementation. Battele—Pacific Northwest Laboratories, Richland, Wash., Tech. Rep. BNWL-1713, 32 pp.

Topp, C.G. and Miller, E.E., 1966. Hysteric moisture characteristics and hydraulic conductivities for glass-bead media. Soil Sci. Soc. Am. Proc., 30: 156—162.

Vachaud, G. and Thony, J., 1971. Hysteresis during infiltration and redistribution in a soil column at different initial water contents. Water Resour. Res., 7(1): 111—126.

Youngs, E.G., 1964. Water movement in soils. Proc. 19th Symp. Soc. Exp. Biol., Cambridge University Press, London, pp. 89—112.

Zelen, M. and Severs, N.C., 1965. Probability Functions. In: M. Abramowitz and I.A. Stegun (Editors), Handbook of Mathematical Functions, Natl. Bur. Stand., No. 55, U.S. Government Printing Office, Washington, D.C.

Zemansky, M.W., 1957. Heat and Thermodynamics. McGraw-Hill, New York, N.Y., 484 pp.

[2]

CONTRIBUTION OF GROUNDWATER MODELING TO PLANNING

JOHN E. MOORE

U.S. Geological Survey, Reston, VA 22092 (U.S.A.)

(Accepted for publication May 10, 1979)

ABSTRACT

Moore, J.E., 1979. Contribution of groundwater modeling to planning. In: W. Back and D.A. Stephenson (Guest-Editors), Contemporary Hydrogeology — The George Burke Maxey Memorial Volume. J. Hydrol., 43: 121—128.

The consideration of groundwater in water-resource planning frequently has been neglected because many planners believed that groundwater could not be adequately evaluated in terms of availability, quality, cost of development, or effect of development on the surface-water supply. The development of predictive groundwater models now provides the water planner with tools to evaluate these problems. Highly sophisticated digital models can be used in planning the development of groundwater and the conjunctive use of ground- and surface water. About 250 digital models have been used to evaluate groundwater problems: The models are powerful tools for predicting the response of groundwater systems to stresses. They can also clarify the cause and progress of past stresses. With these developments it is now possible to integrate the utilization of groundwater into water-resource planning with a high degree of confidence.

INTRODUCTION

The role of groundwater in water-resource planning frequently is neglected because many planners believe that groundwater can not be adequately evaluated in terms of availability, quality, cost of development, or the effect of development on the surface-water supply. However, in many areas the groundwater reservoir is the significant part of the hydrologic system, and its utilization offers many alternatives for effective development of the water resource. For example, in areas where the surface-water supply is plentiful, groundwater is commonly overlooked. As the use of surface water increases and cost of development increases, the search begins for groundwater to supplement the supply. However, as groundwater becomes a major part of the total use, one or more of the following problems will probably occur: (1) the groundwater supply is progressively depleted; (2) streamflow is reduced; and (3) water quality deteriorates. These problems are causes for growing concern and they indicate that water planning is necessary to expand or sustain present supplies. The development of predictive groundwater models since the middle 1950's now provides the water planner with techniques for planning groundwater development to anticipate and to resolve these problems.

The purpose of this paper is to describe techniques that are currently being used to integrate adequate consideration of groundwater into water-resource planning. Several examples of the use of groundwater models in planning are given and the hydrologic data necessary for the development and calibration of these models are described. These models will enable the planner to develope the water resource wisely.

MODELS FOR WATER-RESOURCE PLANNING

Electric-analog and digital models have been used for water-resource planning throughout the U.S.A. and in many foreign countries. Models have been developed that portray the groundwater system and predict the changes imposed by man with varying degrees of accuracy. Digital and analog models simulate the hydrologic properties and boundaries of the hydrologic system.

The major difference between the two models is that in the digital model flow equations are solved mathematically and in electric-analog models the solution is simulated with a resistor—capacitor network. The analog model can handle large and complex hydrologic problems, such as those that involve two or more aquifers with varying degrees of hydraulic connection. It also provides a visual display of the aquifer characteristics (transmissivity and storage) and boundaries (the bedrock—alluvium contact and the stream). This display is useful in explaining the operation of the hydrologic system to the layman. The U.S Geological Survey (USGS) established an analog-model laboratory in 1959 to evaluate groundwater problems. More than 100 hydrologic analog models have been constructed by the laboratory in the past twenty years. The digital model commonly requires less time than a comparable analog model for construction, programming data input, and readout of results. Until recently, computer capacity was a serious limitation to the use of a digital model. The development of high-speed and large-capacity computers in the past ten years has made it possible to solve complex hydrologic problems (Appel and Bredehoeft, 1976).

Digital models are presently (1979) in widespread use and they can calculate changes in quality and quantity of water and the effects of economic and legal constraints on water use. Some of the major water-management problems that have been evaluated with models include: effects of irrigation wells on streamflow, dewatering of waterlogged land, groundwater mining, saltwater intrusion, and effect of lock- and dam-construction on the water table, transport of contaminants, subsidence due to groundwater withdrawal, and effects of geothermal development.

Models have been used to simulate groundwater hydrology in a variety of geologic environments. Four hydrologically complex areas where models were used for water-resource planning are Houston, Texas, San Luis Valley,

Colorado, Arkansas Valley, Colorado, and southwest Florida.

In the Houston area, a three-dimensional flow was modeled in a 23,000-km² (9000-mi.²)-area (Wood and Gabrysch, 1965; Jorgensen, 1975; Meyer and Carr, 1979). Large withdrawals of groundwater have caused large declines in the potentiometric surface. The decline of pressure head in the sand beds resulted in compaction of clay layers which caused land surface subsidence. The decline of pressure head has also caused saline-water movement updip in the southern part of the Houston area. The models were used to gain a better understanding of how the system operated, and to predict hydrologic changes that would occur from planned groundwater developments.

In the San Luis Valley of south-central Colorado, an analog and digital model was used to study a unique problem in a complex unconfined and confined aquifer system. More than $2.5 \cdot 10^{12}$ m³ ($2 \cdot 10^9$ acre-feet) of water is stored in the upper 1800 m (6000 ft.) of the aquifer. A model was first constructed for the uppermost or unconfined aquifer, which ranges in thickness from 0 to 36 km (0 to 120 ft.). The model was used to predict the effects of pumping a network of wells to salvage water that is being consumed by evaporation and nonbeneficial vegetation in the central part of the valley (Emery, 1970). The salvage plan, which was proposed by the Bureau of Reclamation, provides for the construction of 129 wells to pump 10^8 m³ 84,000 acre-feet) of water annually for 50 years. The model analysis showed that water-level declines would range from 1 to 30 m (1 to 100 ft.) and that those declines exceeding 3 m (10 ft.) would be confined mainly to the vicinity of salvage wells. It reassured the local irrigators that the salvage project would have little or no effect on water levels in the principal irrigated areas. This model was later expanded to include both the unconfined and confined aquifer system (Emery et al., 1975).

A combination of the best features of an analog and a digital model was used in the Arkansas Valley of Colorado to define the operation of a stream—aquifer system and to predict effects of proposed changes in water management (Moore and Wood, 1967). The results of these hybrid model studies have been used as a basis for modifying water law, for developing groundwater supplies, and for administrating water distribution. The area modeled extends from Pueblo, Colorado, to the Kansas State line, a distance of 240 km (150 mi.). This reach of the valley is underlain by unconsolidated sand and gravel deposits. The development of groundwater for irrigation caused legal disputes between groundwater and surface-water users because well pumping reduced the flow of the Arkansas River. An electric-analog model was developed to simulate the stream and aquifer. Different rates of pumpage at different points were simulated in the analog model to obtain response curves for water levels and stream flow. The model summarized differences in aquifer transmissivity, specific yield, and the effects of boundaries. The results from the analog model were the basis for constructing a digital model of the stream—aquifer system that was used to analyze and optimize water-management plans. The model was used to predict the availability of surface water at successive diversion points downstream on a month-by-month schedule and to show changes in groundwater storage.

Digital models have been developed in the past five years to incorporate some water-quality parameters. It is now possible to predict changes in water quality that might result from changes in water-management practices or to predict the movement of liquid contaminants intentionally or accidentally released to the environment. An investigation made in an 18-km (11-mi.) reach of the Arkansas River Valley of Colorado demonstrated the use of such a model to describe and predict changes in salinity in the alluvial aquifer and in the adjacent stream on a monthly basis (Konikow and Bredehoeft, 1974). The results greatly aided the understanding of water-quality variations and their relation to irrigation practices.

A digital groundwater model was used to simulate changes in the water levels that would result from proposed groundwater developments by the phosphate mining industry in west-central Florida (Wilson, 1978). The model will also be used to determine the regional effects of continued groundwater withdrawal for irrigation and municipal supplies; to evaluate monitor well networks; and to predict the movement of the saltwater—freshwater interface.

Models have been developed that consider physical, economic, and social constraints, and that analyze effects of alternative water-management plans, allowing selection of those that most closely achieve certain specified objectives. For example, such a model describing a stream—aquifer system was developed by the USGS to show how ground- and surface water could be distributed for irrigation to satisfy the greatest number of water rights in their order of priority and to evaluate the economics of various groundwater schemes (Bredehoeft and Young, 1970).

A compilation of 250 digital models that are available for use in evaluating water-resource management problems is given in a report by Bachmat et al. (1978). The report contains a description of the accessibility of models, inadequacies of model data, and inadequacies in modeling. The authors categorized models into prediction, management and data management.

DATA REQUIREMENTS FOR GROUNDWATER MODELS

The data requirements for groundwater modeling depend upon many factors such as the hydrologic complexity of the area, types of water problems, and size of area. Some studies need sophisticated large-scale model analyses while others require only a descriptive evaluation of existing hydrologic data. A summary of the major steps in an intensive study of a groundwater basin are:

(1) Collect and interpret data needed by the hydrologist to describe geology, hydrology, and historical development of water supplies.

(2) Develop a digital model of the area.

(3) Verify model.

(4) Use model to evaluate water problems and predict future changes.

The first phase of a groundwater basin study consists of the evaluation of existing hydrologic and hydrogeologic data to define the physical character of the groundwater system. This evaluation is used to develop a conceptual model of the system, to guide the data-collection program, and to identify major water-supply deficiencies or problems. The field data-collection program includes the collection of information on the hydrogeologic character and operation of the aquifer, wells and withdrawals, surface-water diversions, precipitation and streamflow. Supplementary data might be obtained by test drilling, installation and measurement of observation wells, geophysical logging, pumping tests, installation and operation of gaging stations, and water-quality analyses. These field data are summarized and then interpreted to provide a description of the water resources and an evaluation of the historical development of the water supply.

If a more intensive study is necessary, the next phase of the study might be the development of a hydrologic model of the basin. A model is essential for many projects because of the difficulty of integrating and analyzing large quantities of hydrologic data. The model should at first be used principally to synthesize the data and to test hypotheses of how the hydrologic system functions. The development and testing of the model for accuracy requires quantitative hydrogeologic and water-quality data to describe the operation of the groundwater system and its interrelation with streams. The results of a model analysis are no more reliable than the basic data used to construct the model and to provide hydrologic input.

Hydrologic and hydrogeologic information on the aquifer, streamflow, and recharge are built into the model to simulate the system. The types of field information needed to build a model can be classified into two categories. These are: (1) field data to define the physical framework; and (2) field data to describe the hydrologic stress on the system. A summary of the data requirements for a prediction model is given in Table I. This tabulation is considered to be only a general guide for data collection. Different types of hydrologic data will be needed for various types of groundwater studies.

After the elements of the hydrologic system have been described and simulated, the next step in the evaluation is the testing for accuracy or verification of the model. The hydrologist selects a time period (months or years) and programs the hydrologic stresses, such as recharge from precipitation and applied water, groundwater withdrawal, and evapotranspiration. Some of these stresses can be defined by direct measurement (for example, the pumping history and stream diversions). However, the definition of most of the stresses requires inference by the hydrologist. Preliminary estimates are usually made of the percentage of applied water and precipitation that recharges the aquifer and the rate of evapotranspiration for those areas where the water table is near the land surface. A comparison to the model results and field data is used to evaluate the correctness of the hydrologist's inferences or estimates.

TABLE I

Data requirements to be considered for prediction model

Physical framework:
(1) Hydrogeologic map showing areal extent and boundaries of all aquifers
(2) Topographic map showing surface-water bodies
(3) Water-table, bedrock-configuration, and saturated-thickness maps
(4) Transmissivity map showing aquifer and its boundaries
(5) Transmissivity map of confining bed
(6) Map showing variation in storage coefficient of aquifer
(7) Relation of saturated thickness to transmissivity
(8) Relation of stream and aquifer (hydraulic connection)

Hydrologic stress:
(1) Type and extent of recharge areas (irrigated areas, recharge basins, recharge wells, etc.)
(2) Surface-water diversion
(3) Groundwater pumpage (time—space distribution)
(4) Depth to water map, keyed to evapotranspiration rate
(5) Tributary inflow (time—space distribution)
(6) Groundwater inflow and outflow
(7) Precipitation
(8) Areal distribution of water quality in aquifer
(9) Streamflow quality (time—space distribution)

Model verification:
(1) Water-level-change maps and hydrographs
(2) Streamflow (including gain—loss measurements)

Prediction and optimization analysis:
(1) Economic information on water supply
(2) Legal and administrative rules
(3) Environmental factors
(4) Planned changes in withdrawals, streamflow and land use

The results of the model analysis consist of a map or hydrographs showing changes in groundwater storage and variations in streamflow. The hydrologist compares these results with field data collected for the same time period. If the model does not duplicate the field condition, then he must decide whether to modify the physical framework or make new estimates of the hydrologic stresses, or both. The testing is dependent on the hydrologist's interpretation of the basic hydrologic data. The final results will be dependent on his knowledge of the hydrologic system and the accuracy of the hydrologic data. The model analysis may indicate major data deficiencies and it may be necessary to collect additional field data before the model testing can continue.

When the hydrologist is satisfied that the model simulates the real system, the model may be used to evaluate water problems and to predict future changes. These analyses will require additional information. Major constraints that affect planning are: the cost of pumping groundwater, the cost of surface water, the legal controls on water use, the groundwater and surface-water quality, and the effects of the water use on the natural environment. For example, the model analysis may indicate that water management in some areas could be improved by increased use of groundwater; this increase, however, may result in long-term degradation of the water quality and eventually, in reduced crop yields.

CONCLUSIONS

In summary, 250 highly sophisticated groundwater models have been developed and are available for use in planning the use of groundwater supplies, for the conjunctive use of ground- and surface water, and for optimizing management objectives within specified physical, economic, and social constraints. Because of the development of these models, it is now possible to integrate use of groundwater into water-resource planning with a high degree of certainty.

REFERENCES

Appel, C.A. and Bredehoeft, J.D., 1976. Status of ground-water modeling in the U.S. Geological Survey. U.S. Geol. Surv., Circ. 737, 9 pp.

Bachmat, Y., Andrews, B., Holtz, D. and Scott, S., 1978. Utilization of numerical ground water models for water resource management. U.S. Environ. Prot. Agency, Rep. EPA-600/8-78-012, 177 pp.

Bredehoeft, J.D. and Young, R.A., 1970. The temporal allocation of ground-water — A simulation approach. Water Resour. Res., 6(1): 3—21.

Emery, P.A., 1970. Electric analog model evaluation of a water-salvage plan, San Luis Valley, Colorado. Colo. Ground Water, Circ. 14, 11 pp.

Emery, P.A., Patten, E.P. and Moore, J.E., 1975. Analog model study of the hydrology of the San Luis Valley, South-Central Colorado. Colo. Water Resour., Circ. 29, 21 pp.

Jorgensen, P.G., 1975. Analog-model studies of ground-water hydrology in the Houston District, Texas. Texas Water Dev. Board, Rep. 190, 84 pp.

Konikow, L.F. and Bredehoeft, J.D., 1974. Modeling flow and chemical quality changes in an irrigated stream—aquifer system. Water Resour. Res., 10(3): 546—562.

Meyer, R.W. and Carr J.E., 1979. A digital model for simulation of ground-water hydrology in the Houston area, Texas. Texas Water Dev. Board Rep. (in press).

Moore, J.E. and Wood, L.A., 1967. Data requirements and preliminary results of an analog model evaluation — Arkansas River valley in eastern Colorado. Ground Water, 5(1): 20—23.

Moulder, E.A. and Jenkins, C.T., 1969. Analog-digital models of stream — aquifer systems. Ground Water, 7(5): 19—24.

Pinder, G.F., 1970. A digital model for aquifer evaluation. U.S. Geol. Surv., Tech. Water Resour. Invest., Book 7, Ch. C1, 18 pp.

Prickett, T.A. and Lonnquist, C.G., 1968. Comparison between analog and digital simulation techniques for aquifer evaluation. In: The Use of Analog and Digital Computers in Hydrology; a Symposium. Int. Assoc. Sci. Hydrol., Publ., 81(2): 625—634.

Weeks, J.B., Leavesley, G.H., Welder, F.A. and Saulnier, G.J., 1974. Simulated effects of oil shale development on the hydrology of the Piceance Basin, Colorado. U.S. Geol. Surv., Prof. Pap. 908, 84 pp.

Wilson, W.A., 1978. Simulated changes in ground-water levels resulting from proposed phosphate mining, west-central Florida — preliminary results. U.S. Geol. Surv., Open File Rep. 77-882.

Wood, L.A. and Gabrysch, R.K., 1965. An analog model study of ground water in the Houston district, Texas, with a section on design, construction, and use of analog models, by E.P. Patten, Jr. Texas Water Comm., Bull. 6508, 103 pp.

[2]

APPLICATION AND ANALYSIS OF A COUPLED SURFACE AND GROUNDWATER MODEL

A.B. CUNNINGHAM and P.J. SINCLAIR

Department of Civil Engineering and Engineering Mechanics, Montana State University, Bozeman, MT 59715 (U.S.A.)
Desert Research Institute, University of Nevada, Reno, NV 89506 (U.S.A.)

(Accepted for publication April 27, 1979)

ABSTRACT

Cunningham, A.B. and Sinclair, P.J., 1979. Application and analysis of a coupled surface and groundwater model. In: W. Back and D.A. Stephenson (Guest-Editors), Contemporary Hydrogeology — The George Burke Maxey Memorial Volume. J. Hydrol., 43: 129—148.

A model which couples two-dimensional transient saturated subsurface flow (Boussinesq equation) and one-dimensional gradually-varied unsteady open-channel flow (Saint-Venant equations) is developed and applied to the Truckee River system in northern Nevada. The coupled model involves simultaneous numerical solution of the surface and subsurface flow equations using the Galerkin finite-element method. This paper attempts to address questions regarding the application of such a model to both real and hypothetical hydrologic systems in terms of analysis of parameter sensitivity, predictive uncertainty and uniqueness of model solution. Investigation of parameter sensitivity provides information regarding the relative accuracy with which individual system parameters should be measured in order to insure successful model operation. Model predictive uncertainty is evaluated based on comparison of model output and field observations obtained over a two-year period from the Truckee River study area in northern Nevada. Results indicate fair to good prediction of both high and low river stage and groundwater table elevations with predictive accuracy decreasing with distance away from river channel. A procedure for quantifying uniqueness of model solution is developed to provide insight into the reliability associated with extrapolation of model output beyond the limits of observed field measurements. Based on results for the Truckee River system, such extrapolation appears to be justified.

INTRODUCTION

For many years hydrologists have been interested in examining the relationship between flow in a stream channel and flow in an adjacent aquifer when hydraulic connection exists between the two. Consequently, in this study, equation describing two-dimensional transient saturated subsurface flow and one-dimensional gradually-varied transient open-channel flow were coupled and a simultaneous numerical solution obtained by the Galerkin finite-element method. This modeling technique (referred to herein by the term "coupled model") was first used to solve a hypothetical problem after which it was calibrated and analyzed for a reach of the lower Truckee River system in northern Nevada. Model performance was evaluated by comparison of observed and simulated data sequences generated over a two-year study period.

Although various approaches have been suggested for structuring coupled models, the basic procedure usually consists of simultaneous numerical solution of differential equations which approximately describe transient open-channel and subsurface flow. Specific methodologies (in addition to the approach used in this study) include the work of Pinder and Sauer (1971) who coupled a model for a two-dimensional horizontal aquifer to a one-dimensional stream flow model. It is in fact their technique which is generalized herein. A subsequent study by Freeze (1972) resulted in a greatly expanded model capable of treating three-dimensional transient saturated— unsaturated subsurface flow coupled with one-dimensional channel flow (in this study lack of field data of sufficient quality and quantity precluded the use of a three-dimensional model). Yet another approach has been suggested by Morel-Seytoux (1975) which combines a model for two-dimensional flow in an aquifer with an expression, linear in stage and discharge, for one-dimensional flood wave propagation.

Besides the development of alternative model formulations, previous investigations also demonstrate that the coupled-model approach, when applied within limits of the inherent physical and mathematical constraints, permits investigation of a variety of hydrologic problems involving the interchange between surface and groundwater systems. Examples of such applications include flood routing in channels with bank seepage (Zitta and Wiggert, 1971), investigation of flood hydrograph modification due to bank storage (Pinder and Sauer, 1971), investigation of the influence of aquifer parameters on hydrograph recession (Cunningham, 1977), and, in probably the most comprehensive study to date, definition of the role of subsurface flow in generating surface runoff (Freeze, 1972). As Freeze points out, the coupled-model approach not only has potentially far reaching applications in the study of the internal mechanisms of the hydrologic cycle, but also could greatly aid in the development of deterministic watershed response models. In fact, if carried to fruition, coupled models could ultimately provide an integrated approach to deterministic modeling of the entire land phase of the hydrologic cycle.

While research results to date provide an optimistic outlook for attaining these goals, they also demonstrate that much additional work is needed in order to completely develop the potential for this particular modeling technique, particularly regarding application to real hydrologic systems. For example, a detailed account of the procedure necessary for calibration and validation of a fully deterministic subsurface model to the Reynolds Creek watershed, Idaho, is given by Stevenson and Freeze (1974) wherein the following statement is made:

"It is clear to us . . . that this technique is far from ready for routine application as a hydrologic response model for large regional watersheds. The major items that seem most significant are limitations in theoretical development, . . . computer limitations, and constraints on calibration procedures. A final limiting factor is concerned with the field data available for the model."

In the case of the coupled model there arise a similar set of fundamental questions which should be evaluated thoroughly before the technique can be considered operational. These additional questions include:

(1) What is the significance of numerous physical and mathematical assumptions and approximations used in the development and solution of the differential equations governing surface and groundwater motions?

(2) How can the model best be calibrated to the real hydrologic system and what level of performance, in terms of prediction of system behavior, can be expected?

(3) Which model parameters are most sensitive and therefore most in need of careful measurement in the field?

(4) Can the degree of uniqueness of solution be defined and established quantitatively, if so, what are the implications?

This paper represents an attempt to at least partially address these and other related questions having to do with the application of coupled ground and surface water models. In so doing, emphasis is placed mainly on evaluation of model performance rather than on its development. Accordingly, only a brief outline of the modeling technique is provided. The results of an evaluation of model performance for both real and hypothetical hydrologic systems in terms of an analysis of parameter sensitivity, predictive uncertainty, and uniqueness of solution are given in more detail.

MODELING PROCEDURE

The modeling technique used herein consists of simultaneous solution of the Saint Venant equations for one-dimensional spatially-varied transient open-channel flow and the linearized Boussinesq equation descriptive of two-dimensional transient groundwater flow. These equations are coupled by an expression for flow through the wetted perimeter of the channel. The Galerkin finite-element technique is utilized in the numerical solution algorithm.

Open-channel flow equations

The momentum and continuity equations that approximately describe gradually varied transient flow in a fixed-bed open channel of arbitrary cross-section and alignment, written as utilized in this study, are:

$$\frac{\partial Q}{\partial t} + \frac{\partial (VQ)}{\partial s} + gA \frac{\partial z}{\partial s} - gA(S_0 - S_f) - u_s q_s = 0 \qquad (1)$$

$$T \frac{\partial z}{\partial t} + \frac{\partial Q}{\partial s} - q_s - q_g = 0 \qquad (2)$$

where A is the area of flow normal to the channel invert; g is the gravitational acceleration; q_s and q_g are the lateral surface water and groundwater inflows per unit length of channel, respectively; Q is the channel discharge; s is the arc length along the channel invert; S_0 is the slope of the channel invert; S_f is the friction slope; t is time; T is the top width of flow; u_s is the average component in the s-direction of the lateral surface water inflow velocity; $V = Q/A$ is the average velocity component in the s-direction; and z is the depth of flow relative to the invert. This particular form of the Saint-Venant equations constitutes a slight generalization of the equations solved by Cooley and Moin (1976, p. 760). It is worth noting that the groundwater inflow is normal to the channel. Thus, q_g does not appear in eq. 1; the groundwater flow does not transport momentum in the s-direction into the channel.

The friction slope, which arises due to the presence of the wall shear stress, is estimated using the empirical Manning's equation:

$$S_f = n^2 V^2 / N R^{4/3} \qquad (3)$$

where n is Manning's roughness coefficient, N is a conversion factor taking the value of 1 m$^{2/3}$/s^2 for S.I. units and $(1.486)^2$ (= 2.208) ft.$^{2/3}$/s^2 for English units, $R = A/P$ is the hydraulic radius, and P is the wetted perimeter of the channel.

The area of flow is assumed to be given by a power law relationship of the form:

$$A = az^b + Bz \qquad (4)$$

and likewise:

$$P = dz^e + B \qquad (5)$$

where a, b, d and e are empirically determined functions of s; and B is the bottom width of the channel, which may also be a function of s.

The top width is determined from the usual expression:

$$T = \partial A / \partial z \qquad (6)$$

Because only subcritical flow is examined herein, boundary conditions used with eqs. 1 and 2 must be supplied at both ends of the reach being studied at the upstream boundary, either a stage or discharge hydrograph together with the backward characteristic equation are used to define the dependent variables there. At the downstream boundary, a rating curve in the form of Manning's equation (with S_f set equal to S_0) in conjunction with the forward characteristic equation are utilized similarly. The particular form of the characteristic equations used in this study is:

$$\frac{\partial Q}{\partial t} + Q \frac{\partial V}{\partial s} \pm c \frac{\partial Q}{\partial s} - T(V \mp c) \frac{\partial z}{\partial t} + gA \frac{z}{\partial s} - gA(S_0 - S_f) - u_s q_s$$
$$+ (V \mp c)(q_s + q_g) = 0 \qquad (7)$$

where $c = (gA/T)^{1/2}$. In eq. 7, the upper signs yield the forward characteristic equation, and the lower, the backward characteristic equation. Eq. 7 may be derived from eqs. 1 and 2 following Strelkoff (1970, pp. 225—232). A rationale for using the characteristic equations in the above context may be found in Strelkoff (1970, p. 237), or Cooley and Moin (1976, p. 761).

Groundwater flow equations

If the general equation approximately describing three-dimensional transient flow of groundwater is integrated over the saturated thickness of an unconfined aquifer with an impervious lower boundary, then the result may be written as:

$$\frac{\partial}{\partial x_i} \left[K_{ij}(h - z_1) \frac{\partial h}{\partial x_j} \right] = [S_y + S_s(h - z_1)] \frac{\partial h}{\partial t} - I - v_s \qquad (8)$$

where h is the mean head over the saturated thickness; i,j are indices which take the values 1,2 only; K_{ij} is the hydraulic conductivity tensor which takes an effective value over the saturated thickness; S_s is specific storage; S_y is specific yield; t is time; $x_1 (\equiv x)$, $x_2 (\equiv y)$ are horizontal Cartesian coordinates; I is the vertical infiltration rate; v_s is a (line) source-sink term, positive for a source; and z_1 is the elevation of the aquifer's lower boundary. The summation convention over repeated subscripts is assumed in eq. 8. The details of the derivation of eq. 8 are given in Cooley (1974, pp. 3—8); alternatively, the general principles on the averaging process may be found in the Hantush (1964, pp. 299—301) or Bear (1972, pp. 218—220). Using the Dupuit—Forchheimer assumptions, which are discussed in some detail in Bear (1972, pp. 361—366), it is relatively simple to derive the Boussinesq equation (Bear, 1972, p. 376) which is similar to eq. 8.

To facilitate the numerical solution of eq. 8, it is linearized in the classical manner by replacing the $h - z_1$ terms with $\bar{h} - z_1$, where $\bar{h}$ is the temporal average of h. This replacement may be justified by the following facts. This study, centered on a reach of the lower Truckee River, Nevada, involved river stage fluctuations less than 1 m. Further, the lithologic logs of the piezometer holes drilled for the study suggest that the saturated thickness of the alluvial aquifer is at least 10 m. Thus, $\bar{h} - z_1 \cong h - z_1$ and eq. 8 may be approximated by:

$$\frac{\partial}{\partial x_i} \left(T_{ij} \frac{\partial h}{\partial x_j} \right) = S_c \frac{\partial h}{\partial t} - I - v_s \qquad (9)$$

where $T_{ij} = K_{ij}(\bar{h} - z_1)$, and $S_c = S_y + S_s(\bar{h} - z_1)$.

Recalling that q_g is the lateral groundwater inflow to the river per unit length of channel, and assuming that channel only partially penetrates its alluvial aquifer, then for the part of the aquifer beneath the channel, eq. 9 must be modified to:

$$\frac{\partial}{\partial x_i}\left(T_{ij}\frac{\partial h}{\partial x_j}\right) = S_c \frac{\partial h}{\partial t} + \frac{q_g}{P} \qquad (10)$$

where, as before, P is the wetted perimeter of the channel; and $S_y = v_s = 0$. Note that eqs. 9 and 10 are both of the linear, heat conduction type.

The boundary conditions used with eqs. 9 and 10 must be in the form of known head, known flow, or a combination of both.

Coupling equation

It now only remains to define mathematically the groundwater inflow to the channel which appears in eqs. 2 and 10 and hence couple eqs. 1, 2, 9 and 10. If the channel is considered to be lined with a thin skin of fine sediments of thickness b_c and hydraulic conductivity K_c, then a simple application of Darcy's law to a unit length of these sediments, gives:

$$q_g = PR_c(h - z - z_0) \qquad (11)$$

where $R_c = K_c/b_c$, h is the hydraulic head under the channel, and z_0 is the elevation of the channel invert measured from the same datum as h. The equation is of the same form as that used by Pinder and Sauer (1971, p. 65).

It is interesting to note that even if the lower-permeability sediments were absent, but the stream channel still only partially penetrates its alluvial aquifer, eq. 11 continues to be applicable. However, K_c is now an effective hydraulic conductivity, and b_c the distance between the channel bottom and the point where the mean head occurs (assumed constant). The other quantities in eq. 11 are defined as given earlier. As these new definitions are somewhat vague, it is perhaps better in this situation to consider R_c ($= K_c/b_c$) as an independent parameter. This interpretation of eq. 11 comes from Cooley and Westphal (1974, pp. 3 and 9).

If, on the other hand, the channel fully penetrates the aquifer (and the sediments are still absent) then the stream and aquifer should be coupled by setting $h = z + z_0$ along the stream. This relationship may be obtained by using eq. 11 also, provided K_c takes a value such that $K_c \gg ||[K]||$ where $[K]$ is K_{ij} written as a 2 × 2 matrix, and $||\cdot||$ is any valid matrix norm. For definitions of matrix norms, see, for example, Forsythe and Moler (1967, pp. 2—4) and the references cited therein.

General solution procedure

Based on present knowledge, an analytical (i.e. exact) solution to eqs. 1, 2,

9 and 10 is impossible to obtain. Thus, a numerical solution is effected using the Galerkin finite-element method over the spatial coordinates, and finite-difference discretization over time, with Q, z and h as dependent variables. Galerkin's method, as one of the methods of weighted residuals, is described in general terms in most of the monographs relating to the finite-element method. See, for example, Zienkiewicz (1977, pp. 49—58), or Pinder and Gray (1977, pp. 54—57 and 64—73). The application of this method to the above equations is rather long and hence no attempt is made to include the work herein. However, the mathematical details will ultimately appear in Sinclair (1979). The application of the Galerkin finite-element method to the Saint-Venant equations themselves (eqs. 1 and 2) together with auxiliary conditions (3)—(6) is given in Cooley and Moin (1976, pp. 761—767). It is their solution technique that was incorporated into the coupled model. The Galerkin finite-element method applied to eq. 9 is described, in principle, in Pinder and Gray (1977, pp. 105—107) where triangular elements with first-degree interpolation are used. The coupled model, in contrast, uses quadrilateral elements each constructed from four first-degree triangles.

Once eqs. 1, 2, 9 and 10 have been discretized, the solution technique proceeds as described in Pinder and Sauer (1971, p. 65) for their finite-difference solution. That technique is as follows. Assuming the solution at time t is known, and the solution at $t + \Delta t$ is required, the Saint-Venant equations (eqs. 1 and 2) are solved iteratively with q_g set at its value at time t. The z from this solution is then used in the solution of eqs. 9 and 10, with eq. 11 substituted into eq. 10, to obtain h directly. Using h and z in eq. 11 an improved estimate of q_g may be calculated, followed by a solution to the Saint-Venant equations again using this new q_g. This cyclic process continues until successive values of q_g differ by an acceptably small amount. Once q_g has converged, the values of Q, z and h are considered known at time $t + \Delta t$, and those values at the end of the next time step may be sought by the same process.

Finding the solution at the end of the first time step by the above procedure requires the solution to be known at time zero — the initial conditions. The initial conditions are assumed to be given by a steady-state solution. This steady-state solution is found from eqs. 1, 2, 9 and 10 with the time derivatives omitted, but otherwise the solution procedure for the unsteady case is followed, with the initial estimate of q_g taken as zero.

EVALUATION OF MODEL PERFORMANCE

Because the coupled model is constructed from the differential equations which approximately govern both open-channel and groundwater flow, it is considered to be physically based and therefore has the capability, when properly calibrated, of simulating hydrologic events beyond the range of recorded field data. However, despite the rigorous mathematical development supporting a physically based model, significant drawbacks still exist, particularly in assessing the overall uncertainty involved with modeling a real hydrologic system. Such uncertainty arises from several sources with one very notable source being the numerous physical and mathematical assumptions which are made in the derivation and solution of the governing differential equations. A detailed account of the derivation of the Saint-Venant equations from the viewpoint of hydrodynamics together with assumptions and limitations is given by Strelkoff (1969) and Yen (1973). Since these assumptions preclude accurate analysis of situations in which flow cannot be considered gradually varied, question arises as to the validity of the Saint-Venant equations in many situations commonly encountered in natural open-channel flow. Compromising physical assumptions likewise exist in the development of the Boussinesq equation (see Cooley, 1974, pp. 3—8) and these, together with approximations and assumptions inherent in the numerical-solution procedures, add additional uncertainty to model operation. Furthermore, when the model is applied to a real hydrologic system, yet another component of uncertainty is encountered due to the error involved in actually measuring time and space variation of system input and output, as well as parameter values for use in model calibration.

In reviewing overall model uncertainty, it is obvious that the approximations and limitations encountered are indeed severe enough to cast doubt, at least at this point, as to the reliability of the final product. It was therefore deemed necessary to undertake a systematic and detailed assessment of the capabilities and limitations of the coupled model, for its complexity certainly does not favor blind application to field problems. Accordingly, model analysis was carried out in three phases: (1) determination of the relative sensitivity among the respective model parameters; (2) calibration of the model to the designated study area (lower Truckee River system, Nevada), and based on comparison of observed and predicted values of system output, determination of predictive uncertainty; and (3) evaluation of uniqueness of solution.

Because only one study area has been analyzed, results of model analysis cannot be considered completely general. It is to be hoped, however, that these results will provide insight into the level of performance which can be obtained, as well as the type and quantity of field data required, when applying a coupled model to a real hydrologic system.

Parameter sensitivity

Determining the degree of sensitivity of model output to individual parameter values is desirable before calibrating the coupled model to a particular hydrologic study area. This information is particularly useful in determining the accuracy to which individual parameters must be known in order to insure successful model operation. For the purposes of this study, the term "sensitivity" is defined as: a measure of the degree to which given variations in a particular model parameter alter model output.

Parameter sensitivity was evaluated by defining a sensitivity index, S, as the average of the absolute percentage differences between the predicted and observed output values. Mathematically, S is given by:

$$S = \frac{1}{n} \left(\sum_{i=1}^{n} \frac{|O_i - P_i|}{O_i} \right) \times 100\% \qquad (12)$$

where O_i is the system output value for time step i; P_i is the corresponding model output prediction; and n is the number of time steps considered. Thus, by systematically varying individual parameters (using always the same model input), the relative sensitivity of the model to changes in each of its parameters can be examined by comparison of S values.

The hypothetical hydrologic system, which was modeled for the purpose of determining parameter sensitivity, is that in Pinder and Sauer (1971, pp. 66 and 67). It consists of a flood plain aquifer which extends 40 km along the length of the channel and 1.5 km across the valley; it is surrounded by impermeable material on all sides. The hydraulic conductivity of the aquifer (assumed homogeneous and isotropic) is 265 m/day and the initial saturated thickness ranges from 65 m at the upstream boundary to 25 m at the downstream boundary. The stream flows along the axis of the valley through a straight rectangular channel with constant cross-section and a slope of 0.001. The hydraulic conductivity of the stream bed along the wetted perimeter is 1200 m/day and is therefore not a limiting factor in the amount of water entering the aquifer. The channel is 30 m wide, the initial depth of flow is 6 m, and Manning's n is 0.03.

A rectangular finite-element grid was used to represent the hydrologic system under investigation as shown in Fig. 1. The scale normal to the channel has been exaggerated to provide resolution necessary for visual analysis of water surface elevation contours not included herein (i.e. node points are 3000 m apart parallel to the channel and 150 m apart normal to it). The dots

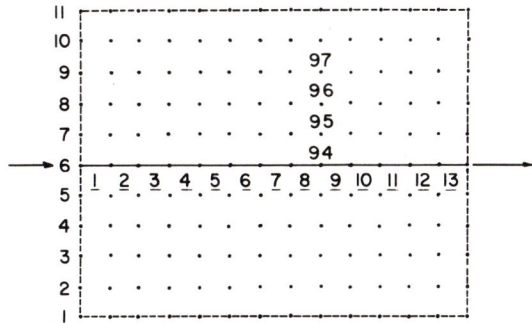

Fig. 1. Finite-element grid for hypothetical river—aquifer system. The method of numerating the nodes is illustrated with *plain numbers*. The channel elements are labelled with *underlined numbers*. The *arrows* show the direction of channel discharge.

in this illustration depict element nodes and therefore represent the locations at which the model provides solution values directly for the variables of interest. The dashed lines bordering the system represent impervious boundaries and thus define the entire groundwater system. There are a total of thirteen river elements for this system and the interchange flow values are the net seepage rates either into or out of the river over each of these elements. This hypothetical hydrologic system was studied in terms of its response to the passage of a flood wave (superimposed on a base flow of 500 m³/s) which has the following characteristics: initial flow is 500 m³/s, peak flow is 750 m³/s, time to peak is 2.0 hr., hydrograph time base in 12 hr.

In carrying out the model sensitivity analysis, an initial model solution was obtained based on the input values listed above. This provided the values of O_i used in eq. 12. Parameter sensitivities were then computed with respect to various model output variables by raising each parameter first by 10% and then by 100% while keeping all other parameters at their original values. Results of this analysis are displayed in Tables I and II. As can be seen, output variables are river discharge and stage (node 94), groundwater–surface water interchange (channel element 12) and water-table elevations at the three arbitrary aquifer nodes 95–97.

Examination of Tables I and II provides insight as to the relative sensitivity of the different model outputs to changes in magnitude of the various parameters. Here it is seen that the model is most sensitive to changes in Manning's n and channel slope for nearly all types of output considered. At the 10% level of parameter change, river stage and discharge along with groundwater–surface water interchange are most sensitive to Manning's n, while water-table elevations are most influenced by channel slope. These results differ only slightly at the 100% level of change in that the river-bottom conductivity be-

TABLE I

Parameter sensitivities at 10% level for various model output variables

Parameter	Water-table elevation node 95	Water-table elevation node 96	Water-table elevation node 97	River discharge node 94	River stage node 94	Groundwater–surface water interchange element 12
Aquifer hydraulic conductivity, K	0.010	0.016	0.0212	0.010	0.009	16.550
Aquifer specific yield, S_y ($\cong S_c$)	0.013	0.021	0.026	0.035	0.022	14.800
River bottom conductivity, K_c	0.010	0.016	0.0212	0.010	0.010	16.512
Manning roughness, n	1.007	1.004	1.003	0.639	6.695	32.366
Stage-area coefficient, a	0.017	0.019	0.016	0.002	0.124	0.180
Stage-area exponent, b	0.023	0.024	0.022	0.003	0.163	0.265
River channel slope, S_0	6.137	6.136	6.134	0.297	3.180	22.423

TABLE II

Parameter sensitivities at 100% level for various model output variables

Parameter	Water-table elevation node 95	Water-table elevation node 96	Water-table elevation node 97	River discharge node 94	River stage node 94	Groundwater–surface water interchange element 12
Aquifer hydraulic conductivity, K	0.052	0.095	0.125	0.063	0.044	165.751
Aquifer specific yield, S_y ($\cong S_c$)	0.108	0.166	0.206	0.298	0.210	122.250
River bottom conductivity, K_c	0.052	0.095	0.125	0.063	0.045	166.710
Manning roughness, n	9.158	9.130	9.114	4.153	61.087	161.166
Stage-area coefficient, a	0.168	0.169	0.168	0.024	1.122	1.789
Stage-area exponent, b	2.534	2.532	2.528	0.571	16.809	45.161
River channel slope, S_0	30.227	30.222	30.217	1.296	12.805	111.73

comes the most sensitive parameter in terms of the computation of groundwater—surface water interchange.

The overall results obtained here are somewhat surprising due to the dominance of river parameters as the most sensitive. For example, it is seen by comparison of sensitivities, that Manning's n is from about 50 to 200 times more sensitive than aquifer parameters (specific yield and hydraulic conductivity) when modeling fluctuations in the water table. This result is indeed welcome, however, since river parameters are, by comparison, more easily estimated than aquifer parameters.

Truckee River system

While parameter sensitivity was studied using a hypothetical hydrologic system, the remainder of the study, namely evaluation of predictive uncertainty and uniqueness of solution, was accomplished by calibrating the coupled model to a real system. The hydrologic system chosen for this particular study consists of a 6-km reach of the Truckee River in northern Nevada as shown in Fig. 2. The groundwater system located immediately south of the river is virtually free from the effects of well pumpage; however, some influence from irrigation and recharge is present. In order to monitor water-table fluctuations in the vicinity of the river, a total of five PVC-cased piezometer holes were drilled at the locations indicated in Fig. 2. The lithology pattern encountered in drilling the piezometer holes basically consisted of assorted combinations of sand, silt and gravel layers. The absence of extended clay lenses suggests the presence of what is essentially an unconfined aquifer system.

Water-table elevations in the five piezometers were monitored on a weekly (and occasionally daily) basis during water years 1975 and 1976. It was soon

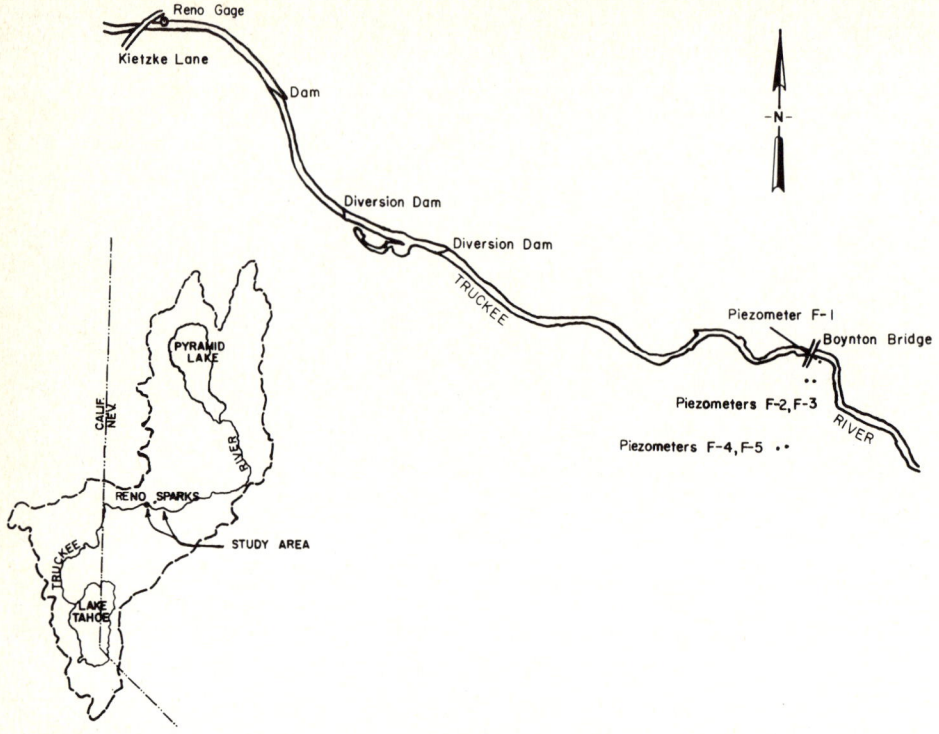

Fig. 2. Truckee River system, Nevada and California. Flow direction is from Lake Tahoe to Pyramid Lake. The *broken line* denotes the watershed boundary.

discovered, however, that piezometers F-4 and F-5 were influenced by irrigation practices to a degree such that they were of little value in an investigation of this type. Thus, model calibration, verification and analysis were carried out using only the data obtained from piezometers F-1, F-2 and F-3. Concurrent river stage and discharge measurements were obtained at the Boynton Lane bridge.

Model calibration

Using measured field parameter values as a guideline, the coupled model was calibrated for the lower Truckee River system. Calibration of the model consisted of repeated trials using alternative combinations of parameter values (taking care to insure that no value exceeded reasonable limits) and comparing model outputs against corresponding system outputs. The model was considered calibrated when no significant increase in accuracy (i.e. the difference between predicted and observed output values) could be achieved by further variation in parameter values. In other words, when the model was calibrated, the above differences were approximately normally distributed about a mean of zero. The final parameter values arrived at in this way are given in Table III.

TABLE III

Final* parameter values used in simulation of lower Truckee River system

Parameter	Value	Units	Method of measurement
Aquifer hydraulic conductivity, K	10.5	m/day	analysis of pump test data
Aquifer specific yield, S_y	0.001	dimensionless	analysis of pump test data
River bottom skin hydraulic conductivity, K_c	10.0	m/day	analysis of river stage and well hydrograph data
Manning's roughness coefficient, n	0.0242	dimensionless	hydraulic survey of lower Truckee River
River channel slope, S_o	0.00329	dimensionless	survey of lower Truckee River

*These adjusted parameter values differ very little (5—10%) from original field observation

Predictive uncertainty

Predictive uncertainty was evaluated based on comparison of observed and simulated river flows and groundwater levels for the Truckee River system. The initial step consisted of a comparison of basic statistical properties computed from both observed and predicted hydrologic data for the 1975 runoff season. Results given in Table IV indicate excellent agreement between predicted and observed mean values for all three variables, a result which serves as a good indication that the model has been calibrated adequately. Standard deviation values as well as first- and second-order serial correlation coefficients are also seen to be in close agreement, while skew values appear to diverge slightly.

TABLE IV

Statistical comparison of observed and predicted hydrologic data (statistics for observed data in parentheses)

Variable	Mean	Standard deviation	Skew	First-order serial correlation coefficient	Second-order serial correlation coefficient
Well F-1	1336.97 (1336.97) (m above M.S.L.)	0.19 (0.22) (m)	0.86 (0.99)	0.84 (0.84)	0.68 (0.68)
Well F-2	1336.97 (1336.97) (m above M.S.L.)	0.19 (0.21) (m)	0.86 (0.56)	0.84 (0.82)	0.68 (0.63)
Truckee River at Reno, 1963 flood	182.4 (193.0) (m^3/s)	144.5 (144.2) (m^3/s)	0.69 (0.91)	0.95 (0.93)	0.88 (0.84)

While comparison of statistical properties provides a general overview of model performance, by far the most straightforward and meaningful method for evaluating predictive uncertainty is that of direct graphical comparison of model predictions with corresponding values of observed field data. Scatter diagrams of this type readily facilitate the construction of confidence bands (Draper and Smith, 1966, p. 24) which in turn allow probabilistic inference to be made regarding the accuracy of model predictions. Results of this graphical analysis (including confidence bands for the 95% probability level) for the lower Truckee River system for water levels in piezometers F-1 and F-2 are

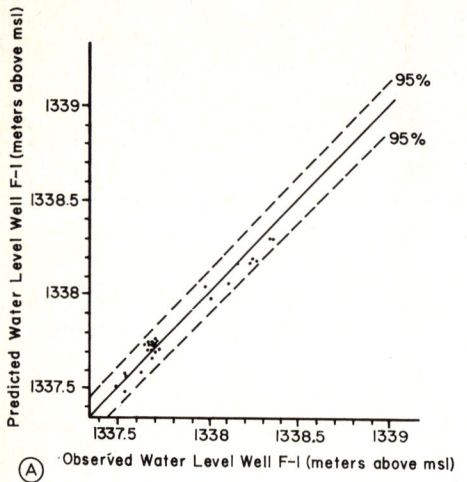
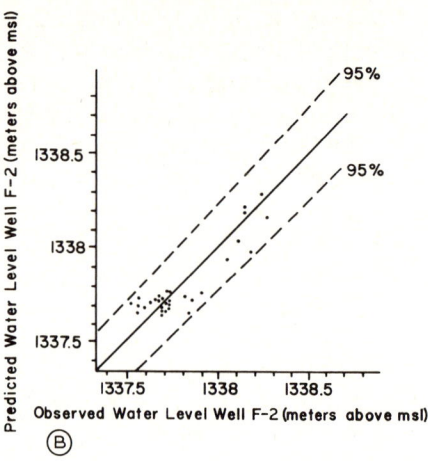
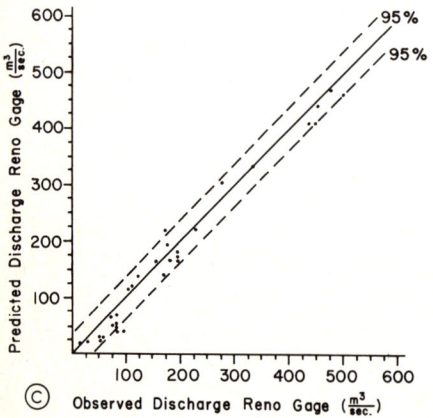

Fig. 3.A. Observations vs. model predictions of water-table elevations, piezometer F-1. The 95% confidence interval is delineated by *broken lines*.
B. Observations vs. model predictions of water-table elevations, piezometer F-2. The 95% confidence interval is delineated by *broken lines*.
C. Observations vs. model predictions for 34 streamflow values (Δt = 4 hr) at Reno gage, flood of February, 1963. The 95% confidence interval is delineated by *broken lines*.

given in Fig. 3A and B while corresponding results for predictions of river flows at the Reno gage are shown in Fig. 3C. The most notable result apparent here is that predictive accuracy decreases markedly with distance away from the river. Comparison of observations and predictions for well F-1 shows the individual points result in a 95% confidence band of about 0.11 m while the more distant F-2 well results exhibit a 95% confidence band of nearly 0.21 m. Thus, predictive accuracy has decreased by a factor of about 2 over a distance of about 100 m.

Results of river flow simulation given in Fig. 3C are the result of modeling the observed flood event of February, 1963 over the 37-km reach of the Truckee River from Farad to the Reno gage. In this particular case it was not necessary to model the groundwater—surface water interchange with any great degree of precision, simply because of the tremendous dominance of river flows compared with interchange flows. Thus, the results of this comparison serve mainly to quantify the accuracy with which river flows can be numerically simulated over long channel reaches. In general, a very high degree of predictive accuracy was observed in the simulation of both high and low river flows. This, together with the unbiased point scatter and narrow 95% confidence bands seen in Fig. 3C, generally indicates that predictive uncertainty is indeed small.

The results shown in Fig. 3 should be taken as the primary measure of the accuracy with which the coupled model simulated the behavior of this particular hydrologic system. Whether or not this level of accuracy is "acceptable" cannot be completely determined here since requirements for predictive accuracy may vary depending on the purpose of the simulation. However, at least in the case of the lower Truckee River system, it can be safely stated that the coupled model accurately reproduced all major features (and most minor fluctuations) of river stage and well hydrographs observed over the two-year study period. Thus, despite the numerous assumptions and approximations inherent in model development and calibration, the coupled model appears to represent a viable tool for hydrologic system analysis.

Uniqueness of solution

It has been stated previously that one advantage of physically based models is their potential for at least limited extrapolation beyond the range of the recorded data base from which they are calibrated. However, reliable extrapolation is in fact possible only if the parameter values determined in the calibration process are unique, and if their use results in accurate simulation over the entire range of observed field data. Thus, an evaluation of uniqueness of model results is fundamental to the comprehensive assessment of overall model performance and reliability.

As has been previously demonstrated, predictive uncertainty of the coupled model appears to lie within tolerable limits. Unfortunately, the above technique for parameter variation neglects the possibility that other combinations of different but compensating parameter magnitudes may in fact yield an equally good (or even better) comparison between model predictions and observed field data. In fact, it is easily conceivable that such a range of parameter values does exist, if only because of the many assumptions made in the derivation of the physical equations of motion governing the movement of ground and surface water.

The problem at hand is, therefore, one of identification of the range over which parameters may vary and still yield statistically similar model results. This range of parameter variation or "parameter slack" will provide a quantitative index to the uniqueness of model solution, and will serve as a basis for making judgments as to the reliability of extrapolated model results.

The procedure by which uniqueness of solution is quantified is to a large degree arbitrary. Basically, the problem is one of identifying some type of index of comparison between model predictions and corresponding observations, and then recording the changes in this index in response to changes in model parameter values. The methodology adopted consists of three parts: (1) variation of model parameters and recording of resulting model predictions; (2) comparison of model predictions with corresponding observations by application of the chi-square test for goodness of fit; and (3) theoretical statistical analysis to determine the range of parameter values for which the model produces statistically similar and hence non-unique results.

In order to avoid an excessive number of computer runs using random combinations of parameter values, a systematic approach to parameter variation was adopted utilizing the results obtained earlier from the parameter sensitivity analysis. Using the calibrated model for the lower Truckee River system, parameter values were changed by designated percentages in such a manner as to collectively provide the maximum increase in the magnitude of model output variables. Using this procedure, model results were obtained for each output variable based on changes in parameter magnitudes of 1%, 2%, 5%, 10%, 20% and 40% from their original values. Here the output variables used in the analysis were the predicted water-table elevations in piezometers F-1 and F-2.

Computation of chi-square. The chi-square "goodness of fit" test was employed to determine probabilistically whether or not the predicted set of data values is statistically indistinguishable from their corresponding observations. Here the chi-square statistic, χ_c^2, is calculated from:

$$\chi_c^2 = \sum_{i=1}^{n} \frac{(O_i - P_i)^2}{P_i} \tag{13}$$

where O_i is the ith observation, P_i is the ith prediction, and n is the number of observations.

In this analysis, the null hypothesis, H_0, which was tested is:

H_0:   model predictions and field observations belong to the same population

In order to test the hypothesis for significance, it was first necessary to compute the chi-square values for model predictions obtained for the various levels of parameter variation. These statistics were computed via eq. 13 and appear in Fig. 4 where they have been plotted against percent parameter change as reflected in piezometers F-1 and F-2. H_0 is then tested by calculating the critical value, C, of χ^2 from:

$$P(\chi^2 \leqslant C) = 1 - \alpha \tag{14}$$

where α is the designated significance level, using a table of the chi-square distribution. Theoretical chi-square values for the 99%, 95%, 5% and 1% significance levels appear in Fig. 4 as horizontal lines superimposed on the computed chi-square curves. Intersections between these horizontal lines and the curves are of fundamental importance because it is these points which allow quantitative inference to be drawn regarding the uniqueness of model results. For example, the 99% theoretical chi-square line is seen to intersect at points corresponding to 19% parameter change for both observation wells. Thus, H_0 is accepted at the 99% level provided the model parameters are not varied more than ±19%. Put another way, the model parameters can be changed up to about ±19% from their original best-fit values and still produce model results which are statistically indistinguishable, and thus non-unique, at the 99% level. Thus, in this case the range of parameter variation of about 19% constitutes the amount of parameter slack at the 99% significance level. This indicates that a reasonably unique solution has been obtained for the lower Truckee River system and, since acceptable predictive accuracy was also demonstrated, simulation beyond the limits of observed data appears justified.

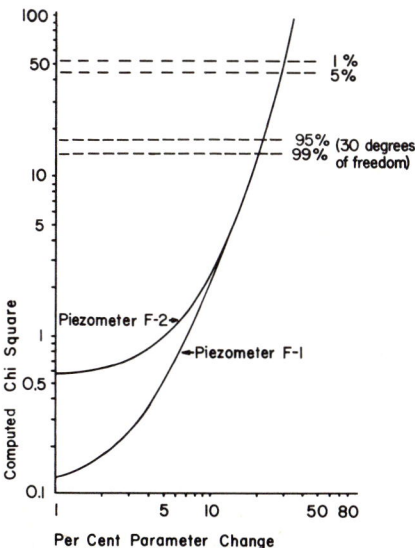

Fig. 4. Chi-square uniqueness test. Theoretical chi-square values for various levels of significance are plotted as *horizontal broken lines*.

In the case of the intersection of the 1% theoretical chi-square line, the interpretation would be to the effect that, if the parameters are not estimated to within about ±30% of their true field values, then there is at least a 99% probability that the predicted water-table elevations will be statistically different from actual observations. Similar statements apply to parameter changes corresponding to the 5%, 95%, or any other desired significance level. These findings, together with results obtained from the analysis of parameter sensitivity, serve as useful guidelines for the calibration of the coupled model to other hydrologic systems.

SUMMARY AND CONCLUSIONS

A coupled surface and subsurface flow model was developed and applied to a hypothetical problem and to a segment of the lower Truckee River system in northern Nevada. Performance of the coupled model was analyzed in three separate but interrelated phases, these being: (1) determination of relative sensitivity of various model parameters; (2) calibration of model to the lower Truckee River system and evaluation of predictive uncertainty; and (3) estimation of uniqueness of solution.

The coupled model was found to be most sensitive to changes in Manning's roughness coefficient and channel slope for virtually all types of output considered. These parameters therefore have the greatest need for accurate field measurement.

Model predictive uncertainty, which was evaluated by several methods, was found to lie within tolerable limits. Good to excellent agreement was found by comparison of statistical properties for model output and corresponding field measurements. Predictive uncertainty was further investigated by direct graphical comparison between model predictions and corresponding field observations. The model was found to predict both high and low river stages and water-table elevations with about the same degree of precision. But model accuracy in terms of predicting groundwater elevations was observed to decrease by a factor of 2 over a distance of about 100 m out from the Truckee River. The major conclusion to be drawn is that despite the numerous physical and mathematical assumptions and approximations inherent in model development and solution, the coupled model appears to represent a viable tool for the investigation of both real and hypothetical systems.

Uniqueness of solution was investigated to determine, at least in part, the degree of model reliability beyond the limits of observed field measurements. The three-part methodology developed was based on the standard chi-square goodness of fit test. Results from the ensuing analysis showed that, for the lower Truckee River simulation, a region of parameter slack exists at the 99% significance level for a range of parameter changes of ±19%. Put another way, the model parameter values can be varied by as much as ±19% from their original best-fit values and still produce model outputs which are statistically indistinguishable at the 99% significance level. Also, if model parameters are not estimated to within ±30% of their true field values, then there is at least a 99% chance that the predicted water-table elevations will be statistically different from actual field observations. The conclusions which follow are: (1) since the magnitude of parameter slack is small and predictive accuracy is adequate, extrapolation beyond the limits of existing data is justified; and (2) there is little to be gained, at least in the case of the lower Truckee River system, in attempting to measure a particular parameter with greater precision than about ±19%. On the other hand, there is justification for attempting to obtain parameter estimates which are within at least ±30% of their true field values.

ACKNOWLEDGEMENT

The authors are indebted to Dr. Richard L. Cooley for his many technical contributions and suggestions throughout the duration of this study. Financial support was provided by the Office of Water Research and Technology, the Desert Research Institute and Montana State University.

REFERENCES

Bear, J., 1972. Dynamics of Fluids in Porous Media. American Elsevier, New York, N.Y., 764 pp.
Cooley, R.L., 1974. Finite element solutions for the equations of ground-water flow. Cent. Water Resour. Res., Desert Res. Inst., Reno, Nev., Tech. Rep. Ser. H-W, Publ. 18, 134 pp.
Cooley, R.L. and Moin, S.A., 1976. Finite element solution of Saint-Venant equations. Am. Soc. Civ. Eng., Hydraul. Div. J., 102(HY6): 759—775.
Cooley, R.L. and Westphal, J.A., 1974. Application of the theory of groundwater and river water interchange, Winnemucca reach of the Humboldt River, Nevada. Cent. Water Resour. Res., Desert Res. Inst., Reno, Nev., Tech. Rep. Ser. H-W, Publ. 19, 74 pp.
Cunningham, A.B., 1977. Modeling and Analysis of Hydraulic Interchange of Surface and Groundwater. Ph.D. Dissertation, University of Nevada, Reno, Nev. (unpublished). Also published in: Cent. Water Resour. Res., Desert Res. Inst., Reno, Nev., Tech. Rep. Ser. H-W, Publ. 34, 89 pp.
Draper, N.R. and Smith, H., 1966. Applied Regression Analysis. Wiley, New York, N.Y., 407 pp.
Forsythe, G.E. and Moler, C.B., 1967. Computer Solution of Linear Algebraic Systems. Prentice-Hall, Englewood Cliffs, N.J., 148 pp.
Freeze, R.A., 1972. Role of subsurface flow in generating surface runoff, 1. Baseflow contributions to channel flow. Water Resour. Res., 8(3): 609—623.
Hantush, M.S., 1964. Hydraulics of Wells. In: Ven Te Chow (Editor), Advances in Hydroscience, Vol. 1, Academic Press, New York, N.Y., pp. 281—432.
Morel-Seytoux, H.J., 1975. A combined model of water table and river stage evolution. Water Resour. Res., 11(6): 968—972.
Pinder, G.F. and Gray, W.G., 1977. Finite Element Simulation in Surface and Subsurface Hydrology. Academic Press, New York, N.Y., 295 pp.
Pinder, G.F. and Sauer, S.R., 1971. Numerical simulation of flood wave modification due to bank storage effects. Water Resour. Res., 7(1): 63—70.
Sinclair, P.J., 1979. Rational Development of a Model Coupling River and Groundwater Hydraulics. Ph.D. Dissertation, University of Nevada, Reno, Nev. (in progress).

Stevenson, G.R. and Freeze, R.A., 1974. Mathematical simulation of subsurface flow contributions to snowmelt runoff, Reynolds Creek watershed, Idaho. Water Resour. Res., 10(2): 284—294.

Strelkoff, T., 1969. One-dimensional equations of open channel flow. Am. Soc. Civ. Eng., Hydraul. Div. J., 95(HY3): 861—876.

Strelkoff, T., 1970. Numerical solution of Saint-Venant equations. Am. Soc. Civ. Eng., Hydraul. Div. J., 96(HY1): 223—252.

Yen, B.C., 1973. Open channel flow equations revisited. Am. Soc. Civ. Eng., J. Eng. Mech. Div., 99(EM5): 979—1010.

Zienkiewicz, O.C., 1977. The Finite Element Method. McGraw-Hill, London, 3rd ed., 787 pp.

Zitta, V.L. and Wiggert, J.M., 1971. Flood routing in channels with bank seepage. Water Resour. Res., 7(5): 1341—1345.

[2]

PROGRESS IN ANALYTICAL GROUNDWATER MODELING

W.C. WALTON

Upper Mississippi River Basin Commission, Twin Cities, MN 55111 (U.S.A.)

(Received March 6, 1979)

ABSTRACT

Walton, W.C., 1979. Progress in analytical groundwater modeling. In: W. Back and D.A. Stephenson (Guest-Editors), Contemporary Hydrogeology — The George Burke Maxey Memorial Volume. J. Hydrol., 43: 149—159.

Analytical models simulating the flow of groundwater to and from wells and streams, land subsidence due to artesian pressure decline, conservative solute transport to and from wells, and heat transport to and from wells have been developed for many aquifer systems, well and stream conditions. These models are useful, though limited in application, for aquifer-test analysis, simple aquifer system evaluation and numerical computer model design and calibration. Accelerated progress in analytical model development and application is called for particularly regarding flow in leaky artesian aquifers, conservative and non-conservative solute transport and transfer, and heat transport.

INTRODUCTION

Analytical modeling has in the past and will continue in the future to serve an important role in groundwater investigations. Although limited in application, analytical models are useful in aquifer-test analysis; evaluation of simple aquifer systems; design, calibration, or verification of numerical computer models; and understanding basic principles of groundwater flow and transport.

Advances in analytical modeling virtually dominated groundwater literature through the 1950's, whereas, in the 1960's and 1970's attention was largely diverted to numerical modeling with the aid of digital computers. Yet, significant progress in analytical modeling continues and the inventory of analytical models is impressive. Additional progress is indicated for the future particularly with respect to solute and low-temperature heat transport and transfer.

The purpose of this paper is to: (1) briefly inventory available analytical groundwater models and describe their usefulness and limitations; (2) identify needed additional progress in modeling; and (3) foster greater use of and appreciation for current models. Emphasis will be placed on models for unsteady-state conditions ivolving well hydraulics. No attempt has been made to cover steady-state models nor to be all inclusive in coverage of available models.

AVAILABLE ANALYTICAL MODELS

A large number of analytical models structured to solve partial differential equations governing the flow of groundwater to and from wells has been developed. These models describe flow conditions in uniformly porous aquifers and aquitards which are homogeneous, infinite in areal extent, and of the same thickness throughout. In some cases, isotrophic conditions are assumed to prevail and in other cases anisotrophy is taken into consideration. Isothermal conditions and no changes or differences in groundwater density and viscosity are assumed and production and injection wells have infinitesimal diameters and no storage capacity or finite diameters and storage capacity. Both fully and partially penetrating wells are considered. Except in the case of flowing wells, the dicharge or recharge of production or injection wells is assumed to be constant. The available analytical models simulating flow to and from wells listed in Table I describe unsteady-state time—drawdown and distance—drawdown in non-leaky, leaky and water-table aquifer systems.

The development of the analytical models discussed in the preceding section was predicated in part on the assumption of infinite areal extent of aquifer systems. These models may be extended to cover finite areal extent situations involving hydrogeologic boundaries (barrier and recharge) based on the image-well theory (Ferris, 1959). The effects of hydrogeologic boundaries on the drawdown in a well are modeled with hypothetical image wells. Hydrogeologic boundaries are replaced by imaginary wells which produce the same disturbing effects as the boundaries. Boundary problems are thereby simplified to consideration of an infinite aquifer in which real and image wells operate simultaneously.

Aquifers are often delimited by two or more boundaries. Two converging boundaries delimit a wedge-shaped aquifer; two parallel boundaries form an infinite-strip aquifer; two parallel boundaries intersected at right angles by a third boundary form a semi-infiite-strip aquifer; and four boundaries intersecting at right angles form a rectangular aquifer. The image-well theory is applied to such cases by taking into consideration successive reflections on the boundaries (Ferris et al., 1962).

A few analytical models involving finite single boundary and multi-boundary aquifer systems have been developed (e.g., see Stallman, 1963; Vandenberg, 1977).

Analytical models simulating land subsidence due to artesian pressure decline are available (Domenico and Mifflin, 1965; Poland and Davis, 1969). These models are based on the theory of one-dimensional consolidation of a linearly elastic soil (see Terzaghi and Peck, 1948), two-dimensional depth-

TABLE I

Available analytical models simulating flow from and to wells

Aquifer system	Isotropic conditions	Well diameter, storage capacity, penetration	Other assumptions	References
Non-leaky artesian	isotropic	infinitesimal diameter; no storage capacity; fully penetrating	uniformly porous aquifer	Theis (1935)
Non-leaky artesian	isotropic	finite diameter; no storage capacity; fully penetrating	uniformly porous aquifer	Hantush (1964)
Non-leaky artesian	isotropic	finite diameter; storage capacity; fully penetrating	uniformly porous aquifer	Papadopulos (1967)
Non-leaky artesian	isotropic	infinitesimal diameter; no storage capacity; fully penetrating	uniformly porous aquifer visco-elastic properties	Brutsaert and Corapcioglu (1976)
Non-leaky artesian	anisotropic	infinitesimal diameter; no storage capacity; partially penetrating	uniformly porous aquifer	Hantush (1964)
Non-leaky artesian	isotropic	infinitesimal diameter; no storage capacity; fully penetrating (flowing)	uniformly porous aquifer	Jacob and Lohman (1952)
Non-leaky artesian	anisotropic; porous blocks and fractures	infinitesimal diameter; no storage capacity; fully penetrating	fractured-rock aquifer	Boulton and Streltsova (1977b)
Non-leaky artesian	anisotropic; porous blocks and fractures	infinitesimal diameter; no storage capacity; production well cased through porous blocks	fractured-rock aquifer	Boulton and Streltsova (1977a)
Non-leaky artesian, partial conversion to water table	isotropic	infinitesimal diameter; no storage capacity; fully penetrating	uniformly porous aquifer	Moench and Prickett (1972)
Non-leaky artesian	anisotropic permeability varies in two horizontal directions	infinitesimal diameter; no storage capacity; fully penetrating	uniformly porous aquifer	Papadopulos (1965)

TABLE I (*continued*)

Leaky artesian	isotropic	infinitesimal diameter; no storage capacity; fully penetrating aquifer wells	negligible aquitard storage and source-bed drawdown changes, uniformly porous aquifer and aquitard	Hantush and Jacob (1955)
Leaky artesian	isotropic	infinitesimal diameter; no storage capacity; fully penetrating aquifer wells (flowing)	negligible aquitard storage and source-bed drawdown changes; uniformly porous aquifer and aquitard	Hantush (1959)
Leaky artesian	isotropic	finite diameter; storage capacity; fully penetrating aquifer wells	negligible aquitard storage and source-bed drawdown changes, uniformly porous aquifer and aquitard	Lai and Chen-Wu Su (1974)
Leaky artesian	isotropic	infinitesimal diameter; no storage capacity; partially penetrating aquitard wells	aquitard storage release; negligible source-bed drawdown changes; uniformly porous aquifer and aquitard	Witherspoon, et al. (1967)
Leaky artesian	isotropic	infinitesimal diameter; no storage capacity; fully penetrating aquifer wells	negligible aquitard storage and source-bed drawdown changes, uniformly porous aquifer and aquitard viscoelastic properties	Corapcioglu (1976)
Leaky artesian	anisotropic	infinitesimal diameter; no storage capacity; partially penetrating aquifer wells	negligible aquitard storage and source-bed drawdown changes, uniformly porous aquifer and aquitard	Hantush (1964)

TABLE I (continued)

Water table	isotropic	infinitesimal diameter; no storage capacity; fully penetrating	uniformly porous aquifer	Boulton (1963)
Water table	anisotropic	infinitesimal diameter; no storage capacity; fully penetrating	uniformly porous aquifer	Neuman (1975)
Water table	anisotropic	infinitesimal diameter; no storage capacity; partially penetrating	uniformly porous aquifer	Streltsova (1974)
Water table	anisotropic	finite diameter; storage capacity; partially penetrating	uniformly porous aquifer	Boulton and Streltsova (1976)
Water table	anisotropic; porous blocks and fractures	infinitesimal diameter; no storage capacity; production well cased through porous blocks	fractured rock aquifer	Boulton and Streltsova (1978)
Water table in aquitard	anisotropic	infinitesimal diameter; no storage capacity; partially penetrating	uniformly porous aquifer and aquitard	Streltsova (1976)

effective pressure head increase diagram areas, and effective pressure head increases which are time-dependent in aquitards and clay interbeds.

Analytical models simulating freshwater injection in a non-leaky artesian salaquifer with differences in density between in situ water and injected water are also available (Esmael and Kimbler, 1967). A model for simulating upconing of salt water below a production well is described by Schmorak and Mercado (1969).

In addition to analytical models simulating groundwater flow to and from wells several analytical models describing flow to and from streams have been developed. These models assume uniformly porous aquifers which are homogeneous, isotropic, finite in areal extent, and of the same thickness throughout. Isothermal conditions prevail and there are no changes or differences in groundwater density and viscosity. Streams and observation wells are assumed to fully penetrate aquifers. The available analytical models describing flow to and from streams listed in Table II simulate changes in groundwater levels caused by stream stage changes, groundwater level declines due to uniform stream discharge, groundwater level changes and stream discharge changes caused by uniform and sudden increments of recharge from precipitation on stream drainage basins, groundwater contribution to streamflow, and bank storage in non-leaky artesian and water-table aquifer systems.

TABLE II

Available analytical models simulating flow from and to streams

Aquifer system	Stream and observation well conditions	Stream stage, discharge or recharge conditions	Boundary conditions	Reference
Non-leaky artesian	stream and observation well fully penetrating	sinusoidal stream stage changes	finite aquifer	Ferris (1951)
Non-leaky artesian	stream and observation well fully penetrating	stream discharging at uniform rate	finite aquifer	Ferris et al. (1962)
Non-leaky artesian	stream and observation well fully penetrating	stream stage suddenly changed	finite aquifer	Ferris et al. (1962)
Water table	stream and observation well fully penetrating	sinusoidal stream stage changes	finite-strip aquifer	Cooper and Rorabaugh (1963)
Water table	two parallel streams and observation well fully penetrating	sudden recharge and subsequent decline in water levels	finite aquifer	Rorabaugh (1960)
Water table	stream and observation well fully penetrating	uniform increment of recharge suddenly added on basin	finite aquifer	Rorabaugh (1964)

Analytical models are available for estimating the rate and volume of stream depletion by nearby production wells (Theis, 1941; Jenkins, 1968) and appraising the cone of depression created by a production well near a stream (Rorabaugh, 1956; Theis, 1963).

Analytical models structured to solve partial differential equations governing solute transport in groundwater have been developed. These models simulate convection and dispersion from a conservative solute-injection well in uniformly porous, non-leaky artesian aquifers which are homogeneous isotropic infinite in areal extent, and the same thickness throughout. Steady-state groundwater flow with or without regional flow components and isothermal conditions are assumed to prevail. Dispersion in one direction is considered dominant. The density and viscosity of the injected solute are assumed to be the same as those of the native groundwater. Solutes of constant concentration are introduced into aquifer at a constant rate through a solute-injection well of infinitesimal diameter with no storage capacity. The discharge of production wells is assumed to be constant. The available analytical models listed in Table III simulate conservative solute transport to and from wells.

TABLE III

Available analytical models simulating conservative solute transport to and from wells

Injection and production well conditions	Groundwater flow conditions	Dominant dispersion	Reference
Single solute-injection well	no regional flow	longitudinal	Hoopes and Harleman (1965)
Single solute-injection well	regional flow	transverse	Lenau (1972)
Single solute-injection well; single production well	no regional flow	longitudinal	Hoopes and Harleman (1967)
Single solute-injection well; single production well	regional flow	transverse	Lenau (1972)

TABLE IV

Available analytical models simulating heat transport from an injection well

Injection and production well conditions	Conduction and convection conditions	Reference
Heated-water injection well	heat convection in aquifer; heat conduction in aquiclude	Laurwerier (1955) Spillette (1972)
Heated-water injection well; production well	heat convection in aquifer; heat conduction in aquiclude	Gringarten and Sauty (1975)
Heat-injection well	heat conduction in aquifer	Carslaw and Jaeger (1959)

Analytical models structured to solve partial differential equations governing heat transport in groundwater have been developed. These models simulate convection without dispersion from a heated water-injection well in uniformly porous non-leaky artesian aquifers which are homogeneous isotropic infinite in areal extent and of the same thickness throughout. Heat conduction in the aquiclude and heat convection in the aquifer are assumed to dominate heat flow conditions. Steady-state groundwater flow without regional flow components is assumed to prevail. The density and viscosity of the injected heated water are assumed to be the same as those of the native groundwater. Heated water of constant temperature is introduced into the aquifer through the heated water-injection well at a constant rate. Wells have infinitesimal diameters and no storage capacity. The available analytical models listed in Table IV simulate transport from an injection well.

USEFULNESS AND LIMITATIONS OF MODELS

Analytical models continue to be most useful in the design and calibration of numerical digital computer models. The general validity of computer modeling results is often verified with analytical models thereby enhancing their credibility.

Most aquifer system parameter evaluation for computer models is accomplished by analysis of aquifer-test data with analytical models. This situation is not likely to change appreciably in the near future even though inverse computer model methods for analysis of aquifer-test data are advancing at a rapid pace.

Analytical models are applied chiefly to relatively uniform aquifers with simple geometry and are used primarily to solve problems involving only parts of aquifer systems or aquifer systems of small areal extent. Computer models are more realistic and adaptable than analytical models. However, the realism and versatility of computer models depend largely upon the availability of abundant and accurate basic data. In some cases basic data are not sufficient to warrant a rigorous description of complex aquifer system conditions and analytical models may be as useful as computer models.

Recognized departures from ideal conditions do not necessarily dictate that analytical models be rarely used. Such departures emphasize the need for sound professional judgment in the application of analytical models to existing hydrogeologic conditions and in properly qualifying results according to the extent of departures. With appropriate recognition of hydrogeologic controls, there are many practical ways of circumventing analytical difficulties posed by complicated field conditions. Many aquifer systems can be highly idealized with little sacrifice in accuracy of analysis (Walton, 1970).

In applying analytical models to field problems, the hydrogeologic boundaries of the aquifers evident from areal studies must be idealized to fit comparatively elementary geometric forms such as wedges and infinite or semi-infinite rectilinear strips. Boundaries ares assumed to be straight-line demarcations. The gross hydrogeologic properties of the aquifer and aquitards, if present, are considered in evaluating the effects of boundaries, and the detailed hydrogeologic property variations are considered in estimating the interference between wells and well fields.

Actual groundwater conditions are simulated with models which have straight-line boundaries and an effective width, length and thickness. The aquifer is sometimes overlain by an aquitard which has an effective thickness. The areal extent of an aquifer often varies laterally or longitudinally, and the construction of models requires the complex aquifer be converted to an equivalent uniform area.

NEEDED PROGRESS

Although the current inventory of analytical models is impressive and there is continued developments in modeling, accelerated progress is called for if voids in models are to be filled in a timely fashion. Based on a careful study of information presented in Tables I—IV and a survey of existing and projected groundwater problems, analytical models for at least the following aquifer systems and discharge and injection conditions need to be developed as soon as possible:

(1) Leaky artesian isotropic uniformly-porous aquifer system with aquitard storage release and negligible source-bed drawdown changes. Finite-diameter production well with storage capacity. Fully penetrating wells.

(2) Leaky artesian anisotropic (permeability varies in two horizontal directions) uniformly-porous aquifer system with aquitard storage release and negligible source-bed drawdown changes. Infinitesimal-diameter production well with no storage capacity. Fully penetrating wells.

(3) Leaky artesian anisotropic (porous blocks and fractures) fractured-rock aquifer system with aquitard storage release and negligible source-bed drawdown changes. Infinitesimal- and finite-diameter production well with and without storage capacity. Fully penetrating wells.

(4) Water-table anisotropic (porous blocks and fractures) fractured-rock aquifer. Finite-diameter production well with storage capacity. Fully penetrating wells.

(5) Leaky artesian aquifer systems with single conservative and non-conservative solute-injection well and without and with production well. No regional flow and with regional flow. Longitudinal or transverse dispersion dominates in aquifer.

(6) Leaky artesian aquifer system with single heated water-injection well and without and with production well. No regional flow and with regional flow. Heat connection dominates in aquifer and aquitard. Longitudinal or transverse dispersion dominates in aquifer.

Additional attention should be directed towards development of analytical models simulating finite aquifer systems. Closed aquifer system analytical models involving shape factors, such as those formulated by petroleum engineers (Earlougher, 1977) but considering both barrier and recharge boundaries, should be developed. Current analytical model theory should be extended to consideration of the effects of changes and differences in density and viscosity. Analytical models should be developed which simulate solute and heat transport from streams to nearby production wells. Finally, application of analytical models to practical field problems is lagging theoretical advances and should be accelerated.

REFERENCES

Boulton, N.S., 1963. Analysis of data from nonequilibrium pumping tests allowing for delayed yield from storage. Proc. Inst. Civ. Eng., 26 (6693): 469—482.

Boulton, N.S. and Streltsova, T.D., 1976. The drawdown near an abstraction well of large diameter under non-steady conditions in an unconfined aquifer. J. Hydrol., 30: 29—46.

Boulton, N.S. and Streltsova, T.D., 1977a. Unsteady flow to a pumped well in a fissured water-bearing formation. J. Hydrol., 35: 257—269.

Boulton, N.S. and Streltsova, T.D., 1977b. Unsteady flow to a pumped well in a two-layered water-bearing formation. J. Hydrol., 35: 245—256.

Boulton, N.S. and Streltsova, T.D., 1978. Unsteady flow to a pumped well in a fissured aquifer with a free surface level maintained constant. Water Resour. Res., 14 (3): 527—532.

Brutsaert, W. and Corapcioglu, M.Y., 1976. Pumping of aquifer with visco-elastic properties. Proc. Am. Soc. Civ. Eng., J. Hydraul. Div., 102, (HY11): 1663—1675.

Carslaw, H.S. and Jaeger, J.C., 1959. Conduction of Heat in Solids. Oxford University Press, New York, N.Y., pp. 261—262.

Cooper, H.H. and Rorabaugh, M.I., 1963. Groundwater movements and bank storage due to flood stages in surface streams. U.S. Geol. Surv. Water-Supply Pap., 1536-S: 343—366.

Corapcioglu, M.Y., 1976. Mathematical modeling of leaky aquifers with rheological properties. Proc. of Anaheim Symposium on Subsidence, Int. Assoc. Hydrol. Sci. Publ. No. 121, pp. 191—200.

Domenico, P.A. and Mifflin, M.D., 1965. Water from low permeability sediments and land subsidence. Water Resour. Res. 1 (4): 563—576.

Earlougher, R.C., 1977. Advances in well test analysis. Soc. Pit. Eng. A.I.M.E., Monogr., 5: 197—203.

Esmael, O.J. and Kimbler, O.K., 1967. Investigation of the technical feasibility of storing fresh water in saline aquifers. Water Resour. Res., 3 (3): 683—695.

Ferris, J.G., 1951. Cyclic fluctuations of water levels as a basis for determining aquifer transmissibility. Int. Assoc. Sci. Hydrol. Gen. Assem. Brussels, Publ. 33, 2: 148—155.

Ferris, J.G., 1959. Ground water. In: C.O. Wisler and E.F. Brater (Editors), Hydrology, Ch. 7, Wiley, New York, N.Y.

Ferris, J.G., Knowles, D.B., Brown, R.H. and Stallman, R.W., 1962. Theory of aquifer tests. U.S. Geol. Surv., Water-Supply Pap., 1536-E: 151—161.

Gringarten, A.C. and Sauty, J.P., 1975. A theoretical study of heat extraction from aquifers with uniform regional flow. J. Geophys. Res., 80 (35): 4956—4962.

Hantush, M.S., 1959. Non-steady flow to flowing wells in leaky aquifers. J. Geophys. Res., 64 (8): 1043—1052.

Hantush, M.S., 1964. Hydraulics of wells. In: V.T. Chow (Editor), Advances in Hydroscience. Academic Press, New York N.Y., 281—442.

Hantush, M.S. and Jacob, C.E., 1955. Non-steady radial flow in an infinite leaky aquifer and nonsteady Green's functions for an infinite strip of leaky aquifer. Trans. Am. Geophys. Union., 36 (1): 95—112.

Hoopes, J.A. and Harleman, D.R.F., 1965. Waste water recharge and dispersion in porous media. Mass. Inst. Technol., Hydrodyn. Lab., Rep. No. 75, 166 pp.

Hoopes, J.A. and Harleman, D.R.F., 1967. Waste water recharge and dispersion in porous media. Am. Soc. Civ. Eng., Hydraul. Div. Proc. Pap. No. 5425, 93 (HY 5): 51—71.

Jacob, C.E. and Lohman, S.W., 1952. Nonsteady flow to a well of a constant drawdown in an extensive aquifer. Trans. Am. Geophys. Union., 33 (4): 559—569.

Jenkins, C.T., 1968. Techniques for computing rate and volume of stream depletion by wells. Ground Water, 6 (2): 37—46.

Lai, R.Y.S. and Cheh-Wu Su, 1974. Nonsteady flow to a large well in a leaky aquifer. J. Hydrol., 22: 333—345.

Laurwerier, H.A., 1955. The transport of heat in an oil layer caused by the injection of hot fluid. Appl. Sci. Res., Sect. A., 5: 145.

Lenau, C.W., 1972. Dispersion from recharge well. Am. Soc. Civ. Eng., Eng. Mech. Div. Proc. Pap., 98: (EM 2): 331—344.

Lenau, C.W., 1973. Contamination of discharge well from recharge well. Am. Soc. Civ. Eng., Hydraul. Div. Proc. Pap. No. 9958, 99 (HY 8): 1247—1263.

Moench, A.F. and Prickett, T.A., 1972. Radial flow in an infinite aquifer undergoing conversion from artesian to water table conditions. Water Resour. Res., 8 (2): 494—499.

Neuman, S.P., 1975. Analysis of pumping test data from anisotropic unconfined aquifers considering delayed gravity response. Water Resour. Res., 11 (2): 329—342.

Papadopulos, I.S., 1965. Nonsteady flow to a well in an infinite anisotropic aquifer. Symp. Int. Assoc. Sci. Hydrol., Dubrovnik.

Papadopulos, I.S., 1967. Drawdown distribution around a large-diameter well. Proc. Symp. on Ground Water Symp., Am. Water Resour. Assoc., pp. 157—167.

Poland, J.F. and Davis, G.H., 1969. Land subsidence due to withdrawal of fluids. Rev. Eng. Geol., 2: 187—269.

Rorabaugh, M.I., 1956. Groundwater resources of the northeastern part of the Louisville area, Kentucky. U.S. Geol. Surv., Water-Supply Pap., 1360—13: 101—169.

Rorabaugh, M.I., 1960. Use of water levels in estimating aquifer constants. Int. Assoc. Sci. Hydrol., Publ. No. 52, pp. 314—323.

Rorabaugh, M.I., 1964. Estimating changes in bank storage and groundwater contribution to streamflow. Int. Assoc. Sci. Hydrol., Publ. No. 63, pp. 432—441.

Schmorak, S. and Mercado, A., 1969. Upcoming of fresh water—sea water interface below pumping wells, field study. Water Resour. Res., 5 (6): 1290—1310.

Spillette, A.G., 1972. Heat transfer duirng hot fluid injection into an oil reservoir. In: Thermal Recovery Tehcniques, Soc. Pet. Eng. of A.I.M.E., Reprint Ser. No. 10. pp. 21—26.

Stallman, R.W., 1963. Type curves for the solution of single-boundary problems. In: Shortcuts and Special Problems in Aquifer Tests. U.S. Geol. Surv., Water-Supply Pap., 1545-C: 45—47.

Streltsova, T.D., 1974. Drawdown in Compressible unconfined aquifer. Proc. Am. Soc. Civ. Eng., J. Hydraul. Div., 100 (HY 11): 1601—1616.

Streltsova, T.D., 1976. Analysis of aquifer—aquitard flow. Water Resour. Res., 12 (3): 415—422.

Terzaghi, K. and Peck, R.B., 1948. Soil Mechanics in Engineering Practice. Wiley, New York, N.Y., pp. 64—64.

Theis, C.V., 1935. The relation between the lowering of piezometric surface and the rate and duration of discharge of a well using ground-water storage. Trans. Am. Geophys. Union., 16th Annu. Meet., Part 2, pp. 519—524.

Theis, C.V., 1941. The effect of a well on the flow of a nearby stream. Trans. Am. Geophys. Union, Part 3, pp. 734—738.

Theis, C.V., 1963. Drawdowns caused by a well discharging under equilibrium conditions from an aquifer bounded on a finite straight-line source. In: Shortcuts and Special Problems in Aquifer Tests. U.S. Geol. Surv., Water-Supply Pap., 1545-C: 10—15.

Vandenberg, A., 1977. Type curves for analysis of pump tests in leaky strip aquifers. J. Hydrol., 33 (1/2): 15—26.

Walton, W.C., 1970. Groundwater Resource Evaluation. McGraw-Hill, New York, N.Y., 644 pp.

Witherspoon, P.A., Javandel, I., Neuman, S.P. and Freeze, R.A., 1967. Interpretation of aquifer gas storage conditions from water pumping tests. Am. Gas Assoc., New York, N.Y., 273 pp.

[2]

CONSIDERATION OF TOTAL ENERGY LOSS IN THEORY OF FLOW TO WELLS

R.L. COOLEY*[1] and A.B. CUNNINGHAM*[2]

Water Resources Center, Desert Research Institute, University of Nevada System, Reno, NV 89506 (U.S.A.)

(Accepted for publication May 28, 1979)

ABSTRACT

Cooley, R.L. and Cunningham, A.B., 1979. Consideration of total energy loss in theory of flow to wells. In: W. Back and D.A. Stephenson (Guest-Editors), Contemporary Hydrogeology — The George Burke Maxey Memorial Volume. J. Hydrol., 43: 161—184.

A coupled numerical solution was developed for unsteady flow in single or multiple confined or semiconfined aquifers and in the well penetrating the system. Analysis of hypothetical problems indicates that, because of friction losses and non-uniform flow in the well bore, a significant region of non-radial flow in the aquifer(s) results whenever aquifer hydraulic conductivity is greater than about 0.015 m/min. and pumping rate is greater than about 1.2 m³/min. Because of this non-radial flow, use of standard aquifer testing equations can lead to significant errors in computed aquifer transmissivity whenever aquifer hydraulic conductivity is greater than about 0.03 m/min. Because of unsteady flow in the well bore, flow may be highly non-radial for the first few seconds of pumping irrespective of hydraulic conductivity or pumping rate. Percentage of open area in the screened or perforated section of well bore, Fanning friction factor, and contraction coefficients characterizing flow through the perforations or slots are parameters that could cause large head losses in the well in some cases. Expressions for parameters B and C in the total drawdown equation $s_w = BQ + CQ^2$ (where s_w is drawdown in the well and Q is pumping rate) were developed dimensionally from expressions used in this study. The development shows that both B and C could theoretically be dependent on Q.

INTRODUCTION

The theory of flow to wells has been expounded upon by numerous authors [see Hantush (1964), Walton (1970) and Lohman (1972) for reviews of the most often used results]. All of the solutions in the references cited above have used as boundary conditions at the well bore the concepts that: (1) the integral of the specific discharges along the well screen equals the total discharge from the well; and (2) the head along the well screen is constant. In

---

Present addresses:
*[1] U.S. Geological Survey, Denver, CO 80225, U.S.A.
*[2] Montana State University, Bozeman, MT 59715, U.S.A.

addition, most of the solutions have assumed radial flow in the aquifer. Implicit in the assumed boundary conditions are the ideas that: (1) all of the head (or energy) loss occurs within the porous medium; and (2) the head in the well is hydrostatic. In reality energy is lost as the water flows through the screen and up the well bore. This idea was considered by Jacob (1947) who postulated that total drawdown in the well is composed of the sum of the theoretical drawdown accompanying flow in the aquifer and a component, called well loss, resulting from flow through the screen and zone near the well bore. This latter component was assumed by Jacob (1947) to be proportional to the square of the discharge. Later, Rorabaugh (1953) extended the idea to include any power of the discharge. However, both of these formulations are empirical and the validity of methods of determining parameters contained in the formulations has been questioned (Mogg, 1968).

Recently, several authors (Peterson et al., 1955; Soliman, 1965; Garg and Lal, 1971; Besbes, 1974) have analyzed the general well loss phenomenon using methods different from those of Jacob (1947) or Rorabaugh (1953). Specifically, Garg and Lal (1971) performed a simplified analysis of steady-state flow to a well in a horizontal non-leaky artesian aquifer where the problem was considered to be composed of three parts: (1) flow in the aquifer; (2) flow through the screen; and (3) flow in the well bore. Energy losses in all three segments of the total flow regime were considered and a combined theoretical solution was obtained. For the purposes of the present paper the most important conclusions of the study by Garg and Lal are: (1) good agreement between theory and experimental results was obtained; and (2) water is not discharged uniformly through the entire length of screen, but instead tends to concentrate near the pump intake.

Based on the above considerations, a description of unsteady flow to wells has been formulated herein to take into account energy loss in the flow system around and within a well. This investigation includes: (1) development of a mathematical statement of the combined energy losses in a well, through the well screen, and in the surrounding porous medium; and (2) solution of selected test cases to allow preliminary assessment of the importance of energy losses in the well in terms of their effect on the flow system around the well. Cases examined herein include unsteady flow due to constant withdrawal from a well under a variety of system conditions.

THEORY

For the purposes of this study the porous medium is idealized as a system of horizontal, uniformly thick aquifers and aquitards (Fig. 1). A well penetrates this system, and screen or slotted casing is set opposite all or parts of the aquifers. The well may or may not be gravel-packed, and drilling mud invasion may or may not be significant.

Well design is simplified as illustrated in Fig. 1. The pump intake is assumed to lie above the uppermost screen so that average flow in the well bore is al-

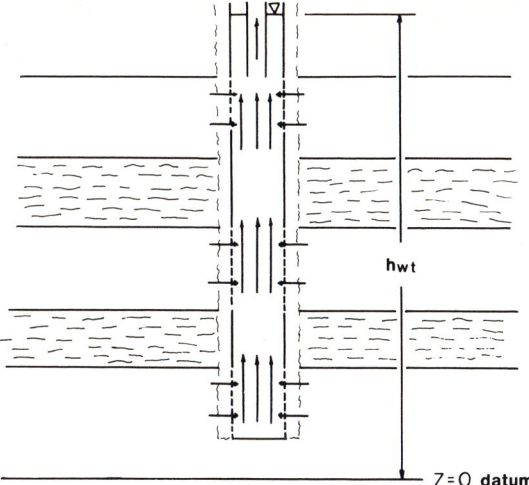

Fig. 1. Simplified well design for a general multi-aquifer system showing screened zones, an altered zone around the well, and the position of the pump intake and water level h_{wt} in the well. Units designated by the *dashed pattern* are aquitards and units not designated by a pattern are aquifers.

ways upward. Because head losses occurring after water enters the pump intake do not affect flow in the well bore or the porous medium, they are not considered in the analysis. It is also assumed that the water level in the well always lies above the uppermost screen, that storage changes in the well due to the changing level of water in the well are negligible, and that water and all well components (pump, etc.) can be considered to be incompressible for the development of the equations for flow in the well bore. The latter two assumptions will, in general, limit the validity of quantitative interpretation of the analysis for early times near the pumping well, and, hence, the numerical experiments presented in this report demonstrate the quantitative effects of energy losses whenever other factors, such as the latter two, are not considered.

All of the conditions enumerated above are contained in the following sections that discuss equations.

Equations for flow in well bore

Equations approximately describing conservation of mass and momentum in the well bore may be derived through a process of integration of general point forms of these equations over a control volume equal to the cross-sectional area of the well bore times differential length of well bore $\delta x_3 \equiv \delta z$. Because water in the well bore is assumed to be incompressible, the point forms of the equations can be written (Mase, 1970, pp. 160–162):

$$\partial u_i / \partial x_i = 0 \qquad (1)$$

$$-\partial(p+\gamma z)/\partial x_i + \partial \sigma_{ji}/\partial x_j = \rho(\partial u_i/\partial t) + \rho u_j(\partial u_i/\partial x_j) \tag{2}$$

where repetition of subscript in a term indicates summation from 1 to 3 on that subscript. Eq. 1 describes conservation of mass, and eq. 2 describes conservation of momentum. The symbol u_i stands for velocity in the x_i Cartesian coordinate direction; p is pressure; γ is specific weight; z is elevation above an arbitrary datum; σ_{ji} is the viscous stress tensor; ρ is mass density; and t is time.

Volume integration of eq. 1 can be performed rigorously or by inspection. Because water in the well bore is assumed to be incompressible, it is easily seen that mass continuity is expressed by the equation:

$$\partial Q/\partial z - q = 0 \tag{3}$$

where $z \equiv x_3$ is the vertical coordinate direction (positive upward), Q is the volumetric discharge given by:

$$Q = \iint_A u_3 \, dx_3 \equiv \iint_A u_z \, dz \tag{4}$$

and q is the lateral inflow per unit length of well bore. In eq. 4, A is the cross-sectional area of well bore. By defining the average velocity as $V = Q/A$, eq. 3 can be written as:

$$A(\partial V/\partial z) - q = 0 \tag{5}$$

Integration of eq. 2 is accomplished in the Appendix. The result is:

$$-\gamma A(\partial h_w/\partial z) - 2\pi r_w \sigma_0 = \rho(\partial Q/\partial t) + \rho[\partial(QV)/\partial z] - \rho v_z q \tag{6}$$

or, using the average velocity, V:

$$\frac{\partial h_w}{\partial z} + \frac{2\sigma_0}{\gamma r_w} + \frac{1}{g}\frac{\partial V}{\partial t} + \frac{2V}{g}\frac{\partial V}{\partial z} - \frac{v_z q}{gA} = 0 \tag{7}$$

where h_w is the mean hydraulic head defined as:

$$h_w = \iint_A (p/\gamma + z) \, dS/A \tag{8}$$

r_w is radius of the well bore, σ_0 is the mean shear stress on the well casing defined by:

$$\sigma_0 = -\iint_{S_3} (\sigma_{1z} n_1 + \sigma_{2z} n_2) \, dS/S_3 \tag{9}$$

v_z is the approximate average vertical component of the inflow velocity, and g is the acceleration due to gravity. In eq. 9, $S_3 = 2\pi r_w \delta z$ is the area of well bore bounding the control volume.

At this point we follow Garg and Lal (1971) and make the following assumptions:

(1) Water enters the well bore through slots or screen openings at right angles to the casing so that $v_z = 0$.

(2) Inflow velocity, u, is given by the orifice law:

$$u = C_v[2g(h-h_w)]^{1/2} \qquad (10)$$

where C_v is the contraction coefficient and h is the hydraulic head in the aquifer just outside of the well bore.

(3) Specific discharge, u_w, corresponding to the inflow velocity can be written:

$$u_w = C_c A_s u = q/2\pi r_w \qquad (11)$$

where C_c is another contraction coefficient and A_s is fraction of well casing (or screen) through which water may flow. The term A_s can be approximated as the product of porosity of the porous medium at the well bore and fraction of well casing (or screen) that is open for water flow.

(4) Mean shear stress may be written in terms of the Fanning friction factor, f, as:

$$\sigma_0 = \frac{f}{4} \gamma \frac{V^2}{2g} \qquad (12)$$

The set of equations to solve for flow in the well bore results from combinations of eqs. 5 and 11, eqs. 7 and 12, and eqs. 10 and 11, which yield, respectively:

$$\partial V/\partial z = 2u_w/r_w \qquad (13)$$

$$\frac{\partial h_w}{\partial z} + \frac{fV^2}{4gr_w} + \frac{1}{g}\frac{\partial V}{\partial t} + \frac{2V}{g}\frac{\partial V}{\partial z} = 0 \qquad (14)$$

$$u_w = C_v C_c A_s [2g(h-h_w)]^{1/2} \qquad (15)$$

Equation for groundwater flow

The equation for conservation of mass referred to coordinates affixed to a porous medium which deforms slightly in response to fluid pressure changes is approximated by (Jacob, 1950, p. 331):

$$-\partial v_i/\partial x_i = S_s(\partial h/\partial t) \qquad (16)$$

where v_i is specific discharge in the x_i direction and S_s is specific storage. Conservation of momentum is approximated by the generalized Darcy's law:

$$v_i = -K_{ij}(\partial h/\partial x_j) \qquad (17)$$

where K_{ij} is the hydraulic conductivity tensor. For a sectionally homogeneous porous medium substitution of eq. 17 into eq. 16 followed by transformation to axisymmetric radial coordinates yields:

$$K_{rr} \frac{1}{r} \frac{\partial}{\partial r}\left(r \frac{\partial h}{\partial r}\right) + K_{zz} \frac{\partial^2 h}{\partial z^2} = S_s \frac{\partial h}{\partial t} \qquad (18)$$

Eq. 18 is assumed to govern groundwater motion for this study. Thus, non-linear flow and influences of inertia, which are omitted from the momentum equation, are neglected.

Boundary and initial conditions

The boundary condition at the well bore is given by eqs. 13—15 coupled with Darcy's law:

$$u_w = -K_{rr}(\partial h/\partial r) \qquad (19)$$

Top and bottom of the aquifer—aquitard system are considered to be horizontal and impermeable so that the boundary condition is $\partial h/\partial z = 0$. The boundary at the outer radius is assumed to be vertical and of constant head, h_e. Finally, the initial head distribution is assumed to be constant at h_e.

In the next section subdomain finite-element solutions are developed for eqs. 13—15, 18 and 19; the solution algorithm is given; and a stability problem is considered. If desired, this section could be skipped with little loss of information pertaining to results of the test problems.

SOLUTION OF EQUATIONS

Direct finite-element solution

Herein, direct (or subdomain) finite-element discretizations of eqs. 13 and 14 are derived to conform to the general direct finite-element solutions for eq. 18 given by Cooley (1974).

The direct finite-element method is explained by Norrie and de Vries (1973, pp. 108—110). Thus, in the present report it is only summarized as applied to flow in the well bore. First, the well bore is divided vertically into sections (subdomains) bounded by node points (Fig. 2). Then, within each subdomain the unknown and known variables are approximated by suitable functions of z. Finally, these functions are substituted into the governing equation which is integrated over each subdomain to produce a set of algebraic equations. These can then be solved using standard procedures.

Linear functions were chosen to describe velocity V in each subdomain. Hence, for the subdomain bounded by nodes k and $k + 1$, V is given by:

$$V = N_k V_k + N_{k+1} V_{k+1} \qquad (20)$$

where

$N_k = (z_{k+1} - z)/(z_{k+1} - z_k)$ and $N_{k+1} = (z - z_k)/(z_{k+1} - z_k)$

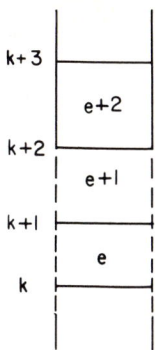

Fig. 2. Section of well bore divided into subdomains (indicated by e, $e + 1$, etc.) bounded by nodes (indicated by k, $k + 1$, etc.).

and the summation convention has been suspended here and henceforth. For application to eq. 13 u_w is assumed to vary linearly in each element also so that we may write for a general subdomain:

$$\int_{z_k}^{z_{k+1}} \frac{\partial}{\partial z}(N_k V_k + N_{k+1} V_{k+1}) \, dz = \frac{2}{r_w} \int_{z_k}^{z_{k+1}} (N_k u_{wk} + N_{k+1} u_{wk+1}) \, dz \quad (21)$$

or:

$$V_{k+1} - V_k = \frac{z_{k+1} - z_k}{r_w}(u_{wk} + u_{wk+1}) \quad (22)$$

To solve eq. 14 h_w is considered to vary linearly over each subdomain, whereas f is constant. Hence, the solution of eq. 14 can be written as:

$$\int_{z_k}^{z_{k+1}} \left[\frac{\partial}{\partial z}(N_k h_{wk} + N_{k+1} h_{wk+1}) + \frac{f}{4gr_w}(N_k V_k + N_{k+1} V_{k+1})^2 + \right.$$

$$\frac{1}{g}\left(N_k \frac{\partial V_k}{\partial t} + N_{k+1} \frac{\partial V_{k+1}}{\partial t}\right) + \frac{2}{g}(N_k V_k + N_{k+1} V_{k+1}) \times$$

$$\left. \frac{\partial}{\partial z}(N_k V_k + N_{k+1} V_{k+1}) \right] dz = 0 \quad (23)$$

or

$$h_{wk+1} - h_{wk} + \frac{f(z_{k+1} - z_k)}{12gr_w}(V_k^2 + V_k V_{k+1} + V_{k+1}^2) +$$

$$\frac{z_{k+1} - z_k}{2g}\left(\frac{\partial V_k}{\partial t} + \frac{\partial V_{k+1}}{\partial t}\right) + \frac{1}{g}(V_k + V_{k+1})(V_{k+1} - V_k) = 0 \quad (24)$$

The time derivatives in eq. 24 are most easily handled by writing them in finite-difference form and evaluating the variables in the remainder of the equation at a weighted mean in the time step. If the weighted mean is defined for V_k for example by:

$$\overline{V}_k = \theta[V_k(t+\Delta t) - V_k(t)] + V_k(t) \tag{25}$$

where $0 \leq \theta \leq 1$, then the finite-difference quotient can be given as:

$$\partial V_k/\partial t \simeq [V_k(t+\Delta t) - V_k(t)]/\Delta t = [\overline{V}_k - V_k(t)]/\theta\Delta t \tag{26}$$

Eq. 24 then becomes approximately:

$$\overline{h}_{wk+1} - \overline{h}_{wk} + \frac{f(z_{k+1} - z_k)}{12 g r_w}(\overline{V}_k^2 + \overline{V}_k\overline{V}_{k+1} + \overline{V}_{k+1}^2) +$$

$$\frac{z_{k+1} - z_k}{2g\theta\Delta t}[\overline{V}_k - V_k(t) + \overline{V}_{k+1} - V_{k+1}(t)] +$$

$$\frac{1}{g}(\overline{V}_k + \overline{V}_{k+1})(\overline{V}_{k+1} - \overline{V}_k) = 0 \tag{27}$$

where $\overline{h}_{wk}$ is defined analogously to $\overline{V}_k$.

Using a process similar to that leading to eqs. 22 and 27, Cooley (1974, pp. 13—15, 20—22) derived a direct finite-element solution to eq. 18 of the form:

$$(G_1 + G_8)_{i,j}(\overline{h}_{i+1,j} - \overline{h}_{i,j}) - (G_4 + G_5)_{i,j}(\overline{h}_{i,j} - \overline{h}_{i-1,j})$$

$$+ (G_2 + G_3)_{i,j}(\overline{h}_{i,j+1} - \overline{h}_{i,j}) - (G_6 + G_7)_{i,j}(\overline{h}_{i,j} - \overline{h}_{i,j-1})$$

$$= (\sum_e \overline{r}_e C_e)_{i,j}[\overline{h}_{i,j} - h_{i,j}(t)] \tag{28}$$

where:

$G_1 = K_{rr1} r_{i+1/2} (z_{j+1/2} - z_j)/(r_{i+1} - r_i)$

$G_2 = K_{zz2}[\frac{1}{2}(r_{i+1/2} + r_i)](r_{i+1/2} - r_i)/(z_{j+1} - z_j)$

$G_3 = K_{zz3}[\frac{1}{2}(r_{i-1/2} + r_i)](r_i - r_{i-1/2})/(z_{j+1} - z_j)$

$G_4 = K_{rr4} r_{i-1/2} (z_{j+1/2} - z_j)/(r_i - r_{i-1})$

$G_5 = K_{rr5} r_{i-1/2} (z_j - z_{j-1/2})/(r_i - r_{i-1})$

$G_6 = K_{zz6}[\frac{1}{2}(r_{i-1/2} + r_i)](r_i - r_{i-1/2})/(z_j - z_{j-1})$

$G_7 = K_{zz7}[\frac{1}{2}(r_{i+1/2} + r_i)](r_{i+1/2} - r_i)/(z_j - z_{j-1})$

$G_8 = K_{rr8} r_{i+1/2} (z_j - z_{j-1/2})/(r_{i+1} - r_i), \quad C_e = S_{se} A_e/(\theta\Delta t)$

$\bar{r}_e$ is the centroidal radius, A_e is the area of the part of element e contributing to the equation for node (i,j) (see Fig. 3), $r_{i\pm1/2} = (r_i + r_{i\pm1})/2$, $z_{j\pm1/2} = (z_j + z_{j\pm1})/2$ and $\bar{h}_{i,j}$, etc., are defined analogously to $\bar{V}_k$.

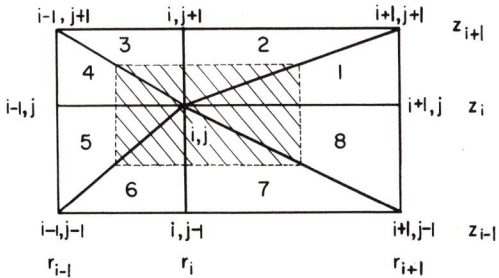

Fig. 3. Part of finite-element mesh centered on node (i,j). The shaded area is the sum of subdomains from elements $1-8$ assumed to contribute to node (i,j). The area of each of these subdomains is A_e.

Special consideration must be given to one of the principal boundary conditions: that discharge from the well is constant. From inspection of eq. 27 it can be seen that, if discharge is specified as constant, then as $\Delta t \to 0$ the approximation for $\partial V/\partial t \to \infty$ on the first time step because V is required to accelerate from 0 to a value sufficient to yield the discharge during this time period. This result is unrealistic because water, the well casing, and all components in the well are compressible and because acceleration of the pump from non-operation to full operation must occur in a finite time interval. Compressibility of the system was ignored in the present study since its effects would probably be significant for only a small length of time and since it would add significant complexity to the solution procedure. Thus, in order to minimize the unrealistic shock due to starting the pump, for most test problems the discharge was allowed to accelerate linearly from 0 to the constant specified value over the first second of time.

Finally, it should be noted that vertical distances to nodes and the values of variables at these nodes are indicated by the index k in equations pertaining to flow in the well bore and by the index j in eq. 28. This signifies that the two node numbering schemes may be different if necessary to produce the most efficient solution scheme. Coupling of the two sets is explained in the next section.

Algorithm for solution of coupled equation system

Solutions of eqs. 15, 22, 27 and 28 together with initial conditions and boundary conditions proceed iteratively at each time step as follows:

(1) An initial $\bar{u}_{wk}$ distribution is calculated. This is obtained by solving

eq. 28 for specific discharge across the well bore; i.e.:

$$\bar{u}_{wj} = [(G_1 + G_8)_{1,j}(\bar{h}_{2,j} - \bar{h}_{1,j}) + (G_2 + G_3)_{1,j}(\bar{h}_{1,j+1} - \bar{h}_{1,j})$$
$$- (G_6 + G_7)_{1,j}(\bar{h}_{1,j} - \bar{h}_{1,j-1}) - (\sum_e \bar{r}_e\, C_e)_{1,j}\{\bar{h}_{1,j} - h_{1,j}(t)\}]/$$

$$[2\pi r_1(z_{j+1/2} - z_{j-1/2})] \qquad (29)$$

For the first iteration at time step 1 trial values of $\bar{h}_{1,j}$ (the heads at the well bore) for use in eq. 29 are obtained from:

$$\bar{h}_{1,j}^{trial} = h_e - Q(r_2 - r_1)/2\pi r_1 T_s \qquad (30)$$

where Q is the discharge pumped from the well and T_s is the transmissivity for the screened or slotted intervals. All other $\bar{h}_{i,j}$ are set to h_e, and the values are used in eq. 29 to produce the $\bar{u}_{wj}$. On the first iteration for all time steps, except the first, values used are those calculated for the end of the previous time step. For all subsequent iterations at all time steps, values of $\bar{h}_{1,j}$ are those obtained from the previous iteration.

As mentioned previously indices j and k stand for different sets of nodes. It was assumed that highly non-linear velocity distributions could occur in the well bore. Thus, allowance was made so each interval could be subdivided by adding intermediate nodes designated by the k system, although nodes j and $j+1$ are also nodes in the k scheme. The intermediate values $\bar{u}_{wk}$ were then obtained by linear interpolation between $\bar{u}_{wj}$ and $\bar{u}_{wj+1}$. An example of this numbering scheme is shown in Fig. 4.

(2) As the $\bar{u}_{wj}$ are computed, total discharge is accumulated so that when the calculations have proceeded through the last node the total discharge from the well is obtained. If this discharge, Q_T, does not equal Q then the $\bar{u}_{wk}$ are adjusted by multiplication by Q/Q_T so that the total discharge is correct. An incorrect value of Q_T also implies that the $\bar{h}_{i,j}$ are incorrect so they are ad-

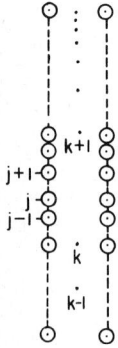

Fig. 4. Schematic diagram of the two node numbering schemes used for the vertical dimension. *Circles* designated by the general index j apply for the porous medium, and *dots* designated by the general index k apply for the well bore.

justed using the theory of superposition by the equation:

$$\bar{h}_{i,j}^{new} = \frac{Q}{Q_T}(\bar{h}_{i,j} - h_e) + h_e \tag{31}$$

(3) Velocity $\bar{V}_{k+1}$ is computed using eq. 22 written in terms of barred variables. The velocity at node 1, $\bar{V}_1$, is a boundary condition and is assumed to be zero.

(4) Hydraulic head in the well bore, $\bar{h}_{wk+1}$, is calculated by employing eq. 27. Again, $\bar{h}_{w1}$ is a boundary condition. It is calculated from eq. 15 rewritten as:

$$\bar{h}_{w1} = \bar{h}_{1,jb} - \bar{u}_{w1}^2/2gC_c^2C_v^2A_s^2 \tag{32}$$

where jb is the node at the bottom of the bottom-most screen. For the first iteration at each time step the value of $\bar{h}_{1,jb}$ used is obtained from the projection:

$$\bar{h}_{1,jb} = \frac{(1-\theta)\triangle t_0 + \theta\triangle t}{\theta\triangle t_0}[\bar{h}_{1,jb}(t-(1-\theta)\triangle t_0) - h_{1,jb}(t-\triangle t_0)]$$

$$+ \bar{h}_{1,jb}(t-(1-\theta)\triangle t_0) \tag{33}$$

where $\triangle t$ is the size of the present time step, $\triangle t_0$ is the size of the previous time step, and the notation $h_{1,jb}(t-\triangle t_0)$, etc., indicate that $h_{1,jb}$ is evaluated at time $t - \triangle t_0$. For all subsequent iterations the value is $\bar{h}_{1,jb}^{new}$ as obtained from eq. 31.

(5) Next, updated values of hydraulic head $\bar{h}_{1,j}$ at the well are calculated by rewriting eq. 15 as:

$$\bar{h}_{1,j} = \bar{h}_{wj} + \bar{u}_{wj}^2/2gC_c^2C_v^2A_s^2 \tag{34}$$

(6) Values of $\bar{h}_{1,j}$ calculated from eq. 34 are used as known head boundary conditions in eq. 28, and an iteration is performed using the SIP iterative matrix solution method (Stone, 1968) as programmed by Cooley (1974).

(7) If the convergence tolerance used in SIP has not been met, steps (1)—(6) are repeated.

Numerical instability at high K

For values of K greater than about 0.283 m/min. a numerical instability appeared which prevented the procedure from converging. At first the flow direction at the bottom of the well reversed so that water discharged into the aquifer. This zone moved up the well bore with each succeeding iteration until it reached the top, at which time the zone then moved to the well bottom again. The cycles continued until the maximum specified number of iterations was exhausted. Several procedures such as underrelaxation and changing mesh and time step spacings were attempted in order to rectify the problem. However, time and budget restrictions prevented thorough analysis which could have resolved the cause of the instability. As a practical matter, the stability problem may not be important because values of K greater than 0.283 m/min. are rare. However, it prevented study of all possible cases in the following section.

RESULTS AND ANALYSIS

A summary of test cases examined is presented in Table I. As can be seen, the cases analyzed cover a wide variety of possible field conditions such as multi-aquifer systems, anisotropy, partial and full well penetration, and responses to systematic variation of key parameters such as pumping rate Q and hydraulic conductivity K.

Effects of varying K and Q

Test cases *9 –20* were analyzed in such a way as to investigate the effects of variation in both hydraulic conductivity and well discharge rate. Fig. 5 shows the aquifer potential field at essentially steady state for various combinations of Q and K. In all cases it is seen that flow to the well is essentially radial near the upper and lower bounds of the aquifer with a region of predominately non-radial flow in between (note here that the terms "radial" and "non-radial" apply only to the case where the flow system is viewed in cross-section; a plan view would reveal no departure from radial flow conditions). This result contradicts, to some degree, the classic assumption that flow to a pumped well is radial throughout a confined aquifer. Furthermore, the degree and extent of the non-radial flow region is seen to vary in proportion to the magnitudes of Q and K. A qualitative explanation for this can be given based on the fact that water movement in the aquifer tends to follow a path such that the total energy loss (i.e. the sum of all energy losses in the well and in the aquifer) is minimized. From Fig. 5 it can be seen that for low K-values the flow paths tend to be more nearly radial than for the cases involving high K-values. This is because for a fixed pumping rate the lower the value of K, the greater the proportion of total head losses that occur in the aquifer. Thus, total energy expenditure is controlled primarily by head losses in the aquifer, which are minimized by water taking the shortest possible flow path, i.e. the radial one. Conversely, if K is large so that head losses in the aquifer are small, flow paths yielding minimum energy loss reflect the larger proportion of head losses in the well. Head losses in the well are minimized if most of the flow is nearest the pump intake (the top of the upper well screen in this study). Therefore, flow paths in the aquifer are directed toward the top of the well, thereby creating a region of non-radial flow. This result appears to be true regardless of the value of Q. However, it is also seen that the degree and extent of the non-radial flow region becomes greater as Q increases. This is probably because head losses vary with Q in the aquifer but vary with Q^2 in the well.

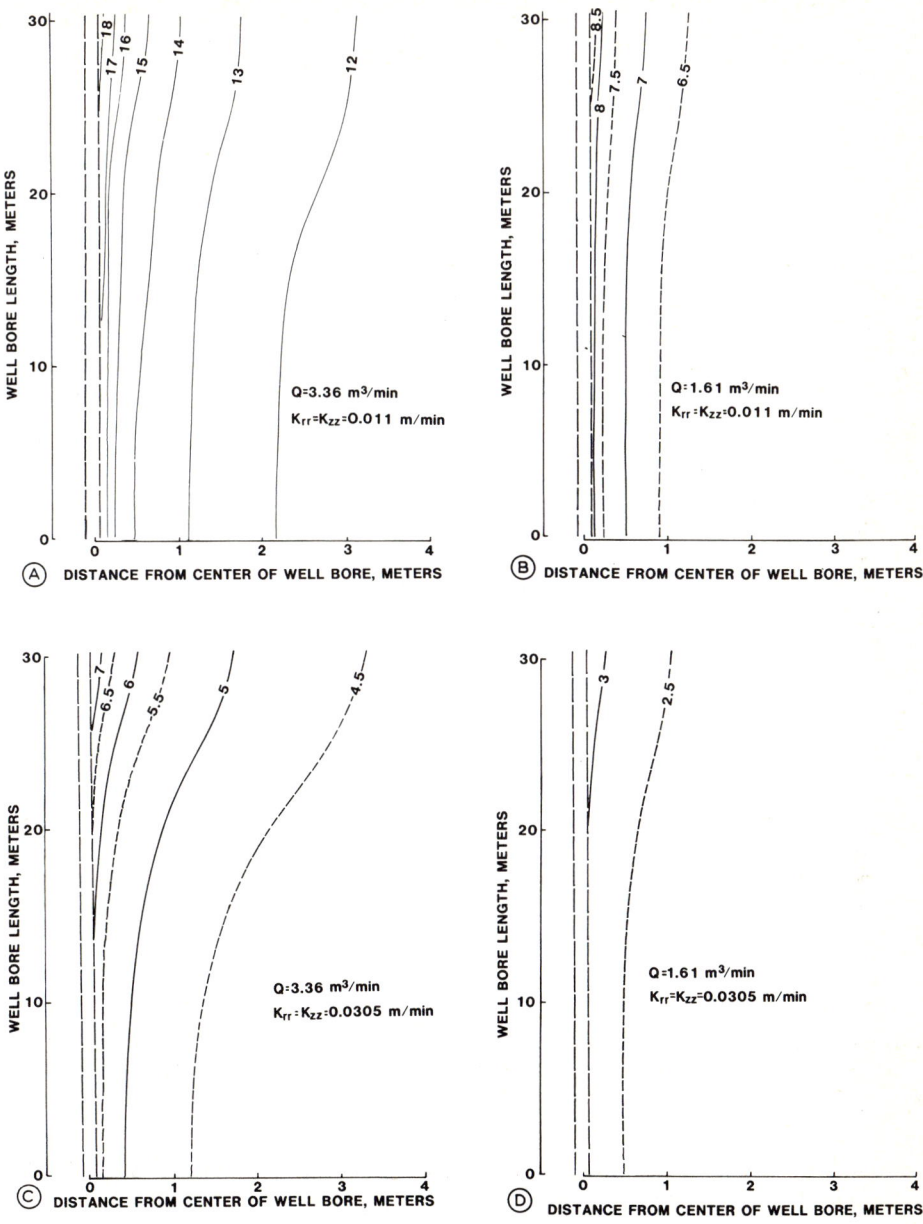

Fig. 5. Distribution of equidrawdown lines adjacent to pumped well (approximately steady-state conditions). *Solid lines* represent 1 m drawdown intervals; *dashed lines* represent 0.5 m drawdown intervals.
A. $K = 0.011$ m/min. and $Q = 3.36$ m^3/min.
B. $K = 0.011$ m/min. and $Q = 1.61$ m^3/min.
C. $K = 0.0305$ m/min. and $Q = 3.36$ m^3/min.
D. $K = 0.0305$ m/min. and $Q = 1.61$ m^3/min.

TABLE I

Summary of cases investigated

Test case	Aquifer description[1]	Aquifer thickness (m)	Storage coefficient ($\times 10^{-4}$)	Hydraulic conductivity (m/min.)	Well discharge (m³/min.)	Well radius (m)	Drawdown at well top/bottom[2] (m)	Specific discharge at well top/bottom (m/min.)
1	single confined homogeneous isotropic aquifer with fully penetrating well	15.25	5	0.00849	0.379	0.152	4.608/4.606	0.0260/0.0259
2	partially penetrating well (12.19 m) in a confined homogeneous isotropic aquifer	15.25	5	0.00849	0.379	0.152	4.937/4.937	0.0326/0.0326
3	three-aquifer system with constant hydraulic conductivity for aquifers, fully penetrating well	10.98	1.10 (for aquifers)	0.00849 (for aquifers)	1.892	0.152	30.897/21.784	0.179/0.183
4	three-aquifer system with gravel packed well, constant hydraulic conductivity for aquifers, fully penetrating well	10.98	1.10 (for aquifers)	0.00849 (for aquifers)	1.892	0.152	25.231/25.015	0.176/0.189
5	three-aquifer system with gravel packed well, formation clogging, constant hydraulic conductivity for aquifers, fully penetrating well	10.98	1.10 (for aquifers)	0.00849 (for aquifers)	1.892	0.152	32.382/32.172	0.178/0.185
6	single confined anisotropic aquifer, fully penetrating well	15.25	5	$K_{rr} = 0.00849$ $K_{zz} = 0.0000849$	0.379	0.152	4.669/4.667	0.0260/0.0260
7	single confined isotropic homogeneous aquifer, fully penetrating well, Darcy "f" for well bore = 0.01	15.25	5	0.00849	0.379	0.152	4.607/4.607	0.0260/0.0260
8	single confined isotropic homogeneous aquifer with fully penetrating well	15.25	5	0.283	0.379	0.152	0.144/0.143	0.0265/0.0258
9	single confined isotropic homogeneous aquifer	30.50	1	0.011	0.102	0.0762	0.531/0.528	0.00702/0.00696
10	single confined isotropic homogeneous aquifer	30.50	1	0.011	0.202	0.0762	1.064/1.055	0.0142/0.0139

TABLE I (continued)

Test case	Aquifer description[1]	Aquifer thickness (m)	Storage coefficient ($\times 10^{-4}$)	Hydraulic conductivity (m/min.)	Well discharge (m³/min.)	Well radius (m)	Drawdown at well top/bottom[2] (m)	Specific discharge at well top/bottom (m/min.)
11	single confined isotropic homogeneous aquifer	30.50	1	0.011	0.403	0.0762	2.142/2.105	0.0289/0.0277
12	single confined isotropic homogeneous aquifer	30.50	1	0.011	0.806	0.0762	4.338/4.189	0.0600/0.0548
13	single confined isotropic homogeneous aquifer	30.50	1	0.011	1.210	0.0762	6.568/6.235	0.0929/0.0810
14	single confined isotropic homogeneous aquifer	30.50	1	0.011	1.613	0.0762	8.886/8.299	0.128/0.107
15	single confined isotropic homogeneous aquifer	30.50	1	0.011	2.667	0.0762	14.927/13.381	0.225/0.171
16	single confined isotropic homogeneous aquifer	30.50	1	0.011	3.360	0.0762	19.399/16.921	0.302/0.214
17	single confined isotropic homogeneous aquifer with fully penetrating well	30.50	1	0.0305	0.102	0.0762	0.193/0.190	0.00715/0.00692
18	single confined isotropic homogeneous aquifer with fully penetrating well	30.50	1	0.0305	1.613	0.0762	3.468/2.903	0.156/0.100
19	single confined isotropic homogeneous aquifer with fully penetrating well	30.50	1	0.0305	2.634	0.0762	6.069/4.614	0.300/0.154
20	single confined isotropic homogeneous aquifer with fully penetrating well	30.50	1	0.0305	3.360	0.0762	8.087/5.775	0.423/0.188
21	single confined aquifer, inefficient well (i.e. percent open area in well screen is reduced from 10% to 1%)	30.50	1	0.0305	3.360	0.0762	8.057/5.793	0.417/0.188

[1] $C_V = 0.97$ and $C_C = 0.63$ in all cases, $A_S = 0.1$ in cases $1-20$, $A_S = 0.01$ in case 21, $f = 0.01$ in case 7, $f = 0.05$ in all other cases.
[2] Drawdown at essentially steady-state conditions. Bt. indicates bottom.

Time—drawdown curves. The time variation of drawdown at both the top and bottom of the well bore was graphed for several test cases (Fig. 6). For cases represented by Fig. 6A and 6D the linear acceleration of pumping rates over the first second, which was explained previously, was not used. In Fig. 6B the effect of the acceleration is represented by the bump in the upper curve. In Fig. 6C the effect does not show because the plot is for drawdown 30 m from the well. In all cases, immediately after pumping begins, drawdown at the well top is seen to be significantly greater than at the bottom, signifying that, subject to the assumptions utilized, flow to wells is initially highly non-radial. However, this initial effect is very shortlived as illustrated by the rapid convergence of the top and bottom drawdown curves as pumping continues. The ultimate degree of convergence of the time—drawdown curves indicates whether or not non-radial flow persists indefinitely. For example, in the case of relatively low values of K and Q (Fig. 6A), the drawdown curves for the top and bottom of the well converge completely indicating that effects of non-radial flow vanish shortly after pumping begins. However, when K and Q assume relatively high values (Fig. 6B) drawdown at the top of the well remains significantly higher than at the well bottom thus indicating the persistence of the non-radial flow region with time.

As indicated previously, the assumptions regarding effects of storage changes in the well and incompressibility of water and well components tend to invalidate quantitative interpretation of early-time results. However, based on the minimum energy-loss argument given above for the formation of non-radial flow, it is the authors' opinion that the highly non-radial flow region would develop at early times with the elimination of the restrictive assumptions, even though the actual shapes of the early-time curves would be different than those predicted by the numerical experiments. Also, elimination of the assumptions should have little effect on the results given in Table I because these results are for systems tending to steady state, and the configurations of steady-state flow systems are independent of the transient conditions leading to them.

Interpretation of pumping test results. The dashed curves in Fig. 6 represent theoretical behavior of time—drawdown in a confined aquifer with finite radius well (Hantush, 1964). Comparison of these curves with the time drawdown curves at the well top reveals how the presence of non-radial flow in the aquifer could cause errors in the interpretation of pumping test data. Of greatest importance are cases where time—drawdown measurements are taken either at or about a meter away from the pumped well, for it is this type of data that most likely would be influenced by non-radial flow. An example of this situation observed in Fig. 6B shows that, due largely to non-radial flow effects, the observed drawdown at the well top would remain consistently greater than the Hantush theoretical curve. In this case, analysis of the observed time—drawdown data by fitting the theoretical Hantush curve would result in a low estimate of aquifer transmissivity and no reasonable

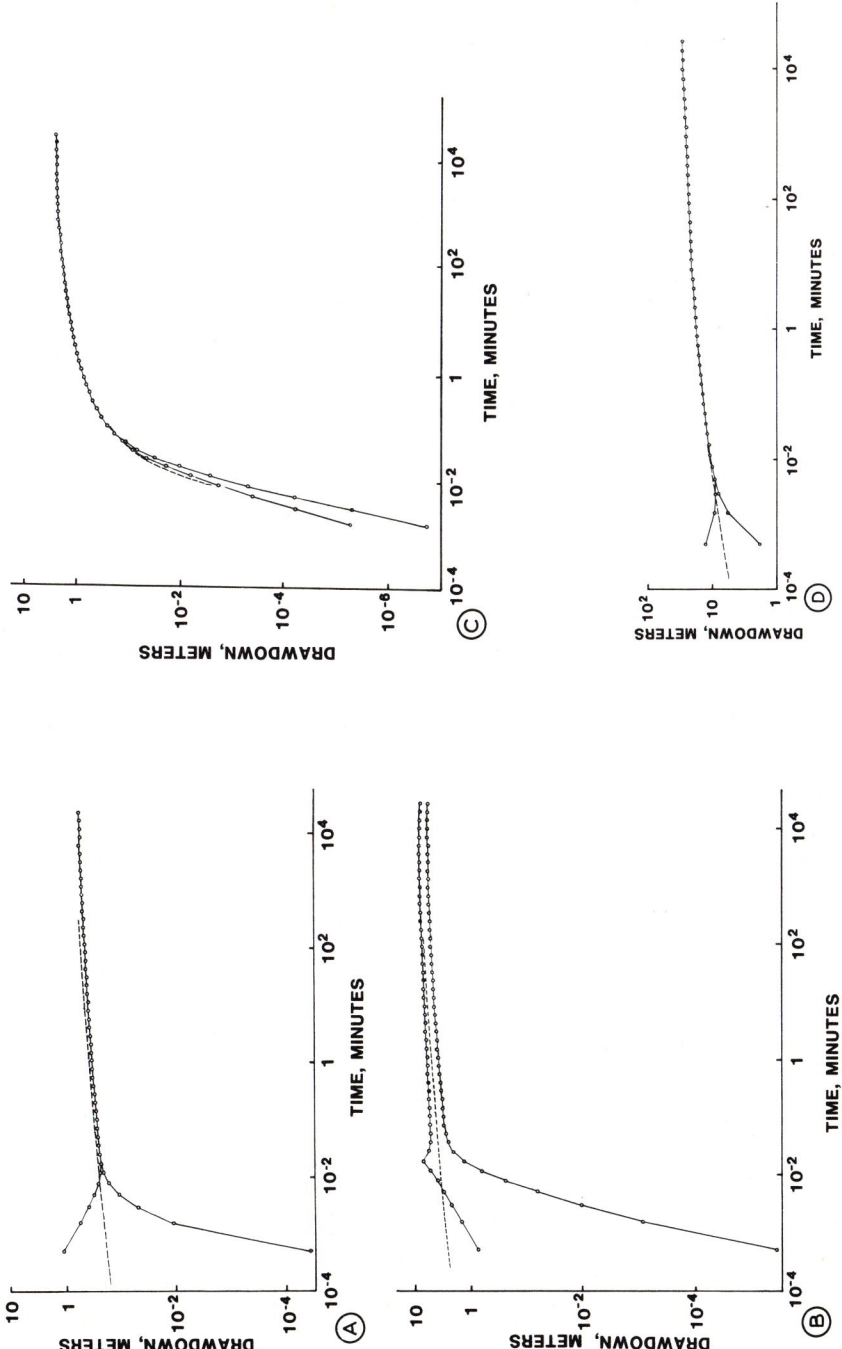

Fig. 6. A. Drawdown vs. time at top (*upper solid line*) and bottom of well bore for $K = 0.011$ m/min. and $Q = 0.102$ m³/min.
B. Drawdown vs. time at top (*upper solid line*) and bottom of well bore for $K = 0.0305$ m/min. and $Q = 3.36$ m³/min.
C. Drawdown vs. time at depths corresponding to top (*upper solid line*) and bottom of well at radius of 30 m from well bore. $K = 0.0305$ m/min. and $Q = 3.36$ m³/min.
D. Drawdown vs. time at top (*upper solid line*) and bottom of well bore for multi-aquifer system with $K = 0.00849$ m/min. (for aquifers) and $Q = 1.892$ m³/min.

estimate of the aquifer storage coefficient. However, this problem is minimized if time—drawdown data are collected at an observation well located a sufficient distance from the pumped well. Fig. 6C represents the time—drawdown curve for the same case as Fig. 6B but with time—drawdown data taken at a distance of 30 m (approximately one aquifer thickness) from the pumped well. Here it is seen that effects of non-radial flow no longer persist because the observed and theoretical drawdown curves tend toward complete convergence. In fact, visual analysis of computer results shows that non-radial flow is virtually non-existent beyond about 10 m from the pumped well after about the first 10 s of pumping.

Effects of variation of other system characteristics

The test cases examined herein revealed very little change in the overall system behavior due to variation in well parameters such as r_w, C_v, C_c, A_s and f. For example, comparison of cases *1* and *7* shows that drawdown at the well is decreased by only about 0.43% as a result of decreasing f from 0.05 to 0.01. Similarly, comparison of results for cases *20* and *21* indicates a very small increase in drawdown resulting from decreasing A_s from 0.1 to 0.01. Changes in r_w generally produced changes in drawdown close to those predicted with standard theory. However, it should be noted that this analysis is by no means complete since the parameters were not studied over all possible values. Indeed, it is obvious that if r_w, C_v, C_c, or A_s were made very small, a significant increase in drawdown would result if the pumping rate remained unchanged.

A similar analysis was made of the effects of variation in aquifer system characteristics such as anisotropy and division into multiple aquifers. Anisotropy was examined in case *6* for which the ratio K_{xx}/K_{zz} was set equal to 100 while all other system characteristics remained the same as for case *1*. Results indicate that anisotropy caused only a slight increase in drawdown at both the top and bottom of the well bore. Specific discharge values were also virtually unchanged. This indicates that in case *6* flow to the well is essentially radial thus allowing the radial hydraulic conductivity term K_{xx} to dominate. Greater drawdown values would be expected if system characteristics were changed so as to cause a more substantial vertical component in the flow entering the well.

Effects caused by multi-aquifer systems were also small. The particular system, as described in case *3*, consisted of three relatively permeable layers of equal hydraulic conductivity separated by low permeability zones (aquitards) of various thicknesses. The results illustrated in Fig. 6D indicate that the plot of time vs. drawdown in the aquifers has the same basic features as does the time—drawdown plot obtained for a single confined aquifer. The theoretical curve in Fig. 6D represents that of a single aquifer with T equal to the sum of the individual transmissivities in the multi-aquifer system. It is seen that this theoretical curve is in virtually perfect agreement with the computed curve, thereby indicating that, at least in this case, neither the presence nor the nature of the multi-aquifer system can be delineated solely on the basis of time—drawdown data in the well.

Final effects studied are those caused by variations in hydraulic conductivity around the well bore (e.g., clogging by mud invasion or increasing K by gravel packing). Both of these effects are predicted almost exactly using the results of Sternberg (1973), which can be written:

$$\Delta h_w = \frac{Q}{2\pi T} \left(\frac{K_s - K}{K_s} \right) \ln \frac{r_s}{r_w} \qquad (35)$$

where Δh_w is the increment in head in the well bore caused by the change in hydraulic conductivity of the aquifer, K, to a value, K_s, inside the annulus bounded by r_s and r_w. Although eq. 35 was derived assuming steady-state radial flow in a non-leaky aquifer, it actually applies as a good approximation to the more general cases studied here. This generality results because $r_s - r_w$ is normally very small compared to the size of the aquifer. Thus, leakage and unsteady flow exert negligible influence on Δh_w.

Comparison with Jacob's theory

Jacob's (1947) equation for drawdown in a well can be written:

$$s_w = BQ + CQ^2 \qquad (36)$$

where coefficient B includes components of drawdown that are linearly related to Q, such as those resulting from flow through the aquifer and any altered zone around the well bore, and C, the well loss coefficient, includes effects related quadratically to Q, such as those caused by turbulent flow in the aquifer and well and flow across the well screen. Rorabaugh (1953) generalized eq. 36 by changing the exponent on Q from 2 to n.

To determine whether or not the solution procedure given here could theoretically produce eq. 36, eqs. 13—15, 18 and 19 are analyzed dimensionally. Define dimensionless velocity V^* as:

$$V^* = V/V_t \qquad (37)$$

where $V_t = Q/(\pi r_w^2)$ and Q, the discharge from the well, is presumed constant. Then using variables:

$$u_w^* = u_w/V_t, \qquad h_w^* = h_w/V_t^2, \qquad h^* = h/V_t \qquad (38), (39), (40)$$

Eqs. 13 15, 19 and 18, respectively, may be written:

$$\partial V^*/\partial z = 2u_w^*/r_w, \qquad \frac{\partial h_w^*}{\partial z} + \frac{f V^{*2}}{4 g r_w} + \frac{1}{gV_t} \frac{\partial V^*}{\partial t} + \frac{1}{g} \frac{\partial V^{*2}}{\partial z} = 0 \qquad (41), (42)$$

$$u_w^* = C_v C_c A_s [2g(h^*/V_t - h_w^*)]^{1/2}, \qquad u_w^* = -K_{rr}(\partial h^*/\partial r) \qquad (43), (44)$$

$$K_{rr} \frac{1}{r} \frac{\partial}{\partial r} \left(r \frac{\partial h^*}{\partial r} \right) + K_{zz} \frac{\partial^2 h^*}{\partial z^2} = S_s \frac{\partial h^*}{\partial t} \qquad (45)$$

Variables given by eqs. 37–40 may be used to write a total drawdown equation of the form:

$$s_w = h_e - h_{wt} = (h_e^* - h^*) V_t + [(h^*/V_t - h_w^*) + (h_w^* - h_{wt}^*)] V_t^2 \qquad (46)$$

where $h_e^* = h_e/V_t$; $h_{wt}^* = h_{wt}/V_t^2$; and h_{wt} is the drawdown at the top of the uppermost well screen, which is the quantity usually measured in a pumping test. Note that the coefficient of V_t in eq. 46 expresses head loss in the aquifer, the first term in the coefficient of V_t^2 expresses head loss across the well screen or slots, and the second term in the coefficient of V_t^2 expresses head loss in the well bore. Using eq. 43, eq. 46 may be rewritten as:

$$s_w = (h_e^* - h^*) V_t + \left[\frac{1}{2g(C_v C_c A_s)^2} u_w^{*2} + (h_w^* - h_{wt}^*) \right] V_t^2 \qquad (47)$$

To the extent that the coefficients of V_t and V_t^2 in eq. 47 remain constant with V_t, eq. 47 is an analog of eq. 36. In opposition to eq. 36, however, the coefficients in eq. 47 vary with vertical position, z. This will not change any of the subsequent conclusions.

To examine the constancy of the coefficients in eq. 47, eqs. 37–45 are examined. From eqs. 37, 39, 41 and 42 it may be seen that, if u_w^* does not depend on V_t, h_w^* and h_{wt}^* do not depend on V_t after the first few seconds of pumping when $\partial V^*/\partial t \to 0$. Under the same assumptions, h^* also does not depend on V_t. However, eq. 43 shows that u_w^* does depend on V_t, which leads to the conclusion that coefficients B and C in eq. 36 may also be functions of V_t. The strength of the dependency of u_w^* on V_t governs the degree of variability of B and C with V_t or Q. Hence, it is possible that an exponent other than 2 in eq. 36 could provide a better fit to test data in some cases. Plots of s_w/Q vs. Q for cases 9–16 and 17–20 shown in Fig. 7 indicate that for the cases examined in this study B and C are not strong functions of Q.

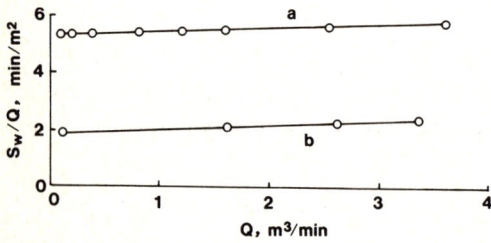

Fig. 7. s_w/Q vs. Q for (a) runs 9–16; and (b) for runs 17–20 at essentially steady-state conditions.

CONCLUSIONS

(1) A numerical solution for coupled unsteady flow in a semi-confined aquifer system and well penetrating it indicates that for values of aquifer hydraulic conductivity greater than about 0.015 m/min. and for pumping rates greater than 1.2 m^3/min. a significant region of non-radial flow resulting from head losses in the well can develop. This region results from friction losses and non-uniform flow in the well. Thus, it persists throughout the pumping period. However, in all the numerical experiments flow was highly non-radial during the first few seconds of pumping. Such potentially important factors as removal of water from storage in the well and compressibility of the well components (pump, etc.) and water in the well would cause the curve shapes to differ from those calculated. However, based on the argument that flow takes place so as to minimize energy losses, it is the authors' opinion that the non-radial flow field would develop even with elimination of the restrictions.

(2) Non-radial flow related to flow in the well can cause significant errors in values of aquifer transmissivity computed from drawdown data from the pumped well if the aquifer hydraulic conductivity is greater than about 0.03 m/min. Values of the aquifer storage coefficient could probably never be computed accurately using drawdown data from a pumped well.

(3) The system parameters C_v, C_c, A_s and f caused little change in system response for wide variations in their values. Furthermore, changes in r_w produced effects nearly predictable using standard theory. However, it is apparent theoretically that if r_w, C_v, C_c, or A_s tended to zero, drawdowns would have to increase dramatically if other parameters were held constant.

(4) Expressions for B and C in Jacob's (1947) formula for total drawdown in a well were developed dimensionally by using the results of this study. It was shown that B and C could vary with Q during a step drawdown test even if no well development occurred. However, they did not vary significantly for the test problems considered in the present study.

ACKNOWLEDGEMENT

The work reported herein was supported in part by the United States Department of the Interior, Office of Water Research and Technology, as authorized under the Water Resources Research Act of 1964, PL 88-379, and in part by funds provided by the Desert Research Institute, University of Nevada System, Reno, Nevada.

APPENDIX

Derivation of eq. 6

Integration of eq. 2 over control volume V and application of the divergence theorem (Boas, 1966, pp. 246 and 247) gives:

$$-\iiint_V \frac{\partial}{\partial x_i}(p+\gamma z)\,dV + \iint_S \sigma_{ji}n_j\,dS = \rho\iiint_V \frac{\partial u_i}{\partial t}\,dV + \rho\iiint_V u_j\frac{\partial u_i}{\partial x_j}\,dV$$

$$= \rho\frac{\partial}{\partial t}\iiint_V u_i\,dV + \rho\iiint_V \frac{\partial(u_i u_j)}{\partial x_j}\,dV \qquad (A\text{-}1)$$

where S is the surface area of V, n_i is the component of the unit outward normal to S in the x_i direction, and the ideas that V is constant with time and $\partial u_i/\partial x_i = 0$ were used.

Only the $x_3 \equiv z$ component of eq. A-1 need to be considered for the present analysis. Thus, with mean hydraulic head defined by eq. 8 and $\sigma_{33} \equiv \sigma_{zz}$ eq. A-1 for $x_i \equiv z$ can be written:

$$-\gamma\frac{\partial h_w}{\partial z}A\delta z + \iint_{S_3}(\sigma_{1z}n_1 + \sigma_{2z}n_2)\,dS - \iint_{S_1}\sigma_{zz}\,dS + \iint_{S_2}\sigma_{zz}\,dS$$

$$= \rho\frac{\partial Q}{\partial t}\delta z + \rho\iiint_V \frac{\partial(u_z u_j)}{\partial x_j}\,dV \qquad (A\text{-}2)$$

where the symbol S_1 indicates the cross-sectional area A at the lower end of δz where $n_1 = n_2 = 0$ and $n_z \equiv n_3 = -1$; S_2 is the cross-sectional area A at the upper end of δz where $n_1 = n_2 = 0$ and $n_z = 1$; and S_3 stands for surface area of well bore contained in V where $n_z = 0$.

The second term in eq. A-2 defines mean shear stress on the well casing as:

$$\sigma_0 = -\iint_{S_3}(\sigma_{1z}n_1 + \sigma_{2z}n_2)\,dS/S_3 \qquad (A\text{-}3)$$

Also, the third and fourth terms can be modified as:

$$-\iint_{S_1}\sigma_{zz}\,dS + \iint_{S_2}\sigma_{zz}\,dS = F_z(z+\delta z) - F_z(z) \qquad (A\text{-}4)$$

where F_z is the normal viscous force in the z-direction. Finally, the last term in eq. A-2 is modified approximately through use of the divergence theorem and the first law of the mean for integrals (Goursat, 1904, p. 152):

$$\iiint_V \frac{\partial(u_z u_j)}{\partial x_j}\,dV = \iint_S u_z u_j n_j\,dS = -\iint_{S_1}u_z u_z\,dS + \iint_{S_2}u_z u_z\,dS +$$

$$\iint_{S_3}u_z(u_1 n_1 + u_2 n_2)\,dS = -u_z(a)\iint_{S_1}u_z\,dS + u_z(b)\iint_{S_2}u_z\,dS +$$

$$u_z(c)\iint_{S_3}(u_1 n_1 + u_2 n_2)\,dS = -V(z)Q(z) + V(z+\delta z)Q(z+\delta z) - v_z q\delta z$$

$$\qquad (A\text{-}5)$$

The terms $u_z(a)$ and $u_z(b)$ are values of the vertical component of velocity in the well bore evaluated at unknown points a and b within the respective areas S_1 and S_2. They are assumed herein to be approximately equal to the average velocities at z and $z + \delta z$, respectively. Similarly $u_z(c)$ is assumed to approximate the average vertical component of the inflow velocity, v_z.

Substitution of eqs. A-3–A-5 into eq. A-2, division by δz, and evaluation of the limit as $\delta z \to 0$ gives:

$$-\gamma A \frac{\partial h_w}{\partial z} - 2\pi r_w \sigma_0 + \frac{\partial F_z}{\partial z} = \rho \frac{\partial Q}{\partial t} + \rho \frac{\partial (QV)}{\partial z} - \rho v_z q \qquad (A\text{-}6)$$

To evaluate $\partial F_z / \partial z$, first note that the term σ_{zz} may be written (Mase, 1970, p. 161):

$$\sigma_{zz} = 2\mu (\partial u_z / \partial z) \qquad (A\text{-}7)$$

where μ is the viscosity of water. Therefore, by integration of σ_{zz} over A it is seen that:

$$F_z = 2\mu (\partial Q / \partial z) = 2\mu q \qquad (A\text{-}8)$$

so that:

$$\partial F_z / \partial z = 2\mu (\partial^2 Q / \partial z^2) = 2\mu (\partial q / \partial z) \qquad (A\text{-}9)$$

Because the value of μ for water is so small, $\partial q / \partial z$ would have to become very large in order for $\partial F_z / \partial z$ to be significant in eq. A-6. It was assumed for the present study that this would not happen, and subsequent numerical experimentation confirmed this assumption. Thus, $\partial F_z / \partial z$ was neglected from eq. A-6 so that eq. A-6 reduces to eq. 6.

REFERENCES

Besbes, M., 1974. Analyse des pertes de charge dans les forages d'eau. Bull. Bur. Rech. Géol. Minières (Fr.), 2ème Sér., Sect. II, 3: 261–270.

Boas, M.L., 1966. Mathematical Methods in the Physical Sciences. Wiley, New York, N.Y., 778 pp.

Cooley, R.L., 1974. Finite element solutions for the equations of ground-water flow. Cent. Water Resour. Res., Desert Res. Inst., Univ. Nevada System, Reno, Nev., Tech. Ser. H–W, Publ. 18, 134 pp.

Garg, S.P. and Lal, J., 1971. Rational design of well screens. Proc. Am. Soc. Civ. Eng., Irrig. Drain Div. J., 97: 131–147.

Goursat, E., 1904. A Course in Mathematical Analysis, Vol. 1. Dover, New York, N.Y., 548 pp.

Hantush, M.S., 1964. Hydraulics of wells. In: V.T. Chow (Editor), Advances in Hydroscience, Vol. 1, Academic Press, New York, N.Y., pp. 281–442.

Jacob, C.E., 1947. Drawdown test to determine effective radius of artesian well. Trans. Am. Soc. Civ. Eng., 112: 1047–1070.

Jacob, C.E., 1950. Flow of ground water. In: H. Rouse (Editor), Engineering Hydraulics. Wiley, New York, N.Y., pp. 321–386.

Lohman, S.W., 1972. Ground-water hydraulics. U.S. Geol. Surv., Prof. Pap. 708, 70 pp.

Mase, G.E., 1970. Continuum Mechanics. Schaum's Outline Series, McGraw-Hill, New York, N.Y., 221 pp.

Mogg, J.L., 1968. Step drawdown test needs critical review. Johnson Driller's J., July—Aug. 1968, pp. 3—11.
Norrie, D.H. and de Vries, G., 1973. The Finite Element Method. Academic Press, New York, N.Y., 322 pp.
Peterson, J.S., Rohwer, C. and Albertson, M.L., 1955. Effect of well screen on flow into wells. Trans. Am. Soc. Civ. Eng., 120: 563—585.
Rorabaugh, M.I., 1953. Graphical and theoretical analysis of step drawdown test of artesian well. Proc. Am. Soc. Civ. Eng., 79, Separate No. 362, 23 pp.
Soliman, M.M., 1965. Boundary flow — considerations in the design of wells. Proc. Am. Soc. Civ. Eng., Irrig. Drain. Div. J., 91: 159—177.
Sternberg, Y.M., 1973. Well efficiency and skin effect. Proc. Am. Soc. Civ. Eng., Irrig. Drain. Div. J., 99: 203—206.
Stone, H.L., 1968. Iterative solution of implicit approximations of multi-dimensional partial differential equations. J. Soc. Ind. Appl. Math., Numer. Anal., 5: 530—558.
Walton, W.C., 1970. Ground-Water Resource Evaluation. McGraw-Hill, New York, N.Y., 664 pp.

[2]

MEASUREMENT OF FLUID VELOCITY USING TEMPERATURE PROFILES: EXPERIMENTAL VERIFICATION

KEROS CARTWRIGHT

Illinois State Geological Survey, Urbana, IL 61801 (U.S.A.)

(Accepted for publication April 25, 1979)

ABSTRACT

Cartwright, K., 1979. Measurement of fluid velocity using temperature profiles: experimental verification. In: W. Back and D.A. Stephenson (Guest-Editors), Contemporary Hydrogeology — The George Burke Maxey Memorial Volume. J. Hydrol., 43: 185—194.

Temperature profiling has been used to predict the rate and direction of groundwater movement. A controlled field experiment was conducted to ascertain the validity of the rate calculations made using this method. The vertical velocity, or leakage, of groundwater between two aquifers was calculated utilizing both hydrologic and temperature measurements in a well drilled into the Paw Paw buried bedrock valley in northern Illinois.

The experiment showed that accurate estimates of leakage can be made in stable boreholes where there are no geologic complications. Estimates utilizing temperature and hydrologic methods produced similar results for one of two aquicludes. However, the methods produced dissimilar results for the second aquiclude. It is speculated that the presence of a thin organic silt caused most of the problem; other complicating factors were lithologic variation and a very low hydraulic gradient. Nevertheless, the method appears to have great promise in many geologic environments.

INTRODUCTION

The vertical distribution of heat in the Earth's materials can easily be measured by the use of a fluid-filled borehole in which the temperature of the fluid is in equilibrium with that of the surrounding rocks. The rate of vertical flow of groundwater can be calculated using the temperature of the borehole fluid if the flow velocity is sufficient to affect the thermal profile, Bredehoeft and Papadopulos (1965) solved the general equation for simultaneous flow of water and heat given by Stallman (1960) and obtained a particular equation describing the steady-state case of both heat and groundwater flow in the vertical direction only. They presented a set of type curves which can be used to compute the rate of vertical groundwater flow, provided the thermal properties of the rock and fluid complex are known. This method has been used by several workers (e.g., Cartwright, 1970; Sorey, 1971; Boyle and Saleem, 1979) to calculate the rate of vertical groundwater flow. The experiment described here is similar to that of Sorey (1971) who studied the temperature profiles of wells in the San Luis Valley, Colorado, and the Roswell Basin, New Mexico. Sorey, as well as the other workers, used estimated values of thermal conductivity taken from the literature (e.g., Clark, 1966); the hydrologic parameters used in Sorey's study were reasonably well known.

The experiment described here was undertaken to check the validity and usability of the method in an unstressed environment where all the basic requirements of the mathematical model are met. The site chosen near Paw Paw, Illinois, appears to meet these requirements: a steady flow of both heat and fluid through the aquicludes in the vertical direction only. It is easily demonstrated that for a change in permeability of two orders of magnitude or greater between two adjacent layers of a porous medium, the direction of fluid flow across the boundary between the two layers will be deflected about 90° (Hubbert, 1940); in this case, from vertical in the aquiclude to horizontal in the aquifers. Test-well data indicate that the groundwater flow is primarily horizontal in the two aquifers; thus, the flow in the aquiclude is vertical. The heat flow by conduction also is presumably vertical.

METHOD OF ANALYSIS

The general differential equation (Stallman, 1960) can be simplified for analysis of steady flow of both heat and fluid in the vertical direction. If the flow is constant in time and space, the time-dependent term and the terms describing flows in the x- and y-directions are equal to zero. The equation then reduces to:

$$\frac{\partial^2 T}{\partial z^2} - \frac{c_w \rho_w v_z}{k} \frac{\partial T}{\partial z} = 0 \qquad (1)$$

Bredehoeft and Papadopulos (1965) solved eq. 1 using the following boundary conditions to describe the problem $T_z = T_0$ at $z = 0$ and $T_z = T_L$ at $z = L$, in which z and v_z are positive downward with the origin at T_0, the top of the aquiclude. T_0 and T_L are the temperature at the origin and distance L from the origin; c_w and ρ_w are the specific heat and density, respectively, of the water; k is the thermal conductivity of the rock—fluid complex; and v_z is the velocity of fluid flow in the vertical direction. Bredehoeft and Papadopulos provide the following solution:

$$(T_z - T_0)/(T_L - T_0) = f(\beta, z/L) = [\exp(\beta, z/L) - 1]/[\exp(\beta) - 1] \qquad (2)$$

where $\beta = c_w \rho_w v_z L/k$. β is a dimensionless parameter that is positive or negative, depending on whether v_z is, respectively, downward or upward. To determine the vertical groundwater velocity, v_z, from the temperature profile, the value of β is obtained by plotting dimensionless depth, z/L, vs. dimensionless temperatures $(T_z - T_0)/(T_L - T_0)$ and matching the plot to the type curves of Bredehoeft and Papadopulos (1965). The value of v_z is then calculated from the equation:

$$v_z = \beta k / L c_w \rho_w \qquad (3)$$

Stallman (1967) suggested that the matching technique could be made more sensitive for low values of v_z by plotting z/L vs. $z/L - (T_z - T_0)/(T_L - T_0)$ and using type curves of z/L vs. $z/L - f(\beta, z/L)$. This modification is necessary when the absolute rate of groundwater flow is less than about 30 cm/yr. (Sorey, 1971).

Kunii and Smith (1961) measured the distribution of temperature in laboratory columns of glass beads and sand through which helium was flowing counter to a heat gradient, in order to determine effective thermal conductivities. A dimensionless plot of z/L vs. $(T_z - T_0)/(T_L - T_0)$ was made with their data by Sorey (1971) and showed close agreement between the flow rate calculated from the temperature distribution and the measured flow rate.

FIELD EXPERIMENT

A field experiment was set up as part of a test-drilling program to define major, underdeveloped aquifers in the greater metropolitan Chicago region. A test well was constructed near the village of Paw Paw, Illinois, in the ancient buried Paw Paw bedrock valley. The exact location of the Paw Paw valley and the nature of the deposits are not completely known. Presumably, the ancient Rock, Troy, and perhaps, Newark valleys merge to the north of Paw Paw and pass through the study area to join the main Mississippi valley near Princeton (Kempton, 1963). There are two major glacial drift aquifers in the valley, a basal pre-Illinoian and an Illinoian aquifer, separated by 3—30 m of glacial till. Wisconsinan tills, along with some aquifers of lesser importance, overlie the main aquifers and form the surface materials. A cross-section of

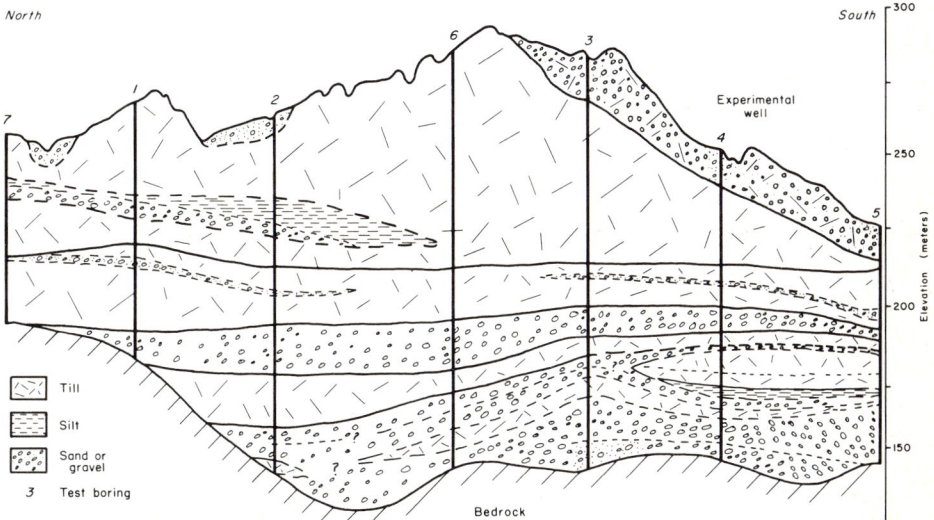

Fig. 1. Geologic cross-section (north—south) across the Paw Paw bedrock valley near the village of Paw Paw, Illinois.

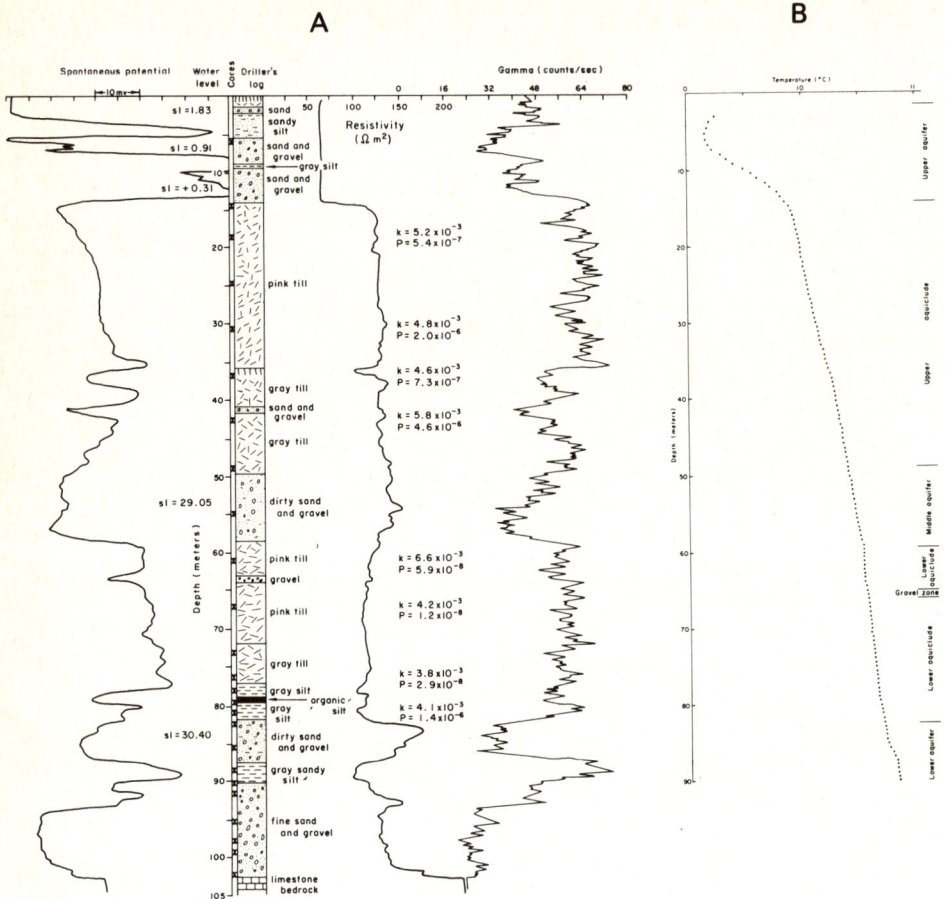

Fig. 2. A. Graphic driller's log, and geophysical logs. B. Measured temperature profile of the experimental well near Paw Paw, located near the SW/c Section 22, T.37N., R.2E. Measurements of static water levels (sl), thermal conductivities (k), and hydraulic conductivities (p) are shown for various units.

the valley which shows the general stratigraphic relationships is shown in Fig. 1.

Only small domestic water supplies are obtained from the glacial deposits in this area. Generally, shallow Wisconsinan or the Illinoian aquifers are tapped. The village of Paw Paw (population 850, 3.5 km north) obtained water from deep bedrock aquifers at the time of the study. The village of Compton (population 400, 10 km west) is the closest large user of water from the valley fill.

The test-well site was selected during the test-drilling project. The site turned out to be more complex geologically than had been anticipated. The well (Fig. 2) was 104 m deep and was systematically cored during drilling. Static water level in each aquifer was determined during drilling by setting a temporary screen and developing the well. The well was constructed of 2-in. pipe with a screen set at 88 m; a bentonite seal was placed around the pipe between the top of the screen and land surface. The screen was later plugged because of the low static water level, and the well filled to within 1.5 m of the surface with water. The well was then allowed to stabilize for about one year prior to conducting any experiments.

The transfer of heat within the borehole by fluid convection is not a serious problem at this well, except near the land surface. Convection of borehole fluids was investigated by Sammel (1968). Water in 2-in. diameter wells should be stable at temperature gradients of 0.1—0.005 °C/m, depending upon the temperature. The well at Paw Paw has a gradient of about 0.013 °C/m at 9—10 °C; this is well below Sammel's upper boundary for stable fluids (water).

Measurements of temperature were made using a Digitec® digital thermistor thermometer, model 1500-47, made by United Systems Corporation. The instrument was specifically modified by the manufacturer to measure a temperature range of 5—25 °C; it has an accuracy of ±0.01 °C and a dial division of 0.005 °C. The thermistor is attached to a weighted 130-m cable which can be lowered down the borehole. According to the manufacturer, the cable will affect the absolute temperature by −0.015 °C at 10 °C, but does not affect the relative temperature difference measured between two points. Temperature measurements in each borehole were made from top to bottom to minimize the effect of measurement on the temperature profiles.

Thermal conductivity measurements were made in the laboratory using a Colora® thermal-conductometer, manufactured in the Federal Republic of Germany, and distributed by Dynatech Research and Development Corporation. Duplicate samples were cut from each core, and two measurements were made on each specimen. The instrument has an accuracy of ±3%; thus a thermal conductivity measurement of 0.0025 is accurate to ±0.0001 cal. s^{-1} cm^{-1} $°C^{-1}$. Repeated measurements on the same specimen generally fell to within this tolerance. The conductivity of duplicate specimens varied as much as ±0.0010 cal. s^{-1} cm^{-1} $°C^{-1}$. Reported values (Fig. 2) are the means of all values measured on each core.

Hydraulic conductivities of the cores were measured using a Soil Test® model K-670 permeameter with a falling-head test. Two or more samples were cut from representative cores. Duplicate measurements of hydraulic conductivity on the same sample varied as much as one-half order of magnitude, and slightly greater between samples from the same core. Hydraulic conductivities reported (Fig. 2) are the means of all measurements on each core.

Vertical temperature profiles were measured in the well at two- to three-month intervals for one year and at yearly intervals for five years. No significant difference in the profiles was noted. A plot of the measured temperature vs. depth is shown in Fig. 2. Note the slight deflection of the measured temperatures from a straight line opposite the two aquicludes. Static water levels for the three aquifers are shown also on Fig. 2.

Nondimensional temperature vs. depth plots were made of the upper (Fig. 3)

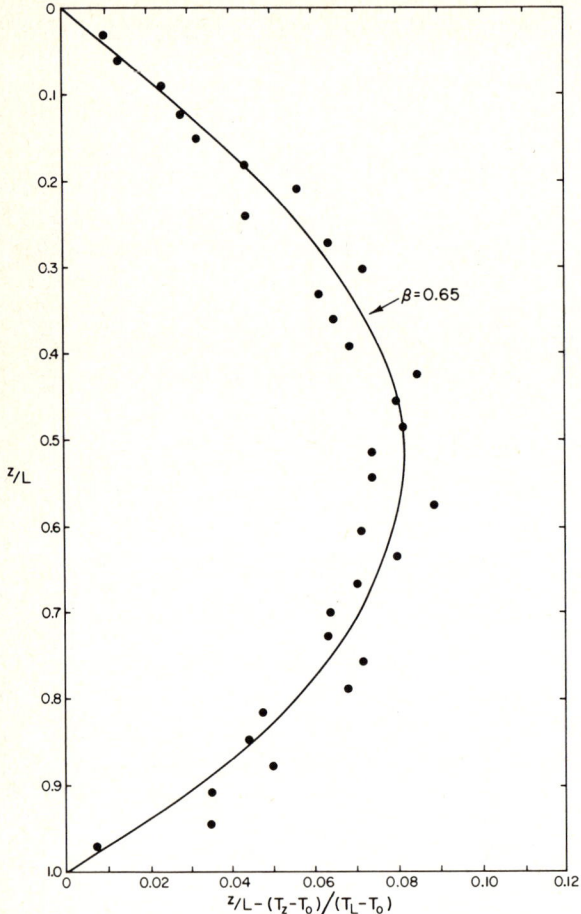

Fig. 3. Nondimensional plot of temperature data between 14 and 49 m (upper confining layer) at the Paw Paw well.

and lower (Fig. 4) aquicludes using the top of each as the origin ($z = 0$). The plots were made using Stallman's (1967) modified form, z/L vs. $z/L - (T_z - T_0)/(T_L - T_0)$. Both plots show considerable scatter, which is not reflected in the measurements of thermal conductivity or hydraulic conductivity.

The upper aquiclude has a large differential head of 0.75 cm/cm. Substituting an average laboratory hydraulic conductivity of $1.96 \cdot 10^{-6}$ cm/s in the Darcy equation, a groundwater flow rate of $1.47 \cdot 10^{-6}$ cm/s (46.4 cm/yr.) is calculated. Inserting $\beta = 0.65$ from the temperature plot (Fig. 3) and an average thermal conductivity of 0.0051 cal. s^{-1} cm^{-1} $°C^{-1}$ in eq. 3, a groundwater flow rate of $9.47 \cdot 10^{-7}$ cm/s (29.9 cm/yr.) is calculated. These values for flow rates are in quite close agreement.

The lower aquiclude has a very low differential head, 0.06 cm/cm. Using

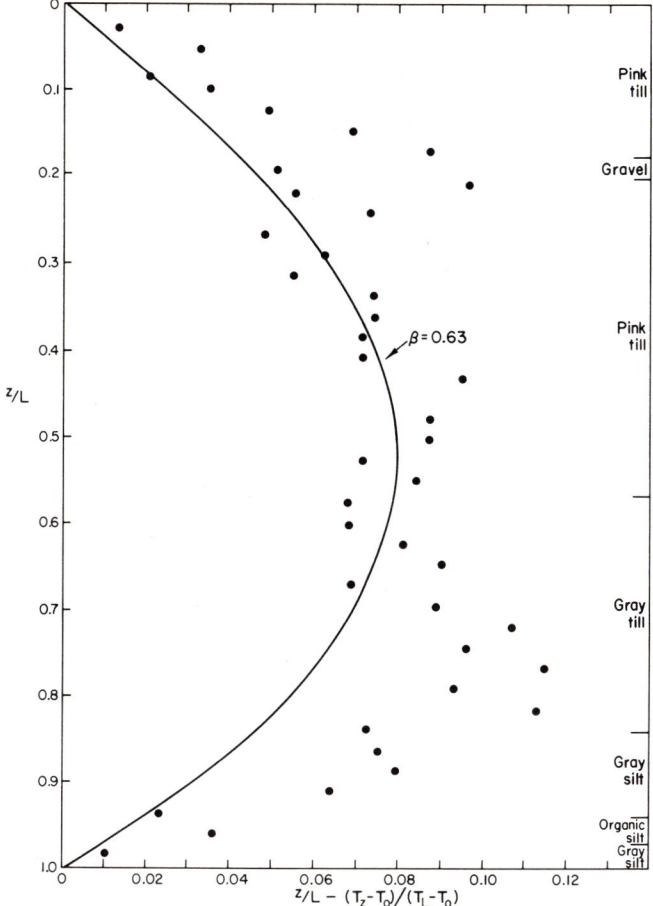

Fig. 4. Nondimensional plot of temperature data between 58 and 82 m (lower confining layer) at the Paw Paw well.

the average laboratory hydraulic conductivity, weighted for the thickness and conductivity of each unit, of $2.91 \cdot 10^{-7}$ cm/s, a flow rate of $1.75 \cdot 10^{-8}$ cm/s (0.55 cm/yr.) is calculated. The nondimensional plot of the temperature vs. depth (Fig. 4) shows considerable scatter, especially near the top and bottom. The best fit of the entire data is to the curve having $\beta = 0.63$. Using this β-value and an average thermal conductivity of 0.0046 cal. s^{-1} cm^{-1} $°C^{-1}$, a groundwater flow rate of $1.21 \cdot 10^{-6}$ cm/s (38.2 cm/yr.) is calculated. These two values for flow rates are two orders-of-magnitude different, nowhere near agreement.

Attempts to divide the lower aquiclude into segments and draw non-dimensional plots for each segment were unsuccessful. There are lithologic breaks (Fig. 2) at the gravel 550 cm below the top of the aquiclude and at the

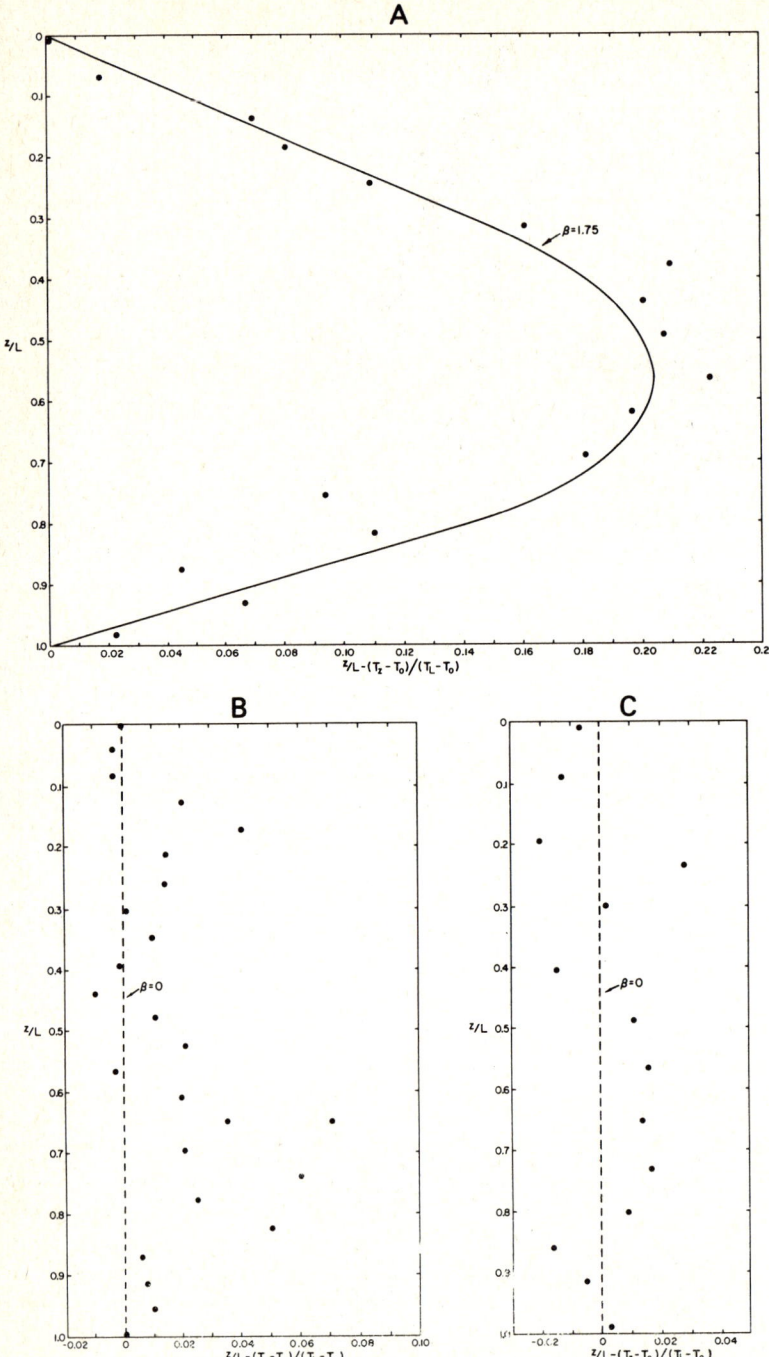

Fig. 5. Nondimensional plot of temperature data for the lower confining layer between the top and 5.76 cm (A), between 575 and 1500 cm (B), and between 1500 cm and bottom (C) of the lower confining layer at the Paw Paw well.

organic silt (strong SP peak) 1950 cm below the top. The breaks in the temperature plot do not coincide exactly with recognizable lithologic breaks; they are at $z/L = 0.25$ and $z/L = 0.65$ in Fig. 4 (575 and 1500 cm, respectively). Temperature plots of these segments are shown in Fig. 5. These plots suggest that all of the groundwater flow is occurring between the middle aquifer and the thin gravel. This possibility is also suggested in the depth—temperature profile (Fig. 2), but is not confirmed by any geologic data. Conductivity measurements on the silt were difficult to obtain; the flexure observed in the temperature profile at 79 m could be caused by the observed decrease of 0.0007 in thermal conductivity from the clayey till to silt. More likely it is the result of the increased organic content of the silt at that point.

I assumed that all leakage is from the middle aquifer to the thin gravel 550 cm below, and that the gravel lens is in hydrologic connection to the main lower aquifer and thus has the same hydraulic head. This gives a hydraulic gradient of 0.25 cm/cm, and using a hydraulic conductivity of $5.9 \cdot 10^{-8}$ cm/s, a leakage rate of $1.48 \cdot 10^{-8}$ cm/s (0.47 cm/yr.) is calculated. Using $\beta = 0.95$ (Fig. 5A) from the upper plot, a leakage rate of $1.09 \cdot 10^{-5}$ cm/s (343.9 cm/yr.) is obtained. Agreement between the two calculations of v_z is worse than using the whole aquiclude; again the flow rate calculated from the temperature data is much too great.

DISCUSSION

The field experiment described shows that under favorable conditions the temperature distribution in a fluid-filled borehole can be used to make reasonable estimates of the rate of groundwater movement through aquicludes. For the upper aquiclude the flow rate of 29.9 cm/yr. calculated using temperature data compares well to the rate of 46.4 cm/yr. calculated using hydraulic data. The calculated rates of flow through the lower aquiclude does not compare as well; rates of 38.2 and 0.55 cm/yr. were calculated from the temperature and hydraulic data, respectively.

Under the relatively undisturbed conditions encountered at Paw Paw, considerable scatter occurred in the temperature data. Scatter severely limits detection of small velocities or measurement of flow across relatively thin confining layers, unless the instrument and fluid convection conditions in the well bore are very stable and the material has a very uniform thermal conductivity. Stability of temperature measurements can be increased by filling the borehole with glycerol (Sammel, 1968).

The complexity of the lower aquiclude at Paw Paw caused considerable difficulty in interpreting the temperature data. The temperature data suggest that the thin gravel 550 cm below the top of the aquiclude may be connected to the lower aquifer, but the connection cannot be confirmed by other evidence As calculated from temperature data, the flow rate is unreasonably high. The organic silt at a depth of 79 m probably contributes a large part of the difficulty in analyzing the lower aquiclude. There is a significant shift in temperature profile at that point. Unfortunately core was not available on which to determine the thermal properties of the organic silt, which must be significantly different from those of other materials in the aquiclude. In other parts of Illinois organic silts are thought to be producing methane (Coleman, 1979) by continued decay of the organic matter. While there is no direct evidence that this is occurring in the Paw Paw well, it seems very likely. Such a heat source could produce the large deflection in the temperature profile at 79 m.

It may be concluded from this experiment that detailed temperature profiles may be used to accurately estimate vertical leakage through an aquiclude between two aquifers. However, the geology of the section must be known with reasonable certainty, as is indicated by the apparently high, and incorrect, flow rate calculated for the lower aquiclude. More experimentation is needed to determine in which geologic environments temperature profiling may be used to determine leakage velocities; however, it appears that it will be valid in many, if not most, areas.

REFERENCES

Boyle, J.M. and Saleem, Z.A., 1979. Determination of recharge rates using temperature—depth profiles in wells. Water Resour. Res. (in press).

Bredehoeft, J.D. and Papadopulos, I.S., 1965. Rates of vertical ground-water movement estimated from the earth; thermal profile. Water Resour. Res., 1(2): 325—328.

Cartwright, K., 1970. Ground-water discharge in the Illinois Basin as suggested by temperature anomalies. Water Resour. Res., 6(3): 912—918.

Clark, Jr., S.P., 1966. Thermal conductivity. In: Handbook of Physical Constants (rev. ed.), Geol. Soc. Am. Mem., 97: 459—482.

Coleman, D.D., 1979. The origin of drift-gas deposits as determined by radiocarbon dating of methane. In: Proceeding of the Ninth International Radiocarbon Conference, June 20—26, 1976, University of California at Los Angeles Press, Los Angeles, Calif. (in press).

Hubbert, M. King, 1940. The theory of ground-water motion. J. Geol., Part 1, 48(8): 785—944.

Kempton, J.P., 1963. Subsurface stratigraphy of the Pleistocene deposits of central-northern Illinois. Ill. Geol. Surv., Circ. 356, 43 pp.

Kunii, D. and Smith, J.M., 1961. Heat transfer characteristics of porous rocks, 11. Thermal conductivities of unconsolidated particles with flowing fluid. J. Am. Inst. Chem. Eng., 7(1): 29—34.

Sammel, E.A., 1968. Convective flow and its effect on temperature logging in small-diameter wells. Geophysics, 33(6): 1004—1012.

Sorey, M.L., 1971. Measurement of vertical ground-water velocity from temperature profiles in wells. Water Resour. Res., 7(4): 963—970.

Stallman, R.W., 1960. Notes on the use of temperature data for computing ground-water velocity. Société Hydrotechnique de France, Nancy, Sixth Assembly on Hydraulics, Rep., 3: 1—7 (also in: R. Bentall (Compiler), Methods of Collecting and Interpreting Ground-water Data, U.S. Geol. Surv., Water-Supply Pap., 1544-H: 36—46, 1963).

Stallman, R.W., 1967. Flow in the zone of aeration. In: V.T. Chow (Editor), Advances in Hydroscience, Vol. 4, Academic Press, New York, N.Y., pp. 151—195.

[2]

UTILITY OF A COMPUTERIZED DATA BASE FOR HYDROGEOLOGIC INVESTIGATIONS, LAS VEGAS VALLEY, NEVADA

ROBERT F. KAUFMANN and HERBERT N. FRIESEN

Office of Radiation Programs, U.S. Environmental Protection Agency, Las Vegas, NV 89114 (U.S.A.)
Water Resources Center, Desert Research Institute, University of Nevada System, Las Vegas, NV 89109 (U.S.A.)

(Accepted for publication May 10, 1979)

ABSTRACT

Kaufmann, R.F. and Friesen, H.N., 1979. Utility of a computerized data base for hydrogeologic investigations, Las Vegas Valley, Nevada. In: W. Back and D.A. Stephenson (Guest-Editors), Contemporary Hydrogeology — The George Burke Maxey Memorial Volume. J. Hydrol., 43: 195—216.

Hydrogeologic study of the shallow groundwater zone in Las Vegas Valley, Nevada involved development of an extensive computerized data base consisting of water analyses and water-well logs. The data were manipulated and reduced using a variety of graphical and statistical techniques applicable to analysis of spatial and temporal changes in water quality. Stratigraphic relations, permeability/transmissivity variations, water budgets and ambient quality relative to drinking-water standards were evaluated as part of an overall EPA funded study to determine the sources and extent of groundwater contamination and develop management alternatives to minimize adverse effects. Extension of the data base to include investigations concerned with improved definition of the stratigraphic and structural makeup of the valley and to applied study of water quality, subsidence and water banking seems reasonable.

INTRODUCTION

Since 1944 and particularly from 1962 until the time of his death, Burke Maxey directly influenced a number of diverse water-resources studies in Las Vegas Valley. If not actually conducting his own studies, he managed investigations of the hydrogeologic framework, subsidence due to groundwater withdrawal, conjunctive use of groundwater and Colorado River water, stratigraphic controls on shallow-groundwater quality, and land- and water-use influences on groundwater quality, to name a few. The authors are deeply indebted to Burke for the guidance and review he so freely gave in the course of the EPA project (Land and Water Use Effects on Ground Water Quality in Las Vegas Valley-Grant R800946) which gave rise to this paper. Burke was instrumental in drafting the original scope of work in 1969 and in periodically realigning the study output with the evolving water-management picture in the years to follow.

Area description

With the influx of population to Las Vegas Valley in the 1940's, the hydrologic regime has undergone significant change. The pristine desert environment of the valley gave way to sprawling urban and suburban areas, sewage treatment plants, a large industrial complex, numerous golf courses, and an extensive marsh system created by point and non-point sources of return flow. There has also been widespread dewatering of the deeper highly-transmissive aquifers and water logging in much of the near-surface water table present in the eastern two-thirds of the developed area.

Las Vegas Valley is a major topographic depression in southern Nevada covering 912 km^2 (Fig. 1). The valley trends NW—SE and is situated along the Las Vegas shear zone characterized by intense structural deformation, primarily consisting of right-lateral movement. Along this zone extensive erosion of bedrock units created a basin at least 1000 m deep in the central part of the valley. Filling the bedrock depression are thick deposits of sand, gravel, silt and clay. The deposits are generally coarsest toward the mountain fronts. Silt and clay deposits occupy the central portion of the valley.

Extensive groundwater withdrawals occur primarily from the Pleistocene valley fill and possibly from coarse-grained facies of the Muddy Creek Formation (Miocene) which may be present in the western part of the valley. At present, essentially no use is made of groundwater in the very shallow deposits and associated caliche strata, or the underlying bedrock units, because of inadequate permeability, excessive depth to water, and/or poor water quality. Extensive groundwater discharge as evapotranspiration and surface water discharge occurs in the Las Vegas Wash area which encompasses about 80 km^2.

Maxey and Jameson (1948) first defined the basic hydrogeology of Las Vegas Valley. The near-surface flow system, which is unconfined except possibly in a few very localized areas, is in hydraulic continuity with underlying saturated sediments. Although perched water is locally present, it is not significant in the context of water-resources management in the valley or relative to the present study. Previous to groundwater development, which effectively began in 1907 and constituted an overdraft situation by the 1940's, recharge was principally by upward movement from underlying aquifers. Because of pumping from deep aquifers, the potential gradient in the near-surface zone in the western and northwestern parts of the valley has been reversed, causing shallow groundwater to move downward into underlying aquifers. Under natural conditions, recharge to the valley fill was believed to be primarily a result of precipitation in the surrounding mountains, principally those to the west and north. Flow from the recharge areas

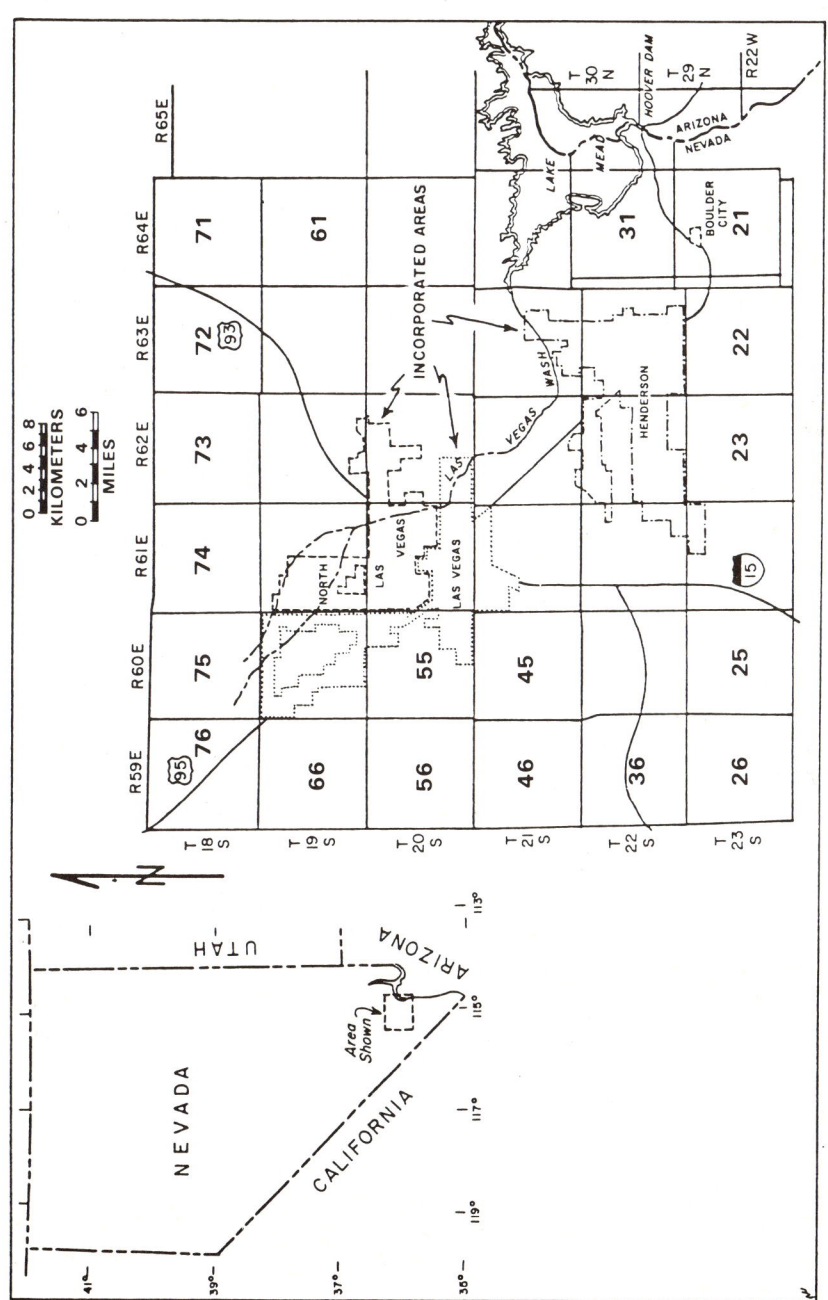

Fig. 1. Location map showing the Las Vegas study area and the matrix notation scheme used to identify townships.

via a deep aquifer system involved an easterly or southeasterly path incorporating lateral and then upward movement in the valley fill either as diffuse seepage or as localized flow towards springs discharging in close proximity to fault planes.

Objectives and scope

The present paper is an abstraction from and extension of larger reports on the shallow groundwater resources of Las Vegas Valley (Kaufmann, 1978; Patt, 1978). These were prepared in the period 1974—1977 by the Desert Research Institute (DRI), University of Nevada System, under a grant from the U.S. Environmental Protection Agency (EPA). Additional research proposals and innovative ideas on the part of the DRI staff also were drawn upon where they relate to the development or application of the data base to water-related problems and issues in the valley.

Water quality in parts of the near-surface system of aquifers and aquitards is greatly affected by urban and industrial land- and water-use practices, chief of which are liquid waste disposal, overdraft of deeper aquifers, irrigation return flows, infiltration of overland flow, and disruption of natural soil conditions (Kaufmann, 1977). The relationship between such land- and water-use patterns and basin-wide water-quality management, particularly the groundwater aspects, should be obvious to responsible water managers and is indeed relevant to the mandates for groundwater protection embodied in Public Law 93-523 (Safe Drinking Water Act) and the water planning efforts under section 208 of Public Law 92-500 (Federal Water Pollution Control Act Amendments of 1972).

DATA-BASE CONSTRUCTION

Data types and sources

Data requirements for the study primarily included water analyses and water-well logs. Water-quality data were not available in computerized form. The U.S. Geological Survey (USGS) had compiled a substantial file of well inventory records which were on magnetic tape (Thordarson and Robinson, 1971). Review and comparison of available data vs. project requirements led to the decision to computerize all available water analyses and well log data, with exception of the information already on the USGS data tape. Well inventory schedules that had been entered into the computer file were complete through 1967, but hundreds of new wells drilled since then necessitated coding of additional inventory schedules. Table I is a partial list of the parameters included on both the USGS and DRI well inventory schedules as of 1973. Fig. 2 shows a portion of the water-analysis data coding form (Friesen, 1970).

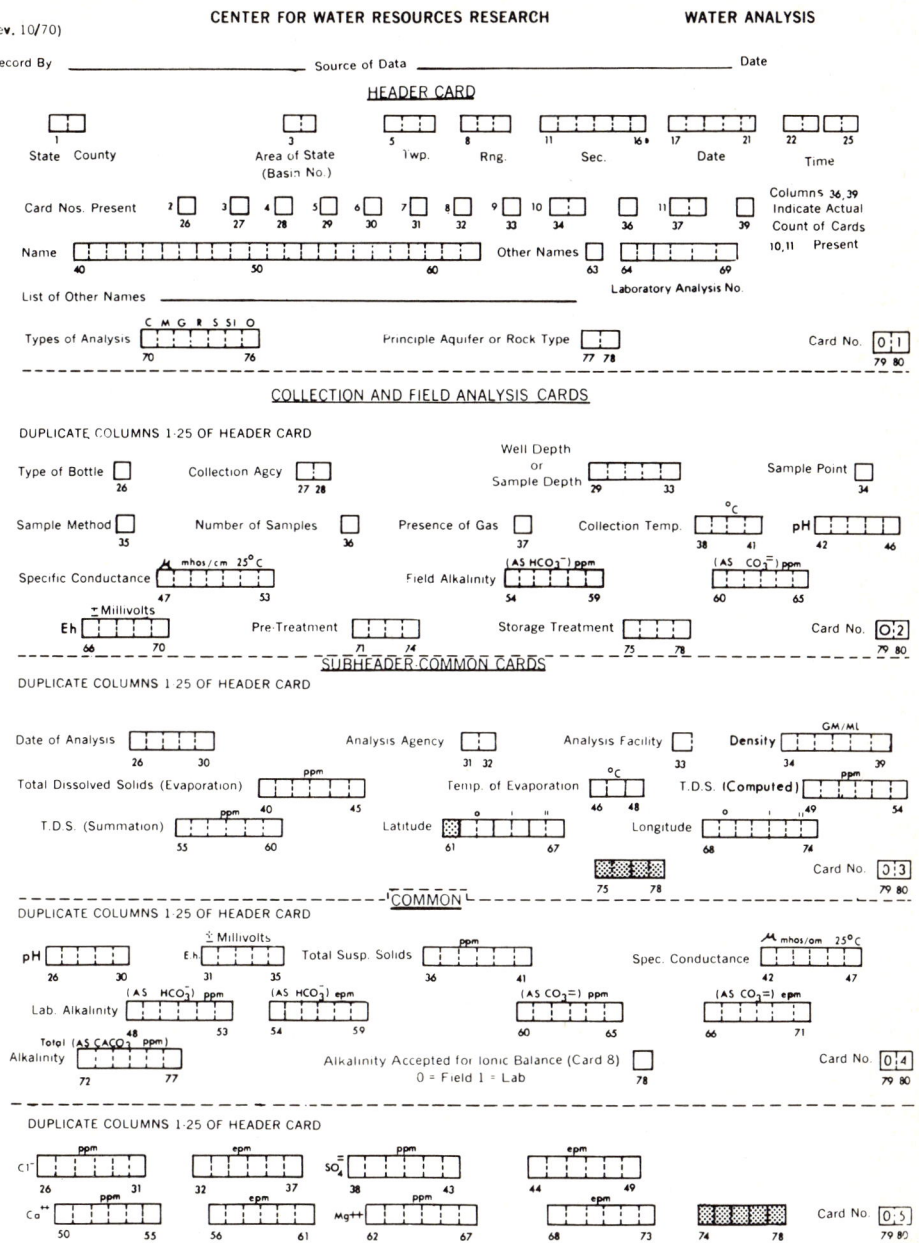

Fig. 2. Center for Water Resources (DRI) water analysis coding form.

TABLE I

Partial list of data items contained on USGS and DRI well inventory schedules

Item	USGS	DRI	Item	USGS	DRI
State and County	X	X	Power	X	X
Latitude and Longitude*	X	X	Altitude*	X	X
Township, range, section	X	X	Water level*	X	X
Name and/or address	X	X	Date measured	X	X
Ownership	X	X	Well*	X	X
Water use	X	X	Drawdown*	X	X
Well use	X	X	Pumping period	X	X
Well data	X	X	Quality of water (7 items)	X	X
Frequency of water level			Physiography	X	X
data	X	X	Drainage basin	X	X
Field characteristics	X	X	Topographic setting	X	X
Hydrologic lab data	X	X	Major aquifer (8 items)	X	X
Quality-of-water analyses	X	X	Minor Aquifer (8 items)	X	
Frequency of quality			Depth to consolidated rock*	X	
sampling	X	X	Depth to basement*	X	
Pumpage inventory	X	X	Surface material	X	X
Aperture cards	X	X	Coefficient of trans-		
Typed log date	X	X	missivity	X	X
Depth of well*	X	X	Coefficient of storage	X	X
Depth cased	X	X	Flow direction		X
Diameter	X	X	SEO log number		X
Well finish	X	X	Water right permit number		X
Method drilled	X	X	Perforated interval (PED)		X
Year drilled	X	X	Infiltration		X
Pump setting	X	X	Uppermost bedrock for-		
Method of lift	X	X	mation		X
Deep or shallow	X		Specific capacity		X

*Starred items have accuracy or qualifier code fields

Most of the water-quality data were obtained from the Clark County District Health Department (CCDHD) and the Nevada Bureau of Environmental Health. Approximately 2000 additional analyses came from surface and groundwater sampling performed by DRI as part of the EPA study. Lithologic data came from drillers' logs on file with the State Engineer and the CCDHD.

Number of records by type

When data collection was completed, the master data tapes contained 139,693 records, each of which was 110 characters long. Following is a summary of the data records on the master tapes:

Total number data points = 9468   Total water analyses = 6700
Total wells inventoried = 5518   Total matched points = 4830
Total lithologic logs = 5617

Extensive effort was devoted to matching water analyses to either well inventory or lithologic log records. The inability to match points was due to lack of a common basis of unique location identification between different record types. For example, among the USGS well inventories the extreme case had 43 wells with the same township, range, section and quartering (TRSQ). Only the sequence number assigned to the latitude and longitude made each record set unique. Street addresses did not appear on the inventory records. On the other hand, the water-analysis records contained street addresses and TRSQ data but no latitude and longitude data, and no well description information. The lithology records had street addresses and TRSQ data but no latitude and longitude. Well descriptions were used for matching to USGS inventories. Efforts to match data records between sources were only partially successful. As indicated in the summary above, there were 6700 water analyses, but only 4830 could be matched with specific well logs. Perhaps 1500 additional matches could have been assigned on a speculative basis, but the decision was made that there had to be a positive match involving location and at least two other descriptive items such as well depth and diameter, year drilled, casing depth and diameter, street address, owner name, date water sampled, and log or laboratory number.

Data-point location system

There are several methods currently in use to sort data points by location. Each has its own advantages and disadvantages depending upon the intended use of the data. The well inventory file originally obtained from the USGS (Thordarson and Robinson, 1971) was sorted by longitude within latitude. A more recent USGS groundwater site inventory (GWSI) code sorts by a number of location parameters. Sorting by longitude within latitude can scatter data for a given township throughout a file, thereby reducing retrieval efficiency and creating undue problems in scanning data from a limited contiguous area. Sorting by township, range and section will produce a file with data points from a contiguous township but the section numbers zigzag so there is little logical relationship between section numbers and real points on the ground.

For purposes of this study, and to achieve a useable unique location identifier for each data point, a new location system was devised, as shown in Fig. 3. The system is basically a series of matrix notations such that any point can be represented by a 12-digit number. The 12 digits are paired off and the following meanings attached. Assume 12 digits in the form:

$i_1 i_2 j_1 j_2 k_1 k_2 l_1 l_2 m_1 m_2 nn$

then,

i_1 = row, i_2 = column of township within the hydrographic basin matrix
j_1 = row, j_2 = column of a section within a township 6×6 matrix
k_1 = row, k_2 = column of a 1/4 section within a section 2×2 matrix

l_1 = row, l_2 = column of a 1/16 section within a section 4 × 4 matrix
m_1 = row, m_2 = column of a 1/64 section within a section 8 × 8 matrix
nn = sequence number from 1 to 99 distinguish between points having identical preceding 10 digits

This procedure allows up to 99 different data points within a given level of location precision. The worst case encountered generated sequence number 48 with two levels of quartering. The data points involved were primarily 20.3-cm diameter wells drilled to 30 m during the mid 1950's and early 1960's in an area in the NW portion of the valley. Most of the wells were

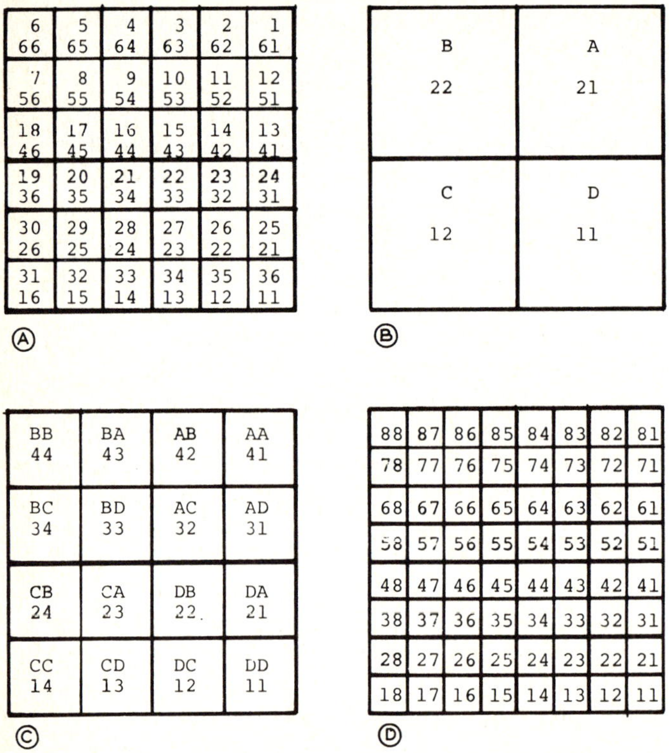

6	5	4	3	2	1
66	65	64	63	62	61
7	8	9	10	11	12
56	55	54	53	52	51
18	17	16	15	14	13
46	45	44	43	42	41
19	20	21	22	23	24
36	35	34	33	32	31
30	29	28	27	26	25
26	25	24	23	22	21
31	32	33	34	35	36
16	15	14	13	12	11

(A)

B	A
22	21
C	D
12	11

(B)

BB	BA	AB	AA
44	43	42	41
BC	BD	AC	AD
34	33	32	31
CB	CA	DB	DA
24	23	22	21
CC	CD	DC	DD
14	13	12	11

(C)

88	87	86	85	84	83	82	81
78	77	76	75	74	73	72	71
68	67	66	65	64	63	62	61
58	57	56	55	54	53	52	51
48	47	46	45	44	43	42	41
38	37	36	35	34	33	32	31
28	27	26	25	24	23	22	21
18	17	16	15	14	13	12	11

(D)

Fig. 3. Matrix notation scheme used to identify data-point locations within township and section.
A. Sections in township. The top number in each square is the section number, the bottom number is the matrix notation.
B. 1/4ths in a section. The letter code is the usual quartering designation, the numbers are the matrix notation.
C. 1/16ths in a section. Letters and numbers as in (B).
D. 1/16ths in a section. Only the matrix notation is shown.
If the data file contains mixed levels of accuracy the data records must allow for all necessary pairs of matrix numbers, however, if only one level of accuracy is present then only one matrix pair is requested.

drilled solely as an investment to gain title to Federal lands sold to the public under the Desert Land Entry Act. In many cases there were no street addresses, and land ownership changed several times in the period up to 1967; hence names and addresses were not adequate for further distinction. Aside from this problem, which was not restricted to any one part of the valley, provisions for sequencing were adequate for the project.

The five levels of matrix notation all have row 1, column 1 in the lower right corner so the data order in the sorted file is very similar to a file sorted one section at a time by latitude and longitude. A very useful advantage of this system is the ease of specifying the boundaries of an irregularly shaped area for retrieval purposes. This feature was extensively used in retrieving static level, perforated interval and specific capacity information for the groundwater model (Westphal, 1978) of several subareas, each with an irregular boundary.

Input data are received from a variety of sources, and location information is specified with varying levels of precision. Many of the data sets assembled for use in the Las Vegas Valley study required matrix notation for three levels of quartering within a section. If all data points in a file have quartering to the same precision level, the more detailed level of matrix notation (Fig. 3C vs. B or Fig. 3D vs. C) adequately expresses the location.

Initially all data processing was performed on a Xerox® Sigma 7 at the University of Nevada at Reno. Changeover by the Univrsity to a CDC® 6400 and access to a second CDC® system at the U.S. Department of Energy in Las Vegas made it necessary to convert all programs. All programs executed within 130-K (octal) available memory. The CDC® 6400 was configured with seven 7-track tape drives, two 9-track tape drives and a 30-in. CALCOMP® plotter. Extensive use was made of the plotter and sort/merge software.

PREVIOUS APPLICATIONS

Spatial trends in groundwater quality

An extensive water-sampling program of the District Health Department resulted in several thousand water analyses for the period 1968—1973. Prior to the study reported on herein there was no attempt to reduce and interpret the data for hydrogeological purposes. The Las Vegas Valley Water District and the city of North Las Vegas have additionally monitored their wells in the western part of the valley, although few data prior to 1969 are available.

A trend-surface analytical technique, in conjunction with the data bank, was implemented to reduce and portray water-quality and well log data. Manual retrieval of analyses and logs and cross-referencing would have been extremely inefficient considering the amount of data available. Trend surfaces have the attribute of portraying broad, overall trends rather than detailed but complex variations of raw data. Statistics such as the correlation coefficient relate how well the surfaces fit the raw data and therefore provide a measure of how well the trends estimate spatial variability for the system and parameter considered.

The trend-surface technique involves fitting polynamial surfaces to map data by means of a general linear model incorporating a least-squares fit of a planar or curvilinear surface to the observed data. More complex surfaces involved higher-order polynomials and more terms in the equation collectively relating each datum (Z) to its location (X, Y) in the area of consideration. In the past, trend-surface analysis has primarily been applied to stratigraphic, structural and sedimentation problems involving such topics as analysis of lithofacies variations, source areas for heavy-mineral assemblages, and evaluation of economic deposits of oil, gas and ore. Davis et al. (1969) applied trend-surface analysis to a problem involving groundwater use, replenishment and aquifer characteristics in Indiana. For the present study, trend surfaces were relied upon to portray chemical quality of groundwater for several reasons. From a research standpoint, the method had not yet been adequately tested and applied to groundwater-quality problems. Another consideration was the need to reduce and generalize a great mass of chemical data, some of which were of questionable veracity. Retrieval of raw data for manual plotting and contouring would have been extremely time consuming and inefficient. Finally, broad overall trends are analytically more useful in describing variations in Las Vegas Valley.

For three separate depth intervals or "slices", 0—15.2, 15.2—30.5 and 30.5—91.5 m, trend surfaces and their associated statistics proved very useful for evaluating several hundred water analyses selected to depict the nature of lateral variations so as to clarify regional and local trends (Kaufmann, 1978). By simplifying otherwise complex patterns of ion distribution, variations attributable to natural and man-related sources of water-quality deterioration were identified. Influences of natural and man-related sources of pollution are most evident from the tritium, chloride and nitrate data and, to a lesser extent, from variations in TDS. Tritium and NO_3^- in particular, were useful indicators of return flows associated with irrigation and effluent disposal (Kaufmann, 1977).

Spatial variations in groundwater quality, particularly for depths below 15.2 m, were determined using analyses from January 1, 1968 up to 1972 for the depth intervals previously indicated, providing the anion/cation ratio was in the range 0.9—1.1. From card decks of acceptable analyses, plots of sampling point locations were prepared to determine density of data points in the study area for the three depth intervals elected. In areas of high density, analyses for depth intervals were averaged to one value per quarter section to avoid undue weighting of the polynomial equations describing the trend surfaces.

Three chemical data bases were utilized: (1) historical analyses from 1909 to 1964 (supplemented by resampling of the same wells or substitutes, where possible); (2) chemical data on file at the District Health Department and representative of Las Vegas Valley-wide sampling of domestic and municipal wells from 1968 to 1972; and (3) approximately 2000 water analyses of shallow groundwater, effluents and potable water sampled as part of the study in the period 1970—1973. The information presented herein is primarily based on item (2) of the foregoing.

Data availability varied with the depth interval considered. In the case of the interval of 0—15.2 m, only TDS, Cl^- and NO_3^- were considered as the water was clearly not potable and these parameters were considered most indicative of return flows from urban and agricultural areas. In addition to the trend surfaces for each parameter, CALCOMP plots of actual values by location were made to enable a visual scan of the raw data and selection of the contour interval and reference contour for the trend.

The data base was used to produce hydrochemical facies maps for the depth interval from 30.5 to 91.5 m. A plotting routine was used in combination with a program which categorized each water analysis into one of 16 facies or classes of water quality, resulting from different sediment composition and source areas, variations in residence time within the flow system and effects of ion exchange, particularly in the fine-grained sediments. Dominant classes or facies were mapped to show trends throughout the area where data were available.

Spatial trends in Cl^- at three different depth intervals illustrate one use of the data base and retrieval programs. Cl^- trend surfaces for depths of 0—15.2 m had low coefficients of correlation, indicating numerous local variations present in comparison to broad valley-wide trends for TDS or SO_4^{2-}. Distribution of Cl^- in groundwater for the interval from 30.5—91.5 m is also poorly described by the trend surfaces. As in the case of NO_3^-, numerous local variations are superimposed on the first-order regional trend (Fig. 4). The latter depicts a concentration gradient of approximately 6 mg l^{-1} km^{-1}. Actual concentration in the SE and NW corners of the study area are 185 and 10 mg/l, respectively. In the NW part of the valley, low concentrations of Cl^- suggest that recharge to the alluvial fill comes from source area(s) low in Cl^- and is characterized by short residence time, or relatively short flow paths, or both. Thus, recharge to the alluvium may be associated with movement in the carbonate aquifers rather than only in the alluvial aprons flanking the carbonates (Kaufmann, 1978).

It is apparent from the first-order Cl^- trends (Fig. 4) that absolute concentrations at any point and concentration gradients across the valley are very dissimilar for the three depth intervals considered. Markedly more saline conditions prevail in the interval from 0 to 30.5 m and largely reflect natural conditions, primarily concentration by evapotranspiration and the presence of saline soils. Both influences are apparent in the interior portion of the valley characterized by shallow depths to groundwater, phreatophytes and fine-grained playa-facies sediment types. However, man-related factors such as waste disposal are also operative and are dominant influences in the eastern part of the valley (Kaufmann, 1977). The fourth-degree trend for Cl^- at depths of 0—15.2 m (Fig. 5) shows concentrations increasing from about 50 mg/l NW of the urban area to between 500 and 1000 mg/l in the upper reaches of Las Vegas Wash.

Cl⁻ concentrations at certain stations near Las Vegas Wash are considerably higher than the regional trend due to the presence of industrial wastewater return flows in the shallow aquifer. From 1971 through 1973, waste discharges from the industrial complex exhibited a wide range in Cl⁻ concentrations (178—168,310 mg/l). An average value is difficult to define but is probably at least 5000 mg/l and, therefore, well above background. This influences the concentration gradients for Cl⁻ in the very shallow aquifer in the SE part of the valley.

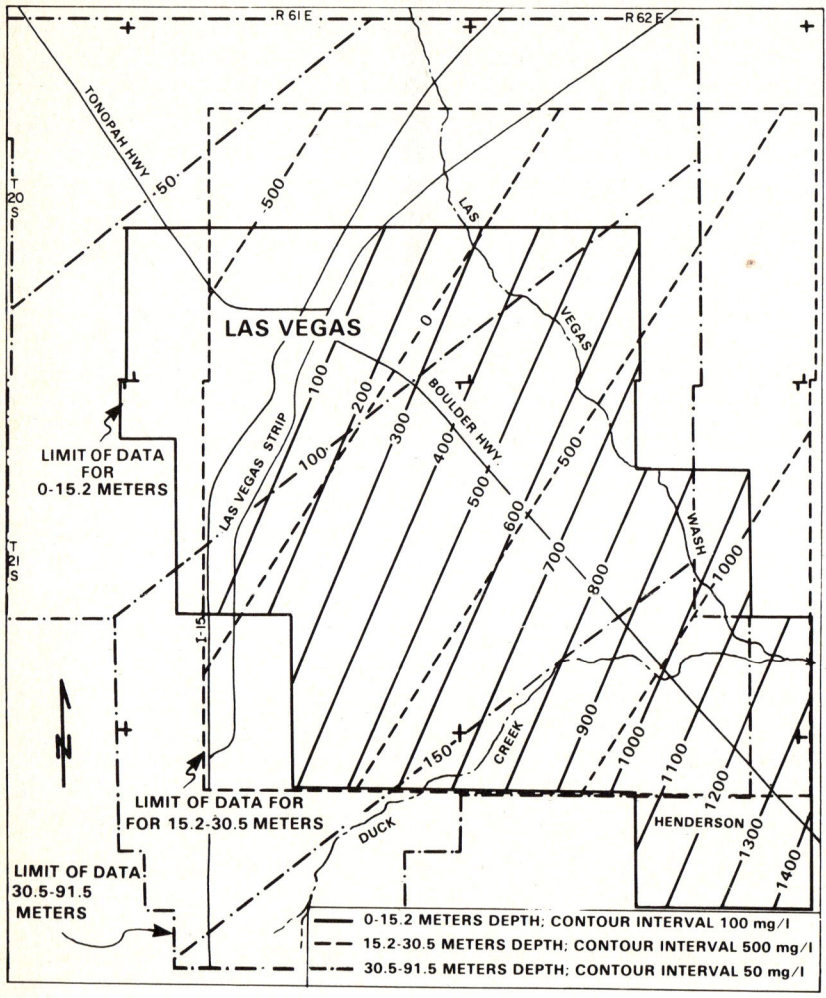

Fig. 4. First-degree trend surface for Cl⁻ in groundwater at depths of 0—15.2, 15.2—30.5 and 30.5—91.5 m.

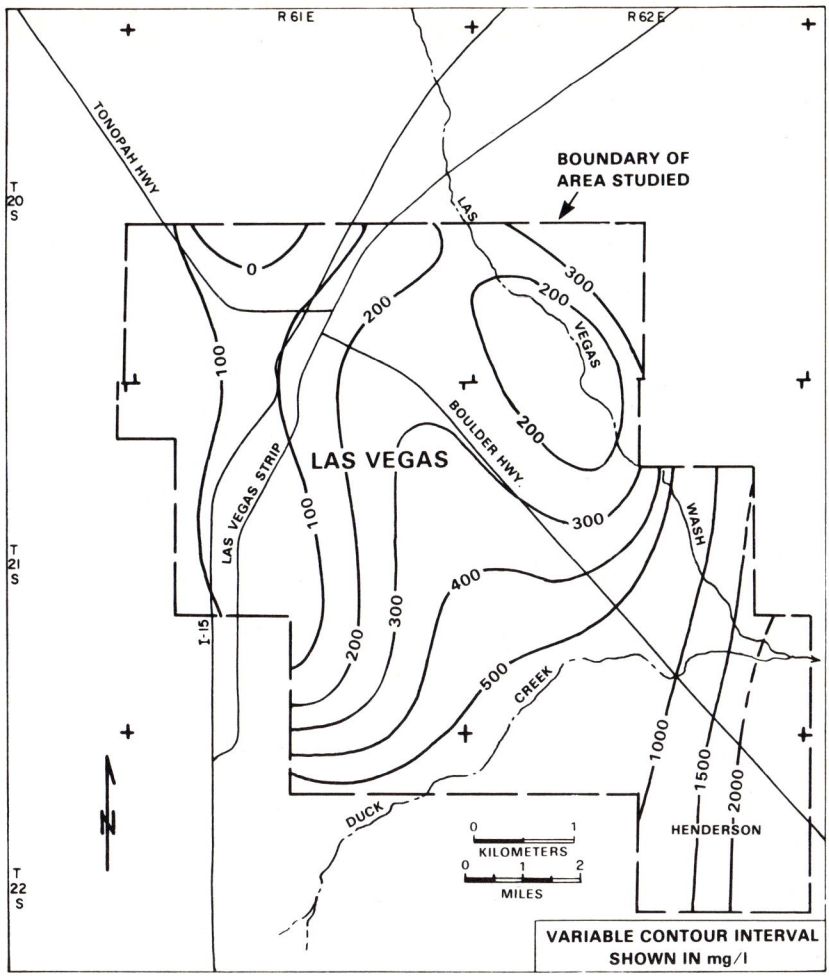

Fig. 5. Fourth-degree trend surface for Cl⁻ in groundwater at depths of 0—15.2 m.

The trend surfaces, particularly for TDS, at depths of 30.5—91.5 m reveal and E—W trending zone of relatively good-quality water extending across the northern portion of the study area to the central portion of Township 20 South, Range 62 East (*20/62*), and then southward toward Boulder Highway. This is believed indicative of favorable conditions for groundwater development due to rather permeable sediments and/or significant inflow. This zone was not identified in previous studies.

Utility of the trend-surface technique

Trend surfaces and their related statistics for various depth intervals proved useful for evaluating and portraying lateral groundwater quality variations and the degree of variability. Judgments were made in rejecting anomalous raw data and in comparing plots of raw data to determine what order surface provides the "best fit". Effects of such judgments were apparent, as evidenced by the increased value of the correlation coefficient derived from selected vs. raw data. Nevertheless, if there was poor initial correlation (in the case of Mg^{2+}, Cl^- and NO_3^- for the 30.5—91.5-m interval), little change was effected in the coefficient calculated using only selected data.

The highest correlation coefficient was generally associated with the sixth-order surface, but only in a few instances was this surface judged the best indicator of overall trends. This decision was made on the basis of visual comparison of various surfaces with plots of the actual data. Absolute values of the coefficients were indicative of the order or trend, or lack thereof, in the data.

Several shortcomings in the use of trend surfaces and their statistics to display groundwater-quality variations were apparent. For the present study, at least, there was an inverse relationship between the amount of data the "goodness of fit" for a given surface and parameter. Similar results were reported by Rockaway and Johnson (1967) in their analysis of water-level data. For example, the interval from 0 to 15.2 m had fewer data points, generally resulting in a higher correlation coefficient for given parameters, compared to lower depth intervals. This erroneously indicated there was greater regularity or order in the very shallow system when in fact the opposite was true.

High concentrations of NO_3^- and Cl^- at depths of 0 to 15.2 m were consistently associated with sanitary and industrial wastes in the eastern part of the valley. Away from areas of waste disposal and for most samples collected below this depth, NO_3^- is 5 mg/l or less, and this is considered background. Where well depths exceeded 15.2 m or were unknown, the chemical data were not used in the trend surface portrayals for the interval. The combination of few data points and the association of high NO_3^- in areas of waste disposal, therefore, partially accounted for the high (0.704) correlation coefficient for the 0—15.2-m interval vs. only 0.246 for 30.5—91.5-m interval. The low coefficient of correlation for NO_3^- at the latter depths suggests very little variation, i.e., background conditions, whereas in the shallow aquifer there is obviously a high positive correlation that can be attributed to human activities.

Another weakness clearly demonstrated by some of the trend surface was that the best fit to the available data frequently had unrealistic values in marginal areas where data were scarce or nonexistent. This even resulted in negative values which indicated failure of the surface to "fit" the actual data. Perhaps a final problem of using the technique was the tendency for misunderstanding the difference between a trend surface and maps or surfaces manually contoured from real data. Whereas both approaches showed concentration increase or decrease, trend surfaces did not always clearly express true values and local deviations.

Impacts of return flows on shallow groundwater quality

NO_3^-, Cl^- and TDS concentrations are particularly diagnostic of return flows associated with urbanized portions of the Las Vegas Valley, areas of sewage disposal, areas irrigated with sewage, and the industrial complex in Henderson. Detailed documentation of water use and return flows (Patt, 1978; Malmberg, 1965) reveal returns from sewage effluent, industrial wastes, cooling water, and septic-tank systems infiltrated the near-surface aquifer.

Because numerous sources of NO_3^- are or have been present, concentrations were initially analyzed through use of trend surfaces to ascertain

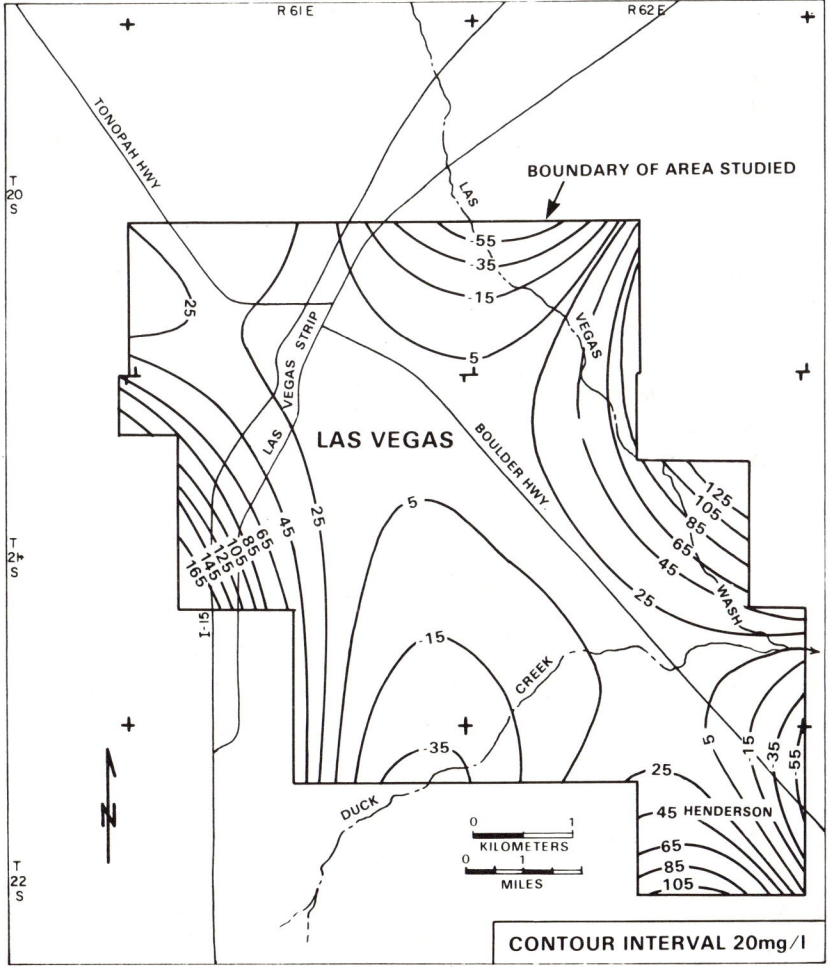

Fig. 6. Fifth-degree trend surface for NO_3^- in groundwater at depths of 0—15.2 m.

variations for the depth intervals considered. The second approach consisted of manually plotting NO_3^- levels equal to or greater than 10 mg/l and comparing the results with known distributions of septic-tank and cesspool waste-disposal systems and areas of sewage disposal. In a negative fashion, trend surfaces for NO_3^- (Fig. 6) proved to be rather revealing. Concentrations above 10 mg/l in the zone from 0 to 15.2 m are common, particularly in the eastern part of the valley. In contrast, NO_3^- at depths of 30.5 m is irregularly distributed and generally quite low in concentration. This implies that waste disposal, and primarily that in the eastern part of the valley, dominates the shallow NO_3^- patterns.

The threshold value of 10 mg/l was selected to distinguish between background and polluted levels of NO_3^-. The value is arbitrary but conservatively so because natural values average less than 5—7 mg/l NO_3^- according to raw-data plots and trends initially generated. By determining deviations from background and relating them to causal factors of land and water use, conclusions can be drawn concerning the impact of septic tanks and other sources of nitrate pollution on shallow-water quality.

Areas of non-potable water

To determine where water quality in the depth interval from 30.5 to 91.5 m does not meet recommended standards for SO_4^{2-} (250 mg/l) and TDS (1,000 mg/l), a plot (Fig. 7) was prepared showing the location of the wells and the parameters involved. An arbitrary limit of 500 mg/l for hardness was set on the basis of comments in McKee and Wolf (1973).

A NO_3^- threshold of 10 mg/l was chosen because concentrations below depths of 30.5 m are normally below 10 mg/l. Thus, although NO_3^- concentrations to date rarely exceed the maximum permissible limit (45 mg/l) in drinking water, high NO_3^- values shown in Fig. 7 are believed indicative of areas where bacterial contaminants or dangerous NO_3^- concentrations are most likely. This is particularly true of areas served by shallow domestic wells and septic tank systems as in Townships *20/61, 20/62, 21/62* and *22/61*.

Hardness, SO_4^{2-} and TDS are the most common parameters exceeding the recommended levels in the valley and are probably the result of natural factors. For example, excessive hardness and SO_4^{2-} generally coincide with the Duck Creek drainage area in Paradise Valley, an area of gypsiferous soils, high evapotranspiration, and poor to moderately permeable sediments. As expected, wells with excessive TDS are located farther eastward, i.e., in the downgradient position of the flow system. In *21/62*, high TDS is associated either with high NO_3^- or is present along the extreme edge of infiltrated-sewage effluent and/or septic-tank leachate, whereas minimal recharge and highly gypsiferous sediments on the flanks of the Frenchman Mountain block could account for hardness—SO_4^{2-}—TDS association.

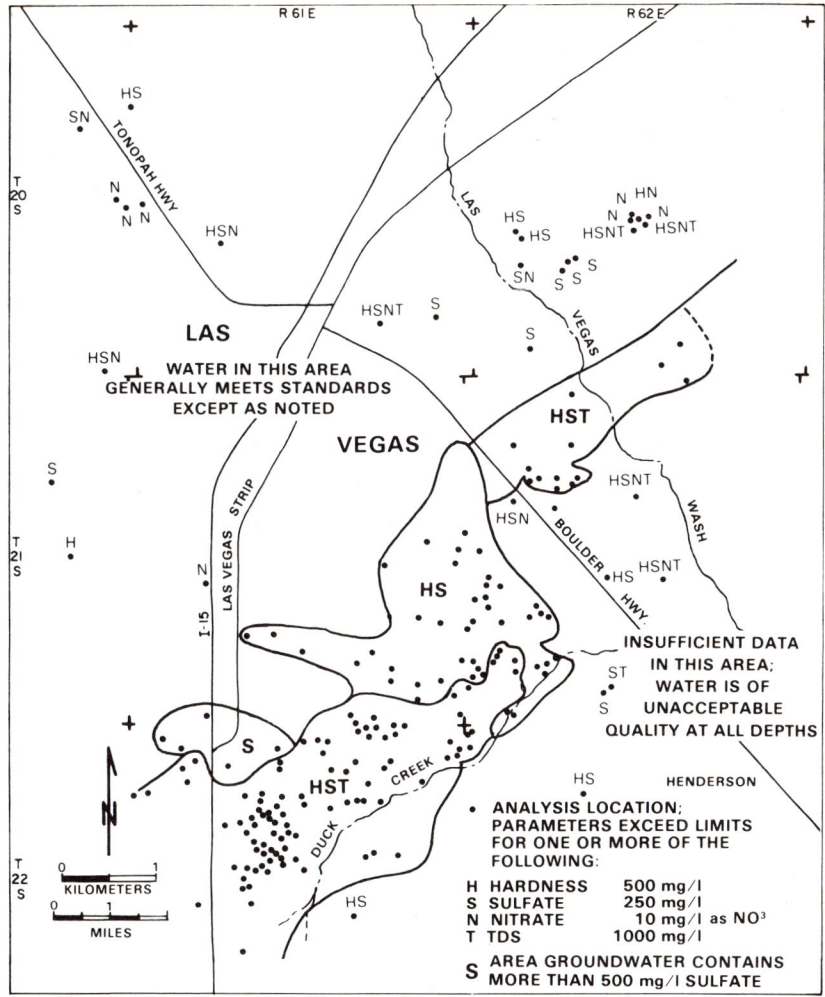

Fig. 7. Locations of wells 30.5—91.5 m deep with inferior water quality.

Hydrochemical facies

Further utility of the data bank created in the course of the study can be seen in Fig. 8 which was generalized from a computer generated plot of water analyses classified as to hydrochemical facies. The hydrochemical facies concept, as used herein, is based on 16 facies identified by relative concentration (in milliequivalents per liter) of the principal ions expressed as a percentage of total anions and total cations. For example, water containing 90% or more of $Na^+ + K^+$ cations and 90% or more $HCO_3^- + CO_3^{2-}$ anions is defined as belonging to the $Na+K-HCO_3+CO_3$ facies.

Lateral variations of facies changes result from lithologic influences in the recharge areas, residence time in the aquifer(s), and changes in the mineral

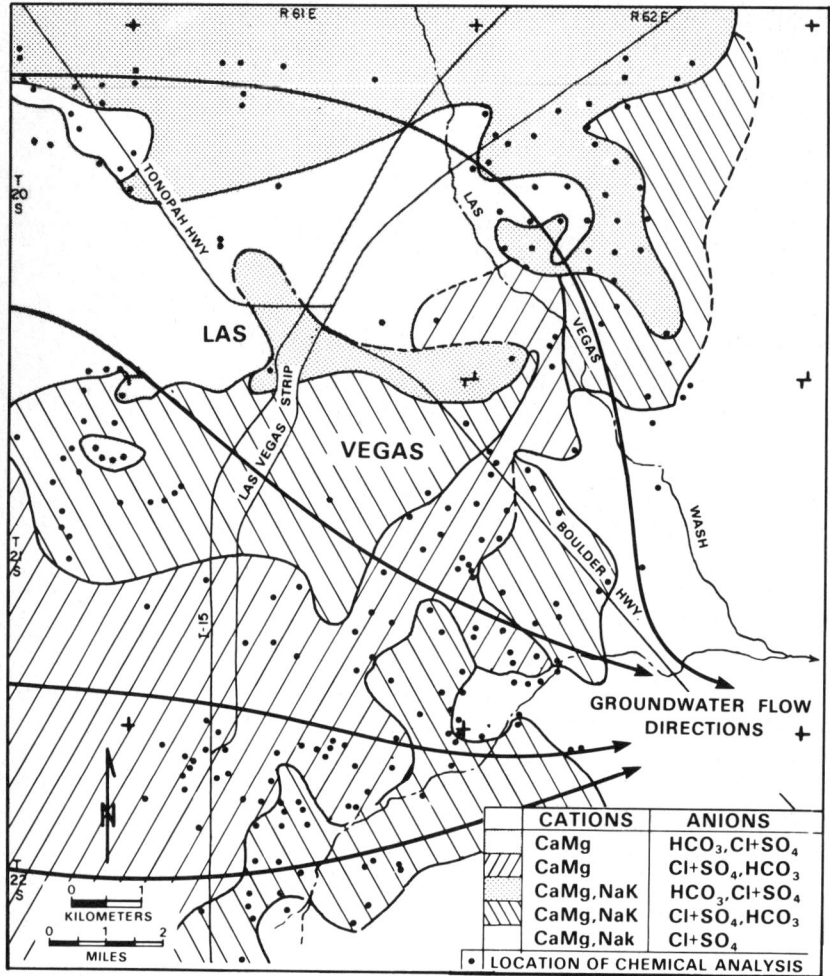

Fig. 8. Hydrochemical facies in groundwater at depths of 30.5—91.5 m.

composition in the aquifer along the direction of flow. Variations in chemical character of water in depth interval from 30.5 to 91.5 m are most apparent along flow paths through the northern and central portions of the valley. The progressive change from Ca+Mg+HCO$_3$—Ca+MgSO$_4$ facies is believed a result of increased gypsiferous sediments, longer residence time, and possible minor recharge from the Frenchman Mountain block. As expected, TDS also increases with distance along the flow path.

Periodic generation of hydrochemical facies maps and comparison of changes in patterns through time can provide an estimate of changing water-quality conditions. The approach is most productive if used in conjunction with time series data for individual parameters (ions, ionic ratios, TDS, etc.) and specific wells or well fields. At present, data from the 0—15.2-m interval partially characterize spatial variations in water quality. Temporal trends are qualitatively known, particularly where waste disposal, waste-water recycling, or other land use is characterized by distinctive return flows.

PROPOSED APPLICATIONS

In excess of 5000 well logs in Las Vegas Valley have been computerized. Drillers' descriptions of penetrated materials are usually brief and there is frequently question as to the veracity of reported lithologic descriptions and thicknesses. Nevertheless, when such logs are the only data available for subsurface hydrogeologic investigations, they should be analyzed for all possible useful information. Of the suggestions that follow, some have already been incorporated into research proposals submitted for funding.

Geologic structure

The geologic structure of Las Vegas Valley includes a number of complexities which influence the occurrence and movement of groundwater and necessitate additional hydrogeologic investigation. The valley is a broad basin surrounded by (normal) fault-block mountains, with significant thrust faulting (Keystone thrust) to the west and a prominent shear system (Las Vegas shear zone), in the NW zone in particular. Many small faults trending N—S along the west side of the valley are the locus of major springs, some of which discharged as late as the 1950's. The faults separate principal aquifers on the west from markedly less productive aquifers and aquitards to the east. Although these basic features are known, our understanding of the detailed effects on underground flowlines, source areas and amount of recharge, and the irregular surface of the static water level largely remain at the level of knowledge gained decades ago. Thorough examination of available logs could lead to a better understanding of how the geologic structure influences the observed hydrologic characteristics.

Research conducted by DRI suggests that valuable information as regards hydrogeologic parameters and lithologic units may be obtained from drillers' reports if they are evaluated as a group rather than on a one-by-one basis (Cooley et al., 1971, 1973). This research dealt with statistical evaluation of groundwater data and included manual compilation of lithologic data into meaningful groupings with regard to hydraulic conductivity. Computerized extension of these earlier efforts would be required in view of the large number of well logs available for processing.

Stratigraphy

Investigation of stratigraphic units would be a natural extension of the geologic structure study. Identification and description of units exerting the most influence on the groundwater regime, and those subject to development, would receive most of the investigative effort. Computer processing of existing data could lead to the capability of generating a stratigraphic cross-section at any desired orientation within Las Vegas Valley. A thorough understanding of the stratigraphy would be necessary to recognize inconsistencies in the raw data and to minimize errors of interpolation and extrapolation.

Two conceptual approaches can be used in computer analyses of well-log data: (1) statistical techniques which basically incorporate all data, good and bad, as mentioned above; and (2) a sorting technique to eliminate the majority of poor data. The first approach is not common and is basically predicated on the assumption that the data are generally valid measures of the environment and that errors will average out. The second approach is predicated on the assumption that important amounts of the data are invalid, and that an accurate portrayal of subsurface conditions can best be established by interpretation of a limited number of data points through deductive interpretation. By deductive reasoning and a few reliable logs, criteria for rejection can be formulated and tested. Ideally, logs would be rejected for increasingly stringent reasons until a point is reached where further rejections would diminish the total information content. At that point basic statistical methods and other tests can be applied to remaining data with more confidence. This work should be done before the permeability)conductivity effort, mentioned below, to minimize processing of unreliable data.

Permeability

Lithologic materials encountered in the valley fill of Las Vegas Valley vary from bouldery gravels to clayey silt, with areally extensive caliche and tightly cemented gravels. Major aquifers are present in the western and northwestern portions of the valley, although domestic wells are feasible in essentially all of the valley except near Las Vegas Wash where water quality is limiting. Static water levels at many points in the valley appear to be inconsistent with expectations based on topography and projected water-level trends. The available data base of lithologic logs, static water levels and water analyses facilitates study of aquifer interconnection and storage capacity of presently unsaturated layers.

Other studies

Worth mentioning are an examination of water banking, aimed at evaluating the long-range costs, benefits and effects of injecting currently unused allotments of Colorado River water into valley fill in Las Vegas Valley. Understanding of geologic structure, stratigraphy and permeability is prerequisite to success in such an endeavor. Essentially any groundwater-quality study or an investigation of subsidence control would make use of the data base currently available. Most water-resources problems in Las Vegas Valley can be categorized as dealing with location, quantity, quality, or a combination of these, and imaginative use of a data bank is a valuable asset.

CONCLUSIONS

Application of an automated data base, consisting of water-quality and water-well logs, and related statistical and plotting routines and hardware facilitated an intensive study of hydrogeologic conditions in Las Vegas Valley, Nevada. This included analysis of shallow stratigraphy, the extent of natural and contaminated water quality, and influences thereon. Through matrix notation for data point location and a variety of retrieval plotting routines, the data were rapidly screened, reduced, and portrayed in graphical or tabular form readily adapted to the present study and to a number of proposed investigations. There are obvious advantages of an ADP system for manipulating well-log and water-quality data for routine administrative matters and in support of hydrogeological objectives. In both cases the utility of such a system is considerable. For the present study, various plotting routines and the trend-surface technique proved particularly useful for spatial pattern analysis. Factor analysis and multiple regression techniques, to name two, are also possible and at costs and speeds not attainable using manual techniques. Maximum use of existing data, although there are limitations in terms of accuracy and spatial distribution, is believed to be a cost-effective alternative to expensive test drilling and water sampling. This is particularly true for extensive reconnaissance and baseline studies.

ACKNOWLEDGEMENTS

The late Dr. George B. Maxey, former Director of the Center for Water Resources, Desert Research Institute, provided valuable counsel and overall support for the work reported on herein, which was conducted under a grant from the U.S. Environmental Protection Agency, Office of Research and Development. Messrs. Jim V. Rouse and Fredric Hoffman served as EPA project officers. Numerous staff of the Desert Research Institute contributed to the effort, but special thanks are extended to Nate Cooper, M.J. Miles, J. Sanders, H. Dewey, and D. Schulke for their administrative and technical assistance. Cooperation and assistance received from the named County, State, and Federal agencies are greatly appreciated.

REFERENCES

Cooley, R.L., Fordham, J.W. and Westphal, J.A., 1971. Hydrology of Truckee Meadows, Nevada. Desert Res. Institute, Univ. Nevada System, Reno, Nev., Cent. Water Resour. Res. Proj. No. 15.

Cooley, R.L., Fordham, J.W. and Westphal, J.A., 1973. Some applications of statistical methods to groundwater flow system analysis. Desert Res. Inst., Univ. Nevada System, Reno, Nev., Cent. Water Resour. Res. Tech. Rep. Ser. H-W, Publ. No. 14.

Davis, L.R., Turner, A.K. and Melhorn, W.N., 1969. Analysis of groundwater use, replenishment, and aquifer characteristics in Bartholomew County, Indiana. Purdue Univ., Lafayetta, Ind., Water Resour. Res. Cent., Tech. Rep. 3, 83 pp.

Friesen, H.N., 1970. Water analysis data system (WADS); a storage and retrieval program for water quality data. Cent. Water Resour. Res., Desert Res. Inst., Univ. Nevada System, Reno, Nev.

Kaufmann, R.F., 1977. Land and water use impacts on ground-water quality in Las Vegas Valley. In: Proceedings: 3rd NWWA-EPA National Ground Water Quality Symposium. Ground Water, 15(1): 81—89.

Kaufmann, R.F., 1978. Land and water use effects on ground-water quality in Las Vegas Valley. U.S. Environ. Prot. Agency, Off. Res. Dev., EPA 600/2-78-179, 215 pp.

Malmberg, G.I., 1965. Available water supply of the Las Vegas ground water basin, Nevada. U.S. Geol. Surv., Water-Supply Pap. 1780, 116 pp.

Maxey, G.B. and Jameson, C.H., 1948. Geology and water resources of Las Vegas, Pahrump and Indian Spring Valleys, Clark and Nye Counties, Nevada. Nev. Dep. Conserv. Nat. Resour., Water Resour. Bull. No. 5, 121 pp.

McKee, J.E. and Wolf, H.W., 1973. Water Quality Criteria. California State Water Resources Control Board, San Francisco, Calif., 2nd ed., 548 pp.

Patt, R.O., 1978. Las Vegas Valley water budget: relationship of distribution, consumptive use, and recharge to shallow ground water. U.S. Environ. Prot. Agency, Off. Res. Dev., EPA 600-2-78-159, 63 pp.

Rockaway, J.D. and Johnson, R.B., 1967. Statistical analysis of ground water use and replenishment. Purdue Univ., Lafayette, Ind., Water Resour. Res. Cont., Tech. Rep. 7, 142 pp.

Thordarson, W. and Robinson, B.P., 1971. Wells and springs in California and Nevada within 100 miles of the point 37°15′N, 116°25′W on Nevada Test Site. U.S. Geol. Surv., Dep. Inter., USGS-474-85 (NTS-227).

Westphal, J.A., 1978. Simulation modeling of the shallow ground water system in Las Vegas Valley. Water Resour. Cent., Desert Res. Inst., Univ. Nevada System, Las Vegas, Nev. Report presented to U.S. Environ. Prot. Agency under grant R800946 (unpublished).

[2]

REGIONAL CARBONATE FLOW SYSTEMS IN NEVADA

MARTIN D. MIFFLIN and JOHN W. HESS

Water Resources Center, Desert Research Institute, University of Nevada System, Las Vegas, NV 89109 (U.S.A.)

(Accepted for publication May 10, 1979)

ABSTRACT

Mifflin, M.D. and Hess, J.W., 1979. Regional carbonate flow system in Nevada. In: W. Back and D.A. Stephenson (Guest-Editors), Contemporary Hydrogeology — The George Burke Maxey Memorial Volume. J. Hydrol., 43: 217—237.

Carbonate rocks include some of the most extensive and productive aquifers in the world. In Nevada, the limited surface water supply has been extensively developed and most of the alluvial groundwater basins are used to the point of estimated perennial yield, thus are closed to further development. Based on the review of over 570 geologic/hydrologic references, 150 petroleum wildcat test-hole records, and cave and carbonate spring data it is believed that deep regional carbonate aquifers exist and that they are potentially favorable for development for water supplies. Approximately the eastern one-third of Nevada (105,000 km^2) is underlain by carbonate rock. Eastern Nevada lies within the miogeosynclinal belt of the cordilleran geosyncline, in which 9,000—12,000 m of marine sediments accumulated during the Precambrian and Paleozoic. Two major periods of deformation have affected the region. Cave, wildcat well and carbonate spring data indicate that the Cambrian and Devonian carbonate strata may generally have the highest permeability. Water quality at depth in general ranges between 300 and 600 mg/l TDS. A "Phase II" project has been designed to further these investigations and will increase our understanding of large regional carbonate aquifers in the Great Basin.

INTRODUCTION

Carbonate rocks underlie ~20% of the land surface of the U.S.A. and generally contain large quantities of groundwater (McGuinness, 1963; Davies and LeGrand, 1972). Carbonate terranes have a large range in permeability including some of the most extensive and productive aquifers in the world (Herak and Stringfield, 1972). Many are proven excellent groundwater supplies, such as in Florida, Texas and Throughout the Appalachian states (N.A.S., 1976).

As indicated in Fig. 1, based on the distribution of bedrock exposure, approximately the eastern one-third of Nevada and the western one-third of Utah are underlain by carbonate rocks (160,000 km^2). Based on background data reviewed, it is believed that the deep carbonate-aquifer systems exist in most of the carbonate rock province in Nevada and possibly Utah.

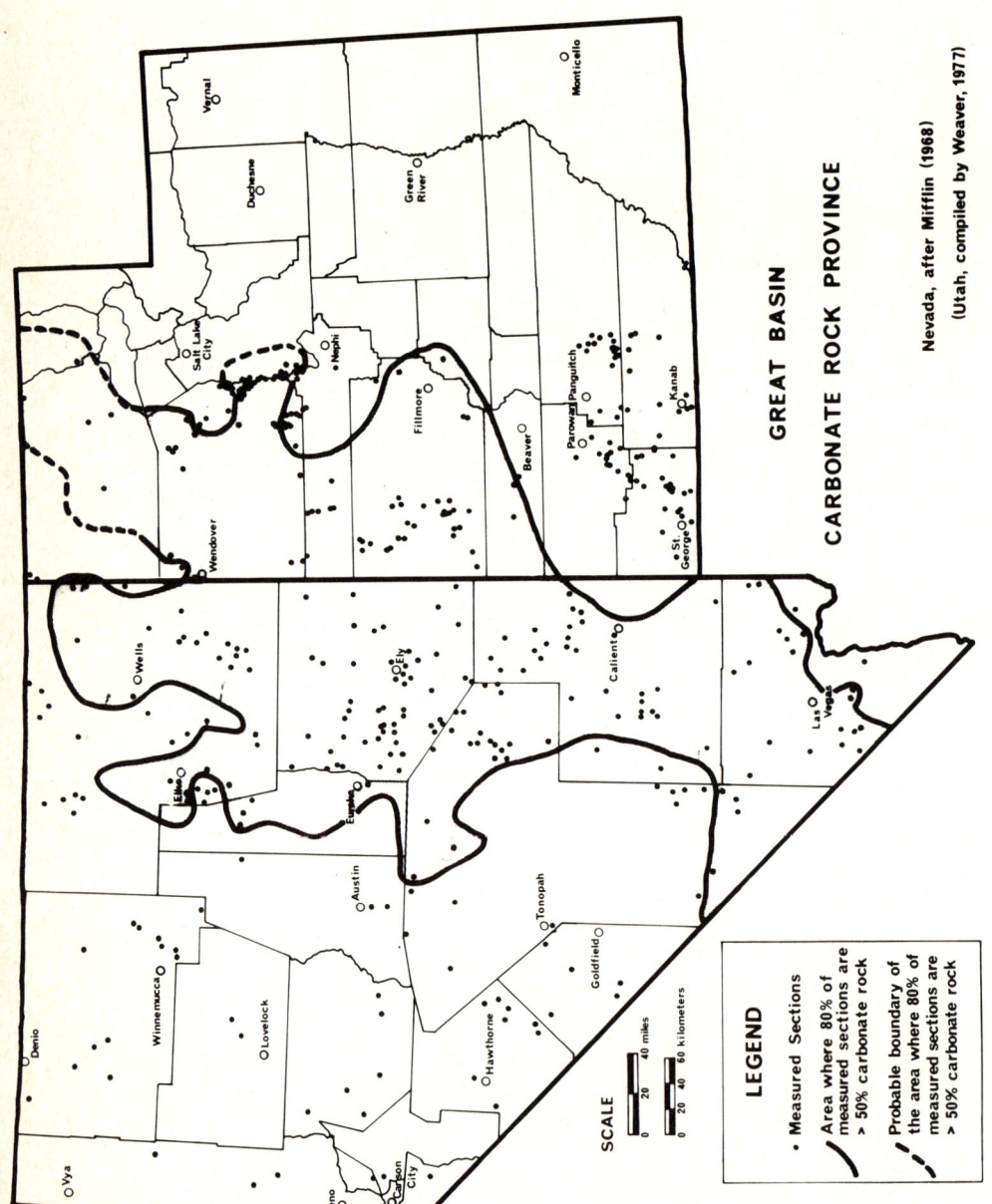

Fig. 1. Great Basin carbonate rock province.

This article is the result of the first phase of a new program of study by the Water Resources Center, Desert Research Institute on the water-supply potential of deep carbonate aquifers in the Great Basin. This initial work was funded by the Las Vegas Valley Water District and was designed to assemble and interpret existing geologic, geophysical, hydrologic and geochemical information available in areas of southern and eastern Nevada believed potentially favorable for development of water supplies from carbonate aquifers. Nothing has been found to reflect negatively on the concept of developing water from the deep carbonate flow systems.

BACKGROUND AND HISTORY

The limited surface water supply of Nevada has been extensively developed. Many of the alluvial groundwater basins are used to the point of estimated perennial yield, thus are closed to further development. As a consequence, a large part of the traditional water resources is no longer available and new uses can be realized only through transfer of water rights from existing uses. In fact, southern Nevada has reached the point in water-supply planning that recognizes no proven additional supplies to meet trends in demand beyond about the year 2000. The deep carbonate rocks of southern and eastern Nevada may represent a potentially "new" source of water. The magnitude of this source may be on the order of millions of cubic meters per year. Little is presently known about this resource in terms of the dynamics of recharge, discharge and physical aquifer characteristics, its total extent, or where it could be developed at a reasonable cost without interfering with existing ground- or surface-water development.

The Great Basin region is characterized by its internal drainage and general aridity. There are few major rivers and streams and thus groundwater has been a key element in water supply. This reliance on groundwater naturally led to major concern about its origins, occurrence and sustained developable volumes. Over the years the general physiography, geological structure and lithology have exercized considerable influence on where and how the groundwater has been developed, on estimates of how much groundwater is available, and on how groundwater flows within the system.

Carbonate rock aquifers are the least understood and exploited type of aquifers for their extent in the Great Basin. As indicated above, approximately the eastern one-third of the State of Nevada is underlain by carbonate rocks. These rocks, over long periods of time, have developed secondary permeability in localized zones through the solution of carbonate minerals by percolating water. Some of the highest known water well yields are obtained in such permeable zones of carbonate rocks in many parts of the world.

There is abundant evidence in Nevada that groundwater does indeed occur in such zones, at least locally, in many areas of the region underlain by carbonate rocks (Maxey and Mifflin, 1966; Mifflin, 1968; Winograd and

Pearson, 1976). The many wells drilled into carbonate rocks at the Nevada Test Site in southeastern Nye County amply demonstrate local occurrence of deep aquifers (Winograd and Thordarson, 1975). Another type of evidence is the occurrence of many large springs associated with these rocks. Unbalanced water budgets in large areas of Nevada indicate substantial flows through carbonate rocks (*Nevada Hydrologic Atlas*, 1972; Mifflin, 1968). Large interbasin groundwater flow systems exist in much of this region within the carbonate rocks (Eakin, 1966; Mifflin, 1968; Naff et al., 1974). Many of the intermountain basin alluvial aquifers may also be related to these larger carbonate flow systems (Bateman et al., 1972, 1974).

Basically the structural framework of the region can be characterized as a series of N—S-trending mountain ranges with general lithology ranging from carbonate rock sequences to volcanic and granitic rocks. These rock types also underlie the valleys at depths ranging from several hundreds to several thousands of meters. The valleys are filled with unconsolidated alluvium and lake sediments of Pleistocene or younger age, and in some basins with semi-consolidated Tertiary sediments. The upper levels of the valley fill, consisting of layers of sands, gravels, silts and clays, have traditionally been considered as the "groundwater basin". Historically, attention has been focused on these alluvial basins for estimating groundwater availability and developing water supplies.

During the early 1950's the Nevada District Office of the U.S. Geological Survey, in a cooperative program with the Nevada Department of Conservation and Natural Resources, developed a technique for estimating the average annual natural recharge to these valley fill groundwater basins. The cooperative program was a concerted effort to develop water supply estimates for each basin during the 1960's and 1970's. The estimating technique used was basically that of developing a valley water budget wherein total natural inflow to each valley must equal total outflow plus changes in storage. In many valleys it was found impossible to balance these water budgets because there was either too much or not enough discharge, and the differences were significantly larger than the probable errors in inflow estimates. These observations and renewed interest in flow system theory led to the studies by personnel at Desert Research Institute (DRI) (Maxey and Mifflin, 1966; Mifflin, 1968) and the U.S. Geological Survey (USGS) (Eakin, 1966) into the reasons for these imbalances. The general conclusions were that while the valleys are topographically separated, there are instances where there is significant groundwater flow between the valleys beneath the mountain ranges. These interbasin groundwater flows (regional flow systems) were found to occur primarily in that portion of Nevada where a significant portion of the geologic section consists of carbonate rock lithologies (Mifflin, 1968; *Nevada Hydrologic Atlas*, 1972). Supporting evidence for the regional flow systems concept was found in the valley water budgets, the occurrence of major springs issuing from or near carbonate rocks, regional valley fill aquifer water levels and regional groundwater chemistry (Naff, 1973).

The regional groundwater flow system concept was generally accepted by the mid to late 1960's. However, early in the development of the concept there was little information on the depths at which these flow systems existed, the effective permeability and porosity of the hydrostratigraphic units or whether these "aquifers" were economically developable. The early direct evidence concerning these latter unknowns came about as a result of the 1963 Limited Test Ban Treaty, which called for all nuclear testing to be done underground. An early concern with such testing was the possibility of radioactive contamination of groundwater. The presence of groundwater was also a possible factor in the successful containment of subsurface detonations. As a result, the U.S. Atomic Energy Commission (now U.S. Department of Energy, DOE) contracted with DRI, USGS, and other research groups to evaluate hydrologic factors associated with this testing. As part of such hydrologic programs, numerous hydrologic test holes were drilled in the vicinity of the Nevada Test Site (NTS) and the Central Nevada Test Area in Hot Creek Valley. These test holes penetrated the valley fill and entered the underlying volcanic and carbonate rocks to significant depths. Some holes were as much as 2500 m deep, and encountered significant permeability and large volumes of water in the carbonate rocks. These provided the first direct hydrologic documentation that groundwater moved through the deep-lying carbonate rocks under the NTS to other closed basins (Winograd, 1962). At the NTS, hydrologic test holes, nuclear event observation hole and event emplacement hole data indicated that as much as 290 km^3 groundwater may be stored and slowly moving through the carbonate rocks (data from Winograd and Thordarson, 1975). The quantity of water under the NTS would be enough, for example, to provide all of Nevada's water needs for approximately 40 years at the present rate of consumption. However, location of this water, probable cost of development for use in other areas of Nevada, and many other considerations suggested a search for similar, and possibly more useable deep carbonate rock aquifers for future development.

From the early 1970's until his untimely death in early 1977, Dr. George B. Maxey and others armed with the available information, worked to obtain funding for the deep carbonate aquifer research project. Research objectives were clarified, proposals developed, and funds were diligently sought to begin the program of research believed necessary to evaluate this "new" water-resource alternative for the State.

In August, 1977, the Las Vegas Valley Water District Board of Directors made it possible to begin initial "Deep Hole" work with money to support the first phase of the research (Hess and Mifflin, 1978). The questions of where to drill, and how much exploratory drilling would be necessary before production wells could be developed were primary. Another important question was whether or not such groundwater can be considered and managed as a "separate" water resource, that is, as water so far removed from the conventionally developed groundwater in the alluvial basins that it can be developed and managed on a different basis.

An adequate "Phase II Deep Hole" research program, which would involve actual drilling and aquifer testing to address the most important questions, would cost millions of dollars over a number of years. The high cost is due mostly to the very expensive drilling, similar to petroleum exploration, using large drilling rigs and constructing test holes that could range up to 3000 m in depth. Research drilling costs are probably many times higher than possible future production well drilling costs due to the hydrologic testing that is desirable and necessary when each permeable zone is penetrated. This testing is coupled with the need to penetrate to much greater depths to more fully explore and understand the flow systems that may eventually be developed. The previously discussed questions of the relation of deep carbonate aquifers to conventionally used groundwater resources demand carefully developed data that does not exist at the present time.

The test-drilling phase of the research program is believed justified because of the potential economic impact to Nevada's economy if the deep carbonate aquifers can be used. In our present state of knowledge, it has yet to be demonstrated that such aquifers could be developed for large-scale water-supply purposes and, if so, what the cost of the water would be. If, for example, the cost of developing deep aquifers for water supply is less than importing water or reclaiming waste water, then this water could have tremendous impact on Nevada's future development. Over one-third of Nevada is underlain by these carbonate rocks, and most of the surface water and conventional groundwater is already in use. Generally, localized water supplies sufficient to permit major new developments are unavailable without radical modification of the present water uses. In some basins, there are essentially no conventional water supplies.

In the Las Vegas area, there is no proven additional source of new water supply once Nevada's share of the Colorado River is fully utilized, a condition that may be reached within 20—25 years. At that time any growth will need to be based on water conservation, intensified mining of conventional groundwater, or groundwater imported from other valleys at the expense of existing water uses in those valleys. This has been the planning picture to present, without recognition of the potential alternative represented by the deep carbonate aquifers. In view of these realities in southern Nevada, the incentive is becoming sufficiently strong to explore and evaluate the carbonate aquifer alternative in spite of the high costs for test drilling.

PHYSIOGRAPHY AND CLIMATE

The carbonate rock province in Nevada generally lies in eastern and southern Nevada, east of longitude 117°. It encompasses approximately 105,000 km² of Clark, Elko, Lincoln, Nye and White Pine counties. This area is within the central Great Basin section of the Basin and Range physiographic province. The area is characterized by isolated, elongate, subparallel block-faulted mountain ranges and broad intervening, nearly flat-floored alluvial valleys or basins. The mountains tend to run north or northeast with many rather regularly spaced between 25 and 40 km apart. They are 30—160 km long, 8—25 km wide and many have elevations between 2400 and 3000 m above mean sea level (M.S.L.). The basins are filled with varying thicknesses of alluvium derived from the surrounding mountains. Basin sediment thicknesses range from a few hundred meters to greater than 3000 m. Valley floor elevations range from approximately 600 m M.S.L. in Las Vegas Valley to about 2100 m M.S.L. in the central part of the area.

The local climates of the region are extremely variable, depending primarily upon altitude. The valleys are arid to semiarid, characterized by low precipitation and humidity and extreme diurnal variation in temperature. The mountains are semiarid to subhumid, receiving about one half of their precipitation as snow during the winter. Thunderstorms account for much of the summer precipitation. Annual precipitation is quite variable, with mean annual precipitation range from as little as 8 cm in some of the southern valleys to more than 76 cm in the highest mountain ranges. The majority of the basin and foothill areas receive 13—30 cm.

STRATIGRAPHY AND STRUCTURE

Eastern Nevada is stratigraphically complex. It lies within the miogeosynclinal belt of the cordilleran geosyncline, in which 9000—12,000 m of marine sediments accumulated during the Precambrian and Paleozoic eras (Roberts, 1964; Stewart, 1964). Precambrian and lower Paleozoic Carbonate and Transitional Assemblages consist of limestone, dolomite and minor amounts of shale, siltstone, sandstone and quartzite.

Upper Paleozoic carbonate and siliceous detrital rocks include thin sequences of conglomerate, siltstone and limestone within the Antler Orogenic Belt, relatively thick sequences of shale, siltstone, sandstone conglomerate, sandy limestone and limestone along the eastern margin of the Antler Orogenic Belt or in the foreland basin to east, and moderately thin to thick sequences of carbonate rock in the foreland basin or on the shelf.

Eastern Nevada is characterized by extremely complex geologic structures as a result of numerous episodes of deformation. Probably three periods of deformation have affected the carbonate rock sequences within the region of interest. In some areas Permian—Triassic deformation seems probable, with attendant metamorphic and igneous activity. Folding and thrust faulting of the Upper Precambrian and Paleozoic rocks occurred during the Late Mesozoic—Early Tertiary Laramide orogeny. Normal block faulting, which produced the present-day basin-and-range topography, began during the Oligocene and reached a maximum in the Miocene with some activity extending to the present time. Strike-slip faults and shear zones have been active during the Laramide and block faulting tectonic periods. Movement along the Roberts Mountain Thrust was as much as 140 km (Wallace, 1964). Displacements of from 40—65 km have been mapped along the Las Vegas Valley shear zone, in Death Valley, the Amargosa Desert, the NTS and in the Las Vegas Valley.

HYDROLOGY

The hydrogeologic setting provides the boundary conditions determining the flow characteristics of an aquifer. Such geologic parameters as stratigraphy, lithology and structure are important in determining the nature of the porosity and permeability of the aquifer, type of aquifer, location of recharge and discharge areas and the water quality.

An overview of the hydrogeology of carbonate aquifers is presented by Stringfield and LeGrand (1969). White (1969, 1977) discusses the effects of structural and stratigraphic controls on carbonate aquifer systems of low to moderate relief. The hydrogeology of carbonate aquifers in folded and faulted rocks is discussed in Parizek et al. (1971).

Various aspects of eastern Nevada geology, hydrogeology and hydrology have been discussed by many authors. The bibliography in Hess and Mifflin (1978) contains over 570 entries and is a compilation of the many references which contain information of potential utility.

Permeability

Carbonate rock permeability is of three types: (1) primary porosity and permeability due to the presence of the intial communicating pore spaces; (2) permeability due to a network of joints, fractures, and bedding planes; and (3) permeability due to cavernous or solution openings. A carbonate aquifer can have high secondary permeability due to solutional development of an extensive system of interconnected cave conduits ranging in size from hundredths of a square millimeter to hundreds of square meters in cross-sectional area.

In the absence of direct information on permeability of deep-lying carbonate strata and in an attempt to identify permeable zones in the carbonate strata, four different types of data sources have been used in this study: (1) number and mapped length of solution caves; (2) carbonate springs; (3) petroleum wildcat wells; and (4) mines. This information has been compiled by geologic age, and is discussed below.

Caves

Solution caves are an extreme example of secondary permeability in carbonate rocks. They are useful features as indicators of which carbonate strata are most susceptable to solution and in the interpretation of the paleo-hydrology of an area. However, it must be kept in mind that lithology is not the only control on cave development. Structure, stratigraphy and history also play major roles. Which of these factors is the most important cannot be determined at this time. In any case, caves are used in this study as gross indicators of which carbonate units have a potential for high solutional permeability.

The number of caves and total mapped cave length by geologic age (McLane, 1972) are summarized in Table I. Cambrian carbonate units contain the most caves (39) and the greatest total mapped cave length (7000 m). Devonian carbonates follow with 14 caves and 1500 m. These two groups of carbonate strata contain 68% of the caves and 74% of the cave length. Thus the cave data suggest that Devonian and Cambrian strata have a high potential for solutional permeability. Appendix III in Hess and Mifflin (1978) lists by county the known solution caves in Nevada longer than 30 m and includes the elevation, mapped length and lithologic unit. These cave loations are shown in McLane (1974).

The comparison of the most cavernous units in Table I suggests that the hydrogeologic history of the area is an important control on cavern development. However, it appears that within a given area certain lithologic units are the best cave formers.

TABLE I

Cave summary for Nevada*

Age	Number of Caves	Approximate total length (m)	Comments
Cenozoic	2	200	
Cretaceous	0	0	
Jurassic	0	0	
Triassic	3	600	
Permian	4	550	
Pennsylvanian	1	300	
Mississippian	5	500	Monte Cristo Group in Clark County
Devonian	14	1,500	Devil's Gate Limestone in Elko County
Silurian	0	0	
Ordovician	10	750	Pogonip Group in Nye County
Cambrian	39	7,000	Pole Canyon Limestone in White Pine County

*Data compiled from a personal communication of A. McLane (1972).

Wildcat petroleum wells

The records at the Nevada Bureau of Mines and Geology, including Reports 18 and 29 and Bulletin 52 (Lintz, 1957; Schilling and Garside, 1968; Garside et al., 1977) were searched for information about the petroleum wildcat wells in eastern Nevada concerning zones of lost circulation, water production and fractures in the carbonate bedrock. In addition, contacts with major petroleum exploration companies were made to obtain additional information on selected wells of special interest. The petroleum wildcat wells are plotted on plate IV in Hess and Mifflin (1978). Appendix IV in Hess and Mifflin (1978) list these wells, including; surface elevation, elevation of the top of the carbonate, thickness of the alluvium, total depth of the well, formation or age of the top of the carbonate, thickness of carbonate rocks penetrated and indications of permeability.

Table II summarizes the geologic age for indications of high permeability. The lack of permeability indications in the Silurian, Ordovician or Cambrian probably reflects a lack of data.

TABLE II

Number of petroleum wildcat wells with permeability indications in eastern Nevada

Age	Number of wells	Age	Number of wells
Permian	3	Silurian	0
Pennsylvanian	6	Ordovician	0
Mississippian	7	Cambrian	0
Devonian	3		

Carbonate springs

The available lithologic data for carbonate springs have been summarized in Table III. The rock unit associated with the discharge point of the spring does not necessarily indicate the strata in which the water is flowing at depth. It does, however, indicate permeability in that particular unit, at least at the land surface. Many of the same constraints as with the cave data must be kept in mind when evaluating the spring data. Structure and history are also important factors controlling stratigraphic location of spring discharge.

TABLE III

Geologic age of rock units associated with carbonate springs in eastern Nevada

Age	Regional springs	Local and small-local springs	Total
Permian	1	4	5
Pennsylvanian	4	10	14
Mississippian	2	4	6
Devonian	15	23	38
Silurian	3	1	4
Ordovician	1	6	7
Cambrian	11	20	31

Table III indicates that the largest number of springs are associated with the Devonian carbonates. Fifteen regional and 23 local or small-local springs discharge from Devonian rocks. Cambrian units have 11 regional and 20 local or small-local springs associated with them. As with the cave data, these two geologic rock units have the highest permeability as indicated by springs. They represent 70% of the regional springs and 66% of the total number of springs.

Mines

During the search of the geologic literature, many references were found that indicated caves or water in mines. No attempt was made to summarize the data due to limited specifics. However, water was a problem in many mines and many caves were encountered, indicating not only the presence of water at depth, but relatively high permeabilities in the area of the mines. Some dewatering attempts have failed due to high permeability in carbonate rocks.

Summary

The cave, wildcat well and spring permeability data have been summarized in Fig. 2. A number of permeability indicators have been plotted against geologic age in a bar graph. The Cambrian and Devonian carbonate units are most frequently associated with permeability indicators selected based on the data available. This is compatible with the hydrostratigraph classification for southern Nevada developed from NTS studies (Winograd and Thordarson, 1975). This lower carbonate aquifer includes the rocks between the Cambrian and Devonian. It is described as a complexly fractured aquifer with coefficients of transmissibility ranging from ~ 4 to 4000 m^2/day.

Flow systems

Studies such as Eakin (1966), Maxey and Mifflin (1966), Mifflin (1968) and Winograd and Thordarson (1975) have established some of the fundamental concepts concerning groundwater flow systems in Nevada. Direct data on these systems are extremely sparse, and the details of the flow systems are not generally known. For purposes of this article, *groundwater flow systems* are defined as follows: a *"regional* groundwater flow system" is defined as a large groundwater flow system which encompasses one or more topographic basins. A regional system may include within its boundaries several groundwater basins; interbasin flow is common and important with respect to total volume of water transferred within the system boundaries; lengths of flow paths are relatively great when compared to lengths of flow paths of "local" groundwater flow system. A *"local* groundwater flow system" is generally confined to one topographic or groundwater basin; interbasin flow is not important with respect to total volume of water transferred within the system; the majority of flux of

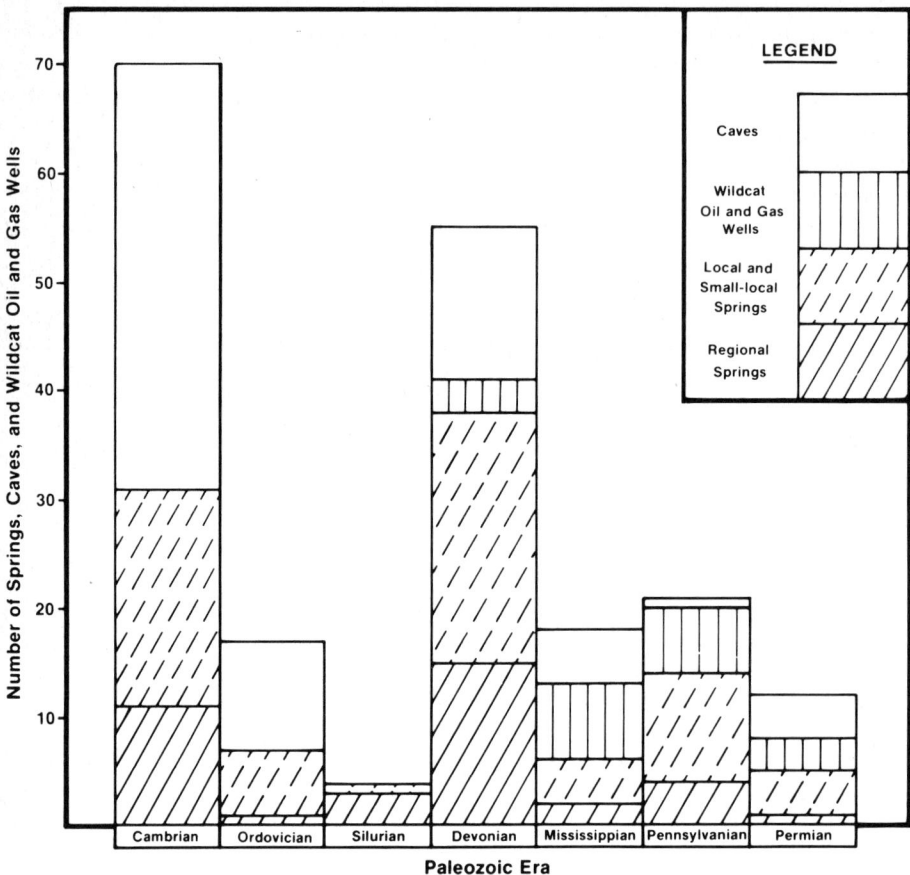

Fig. 2. Permeabilty data sources by geologic age for eastern Nevada.

water within the system discharges within the associated groundwater basin; flow paths are relatively short when compared to regional systems.

Previous investigators (Toth, 1962; 1963; Freeze and Witherspoon, 1966, 1967) have examined theoretical aspects of groundwater flow systems and modeled hypothetical systems in two dimensions. It has been found that, given certain boundary conditions, usually imposed by permeability, topography and available recharge, it is possible for what have been called local, intermediate and regional systems to exist (Toth, 1963).

Models provide important points of departure in attempts to delineate naturally occurring groundwater flow systems. The importance of models is shown in Freeze and Witherspoon (1967) because a commonly assumed criterion for a system boundary is a region of high groundwater potential. These regions of high potential are usually recognized in nature by the configuration of saturation, and fluid potentials at depth are rarely known. These models clearly suggest that such regions of high fluid potentials are not necessarily perfect boundaries to the system as a whole, and that flow can occur at depth from one "cell" of the system to another.

Most important to flow-system delineation are hydrogeologic conditions of relatively high permeability at depth such as exists in carbonate rocks. In such situations, models suggest that large quantities of water can move from one "cell" to another. Thus, it may be insufficient to map only the surficial or "shallow" fluid potential field of a groundwater system if full and reliable identification of the system is to be accomplished in terrane which many be underlain by rocks of high relative permeability. Most hydrologic data of groundwater potentials are limited to the top of the zone of saturation, and hence even a detailed knowledge of the configuration of saturation may be inadequate to ascertain where some waters leave the system, or where they enter the system in areas underlain by permeable zones at depth. There is good evidence for extensive permeable zones at considerable depths within the carbonate rocks in eastern and southern Nevada.

Flow-system delineation in carbonate terrane

The evidence available suggests that thick sequences of carbonate rocks underlie most of the alluvial basins and much of the volcanic rock sequences of eastern Nevada. Deep petroleum wildcat drilling indicates that intervals of cavernous carbonate rock exist to depths perhaps greater than 3000 m, as many test holes experienced extreme circulation difficulties, and a few have experienced dropping bits upon encountering caverns.

A prime hydrologic evidence of extensive zones of permeability in carbonate bedrock is provided by the number of large and important groundwater discharge areas (Mifflin, 1968). Many basins in eastern and southern Nevada also are believed to lose groundwater by interbasin flow, but their smaller size or the limited availability of moisture for recharge makes the phenomenon of interbasin flow less impressive and, in some cases, less certain.

Geologic considerations. Structural history relates to at least four important areas of consideration in the study of deep carbonate flow systems. The first and most fundamental is the distribution of the carbonate rocks in the subsurface. The history of faulting, folding and erosion controls the basic distribution of the rock types of interest. In much of the region, knowledge of this distribution is largely restricted to bedrock exposures, due to the extremely complicated history of deformation, as demonstrated by those areas mapped in detail. Presence of many low- and high-angle thrust faults, associated folding, and superimposed normal faulting greatly complicates the prediction of distribution of formations in the subsurface, particularly in the basin areas. Until demonstrated otherwise, the complicated structural relationships displayed in outcrop areas of the mountains are assumed to continue in the basement rocks underlying the alluvial basins. Changes in structure from range to range are often the case, and therefore interpretation of structure hidden under the basins between the ranges is generally difficult.

Secondly, structural development has rearranged rocks to the extent that permeable and impermeable lithologies are often juxtaposed. This can in turn create corridors of permeability in any given direction, and likewise generate relatively impermeable boundaries on a localized, and perhaps regional, scale. Considered in terms of flow system configuration and size, this aspect has special significance. It is well demonstrated on a local basis that faults and stratgraphic control on permeability can determine general pattern of groundwater flow. What is less evident in documented cases is the possible control that structure may exert in the configuration of large regional flow systems. Further, major linear fault zones, such as regional shear zones and the range front fault zones that sometimes extend for great distances must have, in at least some areas, very profound influence on the movement of groundwater. Little firm evidence exists in the literature as to the importance of such zones throughout the carbonate rock province.

The third aspect of structural history of considerable importance is the direct impact on permeability. Most carbonate rocks deformed at shallow depths within the crust behave in a brittle manner, and thus shear and fracture during deformation. Concentration and extent of fractured carbonate rocks is important in the degree of the development of secondary and tertiary permeability. Fracturing of relatively youthful age in a geologic sense is believed most important with respect to permeability, as fractures in carbonate rocks may heal over prolonged periods of time due to recrystallization of carbonate minerals.

The fourth aspect of structural history that may control permeability in the carbonate rocks relates to present and past patterns of groundwater flow due to position in past circulation systems of groundwater. From at least Miocene time and perhaps the Early Tertiary, the relative position of carbonate rocks in the Basin and Range Province has been changing with respect to recharge areas, discharge areas, and normal flow paths of groundwater. Depending on this relative position at any given time, the potential amount of solution in a fractured carbonate rock may vary greatly due to the thermodynamic environment and associated flux of groundwater.

This type of structural history may have important influences over the distribution of permeability in the subsurface. In some parts of Nevada there is considerable evidence that basin-and-range configuration may not have been constant since the block faulting became a dominant pattern of deformation. Some carbonate rocks now lying buried beneath basin fill deposits did not always occupy such structural and hydrogeologic environments. Conversely, other zones of carbonate rocks may have maintained their relative positions with respect to antecedent groundwater circulation patterns.

Regional movement of groundwater may be strongly influenced by the Late Mesozoic—Early Tertiary deformation of the Upper Precambrian and Paleozoic miogeosynclinal rocks, their subsequent erosion, and the faulting that took place during the Late Cenozoic orogeny.

Groundwater chemistry of large springs. There has been much work on the geochemistry of carbonate waters in recent years. The spatial variations have been investigated by such workers as Back (1966), Back and Hanshaw (1970), Jacobson and Langmuir (1970), and Langmuir (1971). Temporal variations have been studied by Shuster and White (1971, 1972), Thrailkill (1972), Jacobson (1973), Hess (1974) and Hess and White (1974). The geochemistry of carbonate waters is of interest in its own right and is also a means of investigating the hydrogeology and physical hydrology of an aquifer, including flow-system delineation.

Maxey and Mifflin (1966), and Mifflin (1968) have demonstrated the apparent utility of water chemistry from large-discharge springs associated with flow systems in eastern Nevada (Fig. 3). The quality of groundwater flowing through carbonate terrane will vary with length of flow path and

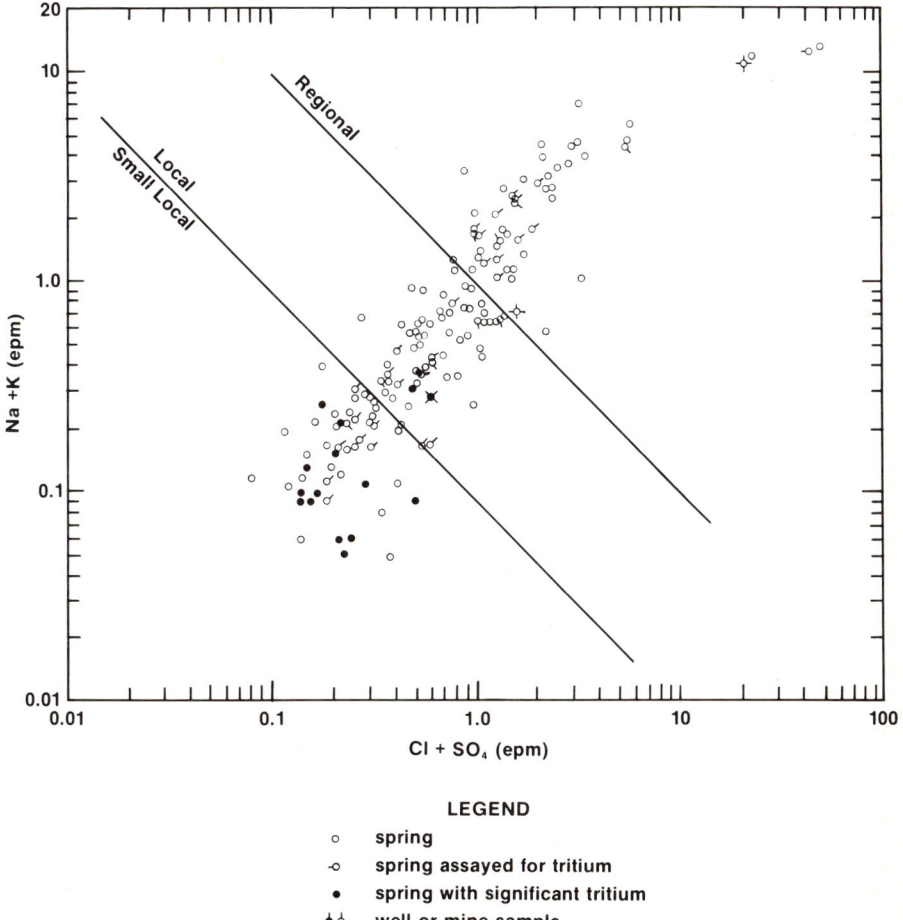

Fig. 3. Plot of relation between water chemistry, tritium and spring classification in eastern Nevada (modified from Mifflin, 1968).

residence time. The minerals calcite ($CaCO_3$) and dolomite [$CaMg(CO_3)_2$], which form the major carbonate rock types of limestone and dolomite, are soluble in water, but not as soluble as other minerals that might be associated with carbonate rocks such as gypsum ($CaSO_4$) and halite (NaCl). Concentrations of the various ions formed by solution of the above minerals will increase in the groundwater with length of flow path and time. However, Maxey and Mifflin demonstrated that water chemistry of springs believed to be associated with regional flow systems characteristically illustrated increased concentrations of Na^+, K^+, Cl^- and SO_4^{2-} with increased lengths of flow paths.

At each large spring associated with carbonate terrane, an indication of distance water has traveled, as well as the fluid potential, temperature and character of discharge, can be obtained. This provides a powerful tool in the absence of widespread fluid potential data. Further, the temperature of water gives an indication of probable depth of circulation immediately upgradient from the spring. The elevation of the spring gives an indication of fluid potential at that point in the system. Characteristics of discharge (i.e., either variable or constant) give indications of proximity of significant recharge areas. All these attributes, if considered together, offer qualitative characterization of the involved flow system at that geographic point.

Tritium in carbonate springs. Tritium, a radioisotope of hydrogen, can be a useful tool in determining age and source of groundwater. Use is made of environmental tritium as a tracer.

Reconnaissance sampling for tritium in large carbonate springs was used by Mifflin (1968) to further investigate the character of carbonate rock flow systems. He found that the concentration of Na+K forms a reliable criterion for predicting tritium concentration in springs. Significant amounts of tritium were found in all sampled springs that contained less than 3.8 ppm Na+K. No significant amount of tritium was found in any sampled spring that contained more than 8 ppm Na+K.

Flow-system classification by chemistry and tritium. Variations in water chemistry and tritium in large springs associated with flow systems in carbonate terrane aid in flow-system delineation. A classification was applied by Mifflin (1968) that divided springs into three general groups: (1) small local; (2) local; and (3) regional flow systems. The approach was to consider water chemistry in springs known to be associated with systems that are interbasin in configuration (regional with long flow paths) and water chemistry in large springs which are intrabasin in configuration (local with short flow paths). Further, the occurrence of tritium in significant quantities in some springs permits a third classification with limits based on tritium.

On the basis of water chemistry and independent hydrologic data, relative paths of flow or lengths of flow systems in carbonate rock terrane were divided into two broad categories, local and regional systems. Occurrence of tritium in significant concentrations in waters with low concentrations of Na^+, K^+, Cl^- and SO_4^{2-}, and consistent absence of significant tritium in waters with higher concentrations of these ions strongly support the characterization of flow systems into local and regional systems. It also indicates that very little or no recharge occurs near points of discharge of large springs associated with regional flow systems. Springs which contain significant concentrations of tritium may be separated further on that basis as being related to "small local" flow systems.

Fig. 3 (modified from Mifflin, 1968) is a logarithmic plot of concentrations of $Na^+ + K^+$ ions compared with $Cl^- + SO_4^{2-}$ ions found in large springs associated with the carbonate rock terrane. Also shown are the discussed boundaries of flow system classification, the springs assayed for tritium, and the springs that displayed significant tritium. Appendix II in Hess and Mifflin (1978) is a compilation of physical and chemical characteristics of carbonate springs in eastern Nevada.

Flow-system boundaries in southern and eastern Nevada

The flow-system boundaries in southern and eastern Nevada were developed by Mifflin (1968) on the basis of both conventional hydrologic data and system classification studies of the large springs. Even with the combined approach, delineation of flow systems in this region is believed subject to major error, and truly confident delineation awaits the proof provided by carefully collected fluid potential data from deep wells in key areas.

Most flow systems in southern and eastern Nevada are interbasin in configuration of flow. Confident delineation of flow-system boundaries in this region cannot be accomplished in detail with available data; however, the general aspects of delineation shown in Mifflin (1968) are more or less valid. The greatest problem of delineation in this region is location of flow-system boundaries in areas where important flow occurs at depth in carbonate rock. Shallow fluid potential data may be misleading as to the location of significant boundaries.

The importance of flow-system delineation to the concept of potential development of carbonate aquifers for water supply is in the disciplines of economics, management and legal considerations. For example, it is known from radiometric age dating of some spring waters (Haynes, 1967; Mifflin, 1968; Winograd and Pearson, 1976) that the regional flow-system waters have rather long average travel times, as would be expected from other lines of evidence. Therefore, the great distances and the slow travel times indicate that even though some conventionally developed groundwater basins may be supplied by the regional flow system, the time frame of connection is so great that thousands of years would be required for development in one area of the system to be manifested in a distant part of the system. Should it be decided to use these waters from the carbonate aquifers, it becomes important to find the most economical places to develop the waters without creating undesired impacts on existing groundwater development. Such considerations as pumping lifts, transfer distances, and perhaps aquifer exploration costs depend on reliable flow-system delineation.

Bedrock—alluvium interaction

An important area of concern in the development of water supplies from the regional carbonate aquifers is the degree of communication between the alluvial aquifers, some presently used for water supply, and the carbonate aquifers. In at least one area of Nevada where there has been development of the alluvial aquifers, there is a demonstrated connection between the two systems. In Ash Meadows spring discharge decreased and water levels in Devil's Hole dropped in response to pumping of the alluvial aquifers (Bateman et al., 1972, 1974). On the other hand, development of the alluvial aquifers in Moapa Valley have not changed the discharge of the Muddy River springs. There does not appear to be a connection that permits transmission of fluid potential changes between the two systems in that valley.

Water quality

Water quality in the carbonate flow systems is in general good. It is low in total dissolved solids (TDS), based on carbonate spring analyses. Hess and Mifflin (1978) gives the chemical quality of carbonate springs in eastern Nevada. Fig. 4 gives the distribution of TDS for the regional springs. A majority of springs have a TDS between 300 and 600 mg/l. The average TDS

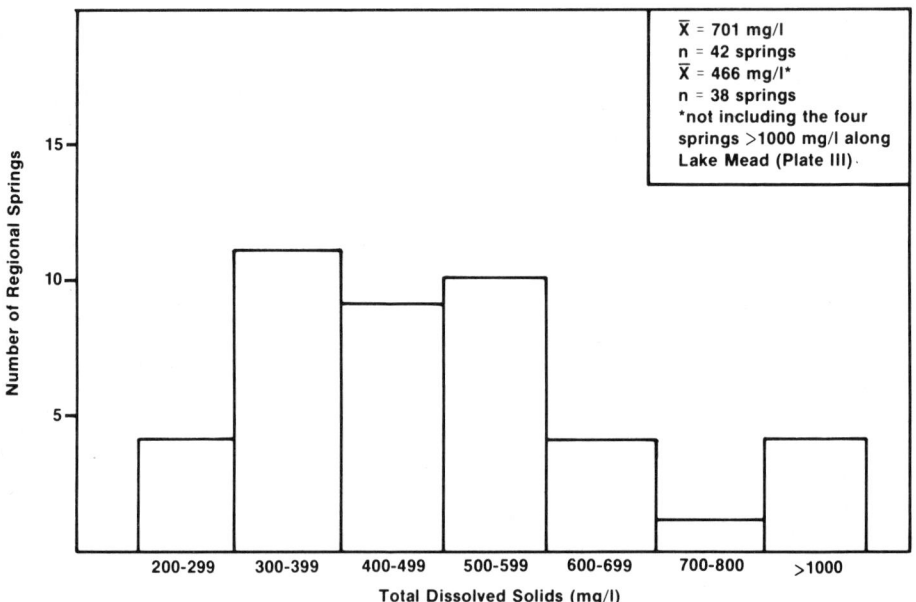

Fig. 4. Distribution of water quality in regional carbonate springs in eastern Nevada.

for the 42 regional springs within the boundaries of flow systems in carbonate rock is 701 mg/l. The water quality is even better if the four springs with a TDS greater than 1000 mg/l are eliminated from the average. They are an anomalous group of springs near Lake Mead in Clark County whose chemistry is influenced by highly soluble minerals in the bedrock in the discharge area. With their removal the TDS average is 466 mg/l.

Data from Nevada Test Site wells (Winograd and Thordarson, 1975) indicate good water quality at depth. Total dissolved solids ranged from 328 to 439 mg/l in four wells at depths between 240 and 600 m. However, data from two wildcat oil and gas test wells in White Pine County indicate higher TDS. One well had a TDS of 1880 mg/l and the other up to 8700 mg/l. Therefore, water quality at depth in the carbonate rocks in general should be good if the well penetrates the zone of active circulation.

SUMMARY

The results of the "Phase I" study of the deep carbonate aquifer systems in Nevada support the hypothesis that the deep regional carbonate flow systems exist and that they are potentially favorable for water-supply development. At the present time, the evidence is strong enough to say that the large regional flow systems exist, the discharge areas are generally known, and there are many areas where carbonate aquifer groundwater could probably be intensely developed without short-term (200-yr.) impact on conventional groundwater supplies. What is not known is just how close to conventional groundwater supplies such intense development might be possible without serious impact. Some very significant economic advantages could be overlooked if adequate flow system studies are not entered into prior to basic water-resource development decisions in southern and eastern Nevada. A "Phase II" project has been designed to further these investigations and will increase our understanding of large regional carbonate aquifers in the Great Basin.

ACKNOWLEDGMENTS

This work was funded by the Las Vegas Valley Water District. The authors would like to thank the several persons who provided valuable assistance during the course of this study. The cooperation and suggestions from various petroleum industry representatives is also appreciated.

REFERENCES

Back, W., 1966. Hydrochemical facies and ground-water flow patterns in northern part of Atlantic Coastal Plain. U.S. Geol. Surv., Prof. Pap. 498A, 42 pp.
Back, W. and Hanshaw, B.B., 1970. Comparison of chemical hydrogeology of the carbonate peninsulas of Florida and Yucatan. J. Hydrol., 10: 330—368.

Bateman, R.L., Mindling, A.L., Naff, R.L. and Joung, J.M., 1972. Development and management of ground-water and related environmental factors in arid alluvial and carbonate basins in southern Nevada. Desert Res. Inst., Water Resour. Cent., Proj. Rep. 18, 43 pp.

Bateman, R.L., Mindling, A.L. and Naff, R.L., 1974. Development and management of ground water in relation to preservation of desert pupfish in Ash Meadows, southern Nevada. Desert Res. Inst., Water Resour. Cent., Rep. HW 17, 39 pp.

Davies, W.E. and LeGrand, H.E., 1972. Karst of the United States. In: M. Herak and V.T. Stringfield (Editors), Karst: Important Karst Regions of the Northern Hemisphere, Elsevier, Amsterdam, pp. 467—505.

Eakin, T.E., 1966. A regional interbasin ground-water system in the White River area, southeastern Nevada. Water Resour. Res., 2: 251—271.

Freeze, R.A. and Witherspoon, P.A., 1966. Theoretical analysis of regional ground-water flow, 1. Analytical and numerical solutions to the mathematical model. Water Resour. Res. 2: 641—656.

Freeze, R.A. and Witherspoon, P.A., 1967. Theoretical analysis of regional ground-water flow, 2. Effect of water-table configuration and subsurface permeability variation. Water Resour. Res., 3: 623—634.

Garside, L.J., Wemer, B.S. and Lutsey, I.A., 1977. Oil and gas developments in Nevada, 1968—1976. Nev. Bur. Mines Geol., Rep. 29, 32 pp.

Haynes, V.C., 1967. Quaternary geology of the Tule Springs area, Clark County. In: Pleistocene Studies in Southern Nevada. Nev. State Mus. Anthropol., Pap., 13: 17—104.

Herak, M. and Stringfield, V.T., 1972. Conclusions. In: M. Herak and V.T. Stringfield (Editors), Karst: Important Karst Regions of the Northern Hemisphere, Elsevier, Amsterdam, pp. 507—518.

Hess, J.W., 1974. Hydrochemical investigations of the central Kentucky karst aquifer system. Ph.D. Thesis, Pennsylvania State University, University Park, Pa., 418 pp.

Hess, J.W. and Mifflin, M.D., 1978. A feasibility study of water production from deep carbonate aquifers in Nevada. Water Resour. Cent., Desert Res. Inst., Publ. No. 41054, 125 pp.

Hess, J.W. and White, W.B., 1974. Hydrograph analysis of carbonate aquifers. Inst. Res. Land Water Resour., Pa. State Univ., Res. Publ. No. 83, 63 pp.

Jacobson, R.L., 1973. Controls on the quality of some carbonate ground waters: Dissociation constants of calcite and $CaHCO_3$ from 0 to $50°C$. Ph.D. Thesis, Pennsylvania State University, University Park, Pa., 131 pp.

Jacobson, R.L. and Langmuir, D., 1970. The chemical history of some spring waters in carbonate rocks. Ground Water, 8: 1—7.

Langmuir, D., 1971. The geochemistry of some carbonate groundwaters in central Pennsylvania. Geochim. Cosmochim. Acta, 35: 1023—1045.

Lintz, Jr., J., 1957. Nevada oil and gas drilling data, 1906—1953. Nev. Bur. Mines, Bull. 52, 80 pp.

Maxey, G.B. and Mifflin, M.D., 1966. Occurrence and movement of ground water in carbonate rocks of Nevada. Natl. Speleol. Soc. Bull., 28: 141—157.

McGuinness, C.L., 1963. The role of ground water in the national water situation. U.S. Geol. Surv., Water-Supply Pap. 1800, 1121 pp.

McLane, A., 1974. A bibliography of Nevada caves. Desert Res. Inst., Water Res. Cent., Misc. Rep. 16, 99 pp.

Mifflin, M.D., 1968. Delineation of ground-water flow systems in Nevada. Desert Res. Inst., Water Resour. Cent., Rep. H-W 4, 111 pp.

Naff, R.L., 1973. Hydrogeology of the southern part of Amargosa Desert in Nevada. M.S.Thesis, University of Nevada, Reno, Nev., 207 pp.

Naff, R.L., Maxey, G.B. and Kaufmann, R.F., 1974. Interbasin groundwater flow in southern Nevada. Nev. Bur. Mines Geol. Rep. 20, 28 pp.

N.A.S. (National Academy of Science), 1976. Water in carbonate rocks. In: U.S. Progress in Perspective, 36 pp.

Nevada Hydrological Atlas, 1972. State of Nevada, Carson City, Nev., 22 maps.

Parizek, R.R., White, W.B. and Langmuir, D., 1971. Hydrogeology and geochemistry of folded and faulted carbonate rocks of the central Appalachian type and related land use problems. Geol. Soc. Am., 184 pp.

Roberts, R.J., 1964. Paleozoic rocks. In: Mineral and Water Resources of Nevada. Nev. Bur. Mines, Bull. 65: 22—25.

Schilling, J.H. and Garside, L.J., 1968. Oil and gas developments in Nevada 1953—1967. Nev. Bur. Mines, Rep. 18, 43 pp.

Shuster, E.T. and White, W.B., 1971. Seasonal fluctuations in the chemistry of limestone springs: A possible means for characterizing carbonate aquifers. J. Hydrol., 14: 93—128.

Shuster, E.T. and White, W.B., 1972. Source areas and climate effects in carbonate ground waters determined by saturation indices and carbon dioxide pressures. Water Resour. Res., 8: 1067—1073.

Stewart, J.H., 1964. Precambrian and lower Cambrian rocks. In: Mineral and Water Resources of Nevada. Nev. Bur. Mines, Bull. 65, 21 pp.

Stringfield, V.T. and LeGrand, H.E., 1969. Hydrology of carbonate rock terranes: an review. J. Hydrol., 9: 349—417.

Thrailkill, J.V., 1972. Carbonate chemistry of aquifer and stream water in Kentucky. J. Hydrol., 16: 93—104.

Toth, J., 1962. A theory of groundwater motion in small drainage basins in central Alberta, Canada. J. Geophys. Res., 67: 4375—4387.

Toth, J., 1963. A theoretical analysis of groundwater flow in small drainage basins. J. Geophys. Res., 68: 4795—4812.

Wallace, R.E., 1964. Topography, In: Mineral and Water Resources of Nevada. Nev. Bur. Mines, Bull., 65: 11—12.

White, W.B., 1969. Conceptual models for limestone aquifers. Ground Water, 7: 15—21.

White, W.B., 1977. Conceptual models for carbonate aquifers: revisited. In: R.R. Dilamarter and S.C. Csallany (Editors). Hydrologic Problems in Karst Regions, Western Kentucky University Press, Bowling Green, Ky., pp. 176—187.

Winograd, I.J., 1962. Interbasin movement of groundwater at the Nevada Test Site, Nevada. U.S. Geol. Surv., Prof. Pap., 450C: C108—C111.

Winograd, I.J. and Pearson, F.J., 1976. Major carbon-14 anomaly in a regional carbonate aquifer: possible evidence for Moga scale channeling, south-central Great Basin. Water Resour. Res., 12: 1125—1143.

Winograd, I.J. and Thordarson, W., 1975. Hydrogeologic and Hydrochemical framework, south-central Great Basin, Nevada—California, with special reference to the Nevada Test Site. U.S. Geol. Surv., Prof. Pap. 712-C, 126 pp.

[2]

GROUNDWATER FLOW SYSTEMS IN THE WESTERN PHOSPHATE FIELD IN IDAHO

DALE R. RALSTON and ROY E. WILLIAMS

Mining and Mineral Resources Research Institute, University of Idaho, Moscow, ID 83843 (U.S.A.)

(Accepted for publication May 10, 1979)

ABSTRACT

Ralston, D.R. and Williams, R.E., 1979. Groundwater flow systems in the western phosphate field in Idaho. In: W. Back and D.A. Stephenson (Guest-Editors), Contemporary Hydrogeology — The George Burke Maxey Memorial Volume. J. Hydrol., 43: 239—264.

The complex geologic setting of the western phosphate field in Idaho provides the environment for equally complex groundwater flow systems. This research was initiated in 1974 to provide general and detailed hydrologic data on specific areas to aid in understanding the water-resource systems in the western phosphate field. Geologic, hydrogeologic and hydrologic data were collected in Little Long Valley and Lower Dry Valley in the Blackfoot River basin.

Two groundwater flow systems are important in relation to the present and proposed mining in Little Long Valley: (1) the local shallow flow systems in the western ridge; and (2) the intermediate flow system in the Dinwoody Formation on the eastern ridge. The baseflow of Angus Creek in Little Long Valley is dependent on the groundwater flow systems in both ridges during the first one or two months following snow melt. The baseflow of the stream is almost completely dependent on the intermediate flow system in the eastern ridge for the rest of the year.

Three groups of groundwater flow systems have been delineated in the Lower Dry Valley study area: (1) flow systems in the Thaynes and Dinwoody Formations along Schmid Ridge; (2) local flow systems in the shallow unconsolidated material on the east slope of Schmid Ridge; and (3) a Dry Valley—Slug Creek Valley flow system in the Wells Formation. The locations of the springs that discharge from the Thaynes and Dinwoody Formations are largely controlled by the axis of the Schmid Syncline and by small E—W-trending faults. The interbasin Dry Valley—Slug Creek Valley flow system is postulated based upon the geologic configuration formed by the Schmid Syncline, the loss of surface water on the floor of Dry Valley, the downward gradient in groundwater potential in the Phosphoria Formation in Lower Dry Valley and the existence and flow pattern of springs in the Slug Creek Valley.

Care must be taken to avoid creating groundwater flow systems in any waste piles constructed as part of the mining activity. Drainage from a waste-pile flow system would probably be poorer quality than water from natural flow systems in the area.

INTRODUCTION

Much of the western phosphate field in the U.S.A. is located in southeastern Idaho in an area of complex groundwater and surface-water flow systems. Ore is mined from the Permian Phosphoria Formation along scattered outcrops of the Meade Peak Member that are exposed in a complexly folded and faulted stratigraphic sequence (Fig. 1). The complex geologic setting of the western phosphate field provides the environment for the equally complex groundwater—surface water flow systems. For example, recharge may occur in the outcrop area of a permeable rock unit in one valley with water transfer through a synclinal axis to a discharge area at the formation outcrop in an adjacent valley. Knowledge of this complex water-resource system is critical in evaluating the range and types of impacts that can be expected as a result of the development, operation and abandonment of open-pit phosphate mines. An understanding of the hydrogeologic system is a prerequisite to the

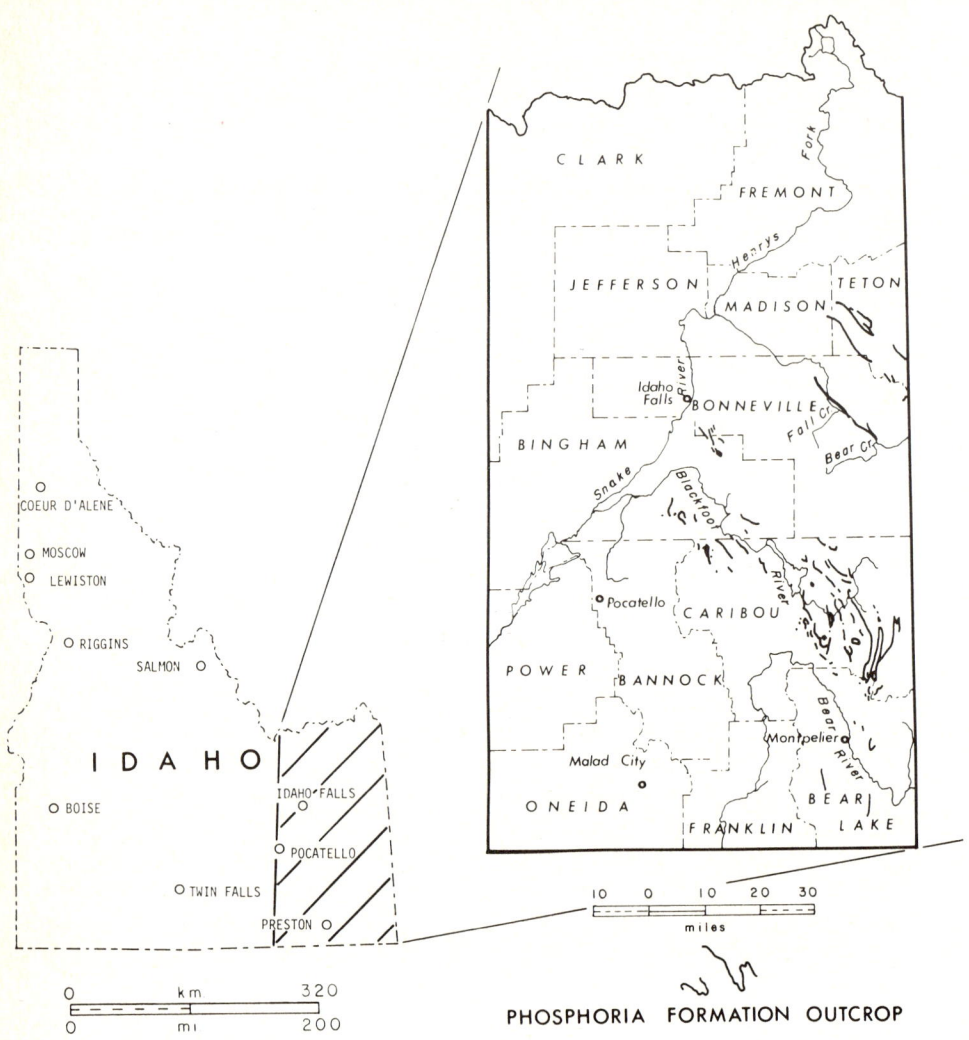

Fig. 1. Location of mining study areas in Upper Blackfoot River drainage.

efficient removal of ore with minimum water-resource impact. This paper is the result of a study that is typical of the hydrogeologic work that must precede efficient and acceptable mining in the Western Phosphate District. These studies are being continually refined and new information leading to changes should be anticipated in future publications.

HYDROGEOLOGIC FRAMEWORK

The phosphate ore occurs in a sedimentary rock sequence that ranges from relatively impermeable mudstones to highly permeable limestones. The Dinwoody, Phosphoria and Wells formations are of primary interest to the hydrogeology of phosphate mining. The geologic and hydrogeologic characteristics of these units are described in Table I. This sequence of sedimentary rocks has been folded and faulted within the Bannock Thrust zone. Thrusting has resulted in a series of roughly parallel NW-trending synclines and anticlines with major parallel and perpendicular fault zones. The topography has been strongly controlled by this structural configuration with many of the valleys located along the axes of the anticlines and the ridges located along the axes of the synclines. This topography facilitates interbasin groundwater flow under the ridges at a number of sites (Fig. 2).

The ubiquitous low permeability of the Phosphoria Formation effectively divides the geologic sequence into two major flow systems. Recharge that occurs to geologic units stratigraphically above the Phosphoria Formation discharges as springs and baseflow from outcrops of these units. Very little of this water appears to move across the Phosphoria Formation into the highly permeable limestones of the underlying Wells Formation. The hydrogeologic characteristics of the Thaynes and Dinwoody formations (stratigraphically above the Phosphoria Formation) are sufficiently continuous to provide spring discharges throughout the Phosphate District at similar stratigraphic positions. For example, a stream which flows stratigraphically downward across this sequence typically will gain flow in the Thaynes Formation, gain additional flow in the Dinwoody Formation, have relatively constant flow across the Phosphoria Formation and lose all its water into the upper member of the Wells Formation. This decreasing potential with depth in the stratigraphic sequence is characteristic of large parts of the phosphate region.

The hydrogeologic characteristics outlined above, are important in the delineation of potential water-resource impacts from and on alternative mining plans. Questions of surface-water depletion by pit construction, flooding of pit operations and possible impact from waste piles may be answered through a general understanding of the hydrostratigraphy. The importance of groundwater flow-system analysis in the phosphate region is shown in the following sections by analysis of the systems at two potential mining sites.

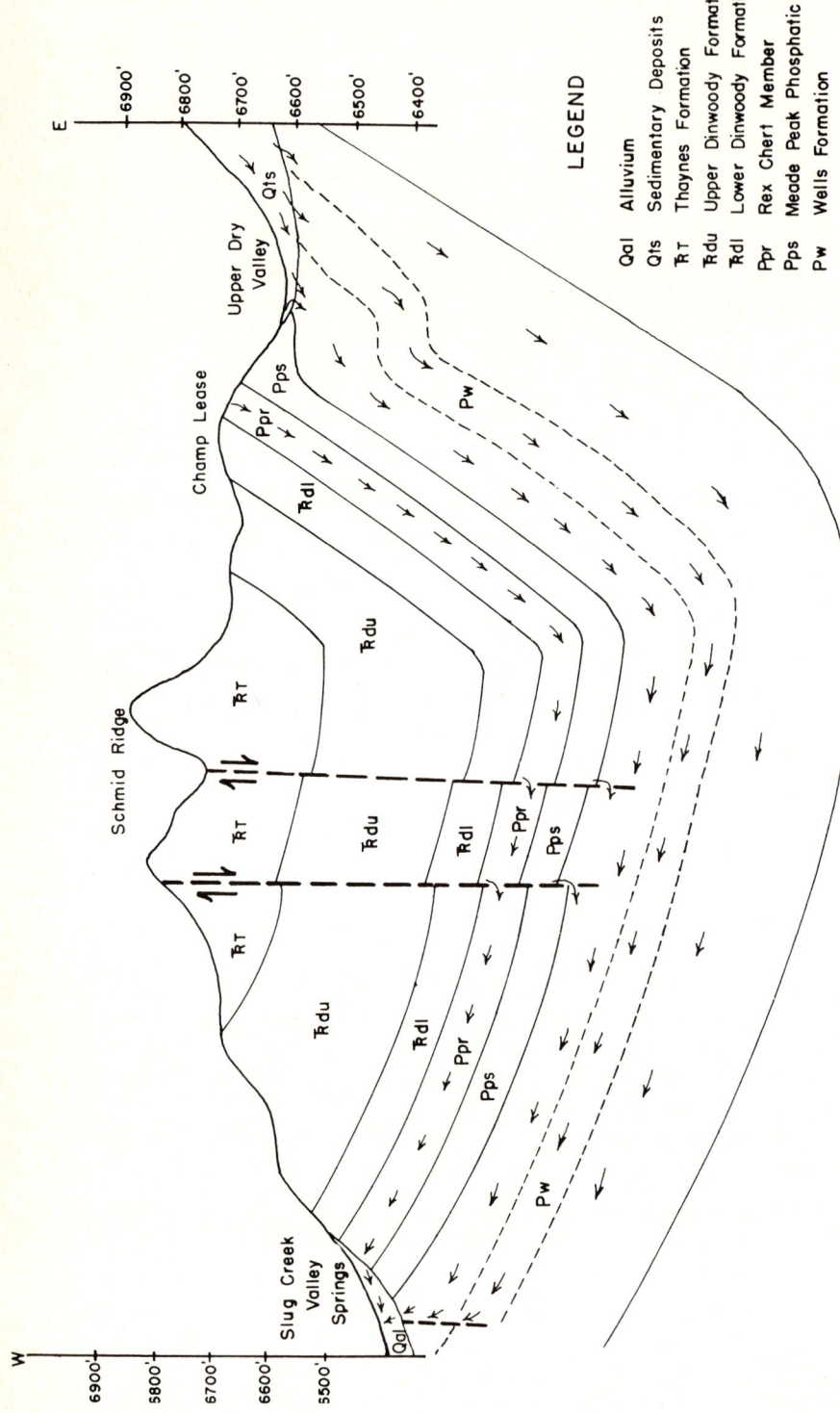

Fig. 2. Hypothetical groundwater flow system from Upper Dry Valley to Slug Creek Valley (section A—A').

TABLE I

Hydrostratigraphic columnar section in the investigated areas in southeastern Idaho

Formation	Member	Thickness* (m)	Lithology	Hydrogeologic characteristics (permeability)
Triassic:				
Thaynes	Upper	274—366 (900—1,200)	limestone and sandstone with some shale siltstone layers	moderate to high
	Middle	607 (2,000)	limestone facies interbedded with greater portion of siltstone and shale	low to moderate
	Lower	607 (2,000)	limestone facies interbedded with greater portion of siltstone and shale	low to moderate
Dinwoody	Upper	274 (900)	interbedded limestone and siltstone with discontinuous shaly zones	moderate for limestone and siltstone, low for shale and silt
	Lower		calcareous shale and siltstone with few thin limestone beds	
Permian:				
Phosphoria	Rex Chert Unit	37—46 (120—150)	chert and cherty limestone, thick bedded	permeable when fractured
	Meade Peak Unit	46—61 (150—200)	phosphatic shale, mudstone and phosphatic rock; some limestone and siltstone	low to very low
Carboniferous:				
Wells	Upper	15 (50)	siliceous limestone	moderate
	Middle	457 (1,500)	sandy limestone, sandstone	high
	Lower		limestone, mostly sandy and cherty	moderate to high
Brazer	Upper	61 (200)	black and white laminated	very low
	Middle	305 (1,000)	thick bedded limestone	high
	Lower	183—305 (600—1,000)	thin bedded limestone	high

*Values in parentheses are in feet.

LITTLE LONG VALLEY

Little Long Valley is a small NW—SE-trending valley located in Townships *6* and *7* south and Range *43* east in the Blackfoot River drainage in southeastern Idaho (Fig. 3). The valley includes an area of approximately 17.5 km^2 (3 mi.^{2}) in the headwaters of the Angus Creek drainage. Angus Creek is a perennial stream at the discharge point from Little Long Valley. Baseflow is provided primarily by eight springs emanating from the eastern ridge. Additional baseflow is derived from the western ridge during the period from snow melt to the end of July.

Little Long Valley is one of a group of NW—SE-trending valleys formed by a series of similarly trending anticlines and synclines. The area ranges in elevation from 1981 m (6500 ft.) on the valley floor to 2326 m (7620 ft.) on the eastern ridge.

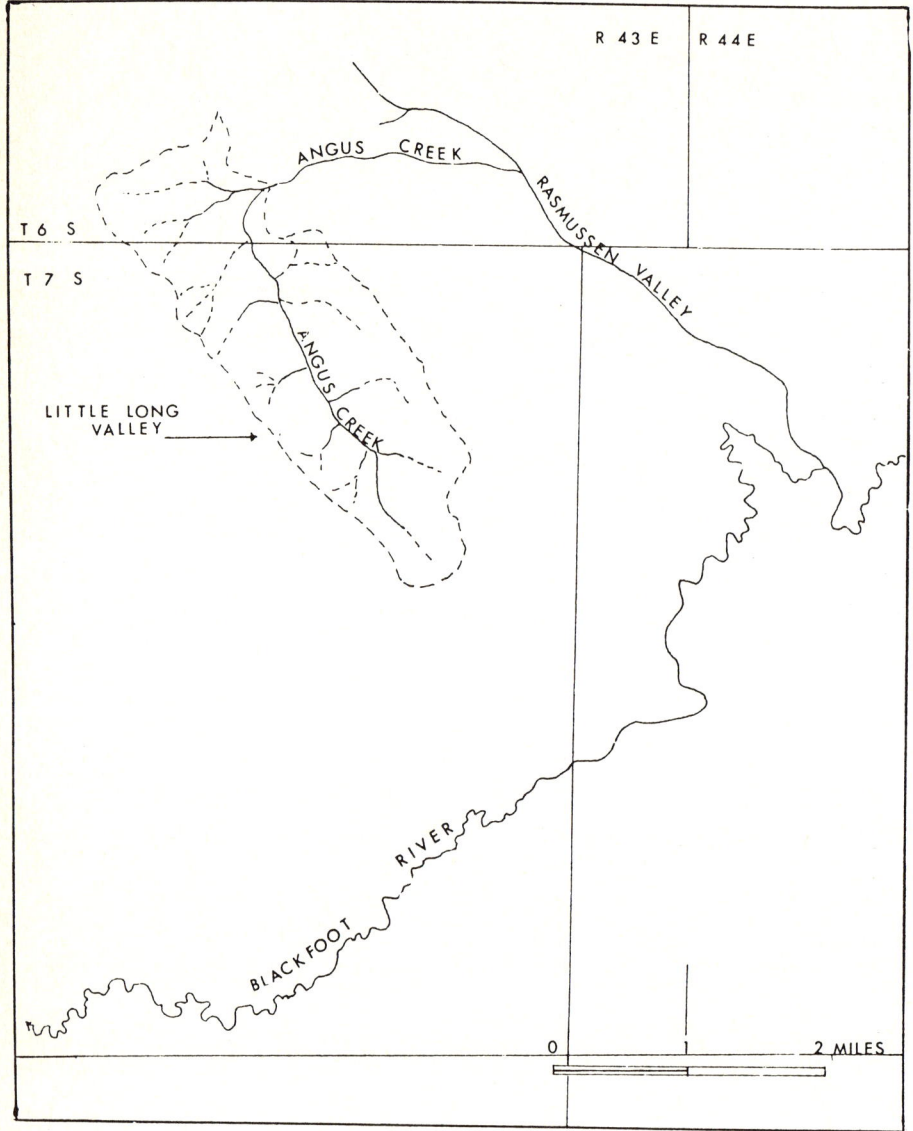

Fig. 3. Location map for Little Long Valley.

Phosphate mining operations began in 1969 in the southern portion of Little Long Valley. Stripping of overburden began in 1976 along a north—south outcrop of the Phosphoria Formation in the west portion of the valley.

The valley includes outcrops of the "Phosphate Sequence" (Dinwoody Formation, Phosphoria Formation and Wells Formation) plus valley alluvium (Fig. 4). The sedimentary rock units in the basin are structurally controlled by

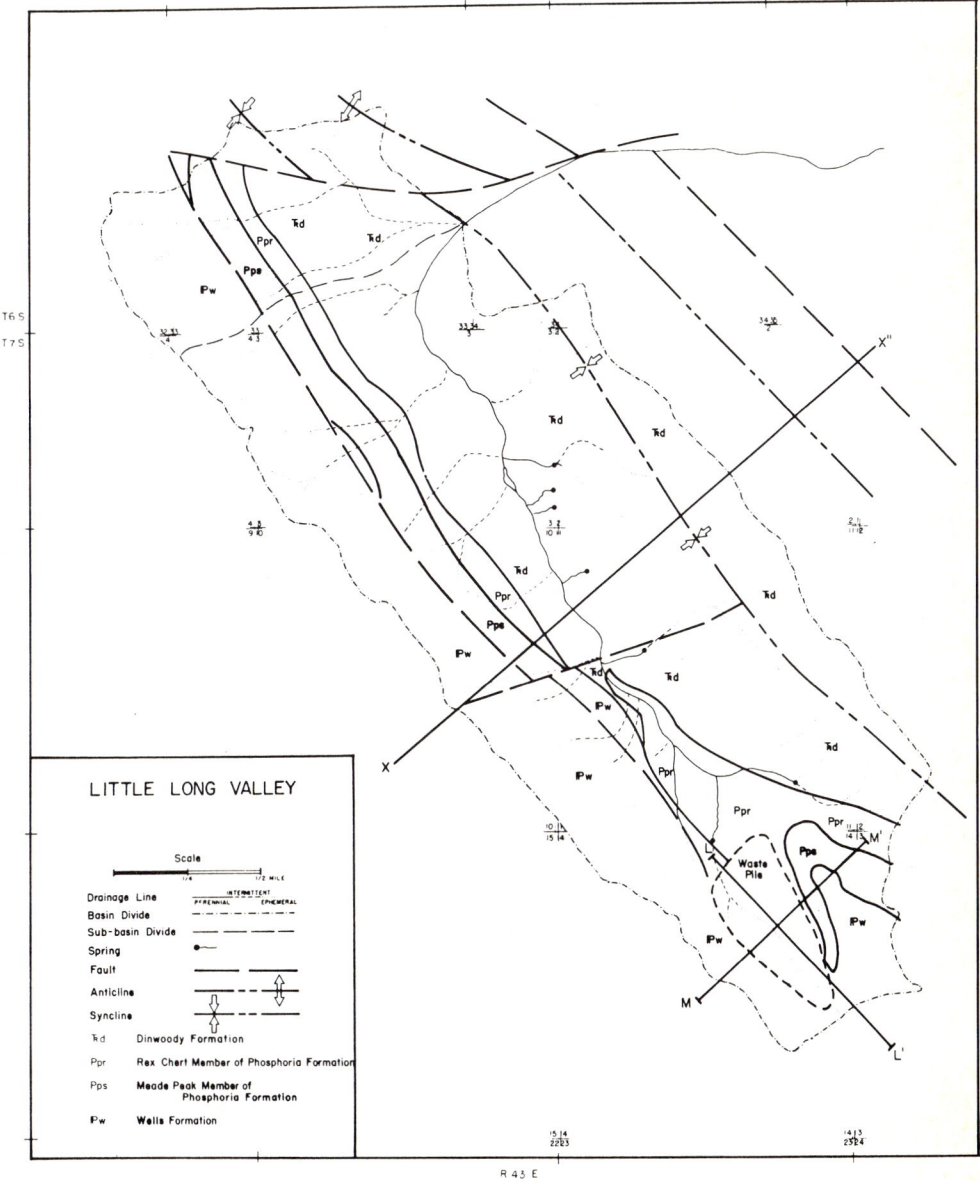

Fig. 4. Geologic map of Little Long Valley.

a syncline whose axis is located along the east ridge of the valley a small anticline in the southern portion of the valley, a major N—S-striking fault along the north side of the valley and several small E—W faults in the northern and central portion of the basin (Fig. 4). The valley floor overlies easterly-dipping sedimentary rocks of the Dinwoody Formation. Mining is progressing along the N—S-trending outcrop of the steeply dipping Meade Peak Member of the Phosphoria Formation part way up the west ridge of the valley.

The structural configuration of the sedimentary rocks provides the dominant control for groundwater flow systems and associated surface-water flow in Little Long Valley. A shallow alluvium—colluvium cover is present over most of the area and is important hydrologically on the west ridge of the valley and in the valley bottom. Generally, the unconsolidated material is relatively thin [less than 15 m (50 ft.)].

The sedimentary rock and unconsolidated material in Little Long Valley can be divided into six general hydrostratigraphic units. The Wells Formation, the Rex Chert Member of the Phosphoria Formation, the upper portion of the Dinwoody Formation and the unconsolidated material act as aquifers in the Little Long Valley area. The Meade Peak Member of the Phosphoria Formation and the lower portion of the Dinwoody Formation both act as aquitards.

Groundwater flow systems in Little Long Valley

Western ridge

Two local groundwater flow systems occur in the western ridge of Little Long Valley. [The terms local, intermediate and regional flow systems are used in the sense of Toth (1963) and Freeze and Witherspoon (1967).] One local flow system occurs on the upper slopes in the top soil and the highly weathered upper few tens of meters of the underlying Wells Formation (Fig. 5, section A—A'). This flow system is recharged at the upper part of the slope and at the top of the ridge. The recharge area encompasses about 2.4 km^2) (0.4 mi.^{2}). The discharge area occupies the topographic lows along the E—W stream channels near the contact between the Wells Formation and the Meade Peak Member of the Phosphoria Formation. This flow system is shallow, within about a meter of the ground surface. Groundwater flow in this system is controlled by the thickness and hydraulic conductivity of the soil and the upper portion of the Wells Formation.

Another local flow system occurs in the lower third of the western ridge in the Rex Chert Member of the Phosphoria Formation and the alluvial material in the valley floor (Fig. 5, section B—B'). This flow system is recharged through outcrops of the Rex Chert Member in relatively elevated areas on the western ridge; the discharge area occupies the valley floor where the hill slope flattens. The depth to water in this flow system ranges from 20.7 m (68 ft.) below land surface as measured in a piezometer on the western ridge, to almost zero on the valley floor.

The two local groundwater flow systems in the western ridge are recharged

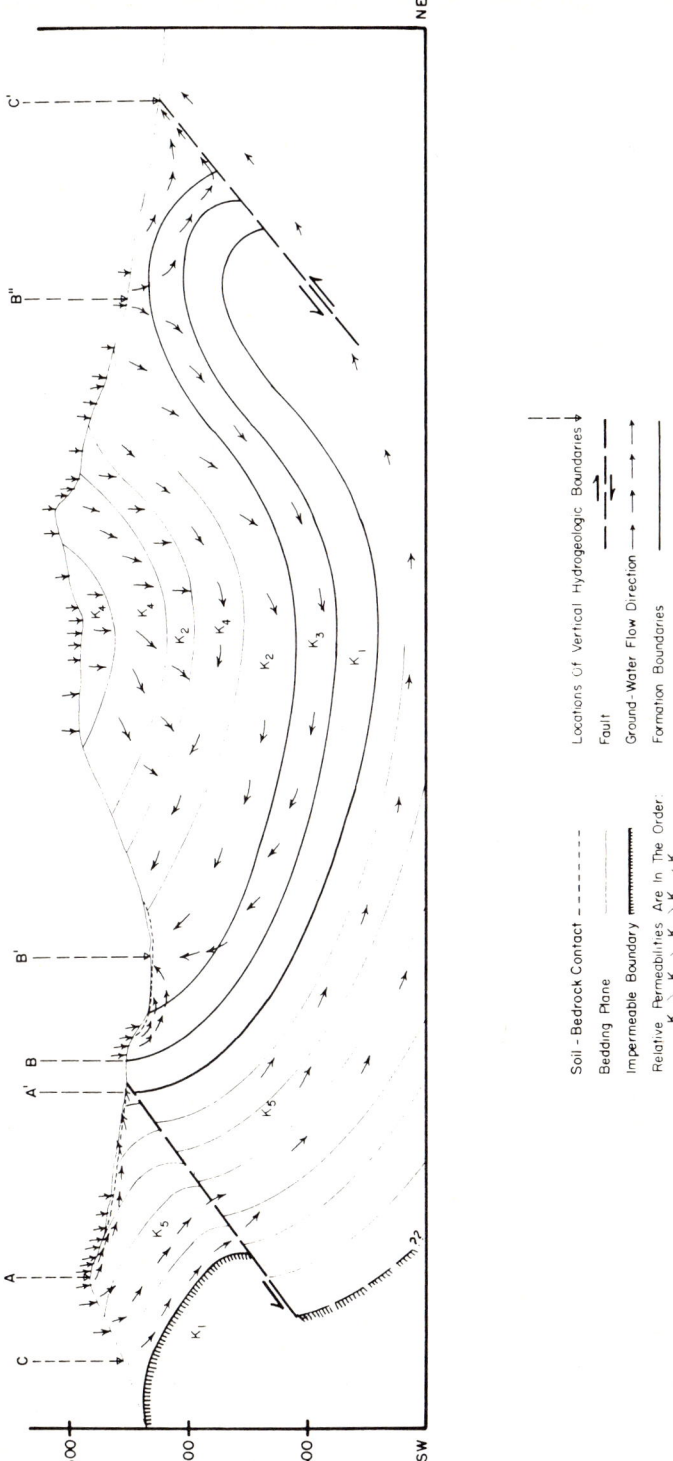

Fig. 5. Hydrogeologic cross-section of the eastern and western ridges of Little Long Valley (section $X–X'$ on Fig. 4).

during the snow-melt season. Discharge from these systems continues for a few months after snow melt but rarely later than the end of July. Subsurface flow to the valley-floor aquifer system may continue to a later time.

An intermediate groundwater flow system is believed to occur in the Wells Formation on the western ridge (Fig. 5, section $C-C'$). The recharge area is the Wells Formation outcrop on the western ridge, extending west of the surface-water divide. Groundwater flow follows the steeply dipping bedding because of greater saturated hydraulic conductivity with bedding than across bedding. The discharge from this flow system is probably located in Rasmussen Valley, east of Little Long Valley. The depth to water in this flow system is unknown.

Eastern ridge

An intermediate groundwater flow system occurs in the Dinwoody Formation on the eastern ridge of Little Long Valley. The recharge area occupies about 6 km^2 (1 mi.^{2}) extending on both sides of the top of the eastern ridge (Fig. 5, section $B'-B''$). The discharge areas are at the base of the ridge in Rasmussen Valley and in Little Long Valley. Most of the discharge from this flow system occurs in Little Long Valley because of the location of the anticlinal axis on the eastern side of the east ridge. The discharge is believed to occur both as springs and as subsurface flow into Angus Creek. Most of the low-flow discharge of the stream is from this flow system. No test holes were drilled on the eastern ridge to determine the depth to water in this flow system. However, one piezometer penetrates the lower portion of the Dinwoody Formation in the valley floor of Little Long Valley. The water level in this hole was measured to be 4.9 m (16 ft.) below land surface on July 12, 1975, which is about equal to the level of Angus Creek in the area. Angus Creek gains from subsurface flow in Little Long Valley. Most of this gain is believed to be from the intermediate flow system in the Dinwoody Formation on the eastern ridge.

Surface water—groundwater interrelationships

The Little Long Valley watershed constitutes the headwaters for Angus Creek. The creek in Little Long Valley is fed by eight perennial springs on the eastern ridge and the nose area and about 19 intermittent discharge areas on the western ridge which flow about one to two months after the end of snow melt (Fig. 6).

Discharge hydrographs for Angus Creek at The Narrows (A) were obtained from recorder charts for a calibrated weir for the two summer seasons of 1975 and 1976 (Fig. 7). The mean daily flow of Angus Creek ranged from 19.9 to 71.0 l/s (315 to 1125 gal./min.) with a mean of 45.4 l/s (720 gal./min.) during the period from July 1, 1975, until September 15, 1975. The total flow for that period was $1.4 \cdot 10^5$ m^3 ($37.2 \cdot 10^6$ gal.) or about 8% of the seasonal yield from snow melt for the year 1975 (Mohammad, 1977). The stream flow after the middle of June, 1975, was mainly from groundwater discharge.

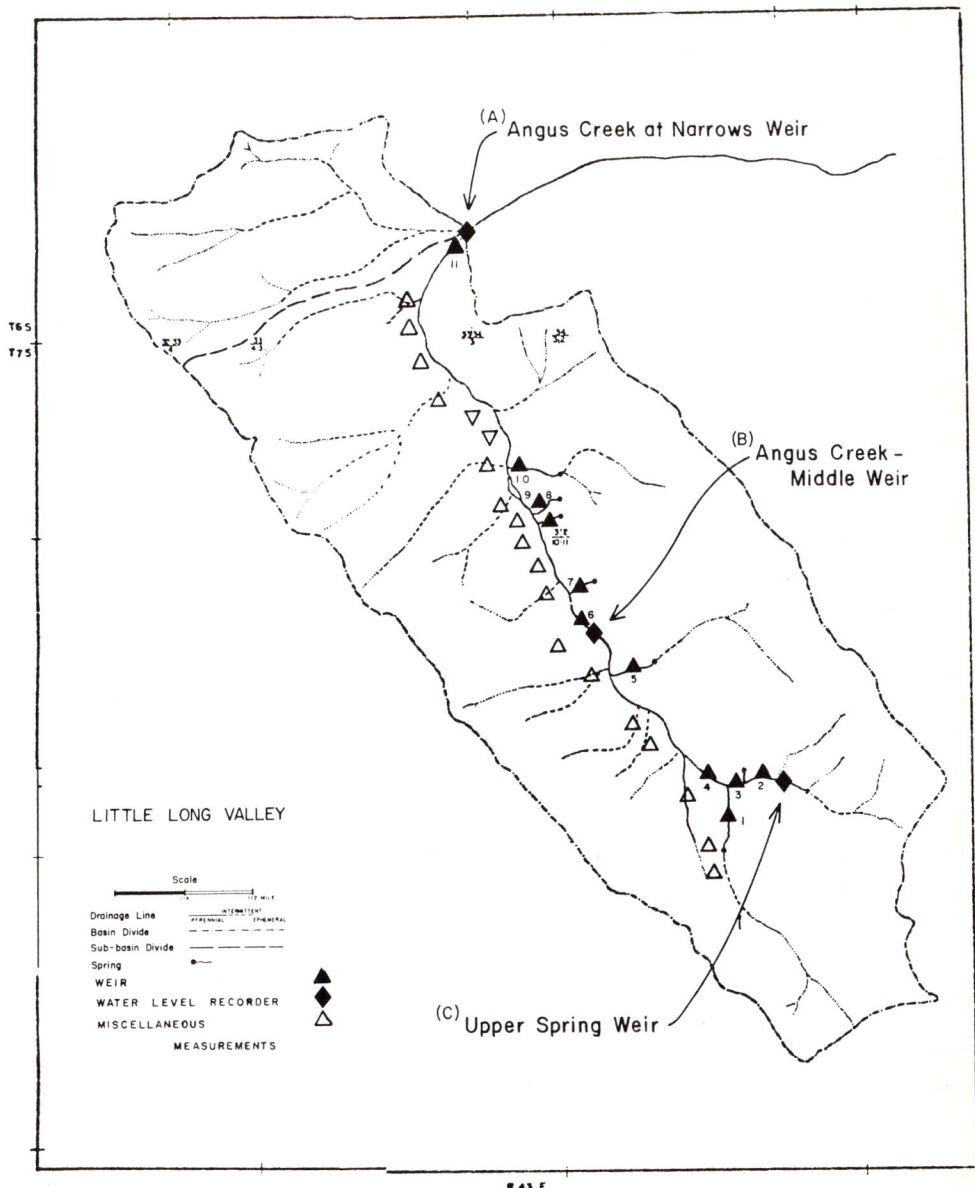

Fig. 6. Location of discharge monitoring stations in Little Long Valley.

Daily fluctuations in gage stage are shown on the recorder charts. These fluctuations are believed to be due to evapotranspiration losses along the stream channel during the day, rather than fluctuation of spring discharges caused by daily variations in barometric pressure. The recorder charts at the Upper Spring (C) did not show such daily fluctuations. Hydrographs for Angus

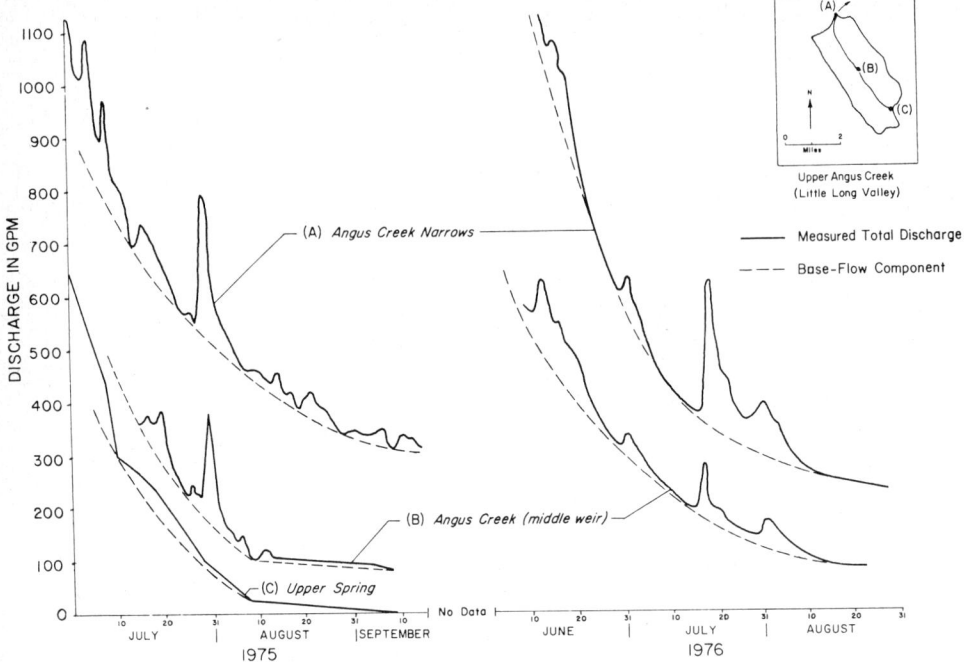

Fig. 7. Discharge hydrographs for Angus Creek in Little Long Valley, June—September, 1975 and 1976.

Creek also were constructed from the stage record at two other weir sites (B and C) and from periodic measurements of discharge for the summer periods of 1975 and 1976 (Fig. 7).

Baseflow recession analysis

Recession curves may be applied to the downward trend of water levels in wells after a period of recharge (Domenico, 1972). Recession characteristics are useful in empirical studies which attempt to relate the geology of a watershed to streamflow parameters. Butler (1957) describes the method used here.

The recession curve of Angus Creek at The Narrows shows two segments of straight lines: lines XY and YZ (Fig. 8). We interpret this to mean that the baseflow is derived from two groundwater flow systems with different recession characteristics. Line XY represents the combined recession of two flow systems: the shallow local groundwater system on the western ridge and the intermittent flow system in the Dinwoody Formation on the eastern ridge. Line YZ has a slope similar to the recession curves for the springs emanating from the east ridge flow system and represents flow almost entirely from that system. The baseflow of the stream is thus supplied from both systems early in the summer while the late summer flow derives entirely from the flow system in the eastern ridge. The recession curve for the western-ridge flow system, line *MN*, was obtained by extrapolating line *YZ* and subtracting values of points on this curve from those on the total recession curve, line *XY*, having the same date (Fig. 8).

The volume of water that is discharged from a flow system during any given time interval may be obtained by integrating the recession equation over the desired time period (Domenico, 1972, p. 49). The monthly discharges of Angus Creek were calculated using this relationship and values of *K* (time increment equal to a log cycle change in *Q*) and Q_0 (discharge at the beginning of the recession period) obtained from the recession curves (Fig. 8). The computed monthly discharges from both ridges are plotted in Fig. 9. The accumulated monthly discharge is from both systems shown on Fig. 10.

These figures show that the stream flow of Angus Creek is primarily from discharge from the eastern-ridge flow system after August, 1975. The surface and subsurface contributions to Angus Creek from the flow system in the western ridge were not significant after this date (see Fig. 10B). The discharge hydrograph in Fig. 9 shows that discharge from the western ridge is significant only during and immediately after the high-flow period.

Total potential discharge, or baseflow storage [defined as the total volume of groundwater that would be discharged during the entire recession if complete depletion were to take place uninterrupted (Meyboom, 1961)] is computed by integrating the recession equation from time zero to time infinity. The resulting equation gives total baseflow storage. Using this equation and the Q_0 and *K* values from Fig. 8, the total baseflow storages for the ground-

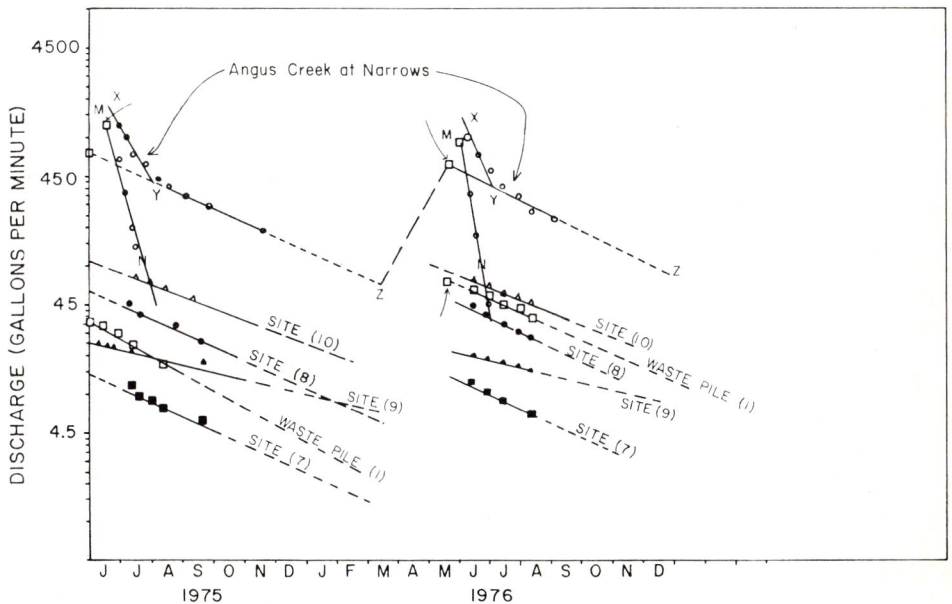

Fig. 8. Semi-logarithmic graph of baseflow recession curves for Angus Creek at The Narrows, and selected springs in Little Long Valley.

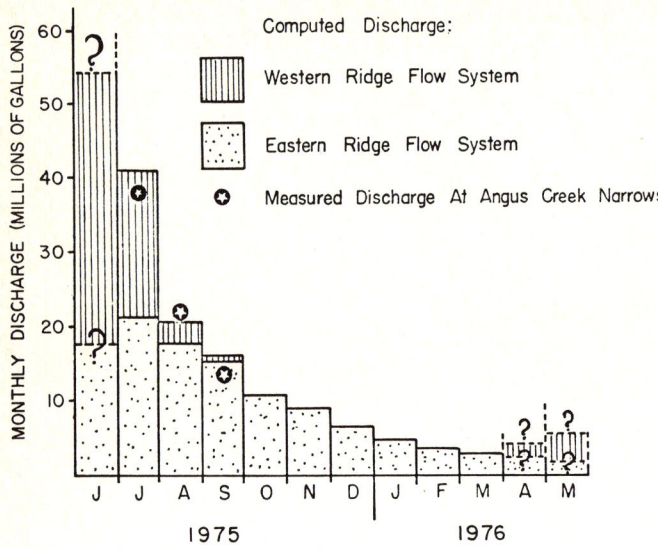

Fig. 9. Computed and measured monthly discharge for Angus Creek at The Narrows, Little Long Valley.

TABLE II

Total baseflow storage for the flow systems in Little Long Valley

Location of flow system	Baseflow storage			
	1975		1976	
	(10^4 m^3)	(10^6 gal.)	(10^4 m^3)	(10^6 gal.)
Eastern ridge	38.9	103	28.4	75
Western ridge	9.8	26	6.8	18

water flow systems on the eastern and western ridges in Little Long Valley were computed from 1975 and 1976 (Table II). The flow system on the western ridge has only about 25% as much baseflow storage as the flow system on the eastern ridge.

The mass curve for the western ridge (Fig. 10B) indicates that 95% of the baseflow storage was discharged during June and July with all storage depleted by the end of September, 1975. No surface flow from this ridge was observed after July, 1975; the rest of the baseflow storage probably was discharged as subsurface flow to the stream, evapotranspiration losses, or deep percolation to another flow system.

The total surface water discharge from Little Long Valley during July and August, 1975 and 1976, as measured at the Angus Creek Narrows, is compared with the calculated baseflow contribution to the creek in Table III.

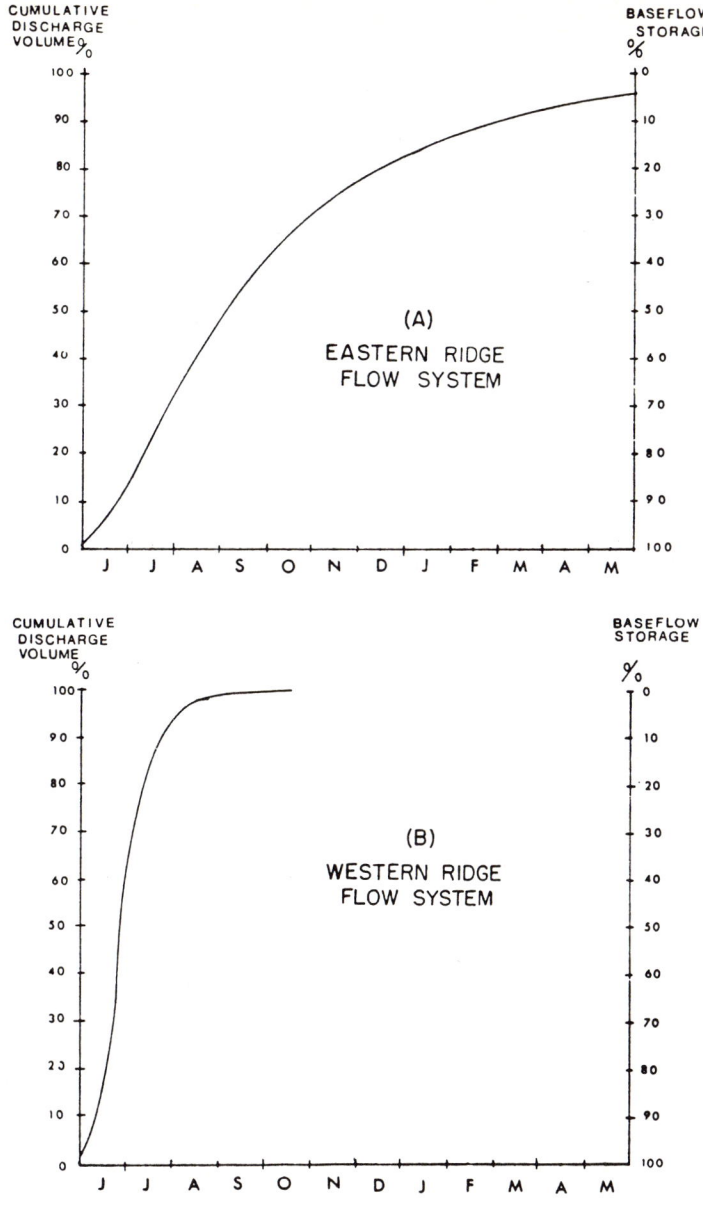

Fig.10. Cumulative discharge volume and remaining baseflow storage (in percentage) for the Eastern Ridge and the Western Ridge flow systems, Little Long Valley, 1975.

The difference between measured discharge and calculated baseflow contribution is believed to be runoff from summer storms. unmeasured subsurface discharge into the stream, mostly from the eastern ridge flow system, and error in recession slope analysis.

TABLE III

Comparison of measured Angus Creek flow with calculated baseflow discharge, Little Long Valley (discharge in 10^4 m^3, values in parentheses in 10^6 gal.)

Time	Calculated discharge			Measured discharge Angus Creek at The Narrows	Difference
	East Ridge	West Ridge	East Ridge + West Ridge		
1975:					
July	7.5 (19.8)	3.1 (8.3)	10.6 (28.0)	13.3 (35.3)	2.8 (7.3)
August	6.2 (16.5)	0.5 (1.2)	6.7 (17.7)	7.4 (19.5)	0.7 (1.9)
Total	13.7 (36.2)	3.6 (9.5)	17.3 (45.7)	20.7 (54.8)	3.5 (9.2)
1976:					
July	5.6 (14.8)	1.1 (2.9)	6.7 (17.7)	8.4 (22.1)	1.7 (4.4)
August	4.3 (11.4)	0.2 (0.4)	4.5 (11.9)	4.8 (12.6)	0.3 (0.7)
Total	9.9 (26.3)	1.3 (3.3)	11.2 (29.6)	13.2 (34.7)	2.0 (5.1)

The groundwater recharge from snow melt on both ridges of Little Long Valley may be computed from the baseflow recession curves for the 1975 and 1976 water years. The baseflow storage in the local flow systems in the western ridge was completely discharged by the end of September, 1975. The groundwater recharge from snow melt during 1975 is thus equal to the baseflow storage, $9.8 \cdot 10^4$ m^3 ($26 \cdot 10^6$ gal.) in 1975 and $6.8 \cdot 10^4$ m^3 ($18 \cdot 10^6$ gal.) in 1976. This indicates an annual recharge of 7.6 and 5.3 cm (3.0 and 2.1 in.) over the 2.94-km^2 (0.5-mi.^{2}) recharge area in 1975 and 1976, respectively.

For the eastern ridge, the recharge from 1976 snow melt is equal to the difference between the remaining potential discharge at the end of the 1975 recession period, and the total potential discharge at the beginning of the 1976 recession. The remaining potential discharge at the end of the 1975 recession can be determined by evaluating the equation for baseflow volume with t_1 equal to 285 days and t_2 equal to infinity. A total of $3.1 \cdot 10^4$ m^3 ($8.3 \cdot 10^6$ gal.) of water remained in storage at the beginning of the 1976 recession.

The recharge which took place between the recessions of 1975 and 1976 should equal the difference between this value ($3.1 \cdot 10^4$ m^3 or $8.3 \cdot 10^6$ gal.) and the total potential discharge at the beginning of the 1976 recession ($28.3 \cdot 10^4$ m^3 or $74.9 \cdot 10^6$ gal.) or a total of $25.2 \cdot 10^4$ m^3 ($66.6 \cdot 10^6$ gal.). This is equal to a recharge of 10.7 cm (4.2 in.) of water over the recharge area of 5.3 km^2 (0.9 mi.^{2}).

The high values of recharge [5.3 cm (2.1 in.) on the west ridge and 10.7 cm (4.2 in.) on the east ridge] may result from the major snow drifts which are located in the recharge areas on both ridges. The snow drifts on the western ridge of Little Long Valley extended for about 4 km (2.5 mi.) along the ridge, with a width up to 42.5 m (140 ft.), and a depth ranging from 4.3 to 6.4 m (14 to 21 ft.) as measured in March, 1975. The snow drifts on the eastern ridge were not measured but should be deeper; the elevation of the east ridge is as much as 137 m (450 ft.) higher than the west ridge.

Conclusions pertinent to Little Long Valley

(1) Two groundwater flow systems are important in relation to present and proposed mining in Little Long Valley: (a) the local shallow flow systems in the western ridge; and (b) the intermediate flow system in the Dinwoody Formation on the eastern ridge.

(2) The baseflow of Angus Creek in Little Long Valley is dependent on the groundwater flow systems in both ridges during the first one or two months following snow melt. The baseflow of the stream is almost completely dependent on the intermediate flow system in the eastern ridge for the rest of the year.

(3) Angus Creek in Little Long Valley is a gaining stream. The stream gained an average of 9.8 l/s (155 gal./min.) between the middle weir and the weir at The Narrows during the period from July to September, 1975. The gain was from subsurface flow from the Dinwoody Formation in the valley floor.

LOWER DRY VALLEY

Dry Valley is a long NW—SE-trending valley located in Townships 7, 8 and 9 south and Ranges 44 and 45 east in the Blackfoot River drainage in southeastern Idaho (Fig. 11). The valley includes an area of about 24.7 km^2 (42 mi.^{2}). It is drained by Dry Valley Creek and streams in Goodheart Canyon and Dry Canyon which are tributary to Slug Creek. Slug Creek flows into the Blackfoot River. The streams are intermittent with surface water in the valley only after periods of snow melt or heavy precipitation. A number of perennial springs exist along the western margin of Dry Valley in Schmid Ridge; however, most of the flow of these springs is lost when the small streams flow out over the valley floor. The flow of Dry Valley is relatively flat with an average elevation of about 2990 m (6550 ft.). Dry Ridge bounds the valley on the east and rises sharply to a maximum elevation of about 2710 m (8900 ft.). Schmid Ridge is located on the west side of Dry Valley and rises to a maximum elevation of about 2290 m (7500 ft.).

Dry Valley has been divided arbitrarily into Lower Dry Valley (northern part) and an Upper Dry Valley (southern part). The area of interest for this report is the western side of Lower Dry Valley. The site includes the valley floor of Lower Dry Valley, the northern portion of Schmid Ridge and extreme eastern side of Lower Slug Creek Valley.

Phosphate mining is proposed for an outcrop area of the Meade Peak Member of the Phosphoria Formation located along the west side of Lower Dry Valley. The proposed mining site extends roughly 12.9 km^2 (8 mi.^{2}) in a N—S

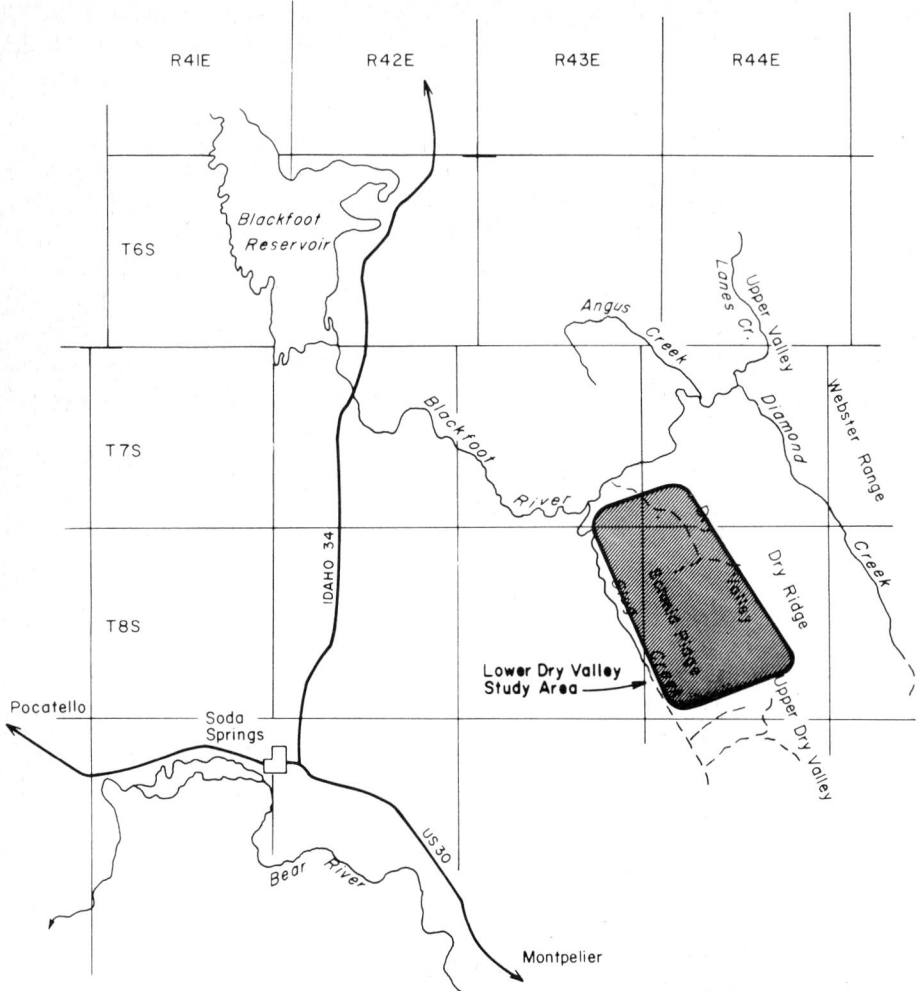

Fig. 11. Location map of Lower Dry Valley study area.

direction near the junction of the valley floor and the east side of Schmid Ridge. Mining operations have not yet begun. An active mining operation exists on the northern portion of Dry Ridge but was not included as part of this report.

The sedimentary rocks that crop out in the Lower Dry Valley site range in age from Mississippian to Quaternary (Fig. 12). They include from oldest to youngest: the Brazer Limestone, the Wells Formation, the Phosphoria Formation, the Dinwoody Formation and the Thaynes Formation. Sedimentary deposits of Tertiary and perhaps Quaternary age form extensive alluvial fans and valley-floor deposits in both Dry Valley and Slug Creek Valley. Alluvial thicknesses along the valley floors are generally unknown. One test hole in

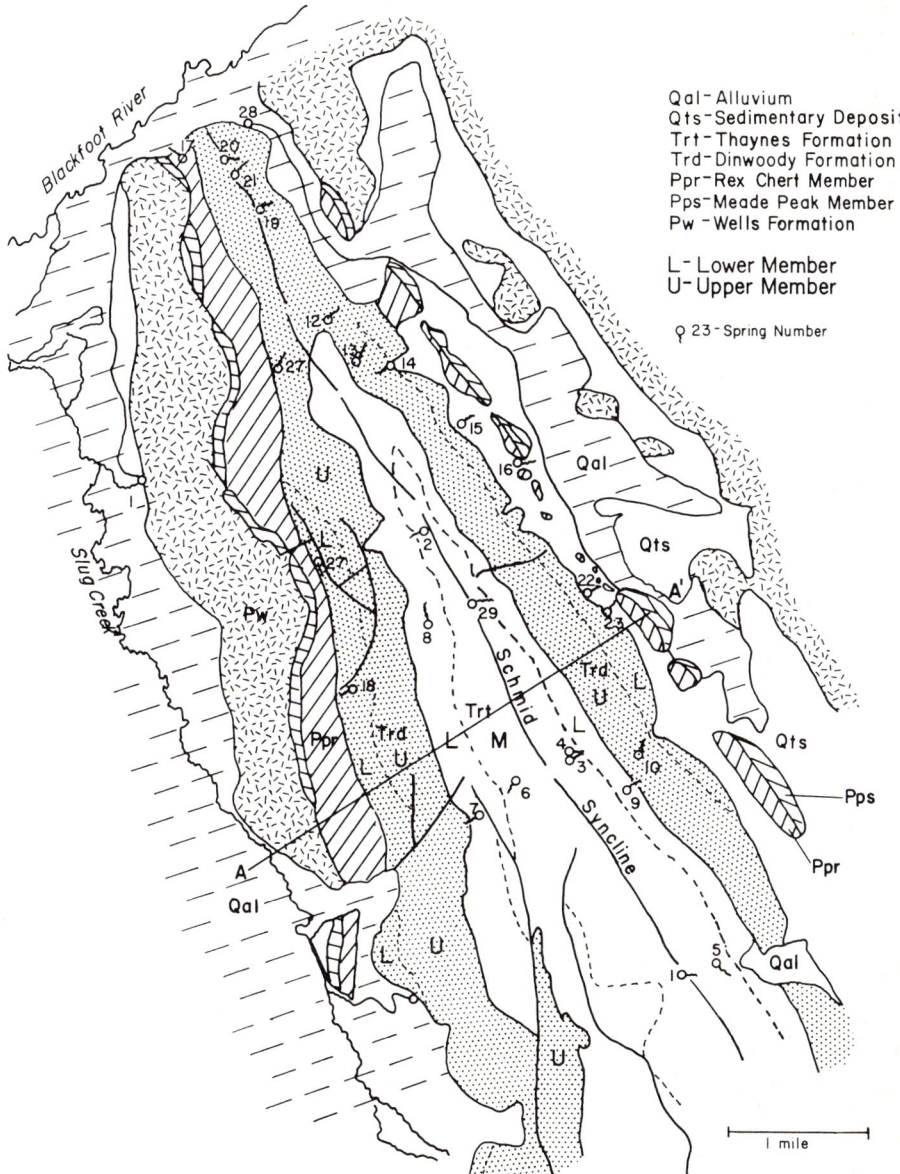

Fig. 12. Geologic map of Lower Dry Valley showing spring locations.

Dry Valley penetrated 122 m (400 ft.) of unconsolidated material without entering bedrock.

The dominant structural features in the area are the Schmid Syncline and the Dry Valley Anticline. The axis of the Schmid Syncline is roughly along the center line of Schmid Ridge (Fig. 12). The axis of the Dry Valley Anticline is located on the west face of Dry Ridge. Both structures trend in a

NW—SE direction and play important roles in the occurrence and movement of groundwater. The sedimentary rock units dip sharply to the west along the proposed mining area in Lower Dry Valley.

Four sedimentary rock units and one unconsolidated rock constitutes the aquifers of the Lower Dry Valley site. Groundwater discharges from springs at sites in the middle portion of the Thaynes Formation, the lower portion of the Thaynes Formation, the Dinwoody Formation, and along the slopes in the unconsolidated alluvial material. In addition, the Wells Formation is believed to be an aquifer for interbasin flow of groundwater.

Groundwater flow systems in Lower Dry Valley

Thaynes and Dinwoody aquifers in Schmid Ridge

A number of groundwater flow systems exist in the Thaynes and Dinwoody Formations in Schmid Ridge. These flow systems are evidenced by springs located along the center and east side of the ridge (Fig. 12). The recharge areas for the flow systems in the Middle and Lower Members of the Thaynes Formation and the Dinwoody Formation are along the top of the ridge at areas of snow accumulation, formational outcrop and intensive fracturing. The locations of the Thaynes and Dinwoody springs are controlled by the axis of the syncline, faulting and formational outcrops. The springs generate small streams that flow only short distances during the summer period.

Unconsolidated aquifers on the east side of Schmid Ridge

Shallow groundwater flow systems are present in the colluvium covering the eastern foothills of Schmid Ridge in the immediate vicinity of the proposed mining area. These sediments overlie most of the Lower Dinwoody, Phosphoria and Wells Formations on the west side of Lower Dry Valley. Recharge to the colluvial aquifers on the west side of Lower Dry Valley occurs in two areas: west or uphill of the outcrop area of the Rex Chert Member of the Phosphoria Formation, and east or downhill of the Chert outcrop. Discharge areas for the flow systems in the unconsolidated aquifers are located in topographic low areas both above and below the Chert outcrops. The discharges are generally only seeps and wet areas, most of which dry up during the summer.

Dry Valley—Slug Creek Valley flow system

An interbasin groundwater flow system occurs in the Wells Formation which allows movement of water from Dry Valley through the trough of the Schmid Syncline to discharge in the Slug Creek Valley (Fig. 13). The recharge area for this flow system is in the valley bottom of Dry Valley where the Wells Formation is present under the valley fill alluvium. Recharge comes from snow melt and from stream loss. The downward pattern of water movement in Dry Valley limits surface-water discharge to periods of snow melt and intense precipitation. The groundwater flow system discharges in Slug Creek

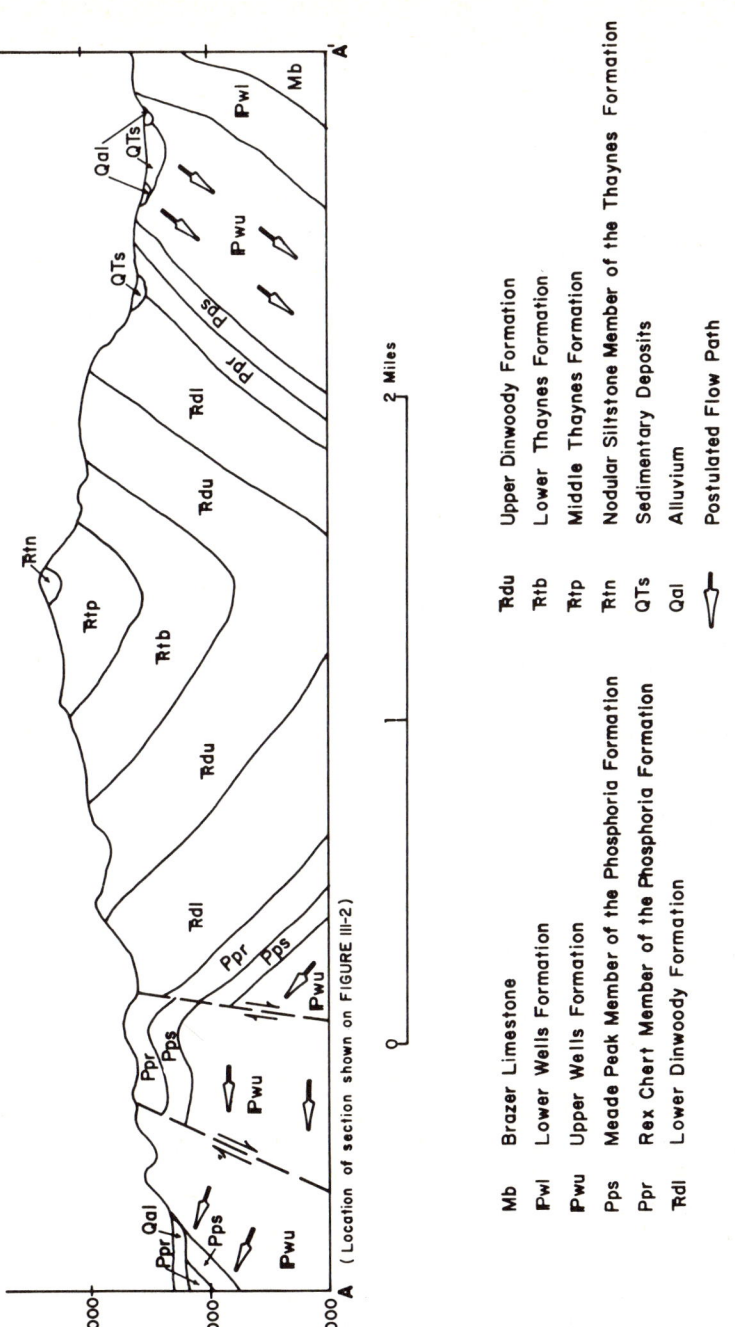

Fig. 13. Postulated Dry Valley—Slug Creek Valley groundwater flow system.

Valley both in the form of springs and as discharge to the alluvial aquifer. Several larger springs exist along the east side of Slug Creek.

Conclusions regarding Lower Dry Valley

(1) Three groups of groundwater flow systems have been delineated in Lower Dry Valley: (a) flow systems in the Thaynes and Dinwoody formations along Schmid Ridge; (b) local flow systems in the shallow unconsolidated material on the east slope of Schmid Ridge; and (c) a Dry Valley—Slug Creek Valley flow system in the Wells Formation.

(2) The Thaynes and Dinwoody Formation flow systems receive recharge on Schmid Ridge from snow melt on formational outcrops and zones of greater fracturing. The locations of the springs that discharge from these systems are largely controlled by the axis of the Schmid Syncline and by small E—W-trending faults.

(3) The flow systems in the unconsolidated materials on the east slope of Schmid Ridge are short and discharge only limited quantities of water. The locations of the small springs, seeps and marshy areas are controlled primarily by the outcrop of a resistant portion of the Rex Chert Member of the Phosphoria Formation.

(4) An interbasin Dry Valley—Slug Creek Valley flow system is postulated based upon the geologic configuration formed by the Schmid Syncline, the loss of surface water on the floor of Dry Valley, the downward gradient in the Phosphoria Formation in Lower Dry Valley and the existence and flow pattern of springs in the Slug Creek Valley.

IDENTIFICATION AND ANALYSIS OF MINING IMPACTS ON GROUNDWATER FLOW SYSTEMS

The geologic, topographic, hydrogeologic and hydrologic factors which control water-resource systems in the phosphate area are subject to changes as a result of mining activities. Mining may impact on both groundwater flow systems and surface streams. Primary emphasis in this discussion is placed upon impacts on groundwater flow systems. This section is a summary of a more detailed discussion presented by Mohammad (1977).

General analysis of potential impacts

Mining impacts on groundwater flow systems may include changes in recharge, discharge, storage, water quality and flow pattern. New recharge and discharge areas may be created at the mining sites with the creation of new groundwater flow systems and/or the modification of originally existing flow systems. Water quality may also be affected by the new exposure of minerals to surface conditions.

Mining waste piles and pits are the most important mining factors in terms of groundwater impacts. New groundwater flow systems may be developed in mining waste piles depending upon the hydrogeologic characteristics of the pile material, the shape of the pile and its location with respect to topography. Mining pits may intersect existing flow systems and create new discharge areas or collect surface water and form new recharge areas. New groundwater flow systems in waste piles may discharge water that is poorer in quality than existing flow systems because of greater surface area exposed in the broken waste material.

The impacts of mining within an area depend upon the following basin characteristics:

(1) *Physical characteristics of the basin*: such as the size, topography, vegetal cover, amount and distribution of precipitation, natural stream flow characteristics, and baseflow component of any streams.

(2) *Hydrogeologic setting of the basin*: which includes the characteristics of any groundwater flow systems occurring in the basin and the location of groundwater recharge and discharge areas.

Mining impacts on groundwater recharge

The effect of the mining pits and waste piles on groundwater recharge depends on the area affected, the change in the rate and quality of infiltration and the location of the mining activity relative to existing flow systems. The mining variables which control the impacts on groundwater recharge are summarized below:

(1) *Waste-pile variables:*
(a) catchment area (include intake area, and any other areas contributing surface-water runoff to the intake area); (b) recharge area (include intake area only); (c) surface topography; (d) vegetal cover; (e) texture and composition of the waste material on the surface of the pile; (f) annual precipitation; (g) occurrence of snow drifts on the pile surface or within its catchment area; (h) effect of secondary structures such as fracturing, piping and compaction; and (i) location relative to existing flow systems.

(2) *Mining-pit variables:*
(a) area flooded by water; (b) depth of water in the pit; (c) period of time for which water stays in the pit; (d) catchment area; (e) occurrence of snow drifts within the catchment area; (f) occurrence of groundwater discharge in the pit; (g) degree of excavation of the Meade Peak rock unit; (h) silt deposits in the bottom of the pit; (i) dip and sequence of geological formations at the pit site; and (j) location of the pit relative to the groundwater flow system.

Mining impacts on groundwater discharge

Mining factors may have a direct or indirect effect on groundwater discharge, depending on whether these factors are associated with a discharge area or a recharge area. Changes in groundwater recharge in a given flow system will ultimately be reflected in changes in discharge in that same system.

Mining pits directly affect groundwater discharge by intersecting existing

flow systems. The change in natural groundwater discharge due to the mining pits depends upon the following factors:
  (1) The configuration of the pit: depth, width and length.
  (2) The location of the pit in relation to existing groundwater flow systems.
  (3) The dewatering program followed during pit construction.

Waste piles may also affect groundwater discharge by creating new discharge areas for either new flow systems or pre-existing flow systems modified by pile construction. The impact of a waste pile upon groundwater discharge depends upon the size, the hydraulic conductivity and thickness of the wastes, the method of construction, the degree of compaction and the age of the pile.

Mining impacts on groundwater storage

The quantity of groundwater in storage may be modified by mining. Changes in recharge and discharge and the creation, modification or elimination of natural groundwater flow systems all will have an impact on the quantity of water in storage. The waste piles will probably increase groundwater storage because the fragmented waste material has a higher porosity than the indurated rock formation. Mining pits will probably cause natural or induced dewatering of the nearby rock units.

Mining impacts on groundwater quality

The greatest impact on groundwater quality will be from flow systems created in mine waste piles. The quality change in a flow system in a waste pile is dependent upon the quantity of water in the flow system, the length of the flow path, the velocity of groundwater movement in the pile and the chemical composition and solubilities of minerals in the waste material. Generally, the water quality of a waste-pile discharge is dependent upon the chemical makeup of the wastes and the time and area of contact of the water and the rock material.

Projected impacts from mining in the study areas

Little Long Valley study area

(1) The proposed mining pit on the western ridge of Little Long Valley will intercept the local flow system on this ridge. This will create a new groundwater discharge area in the pit and eliminate the natural discharge along the lower slopes. This flow system has measurable discharge only in the spring and early summer; impacts from flow-system modification will be noticeable only for this period. A decrease in flow in Angus Creek in the spring and early summer will result from pit construction if the water from the pit is not returned to the stream within the valley.

(2) The mining pit will probably not affect the groundwater flow system in the Dinwoody Formation on the eastern ridge because of the low cross-bedding hydraulic conductivity of the lower Dinwoody Formation and the Rex Chert Member of the Phosphoria Formation. Mining activity will thus

not decrease the flow of the springs that provide the summer baseflow for Angus Creek. The low-flow of Angus Creek should not be decreased by leakage of water into the pit when it is below stream level because of the low cross-bedding hydraulic conductivity of these same units.

(3) The flow system that has been created in the waste pile in the nose area discharges water that is poorer in quality than the other flow systems in the valley. This drainage has not caused a measurable decrease in water quality in Angus Creek to date because of the small discharge rate. Care must be taken to avoid creating flow systems in any new waste piles in the valley to prevent an additional waste pile drainage.

Lower Dry Valley study area

(1) The proposed mining pit on the west side of Lower Dry Valley will intercept the local flow system in the shallow unconsolidated sediments. This will create a new groundwater discharge area in the pit and eliminate the natural discharge along the lower portion of the slope. This flow system has noticeable discharge only in the spring and early summer. Impacts from modification of this flow system will be small and limited to the spring and early summer period.

(2) The construction of the mining pit will not affect the groundwater flow systems in the Thaynes and Dinwoody formations and will not decrease the flow of springs on Schmid Ridge. The stream formed by spring discharges from these units will have to be considered in surface-water diversion plans for the mine.

(3) A major portion of the Phosphoria Formation will be dewatered during pit construction. This will cause little measurable impact on the total water-resource system in the area. Only a very small quantity of water is believed to annually recharge to or discharge from the Phosphoria Formation in this area.

(4) Some dewatering of the valley alluvium may occur if the pit is constructed below the groundwater level in the valley floor in areas where the alluvium is cut by the pit. If the water pumped from the pit from this source is discharged out of the basin, there will be a net decrease in water in the alluvial Wells Formation flow system. This would ultimately decrease the flow of the Slug Creek Valley springs associated with this flow system. The total impact on spring flow would probably be small because of the large quantity of water normally available for recharge from spring snow melt.

(5) Care must be taken to avoid creating groundwater flow systems in any waste piles constructed as part of the mining activity. Drainage from a waste-pile flow system would probably be poorer quality than water from the natural flow systems. A flow system created in a waste pile constructed on the valley floor would probably discharge into the valley alluvium and possibly enter the interbasin Dry Valley—Slug Creek Valley flow system.

REFERENCES

Butler, S.S., 1957. Engineering Hydrology. Prentice-Hall, Englewood Cliffs, N.J., pp. 214—218.

Domenico, P.A.M., 1972. Concepts and Models in Groundwater Hydrology. McGraw-Hill, New York, N.Y., 406 pp.

Freeze, R.A. and Witherspoon, D.A., 1967. Theoretical analysis of regional groundwater flow, 2. Effect of water-table configuration and subsurface permeability variation. Water Resour. Res., 3(2): 623—634.

Meyboom, P., 1961. Estimating groundwater recharge from stream hydrographs. J. Geophys. Res., 66(4): 1203—1214.

Mohammad, O.M.J., 1977. Evaluation of the present and potential impacts of open pit phosphate mining on groundwater resource systems in southeastern Idaho phosphate field. Ph.D. Dissertation, University of Idaho, Moscow, Idaho, 166 pp.

Toth, J., 1963. A theoretical analysis of ground-water flow in small drainage basins. J. Geophys. Res., 68(16): 4806.

[2]

COOLING MECHANISMS AND EFFECTS ON MANTLE CONVECTION BENEATH ANTARCTICA

L.D. McGINNIS

Department of Geology, Northern Illinois University, De Kalb, IL 60115 (U.S.A.)

(Accepted for publication May 28, 1979)

ABSTRACT

McGinnis, L.D., 1979. Cooling mechanisms and effects on mantle convection beneath Antarctica. In: W. Back and D.A. Stephenson (Guest-Editors), Contemporary Hydrogeology — The George Burke Maxey Memorial Volume. J. Hydrol., 43: 265—286.

Temperature in the mantle beneath Antarctica is a function of initial temperatures established during processes of accretion, primarily subduction, of continental nuclei which formed Pangaea. Subsequent near-surface temperature, which is latitude-dependent, has determined the pattern and rate at which the continent has broken up. Subduction and conduction play primary roles in the mantle temperature regimen; whereas, ice-sheet flow systems and deep groundwater flow beneath the ice sheet play secondary roles. All of these effects tend to maintain low mantle temperatures beneath Antarctica. Because of lower mantle temperature, mantle viscosity (η) is several orders of magnitude higher than it is beneath equatorial continents. As η increases, rates of plate motion decrease since strain rate is inversely proportional to η. Since continental accretion ended in late Paleozoic time, mantle temperatures have been increasing due to radioactive decay at rates dependent upon thermal gradients, thus low-latitude fragments of Pangaea were first to separate by rifting and convection. North America was first to drift away from Africa and South America in Triassic time along a rift system which lay along the equator. Greenland rifted from Eurasia and Australia from Antarctica 50—60 m.y. ago. Both rifting episodes occurred in the same time period at the same high latitudes, only one north, the other south. The mantle underlying the remaining high southern-latitude fragment of Pangaea (Antarctica) retains a relatively high viscosity. Indirect evidence for high mantle viscosity beneath Antarctica is seen by the abnormally deep isostatically-uncompensated continental shelves, and an asymmetrically spreading circumpolar ridge. Additional evidence for this is the aseismicity south of 63°S, although aseismicity can also be explained by high, subglacial hydrostatic pressure acting upon bottom melt water. Effective stresses are lowered between adjacent tectonic blocks and accumulating stress is easily dissipated. A present zone of high mantle temperatures beneath the Transantarctic Mountains suggests a possible future spreading center from which additional fragments of the continent might drift; however, the inward-directed circumpolar ridge, high mantle viscosities, and a possible adequate heat loss by volcanism might slow or eliminate the process.

INTRODUCTION

Contemporary tectonics of the antarctic and adjacent ocean lithosphere does not conform to the pattern observed in other regions of the Earth. Antarctica's often-mentioned aseismicity is only one of the region's unsolved enigmas. The aseismicity problem has been explored by Voronov and Klushin (1963) who attribute one of the causes to high mantle viscosity caused by cooling; however, mechanisms of mantle cooling according to their hypothesis are not justified because of the long time required for cold surface temperatures to penetrate to mantle depths by conduction alone. This paper explores various cooling mechanisms and the effect of cooling on mantle viscosities.

According to Dietz and Holden (1970) Pangaea accretion, which terminated in late Paleozoic time, was followed shortly thereafter in Triassic time by progressive continental breakup, with the rifting progressing from low latitudes. According to the Dietz—Holden model all continental fragments have moved away from the poles except for Antarctica. For example, North America moved north, away from Africa 200 m.y. ago (see Phillips and Forsyth, 1972), whereas, Australia parted from Antarctica only 50—60 m.y. B.P. The discussion which follows is an attempt to describe, in a general way, the role of the hydrosphere, lithosphere and asthenosphere on cooling in Antarctica. These ideas rely heavily upon research reported by participants of the Deep Sea Drilling Project (DSDP) Legs 28 and 35 and the Dry Valley Drilling Project (DVDP). DSDP was, of course, a program designed to explore the sediments and basaltic rocks of deep ocean basins; whereas, DVDP was a four-year study, primarily on land, of the McMurdo Sound region of Antarctica. Most of the arguments utilized in this paper are derived from Antarctica and therefore emphasis will be placed on Gondwanaland; however, it can be shown that they apply equally well to the northern hemisphere.

To substantiate arguments presented later, it is necessary to illustrate that cold temperatures have persisted for long periods of geologic time in polar latitudes. These arguments rely on evidence of early glaciations. Intermittent glaciations of Paleozoic age culminate in late Carboniferous time in southeastern Australia (Dott and Batten, 1976), where ice advanced from a center which was located in Antarctica. Glaciers of Gondwanaland persisted contemporaneously with floral assemblages that lived near the perimeter of glacial regions. Evidence of earliest glaciations has not been recorded so much by glacio-marine sediments as it has been by the record displayed on the continents. Following a warm period in Triassic and Jurassic time, in which cold-blooded vertebrates lived in Antarctica only a few hundred kilometers from what is now the pole, Gondwanaland began to break up (Dott and Batten, 1976).

Although evidence for a long cool history is present over much of Gondwanaland the climatic record extracted from the oceans indicates a much later cooling. Oxygen isotope studies on *Foraminifera* indicate ocean surface temperatures during Cretaceous time were similar to those of today in the South Atlantic; however, ocean bottom temperatures were about 10°C warmer (Saito and Van Donk, 1974). Saito and Van Donk found a sharp climate deterioration from middle to late Maastrichtian time (68—66 m.y. B.P.).

Craddock and Hollister (1976) indicate that further cooling of the seas occurred in middle Miocene time between 12 and 26 m.y. B.P. They state that New Zealand detached itself from Antarctica about 81 m.y. B.P.; whereas, Weissel and Hayes (1972) have found the first indications of rifting between Australia and Antarctica occurred about 55 m.y. B.P. The first evidence of ice-rafted glacial sediments occurs in the Antarctic Ocean only 25 m.y. B.P. (Hayes and Frakes, 1975). Webb and Wrenn (in press) indicate that cutting of the deep dry valleys west of McMurdo Sound occurred in Miocene time or earlier.

Hayes and Frakes (1975) state that:

"climate deteriorated markedly during the Cenozoic in the region of the Southeast Indian Ocean, as demonstrated by changes in sedimentation patterns in near-shore and off-shore areas. Culmination of this episode of cooling occurred in the early Pliocene, when the Ross Ice Shelf expanded dramatically and the cold water belt around Antarctica expanded abruptly by about 300 km."

They make the further observation that although antarctic glaciations began only 25 m.y. B.P. the continent has been in a high-latitudinal position for about the last 100 m.y. The cause of a later period of cooling in the oceans is explored in a later section.

Other lines of indirect evidence reflecting on antarctic mantle rheology have been published in the past several years. Weissel and Hayes (1972) found that the Southeast Indian Ridge, separating Antarctica and Australia (Fig.1), has experienced systematic asymmetric spreading since 40 m.y. B.P. in which ocean floor was accreting 30—40% faster on the Australian side of the ridge. They also note, from paleolatitude determinations, that the Indian plate was accreting about 15° north of the Antarctic plate. The latter observation suggests that the ascending limb was very wide. They also suggest that the accreting boundary was migrating away from the "fixed" antarctic plate. Dietz et al. (1972) correctly predicted asymmetric spreading based on the absence of subduction beneath Antarctica. From magnetic inclinations measured on core, Lowrie and Hayes (1975) found the antarctic plate to be fixed for the past 40 m.y.

In West Antarctica, paleomagnetic evidence indicates that the position of the Antarctic plate has remained at high latitudes since at least middle Cretaceous time (Dalziel and Elliott, 1973; McElhinney, 1973). Kemp et al. (1975) require a cooling but temperate climate during early Tertiary time in the McMurdo Sound area and in the Ross Sea.

From the seismicity map of the Earth by Barazangi and Dorman (1969) no earthquakes of tectonic origin have been recorded south of about 63°S. Antarctica is completely aseismic, lacking even intraplate earthquakes common to other continents. For example, 24 earthquakes were recorded in

Australia, an intraplate continent, during the period 1961—1967. Voronov (1964) accounts for the lack of earthquakes beneath the ice sheets of Antarctica and Greenland as being due to three factors which include: (1) a decrease in viscosity of the sub-crust under the ice load; (2) a decrease in temperature of the lithosphere under the ice sheet by some tens of degrees which leads to more difficult plastic deformation; and (3) a binding influence due to ice load acting on lithospheric blocks.

Hayes and Davey (1975) report mean free-air gravity anomalies in the Ross Sea of −12.9 mGal and Bennett (1964) has similarly found the mean anomaly over the Ross Ice Shelf to be −12.0 mGal. Ushakov (1961) too has reported an absence of isostatic equilibrium in several coastal areas in East Antarctica. It has been argued by Bennett (1964) that the negative anomaly and the depression of the continental shelf beneath the Ross embayment of more than 300 m below global averages, are due to glacial isostatic subsidence during an expanded Ross Ice Shelf. Houtz and Meijer (1970), on the other hand, point out that other submarine plateaus lie at similar depths and these have not been ice-covered in recent times. Hayes and Davey (1975) suggest both glacial loading and a possible tectonic cause for the anomalous shelf depths and gravity anomalies.

The south polar historical record, from late Mesozoic time to the present, has been derived primarily from DSDP drilling because of the lack of a Tertiary record on land. Drilling and geophysical studies for geological research have also been used to define the physical, chemical and historical characteristics in the dry valleys in the Transantarctic Mountains of East Antarctica (McGinnis et al., 1972). The study was designed to test several hypotheses, including speculations that melt water at the base of continental ice sheets, produced by sliding friction and geothermal heat (Weertman, 1964), result in groundwater flow systems that have altered the thermal and chemical regimen near glacial peripheries (McGinnis, 1968). Large hydraulic heads produced by the weight of the ice sheet acting on basal melt water supply the energy required to drive deep, cold flow systems which would result in saline groundwater discharges near glacial margins. Since drilling, geophysical logs of the holes (McGinnis et al., in press) and heat-flow analyses (Decker and Bucher, in press) have also been completed. The limited extent of basal melting in Antarctica has been estimated by Budd et al. (1970) and by Drewry (in press); however, the characteristics of aquifers which could play a role in regional subglacial flow have not been defined.

CONVECTION

According to plate-tectonic theory, mantle convection is due to gravitational instability produced by mass differences at ascending and descending limbs of a convection cell. Movement occurs when gravitational forces exceed viscous resistance to mantle flow. Mass differences result from a num-

ber of factors, one of which is thermal expansion or contraction due to differential heating, and this in turn, raises or lowers mantle temperature. Temperature, of course, is inversely proportional to viscosity. Thus a high temperature mantle would most likely be associated with a mantle that is unstable.

Heat energy, produced by radioactive decay, is dissipated by convection, conduction, earthquakes and volcanism. If heat energy in the Earth is produced at a rate faster than it can be lost by conduction then temperature will rise and convection will ultimately result. Evidence favoring a convecting mantle are earthquakes and volcanoes which occur where two plates join. Both occur less frequently within plates where their cause is not completely understood. Inspection of seismicity patterns on Earth (Barazangi and Dorman, 1969) illustrates a latitudinal earthquake distribution which is best illustrated in Antarctica where no earthquakes occur but several volcanoes do. The arctic is also associated with a low level of seismicity. Thus, if it is reasonable to relate earthquakes with plate margins and convection, we can infer that Antarctica is contained within a plate in which mantle viscosity is abnormally high and therefore convection does not occur. Since we have no reason to believe conduction or heat production in Antarctica are different from any other continent, where production exceeds conduction, then we must find an alternate explanation for high mantle viscosity.

From the classical paper on sea-floor spreading by Dietz (1963), continents move with underlying convection cells:

> "until they attain a position of dynamic balance overlying a convergence. There the continents come to rest, but the sima continues to shear under and descend beneath them; so the continents generally cover the down-welling sites."

Toksöz et al. (1971) illustrate the effect of this "down-welling", or subduction as we now call it, on mantle temperature. They have shown how mantle temperatures can be lowered around a down-going slab of lithosphere by as much as 800°C in about 13 m.y., which, needless to say, would cause an orders-of-magnitude increase in mantle viscosity. From reconstructions of Pangaea reviewed by Dietz and Holden (1970), it can be seen that even if it were not in a polar location, the accretionary process dictates a cold mantle beneath Pangaea.

For recent reviews of plate motion and convection in the mantle, see Cross and Pilger (1978), and Marsh (1979). Where subduction occurs, a cold slab of lithosphere will be consumed in a hot mantle. Verhoogen et al. (1970) indicate that heat transfer by convection is globally at least as important as transfer by conduction. With long-term subduction, such as must have been the case for Pangaea, the importance of convection on heat transfer would have been several orders of magnitude greater than conduction.

Elsasser et al. (1979) argue against a substantial increase in viscosity with depth in the mantle. They select as examples, post-glacial uplift of

A. PRECAMBRIAN TO LATE PALEOZOIC
PERIOD OF CONTINENTAL
NUCLEI ACCRETION AND
MANTLE COOLING

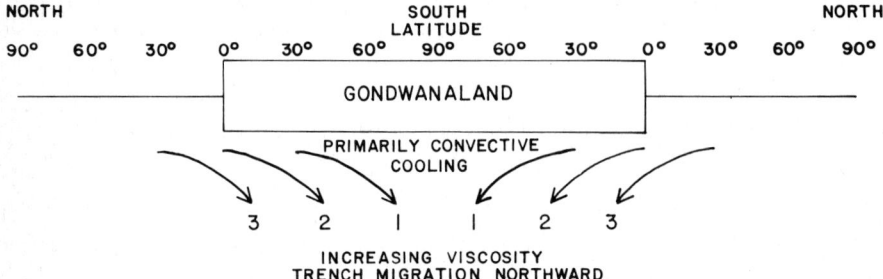

B. LATE PALEOZOIC TO EARLY MESOZOIC
PERIOD OF MANTLE
HEATING

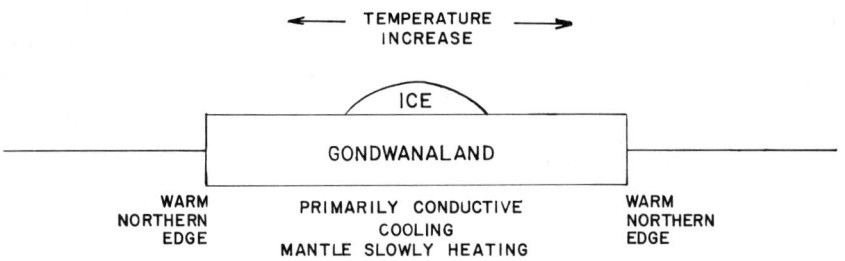

C. EARLY MESOZOIC TO LATE CENOZOIC
PERIOD OF MANTLE INSTABILITY AND
FRAGMENTATION OF CONTINENTAL NUCLEI

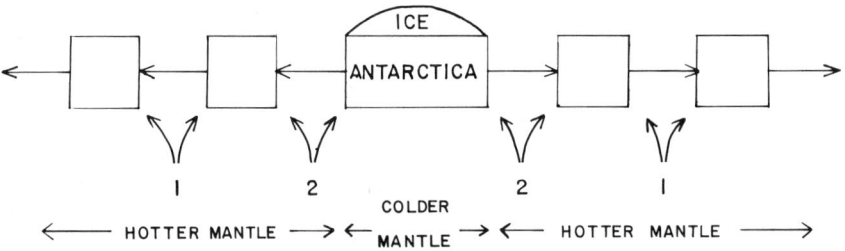

Fig.2.A. Gondwanaland accretion by subduction. The mantle becomes abnormally cold and motion in the mantle is temporarily halted.
B. Mantle temperatures slowly rise in response to the inability of the lithosphere to conduct away all the heat produced by radioactive decay. The rise in temperature is greatest near the northern continental edge where temperature gradients are lowest. An ice sheet forms in the continental interior of Gondwanaland, with advances and retreats being a function of latitudinal position and sea—land transport of moisture.
C. Temperatures of the mantle in the more northerly latitudes rise to produce instability and convection. Fragments of continental nuclei begin to rift, with those more northerly rifting first. The south-polar ice sheet begins to spill off Antarctica into the sea. Deep bottom waters cool by 10°C and the thermal gradients through the lithosphere are increased, producing a lower equilibrium temperature in the upper mantle.

Fennoscandia and the North American Shield, to arrive at a viscosity of about 10^{22} P (poises). A similar value was found by Dicke (1969) from tidal friction; however, Verhoogen et al. (1970) find a value of 10^{26} P from inadequate Earth flattening. Elsasser et al. (1979) do make an exception to homogeneity for the upper mantle down to a depth of 250 km. They state:

> "These top layers are notoriously heterogeneous, whereas the region below about 250-km depth may be assumed to be far more homogeneous."

Thus it is probable that viscosity in the upper mantle shows marked lateral variations.

Accretion of Pangaea can be explained by subduction. Breakup could not begin until long after the accretionary process had ended, since it is associated with rifting and rifting requires a warm and unstable mantle. A differential temperature rise in the mantle beneath accreted continental nuclei must depend upon latitude, all else being equal. Thus rifting will occur first in a region where nuclei accretion is old and where surface temperature is warm. If accretion of all continental nuclei comprising Pangaea occurred nearly contemporaneously, surface temperature will be the most important factor in continental breakup and therefore a latitude effect will be observed. Fig.2 illustrates the successive episodes of cooling by accretion and differential warming by conduction.

CONDUCTION

Heat loss by conduction, averaged over Antarctica, is assumed to be the same as that for other continents. Therefore a cold continent should be cold not only at the surface but also at depth, other factors, such as heat production (ϵ) and thermal conductivity (K) being equal. If steady-state conditions are assumed, i.e. $\delta T/\delta t = 0$, the temperature at the base of the lithosphere is, from Verhoogen et al. (1970), given by:

$$T_L = T_0 + L(Q - \tfrac{1}{2}\epsilon L)/K \qquad (1)$$

where T_L = temperature at the base of the lithosphere; T_0 = temperature at the surface; Q = heat flow; L = thickness of lithosphere; and K = thermal conductivity.

The average value of the quantity $L(Q - \tfrac{1}{2}\epsilon L)/K$ is required to be a constant for all continents in the arguments used in this paper, so T_L is a function only of T_0.

Stacey (1977) shows that a thermal wave imposed on the surface ($Z = 0$) penetrates the Earth as an attenuating wave given by the equation:

$$T(Z) = \Delta T \exp[-(w/2d)^{1/2}Z] \sin[wt - (w/2d)^{1/2}Z] \qquad (2)$$

where ΔT = thermal deviation; d = diffusivity; Z = depth; w = angular velocity = $2\pi/t_0$; and t_0 = period of deviation.

The attenuation length is given by:

$$Z = (2d/w)^{1/2} \tag{3}$$

It is shown in Table I that a period of 10^8 yr. is required for the temperature to penetrate to lower crustal depths; whereas, a period of 10^9 yr. is necessary for the wave to reach the base of the lithosphere. It is clear that a long period of stability of continental nuclei is required for the effects of surface temperature to penetrate into the asthenosphere. Hurley and Rand (1969) have proposed such a stability for pre-Mesozoic continental nuclei. Stable time periods have also been pointed out by Goodwin (1973) in his reconstruction of the distribution of early Proterozoic banded iron formations (1900—2100 m.y. B.P.). From data shown in Table I it is clear that for the effects of surface temperatures to be felt at mantle depths due to conduction

TABLE I

Depth of penetration, Z, and time, t_0, required for a thermal oscillation imposed upon the suface of a semi-infinite medium of diffusivity $(d) = 1.2 \cdot 10^{-6}$ m² s⁻¹

t_0 (yr.)	Z (km)	t_0 (yr.)	Z (km)
1	0.003	10^5	1.098
10	0.011	10^6	3.472
10^2	0.035	10^7	10.982
10^3	0.110	10^8	34.727
10^4	0.347	10^9	109.819

alone, an order of time of 10^9 yr. is required. Differential breakup occurred only after that order of time had elapsed since continental accretion. It is therefore certain that mantle cooling beneath a high-latitude continent is not the key, but it is, on the other hand, more-rapid mantle warming beneath equatorial Pangaea that controlled the on-set of rifting. From the reconstruction of Pangaea by Dietz and Holden (1970), the sinuous boundary separating Laurasia and Gondwanaland lay along the late Permian equator 225 m.y. ago. Thus the mantle beneath Antarctica must still bear the traces of its cold accretionary beginnings.

GROUNDWATER COOLING SYSTEM

Groundwater flow systems of continental proportions which tend to transport heat from the Earth's interior are present throughout the globe. It is probable that extensive circulation and cooling by ocean water is also present in submarine sediments and fractured basalts along submarine spreading ridges (Lowell, 1975). From work recently completed by the DVDP it is inferred that a dynamic groundwater flow system is also present beneath ice sheets (McGinnis et al., in press; Cartwright and Harris, in prep.). In general, regional groundwater flow systems are characterized by salinity gradients, with increasingly saline groundwater following flow lines (Dominico, 1972). These characteristics have not been attributed to polar regions covered by ice sheets because surface expression of their presence was not recognized and subsurface control has been unavailable. However, since subglacial melt water is now known to be present, the increasing salinities along flow lines will permit discharge to occur at temperatures well below freezing. Deep cold circulation of water will have the net effect of lowering upper-crustal temperatures below those which would ordinarily be permitted by conduction. Based on present data it is not possible to estimate the importance of groundwater cooling; however, it is certainly of far less importance than cooling by convection or conduction.

In addition to acting as a coolant, sub-ice water at high pressure will reduce effective stresses in crustal rocks which will, in turn, permit tectonic strain release in a very short time. The process of excess pore-water pressure acting to lower effective stress has been described in the literature by Healy and Pakiser (1971). After release of strain energy the ensuing periods of time will be abnormally aseismic. From Nason and Weertman (1973), weakening at faults can be explained by the suppression of development of an upper yield point which would not permit an unstable state. It is concluded that rather than binding crustal blocks together as suggested by Voronov and Klushin (1963), the presence of an ice sheet and basal melt water will have the opposite effect, i.e. effective stresses will be decreased as the ice sheet thickens because of higher fluid pressures.

In order for a regional flow system to exist in Antarctica all the factors required for regional flow must exist, that is, there must be:

(1) Areas of recharge beneath the ice sheet.

(2) Aquifers of regional areal extent connecting recharge and discharge areas.

(3) Discharge in the periglacial zone on land or in the submarine continental shelves.

(4) A hydraulic gradient.

East Antarctica is a continental shield on which rest relatively undisturbed sediments of Paleozoic—Mesozoic age. The sediments are intruded by thick, often differentiated and extensively fractured, dolerite sheets of early Jurassic age. If a hydrologic flow system were present it would be constrained by the distribution and configuration of these rocks. West Antarctica, on the other hand, contains an archipelago of post-Jurassic volcanic centers and sediments resulting from a convergent plate margin (Craddock and Hollister, 1976). In addition, small continental fragments containing older rocks are found dispersed throughout much of West Antarctica (Ford, 1974). Subglacial groundwater flow in West Antarctica would thus involve aquifers of many origins.

Recharge

Radio-echo soundings in East Antarctica have located seven subglacial lakes west of the dry valley area (Drewry, in press). The lakes occur in elongate, subglacial extensions of the dry valleys. Dimensions of the depressions range from several hundred meters in width to less than 5 km. Discovery of the lakes was made during ten flights oriented N—S parallel to the Transantarctic Mountains, and about 10—15 km apart. It is likely that additional lines along a tighter grid would extend the area of known lakes. The lakes range from about 15—60 km west of the dry valleys and are located beneath 1—2 km of ice. Other areas in both East and West Antarctica are also believed to be underlain by sub-ice lakes (Drewry, in press) and by bottom melting (Budd et al., 1970). These areas along the continental edge can also be characterized as regions of high fluid pressure in the crust and low effective stress.

Regional aquifer

Although rocks of the Beacon Supergroup (Barrett et al., 1972) would form the most likely candidate for a rock unit of regional areal extent they are not located in valley bottoms in the dry valleys where lakes and other evidence of a dynamic hydrologic system are present. Beacon sediments may underlie glacial sediments in McMurdo Sound, but no direct evidence of these rocks has been found at this time. It is certain, however, that Beacon rocks do dip beneath the ice sheet (Crary, 1963), and they probably form a base upon which much of the ice sheet is now resting. Recharge through fractured Beacon sandstone is suggested, although permeability characteris-

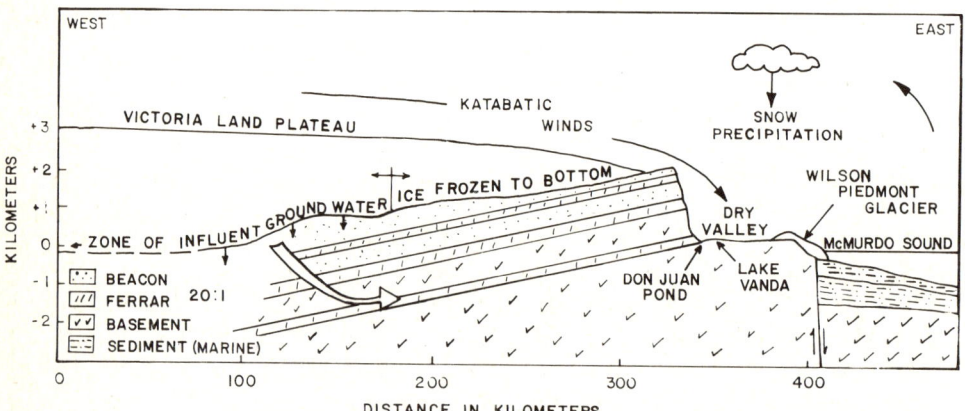

Fig.3. Greatly simplified geologic profile showing possible circulation of the hydrosphere in the East Antarctic ice sheet—dry valley region. Faulting between McMurdo Sound and the dry valleys is based on seismic studies conducted in November 1978. Fluids move from the base of the ice sheet through the Beacon—Ferrar aquifer.

tics of the aquifer have not been determined. In addition to movement of groundwater through the Beacon sandstones, it has been found through drilling that the basement is extensively intruded and fractured in places and therefore contains faults and joints through which fluids can move.

In addition to fractures in the basement, a borehole drilled at Don Juan Pond in Wright Valley encountered a fractured sill of Ferrar Dolerite, approximately 30 m thick, that contained abundant evidence of secondary mineralization. Groundwater at temperature $-16°C$, having salinities over 300 ppt was found to flow from the well. It is assumed, therefore, that beneath the ice sheet, the Beacon—Ferrar aquifer provides access for influent melt water (Fig.3).

Discharge

Three examples of saline groundwater discharge marginal to the ice sheet are either inferred or observed directly. Direct observation includes the flow at Don Juan Pond and the saline discharge at Taylor Glacier in western Taylor Valley. A third example is the Lake Vida borehole where fluids are moving into the hole from a deep fault zone. The volume of waters discharged are probably quite small relative to the size of the ice sheet. A hole drilled 15 km offshore did not encounter unusual pore-water pressures. However, the hole was drilled only 65 m into bottom through permeable sands. A freshwater or ice lens about 30 m thick was found in the sand.

Hydraulic gradient

Because of the low strength of ice, the surface of the ice sheet may be presumed to define a piezometric surface acting on any water present at its base. Flow lines constructed from ice contours for ice flow direction (Budd et al., 1970) also give the direction for the horizontal component of subglacial water flow. A deep, regional flow system will transport heat from bedrock and preglacial sedimentary basins to discharge points near the glacial margin (McGinnis, 1968). Heat transported from beneath the ice sheet in the groundwater flow system will simply be lost to the atmosphere or ocean depending on whether the discharge point is submarine or subaerial. Discharge points can be submarine if regional aquifers outcrop on the continental shelf and periglacial if taliks are present in depressions of glacial valleys or under outlet glaciers, providing salinities are high and mean annual temperatures are sufficiently high. Examples of the latter two situations are believed to exist in the dry valley area. A hydraulic connection from sub-ice lakes to submarine sediments in McMurdo Sound is not immediately apparent. Igneous and metamorphic rocks of the Ross Supergroup and Granite Harbor Intrusives (Warren, 1969) outcrop near the coast and the sediments of McMurdo Sound are not present onshore except in deeply cut valleys. Valley sediments extend continuously from the sea under the Taylor Glacier but they are frozen to basement in the upper reaches of the valley, at least from the site of DVDP No. 12 at Lake Hoare. It is likely that in some regions of Antarctica groundwater discharge occurs offshore.

ICE FLOW

Snow accumulation and ice-sheet flow systems in Antarctica have been described by Budd et al. (1970) and Hughes (1972). The net effect of ice flow on temperature distributions is to lower subglacial temperatures below those which would prevail if ice did not flow. Subglacial temperatures from Budd et al. (1970) are shown in Fig.1. Flow vectors radiate downward and outward away from the ice-sheet center permitting cold surface temperatures to be reached at greater depths, and to transfer geothermal heat to the atmosphere or into the sea. Discharge of relatively cold ice into the world ocean systems will have the further effect of lowering average ocean temperatures. It is likely that during early glaciations (see Holmes, 1965, p.1229) the ice sheet did not have outlet glaciers or ice shelves reaching the sea (Fig.2B). The glacial record has therefore been preserved only on the continents of Gondwanaland.

DISCUSSION

It is assumed that the fundamental motive power for driving plates is radioactive heat. Heat production averaged in large regions is nearly constant and therefore it might be assumed that neotectonics would also be uniform, on average, over the globe. Inspection of global seismicity, volcanism, plate margin distributions, plate movement and slow recovery rates for isostatically depressed continental shelves in the antarctic shows that this is not the case. The polar deficiency in seismicity is only one of the indicators that adequate mechanisms are available to dissipate mantle heat at the poles. The many mechanisms causing heat loss permit lower temperatures to persist, which cause high viscosities to be maintained in the polar asthenosphere.

From data which are not always reconcilable, it is clear that Antarctica has occupied its position near the southern pole of rotation for at least 100 m.y. In fact, it has been associated with Gondwanaland glaciations since at least Permian time. Antarctica has probably been on average, the coldest of the Earth's continents since the early beginnings of Pangaea accretion; however, polar temperatures, as we know them today, may not have persisted continuously. Although ocean cooling of bottom water in the South Atlantic began only 68 m.y. B.P. (Saito and Van Donk, 1974), it is probable that continents were too large prior to break up of Gondwanaland to permit Paleozoic ice sheets to extend out into the oceans to produce cold bottom water.

The most important mechanism of mantle cooling beneath Antarctica was initially by subduction of cold lithosphere. Conduction has played a secondary role in present mantle temperatures, but it has played a primary role in producing global, differential thermal gradients. In the presence of an ice sheet additional factors may be employed in crust and mantle cooling, such as ice-sheet convection and deep cold groundwater circulation. Budd et al. (1970) have shown subglacial temperatures in the interior of Antarctica to be less than −30°C, whereas, Budd et al. (1970) and Drewry (in press) have shown subglacial lakes, at melting-point temperatures, are present in depressions near the margins of the ice sheet. The author proposes that temperatures are lowered by all these mechanisms by an amount sufficient to maintain higher than global average mantle viscosity, which retards strain rate and other secondary effects of mantle convection.

From an inspection of the regional geology of Antarctica one would expect a much more active earthquake regimen, especially in the Antarctic peninsula. The peninsula is a region of arcuate post-Jurassic mountain building which represents an ancient convergent plate margin. In its configuration, along with the South Sandwich Trench and the southern extension of the Peru—Chile Trench, it is similar to the very seismic Carribbean Plate; however, it is aseismic, except for three earthquakes reported in the period 1961—1967 (Barazangi and Dorman, 1969) at its northern tip, north of 63°S. Earthquake frequency in West Antarctica should be similar to that along the Challenger Fracture Zone and the Chile Ridge in the Southeast Pacific (Fig.1). Earthquakes in the East Antarctic shield should occur with about the same regularity as the low-intensity shield seismicity of Australia, where 24 earthquakes occurred during the period 1961—1967.

Over most of the Earth, uniformly cold ocean-bottom water would cause a latitude variation of asthenosphere viscosity to be small; however, land areas of both the arctic and antarctic should display a low degree of neotectonic activity. Upper lithosphere temperatures in Antarctica average 20—30°C less than ocean-bottom temperatures and 50°C less than continents in equatorial regions. Since we can assume that the thermal conductivity of rocks and geothermal heat production in equatorial regions is similar to that in polar regions we can also assume that heat flow is also similar. Thus, if heat flow is comparable in polar and equatorial regions and if thermal conductivity of rocks is also similar, then temperature gradients must also be similar. If temperature gradients are similar then mantle temperatures in Antarctica would average about 50°C less than they do beneath an equatorial continent, assuming sufficient time has elapsed for temperature equilibrium to be reached. For a mantle temperature decrease of only 50°C, Weertman (1970) shows that viscosity increases by about an order of magnitude.

Plates move over, descend into, or are carried along upon an asthenosphere having minimum viscosities of about 10^{22} P (Elsasser et al., 1979) that can convect at rates up to 10 cm/yr. Descent of cold lithosphere beneath Pangaea, would initially cause rapid cooling of the mantle near the subducting limbs. The rate of subduction will decline with cooling as velocity is inversely proportional to viscosity (η) according to Stoke's law. After equilibrium geothermal gradients are established the cold surface temperatures of Antarctica will result in temperatures in the mantle about 50°C colder than equatorial continents.

The significance of the above reasoning has many consequences. Polar areas, especially where they are continental, act as heat sinks. During early accretion of Pangaea plate movements would naturally radiate toward the poles as cold lithosphere descends, causing further cooling and increased viscosity at the poles. With a termination of subduction caused by increasing viscosity, a rise in temperature in the mantle by radioactive heat production will eventually result in rifting, with asymmetric spreading away from polar areas, since the southward-directed limb would be moving over increasingly colder asthenosphere. Weissel and Hayes (1972) have shown that the northern limb of the Southeast Indian Ridge is moving 40% faster than the southern limb (Fig.4).

Finally, a negative mean free-air anomaly of −12 mGal in the Ross Sea, by itself, is probably not a sufficient reason to argue for the cause being due to isostatic depression; however, the negative anomaly along with mean shelf elevations about 300 m below world averages does provide room for conjecture. If asthenosphere viscosity were increased, rebound rate would be greatly slowed.

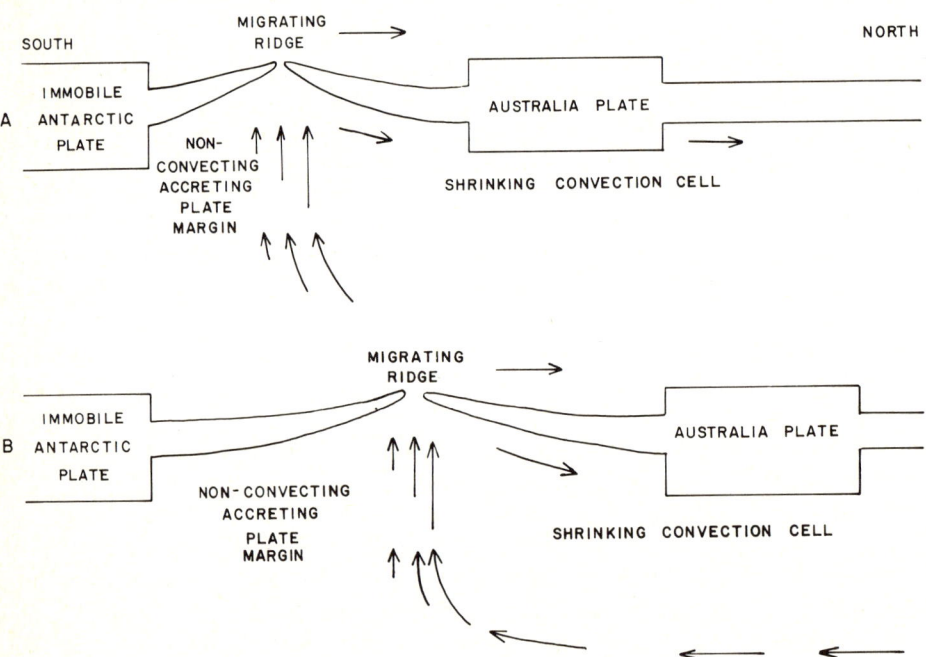

Fig.4. Proposed model for the asymmetric spreading ridge between Antarctica and Australia. High viscosity prevents convection beneath the antarctic plate. A wide, rising, migrating limb is caused by accretionary "plating" onto the immobilized plate edge. Accretion onto the immobilized plate is equivalent to the shrinkage of the convecting plate.

CONCLUSIONS

Continental accretion by subduction and, therefore, mantle cooling on a large scale and at high rates, proceeded beneath Pangaea during Precambrian time. By late Paleozoic time Pangaea extended from the north to the south pole while a broad ocean covered the remainder of the globe. Mantle temperatures gradually increased beneath Pangaea as accretion processes slowed and finally terminated, however, mantle temperatures, viscosities, and therefore convection are a function of surface temperature, which, in turn is a functon of latitude. Rifting, therefore, first began along a sinuous line between North America, Africa and South America in equatorial Pangaea 200 m.y. B.P. Rifting progressed towards higher latitudes until Australia broke from Antarctica and Greenland from Europe 55 m.y. B.P. Continued fragmentation of the last remnant of Gondwanaland, Antarctica, is questionable because it is apparently underlain by a mantle of abnormally high viscosity.

Evidence for a high-viscosity asthenosphere in Antarctica is argued on the basis of:

(1) Lack of earthquakes south of 63°S.

(2) Circumpolar spreading ridge, indicating a latitude effect.

(3) Asymmetric ridge spreading with south-directed rates 40% less than north-directed.

(4) Long-lasting negative mean free-air anomaly along much of the edge of deglaciated East Antarctica.

(5) Abnormally deep continental shelves.

(6) A residence time at high latitudes of sufficient duration to permit asthenosphere temperatures to remain abnormally cold and viscosities abnormally high.

As more and more fragments of Gondwanaland moved north, the nucleus of the continent in East Antarctica was invaded by a relatively warm sea. With temperatures gradually rising in the mantle to accommodate continuing radioactive decay and the new surface temperatures, additional fragments could move north. If this process were to continue the most likely future rift would be along the Transantarctic Mountains. Decker and Bucher (in press) have shown this to be a region of abnormally high heat flow and suggest it is underlain by abnormally hot crust and upper mantle. The problem with additional northward movement is that the region surrounding the Transantarctic Mountains probably has a cold highly viscous mantle and therefore the plates cannot separate. Heat may continue to be lost by volcanism in quantities sufficiently great to permit East and West Antarctica to remain stationary. The long-term decline in global heat production, the long residence time of Antarctica at the south pole, and the shift of the circumpolar spreading ridge northward, may be the beginning of an irreversible process which will lead to continued growth of the region underlain by an immobilized mantle. The consequence of this process may be applied to an understanding of not only the Earth but other planets of the solar system as well.

ACKNOWLEDGMENTS

Support for antarctic research was provided by National Science Foundation Contract C-642. The idea that heat flow is independent of latitude was provided by Ed Decker. The author was introduced to water—earthquake relations by George Maxey. This manuscript has been improved through critical reading by Art Ford, Phil Kyle, Mike Mudrey, and Hans Weertman.

REFERENCES

Barazangi, M. and Dorman, J., 1969. World seismicity maps compiled from ESSA. U.S. Coast and Geodetic Survey, Epicenter Data, 1961—1967, 59: 369—380.

Barrett, P.J., Grindley, G.W. and Webb, P.N., 1972. The Beacon Supergroup of east Antarctica. In: R.J. Adie (Editor), Antarctic Geology and Geophysics, Universitetsforlaget, Oslo, pp. 319—358.

Bennett, H.F., 1964. A gravity and magnetic survey of the Ross Ice Shelf, Antarctica, Univ. Wis., Geophys. Polar Res. Cent., 64-3, 65 pp.

Budd, W., Jenssen, D. and Radok, U., 1970. The extent of basal melting in Antarctica. Polarforschung, Bd. VI., 39: 293—306.

Cartwright, K. and Harris, H.J.H., in prep. Hydrogeology of the Dry Valley region, Antarctica. In: L.D. McGinnis (Editor), Dry Valley Drilling Project. Am. Geophys. Union, Antarct. Res. Ser.

Craddock, C. and Hollister, C.D., 1976. Geologic evolution of the southeast Pacific basin. In: P. Worstell (Editor), Initial Reports of the Deep Sea Drilling Project, Vol. 35, U.S. Government Publishing House, Washington, D.C., pp. 723—744.

Crary, A.P., 1963. Results of the United States traverses in east Antarctica, 1958—61. Int. Geol. Year, Glaciol. Rep., 7: 144.

Cross, T.A. and Pilger, R.H., 1978. Constraints on absolute motion and plate interaction inferred from Cenozoic igneous activity in the western United States. Am. J. Sci., 278: 865—902.

Dalziel, I.W.D. and Elliott, D.H., 1973. The Scotia Arc and Antarctic margin. In: F.A. Stehli and A.E.M. Nairn (Editors), The Ocean Basins and Their Margins, I. The South Atlantic. Plenum, New York, N.Y., pp. 171—245.

Decker, E.R. and Bucher, G.J., in press. Preliminary geothermal studies in the Ross Island—Dry Valley region. In: C. Craddock (Editor), Symposium on Antarctic Geology and Geophysics. University of Wisconsin Press, Madison, Wis.

Dicke, R.H., 1969. Average acceleration of the earth's rotation and the viscosity of the deep mantle. J. Geophys. Res., 74: 5895—5902.

Dietz, R.S., 1963. Ocean-basin evolution by sea-floor spreading. In: S.K. Runcorn (Editor), Continental Drift. Academic Press, New York, N.Y., pp. 289—298.

Dietz, R.S. and Holden, J.C., 1970. Reconstruction of Pangaea: breakup and dispersion of continents, Permian to Present. J. Geophys. Res., 75: 4939—4956.

Dietz, R.S., Holden, J.C. and Sproll, W.P., 1972. Antarctica and continental drift. In: R.J. Adie (Editor), Antarctic Geology and Geophysics, Symposium on Antarctic Geology and Solid Earth Geophysics. Int. Union Geol. Sci., Ser. B, No.1, Scandinavia University Books, Universitetsförlaget, Oslo, pp. 837—842.

Domenico, P.A., 1972. Concepts and Models in Groundwater Hydrology. McGraw-Hill, New York, N.Y., 405 pp.

Dott, R.H. and Batten, R.L., 1976. Evolution of the Earth. McGraw-Hill, New York, N.Y., 504 pp.

Drewry, D.J., in press. Geophysical investigations of ice sheet and bedrock inland of McMurdo Sound, Antarctica. In: C. Craddock (Editor), Symposium on Antarctic Geology and Geophysics. University of Wisconsin Press, Madison, Wis.

Elsasser, W.M., Olsen, P. and Marsh, B.D., 1979. The depth of mantle convection. J. Geophys. Res., 84(B1): 147—155.

Ford, A.B., 1974. Fit of Gondwana continents — Drift reconstruction from the antarctic continental viewpoint. 24th Int. Geol. Cong., Montreal, Que., Sect. 3, pp. 113—121.

Goodwin, A.M., 1973. Plate tectonics and evolution of Precambrian crust. In: D.H. Tarling and S.K. Runcorn (Editors), Implications of Continental Drift to the Earth Sciences. Academic Press, London, pp. 1047—1069.

Hayes, D.E. and Davey, F.J., 1975. A geophysical study of the Ross Sea, Antarctica. In: D.E. Hayes and L.A. Frakes (Editors), Initial Reports of the Deep Sea Drilling Project, Vol. 28, U.S. Government Publishing House, Washington, D.C., pp. 919—942.

Hayes, D.E. and Frakes, L.A., 1975. General synthesis, Deep Sea Drilling Project Leg 28. In: D.E. Hayes and L.A. Frakes (Editors), Initial Reports of the Deep Sea Drilling Project, Vol. 28, U.S. Government Publishing House, Washington, D.C., pp. 919—942.

Healy, J.H. and Pakiser, L.C., 1971. Man-made earthquakes and earthquake prediction. In: Seismology, U.S. Natl. Rep., 1967—1971, 15th Gen. Assem., Int. Union Geod. Geophys., EOS, 52(P. IUGG): 177—174.

Heezen, B.C. and Tharp, M., 1972. Physiographic and tectonic provinces. In: V.C. Bushnell (Editor), Morphology of the Earth in the Antarctic and Subantarctic. Antarct. Map Folio Ser., Am. Geogr. Soc, New York, N.Y., Folio 16, Plate 3.

Holmes, A., 1965. Principles of Physical Geology. Ronald Press, New York, N.Y., 1288 pp.

Houtz, R.E. and Meijer, R., 1970. Structure of the Ross Sea shelf from profiler data. J. Geophys. Res., 75: 6592—6597.

Hughes, T., 1972. Thermal convection in polar ice sheets related to the various empirical flow laws of ice. Geophys. J. R. Astron. Soc., 27: 215—229.

Hurley, P.M. and Rand, J.R., 1969. Pre-drift nuclei. Science, 164: 1229—1242.

Jacoby, W.R., 1970. Instability in the upper mantle and global plate movements. J. Geophys. Res., 75: 5671—5680.

Kemp, E.M., Frakes, L.A. and Hayes, D.E., 1975. Paleoclimatic significance of diachronous biogenic facies, Leg 28, Deep Sea Drilling Project. In: D.E. Hayes and L.A. Frakes (Editors), Initial Reports of the Deep Sea Drilling Project. Vol. 28, U.S. Government Publishing House, Washington, D.C., pp. 909—918.

Lowell, R.P., 1975. Circulation in fractures, hot springs and convective heat transport on mid-ocean crests. Geophys. J. R. Astron. Soc., 40: 351—360.

Lowrie, W. and Hayes, D.E., 1975. Magnetic properties of oceanic basalt samples. In: D.E. Hayes and L.A. Frakes (Editors), Initial Reports of the Deep Sea Drilling Project, Vol. 28, U.S. Government Publishing House, Washington, D.C., pp. 869—878.

Marsh, B.D., 1979. Island—arc volcanism. Am. Sci., 67: 161—172.

McElhinney, M.W., 1973. Paleomagnetism and Plate Tectonics. Cambridge University Press, Cambridge, 372 pp.

McGinnis, L.D., 1968. Glaciation as a possible cause of mineral deposition. Econ. Geol., 63: 390—400.

McGinnis, L.D., Torii, T. and Webb, P.N., 1972. Dry Valley Drilling Project — Three nations are studying the subsurface in the McMurdo Sound Region. Antarct. J. U.S., 7: 7—10.

McGinnis, L.D., Osby, D.R. and Kohout, F.A., in press. Paleohydrology inferred from silinity measurements on Dry Valley Drilling Project (DVDP) Core in Taylor Valley, Antarctica. In: C. Craddock (Editor), Symposium on Antarctic Geology and Geophysics. University of Wisconsin Press, Madison, Wis.

Nason, J. and Weertman, J., 1973. A dislocation theory analysis of fault creep events. J. Geophys. Res., 78: 7745—7751.

Phillips, J.D. and Forsyth, D., 1972. Plate tectonics, paleomagnetism, and the opening of the Atlantic. Geol. Soc. Am. Bull., 83: 1579—1600.

Saito, T. and Van Donk, J., 1974. Oxygen and carbon isotope measurements of Late Cretaceous and Early Tertiary foraminifera. Micropaleontology, 20: 152—177.

Stacey, F.D., 1977. Physics of the Earth. Wiley, New York, N.Y., 414 pp.

Toksöz, M.N., Minear, J.W. and Julian, B.R., 1971. Temperature field and geophysical effects of a downgoing slab. J. Geophys. Res., 76: 1113—1138.

Ushakov, S.A., 1961. The results of the geophysical investigations of the Earth's crustal structure in Antarctica. Izv. Acad. Sci., U.S.S.R., 32: 15—31.

Verhoogen, J., Turner, F.J., Weiss, L.E., Wahrhaftig, C. and Fyfe, W.S., 1970. The Earth. Holt, Rinehart and Winston, New York, N.Y., 748 pp.

Voronov, P.S., 1964. Tectonics and neotectonics of Antarctica. In: R.J. Adie (Editor), Antarctic Geology — Proc. 1st Int. Symp. of Antarctic Geology, Cape Town, 16—21 Sept. 1963, pp. 692—702.

Voronov, P.S. and Klushin, I.G., 1963. On the problem of aseismicity of Antarctica and Greenland. Probl. Arktiki Antarkt., 12: 112—119.

Warren, G., 1969. Terra Nova Bay—McMurdo Sound area geologic map. In: C. Craddock (Editor), Geologic Maps of Antarctica, Folio 12. American Geophysical Society, New York, N.Y., sheet 14.

Webb, P.N. and Wrenn, J.H., in press. Late Cenozoic micropaleontology and biostratigraphy of eastern Taylor Valley, Antarctica. In: C. Craddock (Editor), Symposium on Antarctic Geology and Geophysics, University of Wisconsin Press, Madison, Wis.

Weertman, J., 1964. Profile and heat balance at the bottom surface of an ice sheet fringed by mountain ranges. U.S. Army Cold Reg. Res. Eng. Lab., Res. Rep., 134, 7 pp.

Weertman, J., 1970. The creep strength of the earth's mantle. Rev. Geophys. Space Phys., 8: 145—151.

Weissel, J.K. and Hayes, D.E., 1972. Magnetic anomalies in the southeastern Indian Ocean. In: D.E. Hayes (Editor), Antarctic Oceanology II, The Australian-New Zealand Sector. Antarct. Res. Serv., 19: 234—249.

[2]

MAJOR GEOCHEMICAL PROCESSES IN THE EVOLUTION OF CARBONATE—AQUIFER SYSTEMS

BRUCE B. HANSHAW and WILLIAM BACK

U.S. Geological Survey, National Center 432, Reston, VA 22092 (U.S.A.)

(Accepted for publication May 10, 1979)

ABSTRACT

Hanshaw, B.B. and Back, W., 1979. Major geochemical processes in the evolution of carbonate—aquifer systems. In: W. Back and D.A. Stephenson (Guest-Editors), Contemporary Hydrogeology — The George Burke Maxey Memorial Volume. J. Hydrol., 43: 287—312.

As a result of recent advances by carbonate petrologists and geochemists, hydrologists are provided with new insights into the origin and explanation of many aquifer characteristics and hydrologic phenomena. Some major advances include the recognition that: (1) most carbonate sediments are of biological origin; (2) they have a strong bimodal size-distribution; and (3) they originate in warm shallow seas. Although near-surface ocean water is oversaturated with respect to calcite, aragonite, dolomite and magnesite, the magnesium-hydration barrier effectively prevents either the organic or inorganic formation of dolomite and magnesite. Therefore, calcareous plants and animals produce only calcite and aragonite in hard parts of their bodies.

Most carbonate aquifers that are composed of sand-size material have a high initial porosity; the sand grains that formed these aquifers originated primarily as small shells, broken shell fragments of larger invertebrates, or as chemically precipitated oolites. Carbonate rocks that originated as fine-grained muds were initially composed primarily of aragonite needles precipitated by algae and have extremely low permeability that requires fracturing and dissolution to develop into aquifers.

Upon first emergence, most sand beds and reefs are good aquifers; on the other hand, the clay-sized carbonate material initially has high porosity but low permeability, a poor aquifer property. Without early fracture development in response to influences of tectonic activity these calcilutites would not begin to develop into aquifers. As a result of selective dissolution, inversion of the metastable aragonite to calcite, and recrystallization, the porosity is collected into larger void spaces, which may not change the overall porosity, but greatly increases permeability. Another major process which redistributes porosity and permeability in carbonates is dolomitization, which occurs in a variety of environments. These environments include back-reefs, where reflux dolomites may form, highly alkaline, on-shore and continental lakes, and sabkha flats; these dolomites are typically associated with evaporite minerals. However, these processes cannot account for most of the regionally extensive dolomites in the geologic record. A major environment of regional dolomitization is in the mixing zone (zone of dispersion) where profound changes in mineralogy and redistribution of porosity and permeability occur from the time of early emergence and continuing through the time when the rocks are well-developed aquifers. The reactions and processes, in response to mixing waters of differing chemical composition, include dissolution and precipitation of carbonate minerals in addition to dolomitization.

An important control on permeability distribution in a mature aquifer system is the solution of dolomite with concomitant precipitation of calcite in response to gypsum dissolution (dedolomitization). Predictive models developed by mass-transfer calculations demonstrate the controlling reactions in aquifer systems through the constraints of mass balance and chemical equilibrium.

An understanding of the origin, chemistry, mineralogy and environments of deposition and accumulation of carbonate minerals together with a comprehension of diagenetic processes that convert the sediments to rocks and geochemical, tectonic and hydrologic phenomena that create voids are important to hydrologists. With this knowledge, hydrologists are better able to predict porosity and permeability distribution in order to manage efficiently a carbonate—aquifer system.

INTRODUCTION

During much of his career Burke Maxey was greatly interested in the hydrology of carbonate aquifers. Many of our thoughts and ideas about the geochemical aspects of carbonate aquifer genesis were stimulated by discussions with Burke both in the field during the day and during many indoor sessions long into the night.

Carbonate rocks are among the most productive aquifers even though their permeability and porosity may vary enormously from the time of their origins in the marine environment until their functioning as a freshwater aquifer after emergence. These large variations are related to their mode and environment of origin and to the high chemical reactivity of carbonate minerals. This reactivity causes them to continue to undergo chemical and physical changes from the moment of their biological/chemical origins in the sea, to, and continuing with, their functioning as aquifers on land. In addition, it is this high degree of reactivity that places them apart from all other types of aquifers (e.g., cemented siliceous clastics, basalts, alluvium); indeed, one cannot fully understand the hydrology of carbonate rocks or reconstruct the paleohydrology without an intimate understanding of the chemical processes involved in their formation and evolution.

During the past three decades, comprehension of: the origin of carbonate sediments; processes of deposition and accumulation; and mechanisms of early diagenesis and continuing alteration has increased dramatically. Carbonate scientists can now document a continuum from modern carbonate sediments to ancient carbonate rocks. In this paper we address the pertinent geochemical processes relative to the significance of water chemistry and aquifer formation and transformation in this endless series of changes. Although geochemists are making increasing use of isotopic compositions of both carbonate minerals and dissolved species in groundwater studies, we have not discussed them in this paper.

The sheer volume of material written during the past 30 years that bears on this broad topic would make a full-scale review a truly prodigious task and we have chosen, therefore, to write an overview or synthesis of selected material in a text-book style. In keeping with this approach, we have not included references in the text nor an extensive bibliography, but rather, only a few major references which will serve as a guide.

Although some modern carbonate sediments are deposited in the deep sea, the vast bulk of carbonate deposition occurs in warm shallow (less than 12 m depth) seas in the lower latitudes. Furthermore, the final site of accumulation and initial lithification is typically in the inter- to supra-tidal zone which has often been called the "knee-deep" environment. Therefore, our discussion is limited to the carbonate products from shallow normal marine environments, and we further limited our scope by excluding the hydrology of bedded evaporites, salt deposits and associated subsurface brines.

Much of this paper is based on the work of carbonate petrologists who study recent sediments to understand the processes that form and alter carbonate rocks. Hydrogeologists need to understand many of the same processes in their study of alteration and dissolution of carbonate rocks. In other words, petrologists are more interested in the "construction," whereas hydrogeologists are more interested in the "destruction" of carbonate rocks. For example, a limestone that has not undergone the destructive processes of fracturing, perhaps some folding, extensive weathering, alteration and solution, cannot develop into a major aquifer.

For those topics which are of interest to both sedimentologists and hydrogeologists, such as formation and distribution of porosity and permeability, the approaches are extremely dissimilar primarily because of different perspectives of scale in both time and space. That is, many processes occurring in recent sediments are extremely rapid whereas those occurring in regional aquifer systems are generally exceedingly slow. In addition, carbonate petrologists use the microscope to study porosity of a small sample in a thin-section while hydrogeologists use techniques such as pumping tests to evaluate hydrologic characteristics of an aquifer over several hundred square meters. Even so, this relatively large sample is often inadequate to determine the geohydrologic parameters necessary to describe the aquifer characteristics required for development of predictive models. For many hydrologic problems it is necessary to consider the full spectrum of pore size which can range from small intergranular voids, to larger solution openings and regional fracture systems with attendant differences in permeability distribution. Hydrogeologists can benefit from familiarity with techniques and expertise of carbonate geologists, because the understanding gained from the micro-scale can often supplement that gained from the macroscale in regional flow.

EVOLUTION OF GROUNDWATER CHEMISTRY IN CARBONATE AQUIFERS

Before delving into the origin of marine carbonates and the processes that change them into rocks and aquifers, we briefly summarize what is known about groundwater chemistry in these aquifers. The distribution of chemical species in groundwater is not random; rather, it is controlled by all the processes and reactions of diagenesis including recrystallization, dolomitization and cementation, by structural activity, by dissolution and reprecipitation of minerals during groundwater movement, and by mass transfer of chemical species. The physical manifestations of this interplay of hydrology, sedimentology, mineralogy and water chemistry are seen in landscape modification, karst features, porosity and permeability distribution, and in the establishment and operation of local and regional flow systems. The observed chemical character of water in carbonate aquifers is both a control on the physical parameters just mentioned and a response to them. The chemistry of groundwater is a result of the intimate relationship between mineralogy and flow regime because these determine the occurrence, sequence, rates and progress of reactions. In other words, the mineralogy of an aquifer and its groundwater geochemistry are reflected in each other and change in a systematic and generally predictable manner.

Fig.1 is a conceptual model that depicts the changes in groundwater chemistry of carbonate aquifers from the time of their inception in the shallow marine environment, through their functioning as emergent aquifer systems. These changes are discussed briefly here and referred to throughout the paper.

When carbonate sediments first emerge from the marine environment, they

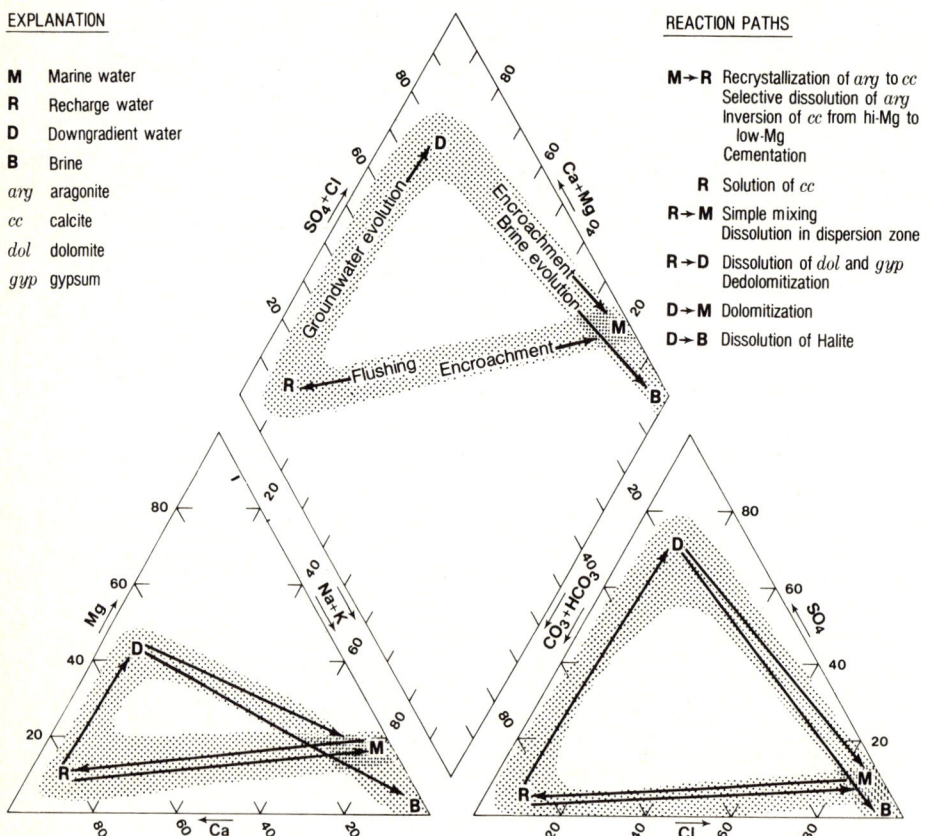

Fig.1. Reaction paths showing evolution of chemistry of groundwater in carbonate aquifers.

undergo flushing of ocean water by freshwater. During this process, some of the most profound mineralogic and petrologic transformations occur; some of these reaction paths are shown on Fig.1 under the heading $M \to R$. Concisely, the dissolved solids content decreases; major ions change from predominantly Na—Cl—Mg—SO_4 to Ca—HCO_3; sediments are recrystallized, selectively dissolved, cemented and perhaps dolomitized.

After carbonate sediments have been transformed into a rock aquifer, many additional chemical changes occur both in groundwater and in aquifer mineralogy. In recharge areas (Fig.1, R), groundwater is typically of the Ca—HCO_3 type. During its movement downgradient (Fig.1, $R \to D$), Mg increases owing to dissolution of dolomite and high-Mg calcite while Ca remains relatively constant; SO_4 increases as gypsum dissolves and HCO_3 remains relatively constant.

Another pathway is also possible from R; this involves marine-water encroachment in a coastal aquifer where recharge occurs close to the coast, which is shown as $R \to M$ (Fig.1). In this case, conservative mixing is the dominant chemical process although selective dissolution may occur in the zone of dispersion.

The schematic model (Fig.2) shows the relative concentration changes of constituents along with the reactions that cause the chemical evolution of groundwater and aquifer minerals for the pathway, $M \to R \to D$, of Fig.1. The box-model conceptualization (Fig.2) of mass-transfer modeling is typical of most carbonate aquifers. The first box of Fig.2 is represented by R on Fig.1. The following two boxes represent chemical reactions and their effects along the pathway $R \to D$ (Fig.1). Although mineralogy of many carbonate aquifers is rather uniform, different reactions occur in different parts of the system because of the flow pattern. In recharge areas, the high concentration of CO_2 and low dissolved solids content cause solution of calcite, dolomite and gypsum. As the concentration of ions increase and their ratios change downgradient (Figs.1 and 2), groundwater becomes saturated with respect to calcite which begins to precipitate. The process of dedolomitization occurs in response to gypsum solu-

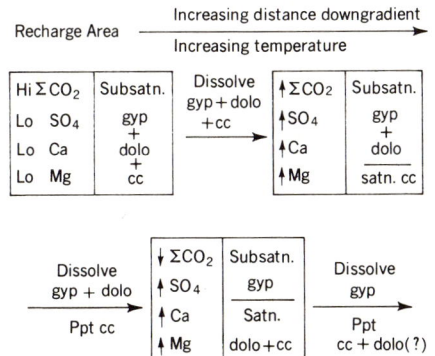

Fig. 2. Schematic model showing evolution of chemical character of water in carbonate aquifers.

tion with calcite precipitation and corresponding effects on the CO_2 and pH. Dedolomitization is the incongruent dissolution of dolomite (incongruent means dissolution during which a solid product remains) to form calcite with a crystalline structure similar to dolomite; calcite formed in this way is called "dedolomite." Hydrologists have demonstrated dedolomitization in the Tertiary Floridan aquifer and in the Pahasapa (Mississippian) aquifer near the Black Hills, South Dakota. It is but a short step from the simple box model (Fig.2) to full-scale computer modeling of reactions in carbonate aquifers.

Where extensive accumulations of evaporite minerals occur, their dissolution results in highly saline brines (Fig.1 pathway $D \rightarrow B$). Another common pathway ($D \rightarrow M$) is caused by subsurface mixing of ocean water that has encroached into the deeper parts of coastal aquifers as in Florida; dolomitization may occur under these conditions.

The remainder of this paper describes major processes that affect the origins of major carbonate aquifer types and groundwater chemistry. However, before going further, it is necessary to describe the mineralogy and crystal chemistry of carbonate minerals needed to comprehend the origin of carbonate minerals in the oceans. This also provides the framework for discussing: the diagenesis of uncemented carbonate minerals to cemented "neo-rocks" after their emergence; how calcium carbonate muds become functioning aquifers; and processes that change their mineralogy, porosity, permeability and water chemistry.

MINERALOGY

Carbonate minerals are those composed of the CO_3^{2-} (carbonate) anion and one or more cations. Approximately 60 carbonate minerals occur naturally, of which calcite ($CaCO_3$, hexagonal-rhombohedral) and its polymorph, aragonite ($CaCO_3$, orthorhombic) predominate in modern marine sediments. In marine carbonate rocks older than Tertiary, calcite and dolomite [$CaMg(CO_3)_2$, hexagonal-rhombohedral] comprise the only major carbonate minerals. Other common rhombohedral minerals include magnesite ($MgCO_3$), siderite ($FeCO_3$), and rhodochrosite ($MnCO_3$). Typical orthorhombic types are strontianite ($SrCO_3$), witherite ($BaCO_3$) and cerrusite ($PbCO_3$). Whereas these other minerals are important in certain sedimentary rocks (e.g., siderite in iron- and peat-bogs) and mineral deposits (e.g., cerrusite in ore gossans), and can have a major influence on the chemistry of some groundwater, they are not an important component in carbonate aquifers and we will, therefore, discuss only aragonite, calcite and dolomite.

In the crystal structure of calcite (Fig.3) the CO_3^{2-} anions (CO_3 groups) may be envisioned as three slightly overlapping O atoms with the small C atom tightly bound in the center. Each Ca has six CO_3 groups as immediate neighbors which are at the corners of an octahedron. However, because the CO_3 groups are triangular rather than spherical, the octahedron is flattened in the direction of the hexagonal C-axis because the CO_3 groups do not readily fit into a regular octahedron. Therefore, the crystal structures that form to accommodate the planar CO₃ groups fall into two distinct types: (1) rhombohedral carbonates (e.g., calcite and dolomite); and (2) orthorhombic carbonates (e.g., aragonite).

The ionic radius of Ca^{2+} is 0.99 Å; Ca^{2+} can be readily accommodated into both the orthorhombic aragonite and the rhombohedral calcite structures (Fig. 4). Because only cations larger than Ca can fit well in the orthorhombic structure, aragonite commonly contains Sr^{2+} which substitutes for Ca^{2+} (i.e., solid solution) in amounts up to 3.5%. Likewise, aragonite will take up sig-

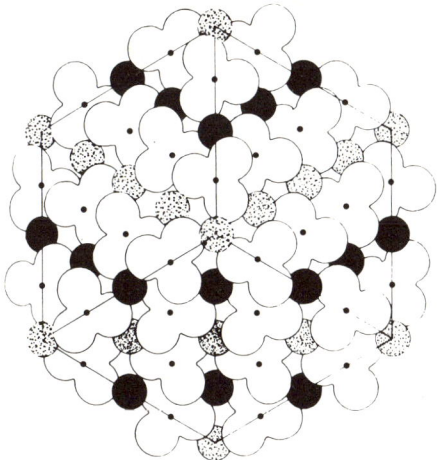

Fig. 3. The crystal structure of hexagonal-rhombohedral calcite. The three *large coalesced blank circles* represent oxygen atoms with a tiny carbon atom (*black dot*) in the center of each planar oxygen trio. The *intermediate-size stippled* and *black circles* represent calcium atoms.

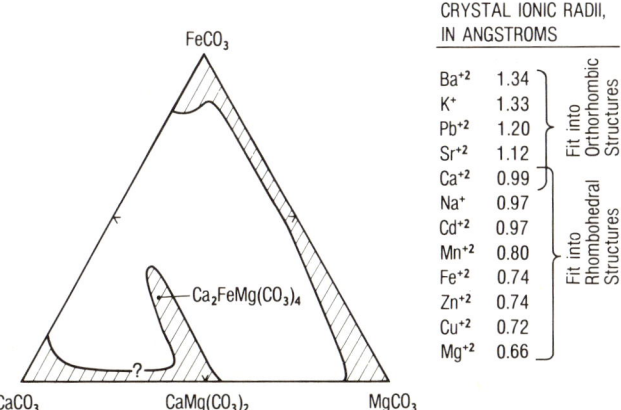

Fig. 4. Compositional ternary phase diagram in the system $CaCO_3$–$MgCO_3$–$FeCO_3$ and crystal cationic radii.

nificant amounts of Ba^{2+}, Pb^{2+}, or K^+ in environments where they are available.

The rhombohedral calcite structure is more compact than the aragonite structure and, therefore, tends to reject the cations larger than Ca^{2+} but readily accepts the substitution of smaller cations, depending, in part, upon their availability in solution. For example, magnesite can accept only about 10% Ca^{2+}, but, on the other hand, calcite may contain up to about 30% Mg^{2+} and still be called calcite, albeit generally preceded by the modifying term "high-Mg".

$CaCO_3$ and $MgCO_3$, when combined in a stoichiometric ratio of 1:1, form the mineral dolomite which has a highly ordered structure. If, in alternating planes of the calcite lattice, the smaller Mg^{2+} were substituted for the Ca^{2+} (i.e., if all the darker cations in Fig.3 were Mg^{2+}) the mineral dolomite is formed for which the rhombohedron has a slightly different angular configuration. Dolomite is typically either Mg-rich or more commonly Ca-rich as shown by the width of the compositional band (Fig.4) rather than being a pure stoichiometric compound with a ratio of Ca to Mg of 1:1. In addition, Fe^{2+} which is intermediate in size between Ca^{2+} and Mg^{2+}, readily fits into the dolomite structure to form a complete solid-solution series between dolomite and ankerite $[Ca_2FeMg(CO_3)_4]$. The typically high Fe^{2+} content of dolomite causes outcrops to weather to a buff-color owing to the oxidation of Fe^{2+} to Fe^{3+} which forms an Fe-oxide staining.

ORIGIN AND ACCUMULATION OF MARINE CARBONATES

Fig.5 is a diagrammatic flow chart which depicts the processes and environments that are associated with the origin, accumulation, transportation, cementation, diagenesis and alteration of carbonate sediments. In this section we start with ocean water in the warm shallow environment and follow the initial formation of carbonate minerals through the paths and processes that lead to carbonate rocks.

Shallow marine environment

Most near-surface ocean water is supersaturated with respect to magnesite, dolomite, calcite and aragonite. Increasing temperature, decreasing pressure and outgassing of CO_2 all lead to increasing saturation. Therefore, carbonate minerals tend to form and accumulate primarily in warm shallow lower-latitude parts of the world's oceans. Despite ocean-water saturation with respect to magnesite and dolomite, these minerals are unknown as primary products in the marine environment. A major reason for this is apparently due to the interaction of Mg^{2+} ions with water molecules; the water molecule has a strong dipole moment owing to the partially ionic nature of its hydrogen—oxygen bonding. Mg^{2+} ions, much more so than Ca^{2+}, have very strongly bound water molecules in solution. In order to be incorporated into a crystal structure such as dolomite or mag-

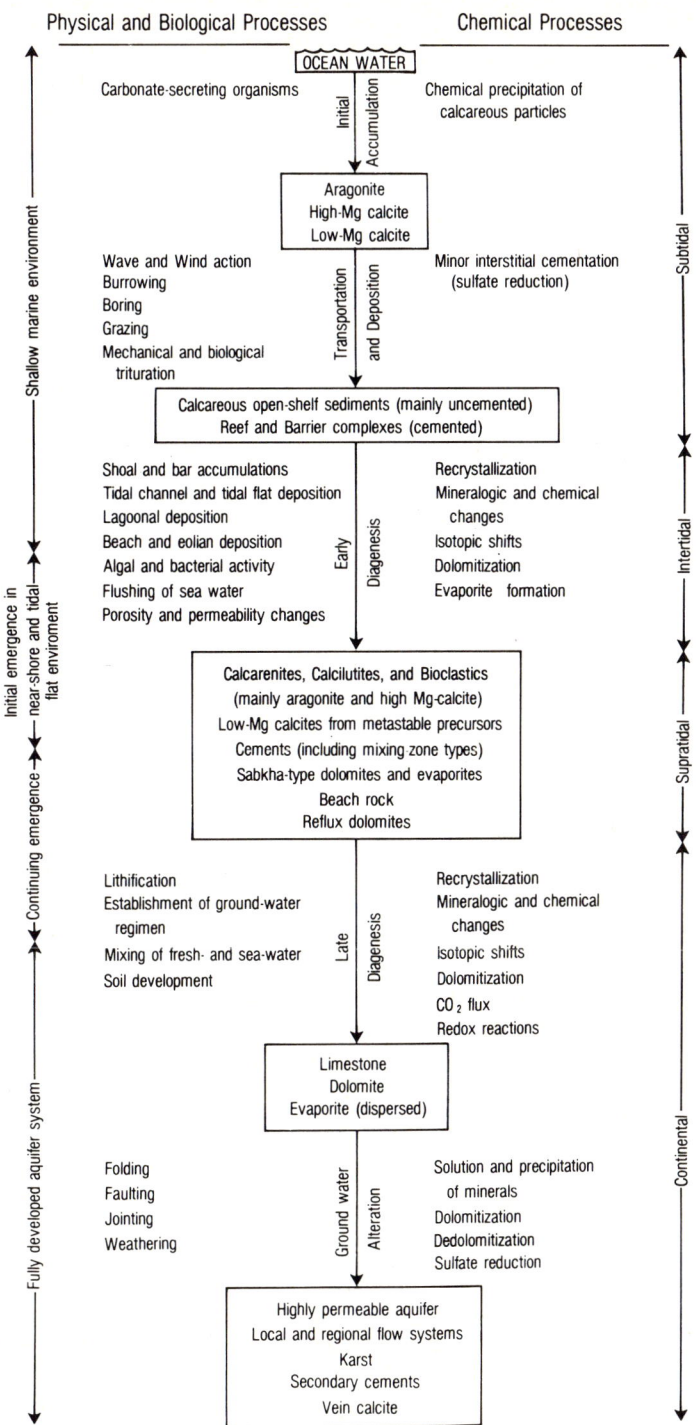

Fig. 5. Processes and environments associated with the evolution of carbonate aquifers.

nesite, the Mg^{2+} ion has to undergo dehydration but the energy requirements for such a process is almost twice that necessary to dehydrate the Ca^{2+} ion. This phenomenon is commonly called the magnesium-dehydration barrier. Carbonate ions (CO_3^{2-}) possess sufficient energy to aid in dehydrating Ca^{2+} ions and, in sufficient concentration, can even displace the water molecules adhering tightly to the Mg^{2+} ions. However, this ability is proportional to CO_3^{2-} concentration which is exceedingly low in ocean water. Only in alkaline continental lake environments does the CO_3^{2-} concentration apparently become high enough to allow dolomite or magnesite to precipitate naturally.

In addition to the Mg-dehydration barrier to primary precipitation of dolomite directly from seawater, another important factor is probably the highly ordered nature of dolomite as noted above. Apparently this requirement of high ordering cannot be satisfied owing to the Mg-dehydration barrier compared to the ease of aragonite or calcite nucleation and precipitation in the ocean-water environment. Therefore, modern marine carbonate accumulations are all initially one or the other mineralogic forms of calcium carbonate.

Although, thermodynamically, calcite should precipitate directly from seawater, it has been shown by both experiments and field observations that only aragonite precipitates inorganically. One might think that the behavior of the Mg^{2+} ion in solution would favor the inorganic formation of calcite because Mg-rich calcites are common; in fact, Mg^{2+} is known to substitute randomly in some organically-derived calcites up to about 30%. On the other hand, aragonite is nearly Mg-free and accommodates only ions larger than Ca^{2+} into its orthorhombic structure (Fig.4). The precise reason for Mg-inhibition of calcite precipitation is apparently not known, but it is well documented experimentally that at molar Mg/Ca ratios greater than about 2, even at low dissolved solids content, aragonite rather than calcite precipitates. Thus, because the Mg/Ca of ocean water is 5.3, only aragonite is precipitated inorganically, although not in large quantities. Some aragonite needles, oolites and larger grapestones may have an inorganic origin but the issue is still hotly debated. Virtually all modern workers agree that organisms are overwhelmingly responsible for the formation of almost all calcium carbonate in present-day oceans.

Under near-surface conditions, most oceanic water is supersaturated with respect to calcium carbonate whereas deep ocean waters are generally undersaturated. Therefore, the greater majority of the biogenic or inorganic calcium carbonate minerals that form near the surface, dissolve in the deep ocean as they sink through the water column. Thus, most calcium carbonate that eventually gets incorporated into the geologic record is formed and preserved in the warm shallow water environment. Various estimates of organic production of calcium carbonate sediment in the shallow marine environment range from hundreds to thousands of grams per square meter year; the average rate in organically productive areas is probably between 500 and 1000 g m^{-2} yr.$^{-1}$.

GENETIC TYPES OF CARBONATE AQUIFERS

Reefs

Reefs are the most spectacular form of carbonate mineral accumulation although they are volumetrically insignificant in modern, and apparently also in ancient areas of carbonate accumulation. Reefs come in a variety of sizes, shapes, lengths and forms, from the Great Barrier Reef with rigid framework off Australia, to the modest sandpile reefs with no demonstrable framework, which are common in parts of the Bahamas and Caribbean. The best known reef-type is the tropical coral reef environment which includes the back-reef lagoons and the adjacent bank areas upon which carbonate mud accumulates. Usually these reefs have a steep drop-off at the shelf boundary and good water circulation which aids in the growth of algae that encrusts the framework coral and cements the reef into a solid massive structure. At the other end of the spectrum are the sandpile reefs. These typically contain individual living corals and exhibit some aspects of coralgal framework reefs when seen in place, yet drilling reveals coral skeletal fragments in an extensive pile of uncemented sand. Virtually all reefs in ancient rocks are good aquifers; they are one of the most sought after types of oil reservoirs owing to their initial high porosity, permeability and up-dip proximity to petroleum source beds. Another important type of reef-like mass, especially in Cretaceous carbonate rocks, is formed largely by giant rudist bivalve mollusks which are generally associated with corals.

On the landward side of a typical reef development, evaporites are common owing to restricted circulation in the back-reef environment and attendant evaporation to the point of precipitation of gypsum and, less commonly, halite. As a result, water from reef aquifers generally has high sulfate content. The geochemical path, $R \rightarrow D$ (Fig.1), develops rapidly; commonly groundwater chemistry evolves quickly to somewhere on the $D \rightarrow B$ pathway of Fig.1 as a result of solution of back-reef evaporites.

Carbonate sand

Carbonate-sand accumulations are volumetrically less important than are carbonate muds, both in areas of modern marine carbonate deposition and in the ancient rock record, yet these cemented sands are commonly prolific aquifers. Sand-size carbonate particles tend to accumulate near their site of origin. Carbonate muds typically contain some sand-size grains which appear to float in a matrix of finer carbonate material. Sands tend to accumulate in higher energy areas than the carbonate muds which typically evolve in, or are transported into, lower energy environments. Further, sands typically accumulate as linear features such as bars, beaches, and upon emergence, as beach ridges. Generally, the clay-sized content is low in carbonate sands owing to the winnowing action of the higher energy environments in which sands evolve and accumulate.

The origins of most carbonate sands may be conveniently divided into two categories: (1) the inorganic oolites; and (2) the skeletal remains of calcareous invertebrates. It is now widely agreed that oolites result from the direct precipitation of aragonite in a high energy environment subjected to strong bottom currents. Oolites, which are smooth, oblate to prolate, spheriods with concentric rings of carbonate, are typically well sorted and range up to several millimeters in diameter. Recent work suggests that the larger sized pisolites may form primarily in the vadose zone after emergence and not in the marine environment.

The second source of sand-size carbonate particles is the skeletal remains of marine invertebrates. Many organisms have sand-size shells; some larger shells readily decompose into sand-size particles. For example, most calcite shells are coarse-grained and break down readily into sand-size material, whereas aragonitic shells typically disintegrate into mud-size particles. Other larger organisms are broken into sand-size by either physical or biological processes.

Perhaps the most prevalent sand-size-appearing materials are the fecal pellets produced by organisms that ingest carbonate mud during the feeding process and excrete enormous quantities of subspheroidal (generally prolate) lumps of mud that are poorly cemented or coated by organic matter. Most pellets crumble when disturbed. Because the pellets are composed primarily of carbonate mud in a matrix of the same material and are poorly cemented, they seldom become incorporated into sand-type deposits. Microscope study of ancient calcilutites (carbonate muds) commonly reveals vague outlines of fecal pellets; apparently much modern and ancient carbonate mud has passed numerous times through the digestive tract of various burrowing and bottom living organisms. Calcilutites that contain large quantities of fecal pellets generally show burrows and tubes or other signs of disturbance. Under certain geochemical conditions, the pellets, burrows, or carbonate mud may be either preferentially recrystallized, dolomitized, or dissolved which will dramatically affect later characteristics of the aquifer, generally toward increased permeability.

Upon emergence, uncemented sand bodies are especially subject to extensive eolian action. Oolites are particularly vulnerable to wind action because of their smooth rounded shape (i.e., the original rolling stones). During times of sea-level lowering, many marine oolite deposits are exposed subareally and are swept into large dune deposits. Unless these wind-formed deposits are subjected to extensive groundwater alteration, they will be poorly cemented and friable as in the Yucatan (Mexico) where spectacular examples of such dunes exist. During groundwater diagenesis, the aragonitic oolite inverts to low-Mg calcite and produces minor, but areally extensive cementation. As a general rule, cementation that occurs in the vadose zone is minor, whereas alteration under phreatic conditions leads to more extensive cementation. Because oolites originate in a high energy environment, finer particles are winnowed out; wind erosion and dune accumulation further enhances the segregation of sizes. In general, therefore, sparingly cemented oolitic rocks have extremely high porosity and permeability and commonly are excellent aquifers with prodigious productivity.

Most carbonate-sand and -mud sediments are minimally influenced by current action and remain at or near their point of origin. These uncemented deposits are highly cohesive and are not greatly disturbed by major storms even in a shallow-water environment. On the other hand, initially well-cemented framework reefs live at or close to the atmosphere—ocean interface and are, therefore, subjected to the full force of major storms. However, the broken fragments are seldom transported far from their place of origin. Thus, most carbonate accumulations are not influenced greatly by storms or even by more normal current action with the exception of beach and bar deposits of sand-size carbonate fragments which are strongly influenced by currents. Typically such deposits record the action of currents when they become lithified and are readily recognized in ancient rocks where they occur as isolated microfacies surrounded by facies that do not exhibit the effects of current. Interpretation of these features by carbonate sedimentologists is important in unravelling paleoenvironment conditions and provides clues to both paleo- and modern-permeability distribution.

Carbonate sands generally are very good aquifers; they commonly originate and accumulate in high-energy wave or wind environments, are well sorted, and contain only minor evaporite minerals. Therefore, their groundwater generally contains low dissolved solids and is typically $Ca-HCO_3$ (Fig.1 R), which may evolve along the path $R \rightarrow D$.

Carbonate mud

Modern carbonate sediments have a strongly developed bimodal grain size distribution and consist of sand- and clay-size (mud) carbonate particles. By far the most predominant size is the mud which is composed of needle-shaped aragonite particles averaging about 3 μm long by 0.5 μm wide. When these carbonate muds become cemented into rocks, called micrite or calcilutite, the individual crystals are approximately 5 μm in diameter owing to recrystallization during lithification.

Carbonate muds have several origins, the most important of which is biological. Calcareous green algae are prolific producers of aragonite needles precipitated around threads of living cells. Upon death of the plant, the aragonite needles are released to become incorporated into the carbonate mud accumulating on a bottom environment remarkably devoid of land-derived sediment. Many investigators believe that two groups of green algae (codiaceans and dasycladeans) were the major source of material for ancient carbonate rocks. Modern representatives of these groups, especially the genera *Halimeda*, *Penicillus*, *Rhipocephalus* and *Udotea* are the major aragonite mud contributors in many Holocene accumulations. In addition to the green algae, some brown and red algae are locally important contributors to carbonate mud. In particular, common turtle grass, *Thalassia*, generally supports a rich encrustation of red algae which is composed of Mg-rich calcite, and minor aragonite-producing bryozoans and serpulid tubes. In some modern areas, these encrustations produce enormous quantities of carbonate mud.

Many other organisms contribute directly to the accumulation of carbonate mud upon disintegrating after death. Larger organisms yield mud-size grains as the result of disintegration caused by wind and wave action, burrowing, boring, grazing and biological abrasion. The latter process may be divided into two parts; mechanical — as when a fish takes a bite of coral and spits out the crushed skeletal carbonate; and digestive — when the carbonate material passes through the digestive system and is abraded internally before being excreted as fine particles. Many of these sources of carbonate mud are either low-Mg calcite (e.g., certain brachiopods, perforate foraminifers, mollusks, coccoliths) or high-Mg calcite (e.g., imperforate foraminifers, holothurians, echinoids, crinoids, some corals). Some subclasses of corals are composed of aragonite and certain organisms produce both low-Mg calcite and aragonite. Nevertheless, owing to the high productivity of aragonite-producing algae, most carbonate muds are predominantly aragonite.

A third source of carbonate mud may be the direct inorganic precipitation of aragonite needles in shallow water, although this is volumetrically unimportant in most environments.

EFFECT OF DIAGENETIC PROCESSES ON GROUNDWATER

Many of the diagenetic processes and reactions that occur in the early development of carbonate rocks have significant influence on the permeability distribution and chemical character of groundwater after the rocks have developed into an active groundwater system. For example, most modern marine carbonate muds are composed of about half aragonite, and half calcite; the latter contains varying amounts of Mg in the lattice. During the groundwater induced process of recrystallization and cementation, aragonite inverts to the stable form of low-Mg calcite. Because Sr^{2+} fits readily in the orthorhombic aragonite structure but not in the rhombohedral calcite structure, Sr^{2+} is rejected from the recrystallizing materials. Dramatic increases in the Sr^{2+} content of groundwater on carbonate islands have been observed as a result of this process.

As marine water is flushed from the sediments (Fig.1 $M \rightarrow R$; Fig.5, *Early Diagenesis*) and fresh groundwater circulation is initially established, profound mineralogic and hydrologic changes occur. For example, most of the carbonate sediment is calcilutite which initially has poor aquifer capabilities. Fig.6 shows a typical diagenetic pathway from unconsolidated fine-grained material to limestone. During the flushing process, fine particles recyrstallize to larger grains, porosity is "collected," aragonite inverts to calcite, and cementation occurs.

During the recrystallization of high-Mg calcite, Mg^{2+} is ejected from the lattice, with a concomitant increase in Mg^{2+} in fresh groundwater. Commonly,

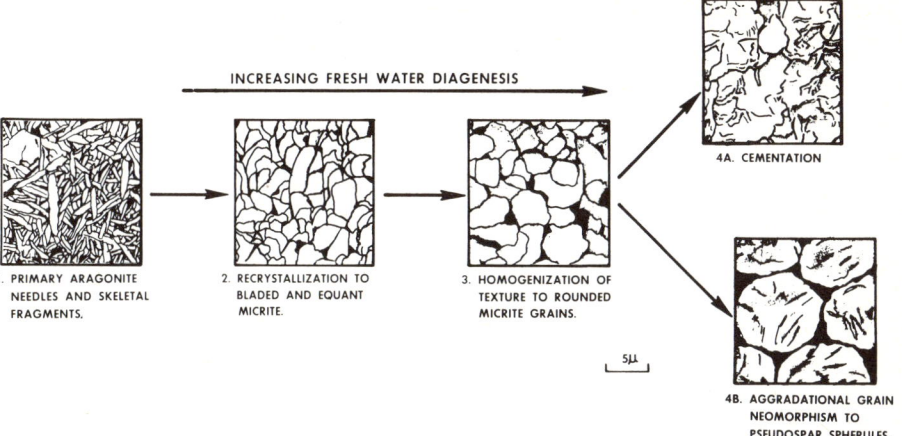

Fig. 6. Typical early diagenetic sequence in transforming carbonate muds into limestones (from Longman and Mench, 1978).

because aragonite is more soluble than calcite, aragonite grains and even entire aragonite skeletal remains of organisms, are preferentially dissolved leaving hollow molds of the material as pores. Typically, adjacent calcite material is not removed but only recrystallized (i.e., dissolved and reprecipitated). Through these processes, freshwater diagenesis leads to redistribution of pore space with greatly increased pore sizes and permeability.

Even though the processes of lithification continue in the groundwater regime, the emphasis is different from those processes of early diagenesis which occurs in the supratidal and near-shore emergent zones. For example, recrystallization is extremely important in both environments; crystal growth and reprecipitation of cement are critical during early diagenesis, whereas, dissolution and removal of material along with continued growth of porosity are of more concern in the groundwater regime. Also many reactions that proceed to the right in the equations below (i.e., forward direction) during early diagenesis, proceed to the left (i.e., reverse direction) during groundwater alteration and solution of minerals. That is, for precipitation or dissolution of calcite:

$$Ca^{2+} + 2HCO_3^- \rightleftharpoons \underset{\text{calcite}}{CaCO_3} + H_2O + CO_2$$

Similarly, for gypsum ($CaSO_4 \cdot 2H_2O$):

$$Ca^{2+} + SO_4^{-2} + 2H_2O \rightleftharpoons \underset{\text{gypsum}}{CaSO_4 \cdot 2H_2O}$$

and for dolomite:

$$Mg^{2+} + 2CaCO_3 \rightleftharpoons \underset{\text{dolomite}}{CaMg(CO_3)_2} + Ca^{2+}$$

where the reverse direction is the process of dedolomitization under certain geochemical conditions in groundwater. Both reactions to the left, solution of gypsum and dedolomitization, increase porosity and permeability whereas the process of dolomitization may either increase or decrease these parameters.

Obviously, many reactions continue to operate in the forward direction whenever local conditions permit the precipitation of minerals. The reduction of sulfate in the presence of organic matter is an example of a reaction that goes only in the forward direction and occurs during diagenesis in the tidal zone and continues to be extremely important in groundwater alteration; that is:

$$SO_4^{2-} + 2CH_2O \rightarrow HS^- + 2HCO_3^- + H^+$$

This reaction not only generates hydrogen sulfide, but also provides an additional source of HCO_3^- which can be significant in causing calcareous cementation in both tidal zones and aquifers.

In addition to the chemical aspects of diagenesis, physical processes also exert a major influence on the evolution of a functioning aquifer from the original carbonate sediments. Physical diagenesis results from many processes of consolidation, in addition to influences of initial uplift and fracturing. Chemical and physical diagenesis continues in the groundwater regime where the amount of solution of carbonate rocks is controlled, in part, by many hydrogeologic factors that affect the source and movement of water. These include structural features, total thickness, bedding thickness and competency of the rock; thickness and permeability of adjacent beds; lithology, texture, distribution of primary and secondary permeability of the aquifer; head distribution and groundwater flow pattern. These factors are significant because they control the surface area and contact time between the rocks and the water, thus affecting its chemical evolution. Competency of carbonate beds is determined by the mineralogy and effects of diagenetic processes; competency, in turn, controls the rock response and the nature of fracturing resulting from tectonic activity. That is, some rocks fold and fracture under the same stress that cause others to form extensive fault systems, which drastically effects the hydraulic response.

Fault zones within carbonate rocks commonly serve as conduits and generate zones of solution greater than the solution along joints or bedding planes. However, the statement must not be accepted as a generalization because some faults have either no influence in the solution of limestone or serve as a hydraulic barrier by transposing less permeable material adjacent to the aquifer and thereby effectively resulting in a no-flow boundary which retards, rather than facilitates, additional solution and permeability generation.

An example of the influence of structure on chemistry and flow of groundwater results from the extensive faulting and uplift of the Edwards and associated limestones in Texas. The "bad-water line" separates potable from non-potable groundwater in the limestone aquifer; its location is partly fault-controlled. Non-potable water is characterized by both high temperature and high concentrations of dissolved solids, especially sulfate and sulfide; and lies along the pathway $D \to M$ (Fig.1). It occurs downgradient where the low permeability is caused in part by the presence of evaporites and carbonaceous material. The position of the "bad-water line" is largely a function of the availability of water during and shortly after the uplift, as the present groundwater regime was being established. The active circulation, with flushing and solution on the upgradient side, developed a major freshwater aquifer with a chemical type that plots near R along the pathway $R \to D$ (Fig.1). Primary reactions were oxidation of sulfides and carbonaceous material, solution of limestone along the fractures, recrystallization of fine-grained micrite to coarser spars, and extensive dedolomitization. In the bad-water zone, little alteration has occurred; the original organic material, sulfide and evaporites have been preserved. Because the mixing zone of dispersion is extremely narrow, and apparently has remained quite stable since development of the present flow pattern, none of the diagenetic effects normally present in mixing zones are found.

GEOCHEMICAL AND HYDROLOGIC SIGNIFICANCE OF THE ZONE OF DISPERSION

Throughout the geochemical evolution of an aquifer, reactions that result from mixing water bodies of different chemical compositions continue to be a major control on groundwater alteration processes including porosity and permeability redistribution. These reactions occur when bodies of saline and fresh water mix (e.g., $M \to R$, $R \to M$, $D \to M$, $D \to B$, Fig.1). Evidence for occurrence and progress of mixing zone reactions in the subsurface is primarily mineralogic; carbonate petrologists are now able to provide a consistent and convincing interpretation for the origin of various cements, textures and crystal morphologies. Mixing owes its significance to the non-linearity of mineral solubility as a function of variables such as salinity, partial pressure of carbon dioxide, temperature and activity (effective concentration).

An example of non-linearity is given in Fig.7 which shows the relationship between ionic strength (a measure of dissolved solids content) and activity coefficient, γ (a factor to correct for ion interaction and complexing). The activity coefficient of a single ionic species such as calcium is defined as:

$$\gamma_{Ca} \equiv \alpha_{Ca}/m_{Ca}$$

where m_{Ca} is the analyzed concentration and α_{Ca} is the effective concentration

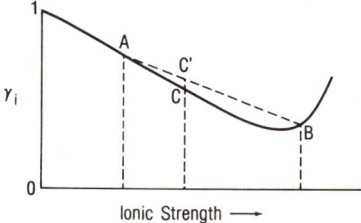

Fig. 7. Graph showing the upward concavity of activity coefficient as a function of ionic strength.

used in thermodynamic calculations. In very dilute solutions α_{Ca} approaches m_{Ca}, and γ_{Ca} approaches unity (Fig. 7). When two solutions of different ionic strength are mixed, the resulting concentration of dissolved solids and of any ion in the mixture (e.g., m_{Ca}) is linear and, therefore, directly proportional to the ratio of volumes of the two solutions. However, the activity coefficient is non-linear as a function of ionic strength and the curve is concave upward (Fig. 7). If a linear relationship existed between these two parameters, the activity coefficient, γ_i, of any ion, i, resulting from mixing solution A with solution B (Fig. 7) to attain a solution with the ionic strength given by the line $C-C'$, would have an activity coefficient corresponding to the point C'. However, owing to non-linearity, the true value is at C. Therefore, because $\gamma_C < \gamma_{C'}$, the resulting activity, α_C, is less than that anticipated from simple linear mixing. Thus, if solutions A and B had been in equilibrium with a mineral such as calcite, the resulting solution C would be undersaturated and, therefore, capable of dissolving additional calcite. In like manner, any other non-linear relationships may result in over- or under-saturation of mineral species, depending upon the shape of the function curve.

The effects of this non-linear relationship resulting in calcite undersaturation, have been demonstrated at a lagoon in Yucatan (Mexico). Regionally, in this area, a thin lens of rapidly moving fresh groundwater overlies a thick body of nearly stagnant saline groundwater with a chemical composition close to ocean water. These two chemically distinct groundwaters are at, or slightly beyond, saturation with respect to calcite. When they mix to form the brackish dispersion zone, the resulting water is undersaturated with respect to calcite and extensive solution occurs in the subsurface whereas only minor solution occurs in the lagoonal environment. Cuspate bays along the carbonate coast of the Yucatan peninsula probably formed along fracture systems by solution resulting from mixing of waters of different chemistry; this may be an important geomorphic process along other carbonate coasts.

Thus far we have restricted the discussion to effects of mixing waters of different chemical composition that are in equilibrium with a mineral to produce a mixture that is undersaturated. Fig. 8 is a hypothetical curve of solubility as a function of salinity and shows that it is also possible for two waters, such as D and E which are supersaturated, to produce an undersaturated mixture, e.g., point F below the solubility curve; or for two waters that are undersaturated, such as A and B, to form a mixture that is supersaturated, e.g., point C above the curve. This increase in salinity may result from either mixing, evaporation, or solution of evaporite minerals.

Even without mixing, any process (e.g., changes in temperature, pH, CO_2 content) that alters the activity of ions will affect the equilibrium conditions of a solution relative to one or more minerals. One such process associated with the zone of dispersion is the CO_2 flux. Any influx of CO_2, such as from root respiration, decomposition of lignite in an aquifer, or decomposition of organic material in intruding ocean water, will cause additional solution of calcite. Conversely, outgassing, which commonly occurs when groundwater discharges at the surface, will prevent further solution and may, under certain conditions, cause precipitation of calcareous cements. Degassing of CO_2 is probably the major process in the formation of beach-rock cements. Mass-transfer calculations show that tidal oscillations of the zone of dispersion can induce degassing of discharging groundwater and cause calcite precipitation. Therefore, it appears that mixing of fresh and ocean water is a dominant control in generating subsaturated solutions and, although mixing can produce supersaturated solutions, the mixing must be accompanied by outgassing to overcome the kinetic factors and permit precipitation of calcite.

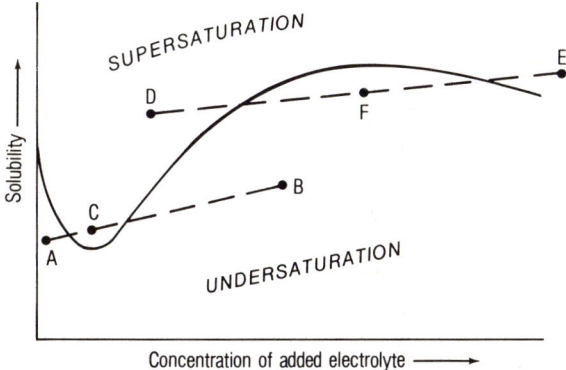

Fig. 8. Hypothetical solubility curve showing how mixing of two waters may result in either undersaturation or supersaturation (from Runnells, 1969).

All aquifers originating as marine limestone have undergone the influence of reactions resulting from mixing of freshwater with ocean water at least once and many aquifers have undergone the influence repeatedly. During emergence, either by tectonic uplift or sea regression, the entire carbonate unit would be subjected to these mixing zone reactions occurring in the zone of dispersion. Within all dispersion zones and in all aquifers throughout their existence, the same group of mixing zone reactions occur. We base this on the inviolate validity of the equilibrium laws of chemical thermodynamics. However, because of other considerations, primarily kinetic, not all reactions proceed equally, nor do they progress to completion. In addition to mineralogic controls, the progress of these reactions is also controlled by many variables which include their sequence, the degree of chemical dissimilarity between two mixing water bodies, time within the dispersion zone, and width of the dispersion zone; these latter two are controlled largely by hydraulic conductivity, rate of uplift (or regression), and head difference between fresh and head difference between fresh and saline water which affects the amount of freshwater discharge.

DOLOMITIZATION

One of the important processes that occurs in the mixing zone is dolomitization. The amount and rate of solution and precipitation of these minerals are major controls on the redistribution of porosity and permeability in carbonate aquifers. Because of the dramatic hydrologic consequences of the processes of dolomitization and dedolomitization and owing to major advances in solving the classic "dolomite problem" in the past decade, we feel that an extensive discussion is justified. The process can range in time from early diagenesis to groundwater alteration and can range in space from near-shore marine-environment to deep-aquifer systems. These processes are recognized by petrologic studies; their importance to hydrologists is that the various minerals found in carbonate aquifers have different solubilities and produce different chemical reaction paths. Furthermore, these studies provide the basis for stratigraphic interpretation and extrapolation of bore-hole data for areal and regional estimates of hydrologic parameters of the aquifer.

During any of the several processes that dolomitizes unconsolidated marine carbonates, porosity and permeability is generally greatly enhanced. The dolomitization process is another that can be thought of which "collects" porosity into larger pores. For example, most marine carbonates are composed of mud-sized particles. Although initial porosity is typically quite large, initial permeability is low. Therefore, as dolomitization proceeds, the small particles of calcium carbonate recrystallize into larger grains (commonly rhombs) of dolomite. In the process, the myriad of small pores are collected into fewer but larger pores. Thus, although total porosity may not change much or even decrease during dolomitization, the resulting dolomite rock may have greatly enhanced permeability.

A typical evolutionary sequence (Fig.9) shows types of dolomite in several diagenetic environments in the Edwards limestone aquifer as the salinity decreases by freshwater flushing. Tidal-flat dolomite forms by penecontemporeneous replacement of soft aragonitic supratidal sediments in a high-Mg hypersaline environment (i.e., the sabkha environment, discussed below). One of the earliest diagenetic processes results in overgrowth of dolomite on pre-exisiting dolomite rhombs. Dolomitized micrite, the most abundant type of dolomite in the Edwards, is not contemporaneous with the hypersaline tidal-flat type, but was formed within the groundwater regime. The fourth type with flame structure, results from crystal overgrowth, perhaps rapid, where there was no competition for space with micrite. The fifth is leached dolomite resulting from both external and internal solution in which the internal leaching is caused by removal of a chemically unstable core. During diagenesis, the salty water was flushed by fresh groundwater and disordered meta-stable dolomite was formed. The last type is composed of clear limpid crystals that grow as a result of slow ordered precipitation from potable groundwater. The types of dolomite that form in these various environments have different textures and morphologies which influence their solubility and stability in different chemical types of water.

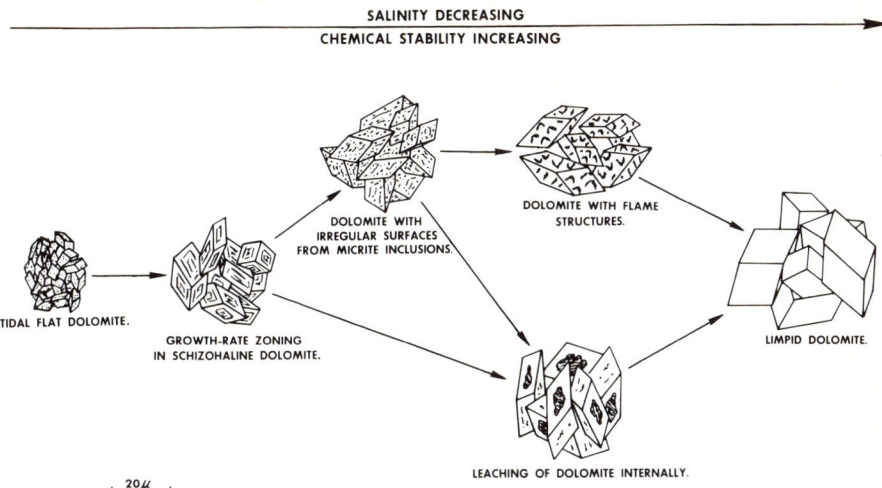

Fig. 9. Typical sequence showing types of dolomite in several diagenetic environments (from Longman and Mench, 1978).

The onshore supratidal and the intertidal areas are profoundly important in the diagenetic history of carbonates because this is where many evaporite minerals are deposited and dolomitization can occur. These are among the most active areas where calcareous sediments first begin the process of becoming rocks — limestones, evaporites and dolomites; in fact, most carbonate-mud aquifers originate here.

In some cases, tidal pools may evaporate to the point where gypsum crystals are deposited. Eventually, some of these crystals may become incorporated as a disseminated minor constituent in carbonate rocks. Another environment where both evaporite minerals and dolomite commonly form is the broad flat extensive supra-tidal zone. Although not widely recognized by geologists and hydrologists, virtually all carbonate-mud aquifers contain disseminated gypsum which is of profound importance to water—rock interactions and commonly is the main driving force of the geochemical system.

In some parts of the world, for example, the Trucial Coast in the Persian Gulf, these supratidal areas are hundreds of kilometers long and tens of kilometers wide. This geochemically and hydrologically active carbonate-mud flat in an arid climate is called a "sabkha." The sabkha environment generally has fresh to brackish groundwater moving under a slight hydraulic gradient toward the sea coast where the water becomes more saline. Solar load on the stark light-colored sabkha plain results in considerable evaporation by capillary action. This evaporation of water from the subsurface causes dissolved solids to increase to the concentration where gypsum precipitates; this process removes Ca^{2+} from solution which increases the Mg/Ca ratio until dolomite forms from a calcium carbonate precursor as described below. Under these conditions, large areas may form dolomite which will have gypsum associated with it.

Some ancient dolomites probably formed in this manner and one of the clues in recognizing their origin is the interspersed evaporite minerals.

Another process that may form dolomite as an early diagenetic product is known as the "reflux model." Under this model, an isolated surface pool of ocean water is evaporated by solar action which produces a high-density brine. Supposedly, under the influence of a density difference between the pool of evaporated water and the less dense underlying ocean water, dolomitization occurs as the heavy Mg-enriched brine migrates downward through the carbonate sediments. The process is thought to be most active in reef environments where the restricted lagoon or pool landward of the reef is the evaporating water body. Although once thought to be an important dolomitizing process, the model is currently not in high favor among sedimentologists.

The environments of dolomite formation cited above should provide diagnostic sedimentary features if the material were preserved in the rock record. These would include association with evaporites and/or shallow water or supra-tidal phenomena, such as mud cracks and broken-shell debris. However, these mechanisms do not account for many of the dolomites of regional extent which are common in the geologic record.

Most modern disordered dolomites have been found in dynamic environments where the molal Mg/Ca ratio reported for interstitial water varies from a low of about 3 to as high as 100. This ratio is important in the kinetics of dolomitization; the higher values may not be necessary for dolomitization to occur, but they do speed up the process. Therefore, the formation of dolomite is at least a three-fold problem: (1) kinetics, i.e., sufficient time; (2) continuing supply of Mg^{2+} ion; and (3) high CO_3^{2-} ion content to overcome the Mg-dehydration barrier. In groundwater systems, sufficient time is generally available; the problem then reduces to determining whether there is enough Mg available. In order to do this, we must know the specific Mg/Ca ratio in solution for the three-phase equilibrium, dolomite—calcite—solution. If this ratio is exceeded in any limestone terrane, then dolomitization should proceed spontaneously, given enough time and CO_3^{2-} ion.

At the three-phase equilibrium, a unique species ratio in solution must exist, provided that the fluctuations of temperature and pressure are kept reasonably small, as in most groundwater environments. For the reaction:

$$CaMg(CO_3)_2 \rightleftharpoons Ca^{2+} + Mg^{2+} + 2CO_3^{2-}$$

it can be shown that:

$$K_d/K_c^2 = \alpha_{Mg^{2+}}/\alpha_{Ca^{2+}} = \text{constant}$$

where K_d and K_c are the equilibrium constants for dolomite and calcite, respectively. Thus, for isothermal three-phase equilibrium, the molal Mg/Ca ratio is fixed at some constant value. A study of water chemistry from various carbonate aquifers suggests that this value is close to unity. The foregoing discussion implies that if calcite is immersed in a solution with a higher ratio, then dolomitization should proceed spontaneously, provided sufficient time and Mg^{2+} ion are available.

One environment which can fit all these requirements and which can also account for regional dolomitization is a groundwater system. However, most potable groundwater systems simply approach three-phase equilibrium and are, therefore, incapable of extensive dolomitizing without an additional source of Mg. Autocannibalization of high-Mg calcite during periods of emergence could account for some dolomites of regional extent; we believe this may have occurred on the Florida platform during Tertiary time. This could account for much of the dolomite found today in cuttings from the potable part of the system.

Underlying the potable water lens in Florida is a zone of brackish water, underlain in turn by an extensive zone of hypersaline water. The volume of Tertiary rocks in contact with non-potable water is much greater than that in contact with freshwater. A diagrammatic cross-section (Fig.10) of the central Florida peninsula illustrates the distribution of hydrochemical facies and Mg/Ca ratios within the groundwater system. Samples from the brackish zone generally contain less than 150 mg/l Mg^{2+} ion; Mg concentrations in the saline zone range from 300 to less than 1000 mg/l and, in the hyper-saline zone, up to 3500 mg/l. Samples from both the brackish and saline zones generally have Mg/Ca ratios greater than unity and this water could, therefore, dolomitize limestone. Mg/Ca ratios from samples in the hypersaline zone vary greatly. The hydraulic connection between these very deep zones and upper parts of the system is poorly known.

Cyclic flow of salt water in the Biscayne aquifer of southeastern Florida may be related to thermal convection within the aquifer. A similar mechanism may cause slow convective movements in salty water underlying all of peninsular Florida. If this is so, then a mechanism is available for brackish and saline waters to dolomitize much of the carbonate material in the aquifer. Such dolomitization is also suggested by the report of dolomite crystals from the highly cavernous Boulder zone, in the Florida aquifer. The Boulder zone exists over much of central Florida and parts of the zone correspond to the zone of brackish water. The cavernous Boulder zone may be caused by solution of limestone brought about by mixing potable and saline water. As discussed previously, when two waters saturated with the same mineral phases are mixed, undersaturation may occur. Thus, mixing of saline with potable water to produce the brackish water zone (Fig.10) may cause solution of limestone. In this

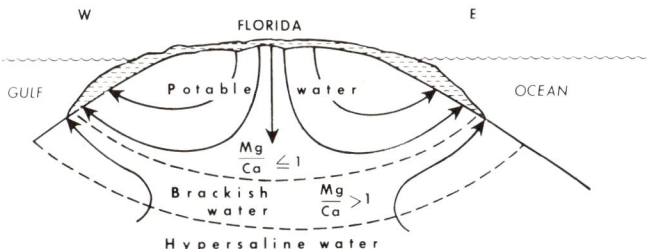

Fig.10. Diagrammatic cross-section of central Florida with *arrows* showing generalized flow paths.

process both Ca^{2+} and CO_3^{2-} would be increased in the solution phase which could cause supersaturation with respect to dolomite. Based on the Mg/Ca ratios in the brackish zone, all the water should dolomitize limestone; it is, in fact, supersaturated with respect to dolomite. Owing to sea-level changes, climatic changes, and/or occasional uplift or downwarp of the Floridan platform, the brackish and saline zones would change position within the system and dolomitize considerable parts of the rock below the potable water lens.

MASS TRANSFER

The opposite of dolomitization is dedolomitization which has been demonstrated by mass-transfer calculations to be an important control on the chemistry of groundwater in a Mississippian carbonate aquifer in parts of the western U.S.A. Basically, it has been found that this aquifer behaves similar to other carbonate—rock systems. The evolutionary pathways and the processes that control its chemistry of water are the same as those shown in Fig.1 and 2. The recharge area has a high total CO_2 and relatively low total-dissolved solids. As groundwater moves downgradient through the aquifer it begins to dissolve calcite, dolomite and gypsum from the mineralogic framework. This results in an increase in the total CO_2 and increases in many of the other chemical constituents, particularly SO_4^{2-}, Ca^{2+} and Mg^{2+}. Eventually, equilibrium with respect to calcite is attained and no more calcite is dissolved from the aquifer. However, the groundwater is still undersaturated with respect to dolomite and gypsum and it continues to dissolve both of these minerals. In so doing, dolomite contributes calcium, magnesium and carbonate species to the solution and gypsum contributes both calcium and sulfate. Ca^{2+} ions from the two sources and carbonate from dolomite combine to precipitate calcite. As a result of the above processes, the total dissolved inorganic carbon decreases while the other constituents all continue to increase. This continues until saturation with respect to dolomite occurs, at which time dolomite may precipitate concurrently with calcite. This process will continue until equilibrium with gypsum or anhydrite is attained. These conceptual models provide a basis for more refined predictive models which incorporate all of the parameters operating in the hydrochemical system.

The development of predictive models for the chemistry of water in hydrologic systems requires detailed information on the reactions occurring in the system. Specifically, one must know what reactions are occurring, the rates of these reactions, and the extent to which these reactions may proceed. The approach now being used by geochemists involves the application of mass-balance and mass-transfer calculations to the observed chemical and isotopic data from the system. These calculations screen out reactions that cannot occur in the system. The successful matching of chemistry from one well to another down the hydraulic gradient implies that the reactions chosen are the correct ones. However, proof that the correct reactions were used is never absolute with such an approach, but, through a series of elimination processes there often remains a single reaction model that is consistent with the chemical and isotopic data observed for the system.

To fully comprehend the functioning of a carbonate—aquifer system requires knowledge of: distribution of head, porosity, permeability and chemical character of the groundwater, including isotopes; boundary conditions; tectonic history; and the sedimentologic and mineralogic framework. As the application of predictive chemical models to practical hydrologic problems becomes more widespread, the expertise of carbonate petrologists and geochemists will become increasingly integrated in the hydrologists' studies of ground—water systems.

ACKNOWLEDGEMENTS

Development of many of our ideas and direction of our thinking was guided by the discussions during field trips of the U.S. Committee on Limestone Hydrology to various parts of the U.S.A., Puerto Rico and Mexico. In addition to Burke Maxey, the committee was composed of V.T. Stringfield, Harry Legrand, Leo Heindl, and Philip E. LaMoreaux. We also have been strongly influenced by discussions with participants of the 1973 G.S.A. Penrose Conference on Water and Carbonate Rocks convened with Tom Freeman. Field discussions in both the Yucatan and northern Mexico with A.E. Weidie and William C. Ward have been most helpful. We appreciate the discussions we have had with many other colleagues and particularly Niel Plummer for enlightening us regarding mass-transfer calculations. We are especially grateful to Craig T. Rightmire and Roger Wolff for the extensive and critical review of the manuscript. We thank Mark Longman, Cities Service Company, and Patricia Mench, Continental Oil Company, for providing and granting permission to use two of their illustrations.

BACKGROUND REFERENCES

Back, W. and Hanshaw, B.B., 1970. Comparison of chemical hydrogeology of the carbonate peninsulas of Florida and Yucatan. J. Hydrol., 10(4): 330—368.
Back, W., Hanshaw, B.B., Pyle, T.E., Plummer, N. and Weidie, A.E., in press. Geochemical significance of groundwater discharge to the formation of Caleta Xel Ha, Quintana Roo, Mexico. Water Resour. Res.
Blatt, H., Middleton, G. and Murray, R., 1972. Origin of Sedimentary Rocks. Prentice-Hall, Englewood Cliffs, N.J., 634 pp.
Folk, R.L. and Land, L.S., 1975. Mg/Ca ratio and salinity: two controls over crystallization of dolomite. Am. Assoc. Pet. Geol. Bull., 59(1): 60—68.
Hanor, J.S., 1978. Precipitation of beachrock cements: mexing of marine and meteoric waters vs. CO_2-degassing. J. Sediment. Petrol., 48(2): 489—501.
Hanshaw, B.B., Back, W. and Deike, R.G., 1971. A geochemical hypothesis for dolomitization by ground water. Econ. Geol., 66(5): 710—724.
Hudson, J.D., 1977. Stable isotopes and limestone lithification. J. Geol. Soc., London, 133: 637—660.

Lippmann, F., 1973. Sedimentary Carbonate Minerals. Springer, New York, N.Y., 228 pp.

Longman, M.W. and Mench, P.A., 1978. Diagenesis of Cretaceous limestones in the Edwards aquifer system of south-central Texas: a scanning electron microscope study. Sediment. Geol., 21: 241—276.

Neumann, A.C. and Land, L.S., 1975. Lime mud deposition and calcareous algae in the bight of Abaco, Bahamas: a budget. J. Sediment. Petrol., 45(4): 763—786.

Roehl, P.O., 1967. Stony Mountain (Ordovician) and Interlake (Silurian) facies analogs of recent low-energy marine and subaerial carbonates, Bahamas. Am. Assoc. Pet. Geol. Bull., 51(10): 1979—2032.

Runnells, D.D., 1969. Diagenesis, chemical sediments, and the mixing of natural waters. J. Sediment. Petrol., 39(3): 1188—1201.

Wilson, J.L., 1975. Carbonate Facies in Geologic History. Springer, New York, N.Y., 471 pp.

[2]

EFFECTS OF KARST AND GEOLOGIC STRUCTURE ON THE CIRCULATION OF WATER AND PERMEABILITY IN CARBONATE AQUIFERS

V.T. STRINGFIELD, J.R. RAPP and R.B. ANDERS

U.S. Geological Survey, National Center 412, Reston, VA 22092 (U.S.A.)

(Accepted for publication May 10, 1979)

ABSTRACT

Stringfield, V.T., Rapp, J.R. and Anders, R.B., 1979. Effects of karst and geologic structure on the circulation of water and permeability in carbonate aquifers. In: W. Back and D.A. Stephenson (Guest-Editors), Contemporary Hydrogeology — The George Burke Maxey Memorial Volume. J. Hydrol., 43: 313—332.

The results of the natural processes caused by solution and leaching of limestone, dolomite, gypsum, salt and other soluble rocks, is known as karst. Development of karst is commonly known as karstification, which may have a pronounced effect on the topography, hydrology and environment, especially where such karst features as sinkholes and vertical solution shafts extend below the land surface and intersect lateral solution passages, cavities, caverns and other karst features in carbonate rocks. Karst features may be divided into two groups: (1) surficial features that do not extend far below the surface; and (2) karst features such as sinkholes that extend below the surface and affect the circulation of water below.

The permeability of the most productive carbonate aquifers is due chiefly to enlargement of fractures and other openings by circulation of water. Important controlling factors responsible for the development of karst and permeability in carbonate aquifers include: (1) climate, topography, and presence of soluble rocks; (2) geologic structure; (3) nature of underground circulation; and (4) base level. Another important factor is the condition of the surface of the carbonate rocks at the time they are exposed to meteoric water. A carbonate rock surface, with soil or relatively permeable, less soluble cover, is more favorable for initiation of karstification and solution than bare rocks. Water percolates downward through the cover to the underlying carbonate rocks instead of running off on the surface. Also, the water becomes more corrosive as it percolates through the permeable cover to the underlying carbonate rocks. Where there is no cover or the cover has been removed, the carbonate rocks become case hardened and resistant to erosion. However, in regions underlain not only by carbonate rocks but also by beds of anhydrite, gypsum and salt, such as the Hueco Plateau in southeastern New Mexico, subsurface solution may occur where water without natural acids moves down from bare rock surfaces through cracks to the beds that are more soluble than carbonate rocks. For example, in the area of Carlsbad Caverns in southeastern New Mexico, much of the water responsible for solution that formed the caverns apparently entered the groundwater system through large open fractures and did not form sinkhole topography. East of the Carlsbad Caverns, however, in the Pecos River Valley where the carbonate rocks are overlain by the less soluble Ogallala Formation of Late Tertiary age, solution began along escarpments as the Pecos River and its tributaries cut through the less soluble cover. As these escarpments retreated, sinkholes and other karst features developed.

Joints or fractures are essential for initiation of downward percolation of water in compact carbonate rocks such as some Paleozoic limestone in which there is no intergranular permeability. Also joints or fractures and bedding planes may be essential in the initiation of lateral movement of water in the zone of saturation. Where conditions of recharge and discharge are favorable, groundwater may move parallel to the dip. However, the direction of movement of water in most carbonate rocks is not necessarily down dip or parallel to the dip. The general direction of movement of both surface and groundwater may be parallel to the strike in a breached anticline. Faults may restrict the lateral movement of water, especially if water-bearing beds are faulted against relatively impervious beds. Conversely, some fault may serve as avenues through which water may move as, for example, in the Cretaceous Edwards aquifer in the San Antonio area, Texas.

Karst aquifers, chiefly carbonate rocks, may be placed in three groups according to water-bearing capacity. Water in aquifers of group *1* occurs chiefly in joints, fractures, and other openings that have not been enlarged by solution. The yield of wells is small. Aquifers in group *2*, with low to intermediate yields, are those in which water occurs in joints and fractures with some cavities and channels enlarged by solution. Aquifers in group *3* are those in which the yield of wells and springs range from intermediate to very large. This group includes five of the most productive aquifers in the U.S.A.

The water-bearing beds of all of these productive aquifers, except the Biscayne aquifer in southeastern Florida, contain buried paleokarst in which the permeability has been reactivated and enlarged by the present circulation system.

INTRODUCTION

The purpose of this paper is to discuss some of the features of physical hydrogeology of karst including geologic structure that affects the circulation of water and permeability in carbonate aquifers. The development of karst, commonly known as karstification, is the result of solution and leaching of limestone, dolomite, gypsum, salt and other soluble rocks. Circulation of water through fractures and other openings enlarges these openings, increases permeability, and hence, the potential rate of circulation. Where cover and other conditions are favorable and sufficient water is available, the process is more or less continuous and depth of circulation, size of openings, and vertical and horizontal permeabilities increase with time. The evolution of the circulation system and resultant very large openings finally give rise to an irregular topography marked by sinkholes, vertical shafts, abrupt ridges, caverns, underground streams, and other karst-related phenomena.

Paleokarst is buried karst that formed when the soluble rocks were at or near the land surface. It may occur at depths ranging from a few feet to many hundreds of feet. In areas where the paleokarst is not too deeply buried, the present circulation system in the overlying carbonate aquifers may reach the paleokarst due to revived secondary permeability, thus forming a very productive aquifer such as the principal artesian aquifer of Tertiary age in southeastern Georgia and Florida. In addition to water resources, paleokarst may also be of economic importance from the standpoint of other natural resources, such as zinc, lead, oil and gas.

Special acknowledgments are due to our colleagues for their helpful suggestions and review of this report.

Karst terminology

The following section on karst terminology is included to aid the reader in better understanding the terms used to describe the hydrology of karst aquifers.

The German word "karst" is from the Indo-European word "kar" meaning rock, and the Slovenian word "kras," as used in the classic karst region in Yugoslavia where carbonate rocks are exposed (Roglic, 1972, p. 2). In Yugoslavia the terms "Krš" and "Carsus" are also used by some authors, as in the book: "*Krš Jugoslavije, Carsus, Iugoslaviae*" (Petrik and Herak, 1969). "Carsus" is from the Italian place name "Carso". The term "Krš" is Serbo-Croatian for "rock," or "stone," and "kras" is the Slovenian word for a bleak waterless place (Williams, 1970, p. 115). These terms were adopted by early investigators of the landforms in carbonate terranes.

Types of karst and the main groups of karst landforms are described by Sweeting (1972). Karst regions of eleven European countries, Jamaica (West Indies) and the U.S.A. are discussed in *Karst — Important Karst Regions of the Northern Hemisphere* (Herak and Stringfield, 1972). Both books contain references to many earlier reports on karst throughout the world. Among the later reports are papers presented at two symposia: *Karst Hydrogeology, 1972* (Tolson and Doyle, 1977), *Karst Geology and Hydrology, 1974* (Rauch and Werner, 1974), and *Hydrologic Problems in Karst, 1977* (Dilamarter and Csallany, 1977). Other reports containing information on the physical hydrogeology of karst include a paper and bibliography by LaMoreaux et al. (1975). That report gives a list of 52 symposia and conferences on carbonate rocks 1948—1975. Also, the report contains a discussion of historical references and the titles of several new books including Petrik and Herak (1969), Jennings (1971), and Gvozdetsky (1972). Glossaries of karst terminology include reports by Monroe (1970), and UNESCO—FAO (1972). An album, *Karst in China* (Chinese Academy of Geological Sciences, 1976) contains a brief discussion of karst of China with photographs of spectacular karst features in that country. A paper by Silar (1965) describes tower karst in China and North Vietnam.

Reports relating chiefly to the chemical hydrogeology of karst include those by Back and Hanshaw (1965, 1970), Thrailkill (1968, 1972), Jacobson and Langmuir (1970), Rauch and White (1970), Langmuir (1971), and Shuster and White (1971). LaMoreaux et al. (1975) reported on the progress of knowledge about hydrology of carbonate terranes with a selected bibliography.

The most extensive annotated bibliographies are those by LaMoreaux et al. (1970) and Warren and Moore (1975) which includes 800 new references. The 1970 report contains 1462 references to the geology, hydrology, geochemistry and geophysics of carbonate rocks and closely related subjects.

In the U.S.A., one of the early references to a karst feature was by Thomas Jefferson who suggested in 1794 that Natural Bridge, Virginia, 22.5 km (14 mi.) SW of Lexington was the result of a great "convulsion of nature" (Thornbury,

1954, p. 337). Jefferson's interest in the bridge caused him to give King George III "20 shillings of good and lawful money" for the bridge and 157 acres of land in 1774. In 1976 the Governor of Virginia proposed that the State of Virginia purchase the bridge from the present owners, a private company, for 4 million dollars (McAllister, 1976).

About a quarter of a century after Jefferson concluded that the Natural Bridge resulted from a great "convulsion of nature", Gilmer (1818) attributed the Bridge to solution of limestone and cited Natural Tunnel about 5 km (3 mi.) north of Clinchport, Virginia, as being similar to the Bridge. Since that time, different ideas have been proposed to explain the Bridge, but all have recognized that solution of limestone was required.

Karst features may be divided into two groups as follows: (1) surficial features that do not extend far below the surface; and (2) karst features such as sinkholes, solution shafts, solution cavities, and caverns that may affect the circulation of water many hundreds of feet below the surface.

The term sinkhole or sink in a karst terrane is used to designate a hole or depression formed by sinking or collapse of the land surface where solution of the underlying rocks has formed cavities or other openings. It does not denote sinking of a surface stream as used for a "swallow hole" (a karst term used to indicate a hole that swallows water) or a ponor, which is a sinkhole into which a surface stream flows. In the U.S.A., some investigators have adopted some of the foreign terms for karst features. Therefore, some of the foreign terms of karst features that are significant in hydrologic investigations are included here.

The Serbian word, "doline", meaning a little dole or valley, is commonly used as the equivalent of sinkhole in karst areas. In a glossary of geology (Howell, 1957, p. 86), "doline" is defined as equivalent to "swallow hole", "sinkhole", or "cockpit". Thornbury (1954, p. 322; 1969, pp. 308 and 309) defined "doline" as a solution-type sink that develops slowly by solution beneath the soil mantle. Monroe (1970, p. K.7) stated that "doline" is a basin or funnel-shaped hollow in limestone, ranging in diameter from a few meters to a kilometer and in depth from a few to several hundred meters. He divided dolines into two groups: solution dolines and collapse dolines. Solution dolines are formed chiefly by solution of the surface of the carbonate rocks. Collapse dolines are formed by the collapse of the surface into a cavern.

Sinkholes can be divided into three general types according to the conditions under which they developed. These include, in addition to the two types of sinkholes (solution dolines and collapse dolines of Monroe (1970), the cenote-type that is formed by collapse of the roof of a cavern, bed by bed, forming a vertical shaft (Stringfield and LeGrand, 1974, 1976). When the shaft finally reaches the surface it forms a steep-sided sinkhole. Where the rocks are overlain by a thick cover of unconsolidated material, the sink may be funnel-shaped above the carbonate rocks. The slope of the funnel will depend on the angle of repose of the material overlying the rocks. The conditions under which sinkholes (including cenotes) form are different from those of vertical solution shafts. Some of the shafts are known as "jamas" or "avens" in some parts of Europe.

Solution shafts, sometimes referred to as natural wells, commonly form along solution escarpments of karst areas. These shafts are caused by water moving downward in the zone of aeration to the zone of saturation. The solution may begin along intersecting vertical joints forming vertical tubular shafts that are enlarged by solution, as thin films of water move down the walls. In a cave where a shaft does not extend to the surface, the roof of the shaft is dome-shaped and the floor is a pit; therefore, the shaft is known as a dome pit (Pohl, 1955).

Some vertical shafts and cenote-type sinkholes are as much as 305 m (1000 ft.) deep in areas of considerable topographic relief. In a table giving the depths of deep caves and deep pits of the Americas, Fish (1972, p. 37) reports El Sotano, Queretaro, Mexico as the deepest pit (shaft) having a depth of 410 m (1345 ft.). The pits or shafts are known as "sótanos" (basement) in the karst region between Mexico City and Monterey, Mexico. A solution shaft may appear to be similar to a cenote-type sinkhole. However, the permeability of the wall of a shaft is much less than that of a cenote-type sink. Such a difference may have a significant effect on the circulation of water in a carbonate aquifer as described by Stringfield and LeGrand (1969a, b, 1971).

A number of sinkholes may in time coalesce by lateral enlargement and form a larger karst basin which is called a "uvala". A uvala is an intermediate form between a sinkhole and a "polje". The karst term polje (which is the Serbian word for flat field) is a large basin or valley formed chiefly by solution of the carbonate rocks. Typical poljes are larger than uvalas and have no surface outlets; drainage is underground. Faulting is involved in the development of some poljes, as for example in Jamaica where poljes are also known as interior valleys.

In some karst areas residual hills of carbonate rocks remain among the sinkholes. Under some conditions the sinkholes are filled with unconsolidated deposits that form a plain among the hills. Topographic relief among the hills may be as much as several hundred meters. One such hill in the Mekong Valley in Laos was reported to be nearly 305 m (1000 ft.) high (White, 1968, pp. 754 and 755). Jennings (1971, p. 190) reported karst towers as much as 533 m (1750 ft.) high in Malayan karst. The karst hills have local names in different countries. For example, they are known as "hums" in Yugoslavia; "pepino hills", "haystack hills", and "mogotes" in Cuba; and "knobs" in Indiana. The shapes of the hills are used to describe the karst in some areas. In Jamaica the karst with cone-shaped hills is called cone karst or kegel karst. Where the hills are shaped like towers, the name tower karst or turm karst is used.

Solution shafts, natural wells, cenote, and other types of sinkholes form in the zone of aeration above the water table in carbonate rocks. However, with changes in base level, such as changes in sea level during Pleistocene time, some of these karst features that were formed in the zone of aeration in coastal areas, such as those in Florida, U.S.A., and Yucatan, Mexico, became submerged in the zone of saturation when the sea rose to its present level. Also solution channels and cavities that formed in the zone of saturation are now in the zone of aeration where declines in base level have lowered the zone of saturation.

SOLUTION IN CARBONATE ROCKS

The water-bearing capacity of karst aquifers depends in part on the primary permeability of the aquifer and on the secondary permeability which is formed by fracturing, solution, and other processes. However, the secondary permeability, due to solution of limestone, dolomite, anhydrite, and gypsum, accounts for much of the large yield of the most productive aquifers in the U.S.A. In order for large permeability to develop by solution, the soluble rocks must be in a groundwater circulation system. The geologic formation must: (1) receive water in an intake area; (2) transmit water; and (3) discharge water. Means of discharge are especially important. Under water-table conditions, circulation tends to be large in a zone just below the water table in the aquifer and to be less at greater depths. However, the depths of circulation depend, in part, on the relative elevations of the discharge area to the recharge area. For example, in the Cahaba Valley, east of Saginaw, Shelby County, Alabama, where surficial karst (a karren field) was exposed by quarrying operations in the Ordovician Newala Limestone (Moravec, 1974, pp. 113—119), groundwater circulation was chiefly surficial, and only shallow solution occurred. A quarry adjacent to the 'karren field' revealed that although joints filled with water extends to a considerable depth below the land surface, solution along the joints was limited because there was no discharge area to provide circulation of the water. A few kilometers farther south, along the strike of the formation, solution of the limestone extends to a considerable depth because there is a discharge area for the circulation system.

The rate of solution depends on many factors, including the rate of circulation of the water, time, volume of solvent, area of contact, concentration and partial pressure of carbon dioxide, and natural acids in the water. However, these factors and related complex karstifications are beyond the scope of this paper. For example, Swinnerton (1949, p. 659) reported that the temperature factor in the ordinary range of groundwater is complicated but is not very significant. Above the water table, however, the variations in temperature may be large enough to be significant. Carson (1964, p. 5) observed that in Tolley's Cave, Virginia, the ratio of solution to deposition varies with the season. During the winter, water entering the cave is cold and contains a large amount of calcium carbonate in solution. As the water becomes warmer inside the cavern, carbon dioxide is released and travertine is deposited. In a discussion of the amount and rate of solution of carbonate rocks, Burdon and Papakis (1963, pp. 61 and 62) cited results of studies by Corbel (1959, p. 117) tending to indicate that the rate of erosion from cold rain is greater than that from warm rain. As stated by Back and Hanshaw (1965, p. 80), solution is a major process by which the permeability of aquifers containing minerals of high solubility, such as calcite and gypsum, is increased. The chemistry of the water is a complex controlling factor. For example, they report that salt-water encroachment in a carbonate aquifer can change the permeability distribution and that, because of the effect of dissolved-solids concentration on the activity of each ion in solution, salty water can dissolve more limestone than freshwater. Although the conclusion that salty water is more capable than freshwater of dissolving limestone under some conditions, other factors are generally more significant. For example, much of the fresh groundwater is undersaturated with respect to calcium carbonate, and in general fresh groundwater circulates much faster than salty ground water. These two factors, in combination, result in a much greater development of permeability in parts of limestone containing freshwater than in parts containing salty water. All limestone terranes considered in this report indicate the importance of freshwater circulation to develop permeability. Two examples are the deep-lying limestones of low permeability in the midcontinent region that have never contained freshwater and, in contrast, the permeable parts of limestones of Jamaica and Puerto Rico that have been elevated above sea level and have not contained salty water since Miocene time (Stringfield and LeGrand, 1969a, p. 376).

DEVELOPMENT AND EFFECTS OF KARST IN CARBONATE ROCKS

No single geologic rock unit ranges more widely with respect to yield and circulation than carbonate rocks. Some carbonate formations rank among the best aquifers, yet others are as unproductive as shale. These differences are due in large part to the extent to which they have been subjected to solvent action of circulating water. There also is a great range in the solubility of carbonate rocks. Under some conditions limestone is among the most soluble rocks whereas under other conditions limestone is among the most resistant of the rocks to solution and erosion. The secondary permeability formed by solution of the rocks accounts for the high permeability of the most productive aquifers.

Important controlling factors responsible for development of karst and permeability in carbonate aquifers include: (1) climate; (2) topography; (3) presence of soluble rocks; (4) geologic structure; (5) nature of underground circulation of water; and (6) base level (Herrick and LeGrand, 1964, p. 30). Another important factor is the condition of the surface of the carbonate rocks at the time the rocks are exposed to meteoric water. A carbonate rock surface with a soil or relatively permeable, less soluble cover is more favorable for the initiation of karstification and solution than bare rocks. Water entering the permeable cover percolates downward to the underlying carbonate rocks instead of running off on the surface. Also, the water may dissolve natural acids and become more corrosive as it percolates through the permeable cover to the underlying carbonate rocks.

The cover may consist of: (1) a mantle or a blanket of unconsolidated, relatively permeable deposits on carbonate or other soluble rocks; or (2) consolidated formations which may include resistant beds, such as sandstone.

Where bare carbonate rocks are exposed to weathering before the initial stages of solution have developed, the rocks become indurated and any soil formed by weathering of the carbonate rocks is likely to be removed by surface erosion especially on steep slopes. In such areas solution of the rocks tends to be confined to the surface. Surface streams crossing the area may cut deep channels or gorges, but carbonate rocks at the surface in interstream areas become indurated and resistant to solution. Any erosion is likely to be more or less sheet erosion on the surface, as, for example, on the chalk of the Selma Group in Alabama. If, on the other hand, the bare rock surface is cut by deep fissures or crevices through which water may enter, some solution will occur along these crevices. However, where the rocks include anhydrite, gypsum, or salt, solution of these rocks may be sufficient to develop a circulation system in and thereby increase the permeability.

In the major karst areas in the U.S.A., solution of the carbonate rocks began as streams cut through the less soluble overlying formations, including resistant formations such as sandstone, exposing the carbonate rocks to meteoric water which percolates into the underlying carbonate rocks forming a zone of saturation and a circulation system. As solution of the carbonate rocks and collapse of the less soluble overlying beds occur, solution escarpments develop and retreat, leaving a sinkhole plain. Most, if not all, of the sinkhole plains in the U.S.A. developed in this way.

Where the less soluble beds include formations resistant to erosion, such as the rocks of Pottsville age or Cypress Sandstone in the Mammoth Cave region of Kentucky, one or more solution escarpments may form as shown in a cross-section of the formations, Fig. 1 (Livesay and McGrain, 1962, p. 9). The Pennyroyal Plateau, a sinkhole plain adjacent to the escarpments, formed as the escarpments retreated by solution and erosion. A plan view of the Chester escarpment and sinkhole plain (Mitchell plain) in southern Indiana, is shown in Fig. 2 (Powell, 1961).

Herrick and LeGrand (1964, pp. 27 and 28) show the relation of the solution escarpment (Hawthorn escarpment) to a sinkhole plain in Baker and Mitchell counties in southeastern Georgia. In that area the retreat of the escarpment was due chiefly to solution and sapping of the carbonate rocks by water from migrating springs at the head of pocket valleys when the zone of saturation stood at a higher level. That escarpment extends into Florida where large solution basins, some of which are occupied by lakes, such as Lake Iamonia near Tallahassee, occur along the escarpment (Stringfield et al., 1977, pp. 10—13).

Where the less soluble beds overlying the carbonate rocks are unconsolidated and permeable, karstification may begin throughout the area when surface streams cut through the cover. These streams serve as outlets for the water that percolates into the carbonate rocks. For example, the blanket sands (Briggs, 1966, p. 60; Monroe, 1966, p. 6, 1968, pp. 75—85, 1974, pp. 159—163) served as a cover on some of the carbonate rocks in the karst region of northern Puerto Rico.

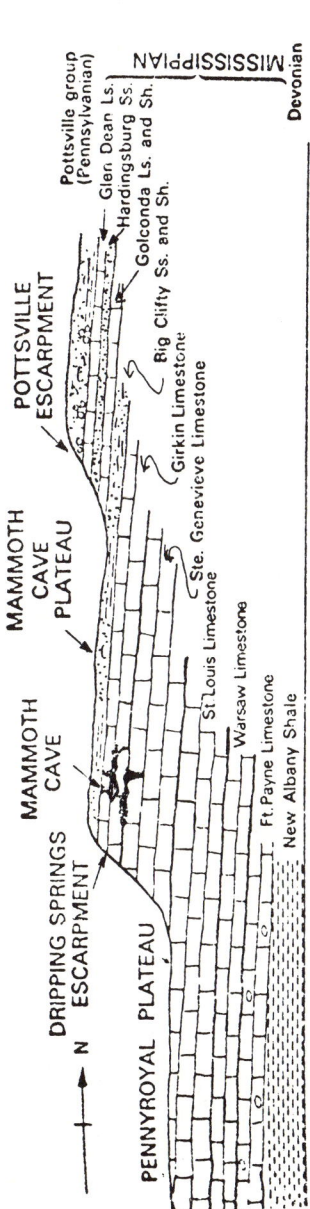

Fig. 1. A north–south cross-section through formations in the vicinity of Mammoth Cave, Kentucky (from Livesey and McGrain, 1962).

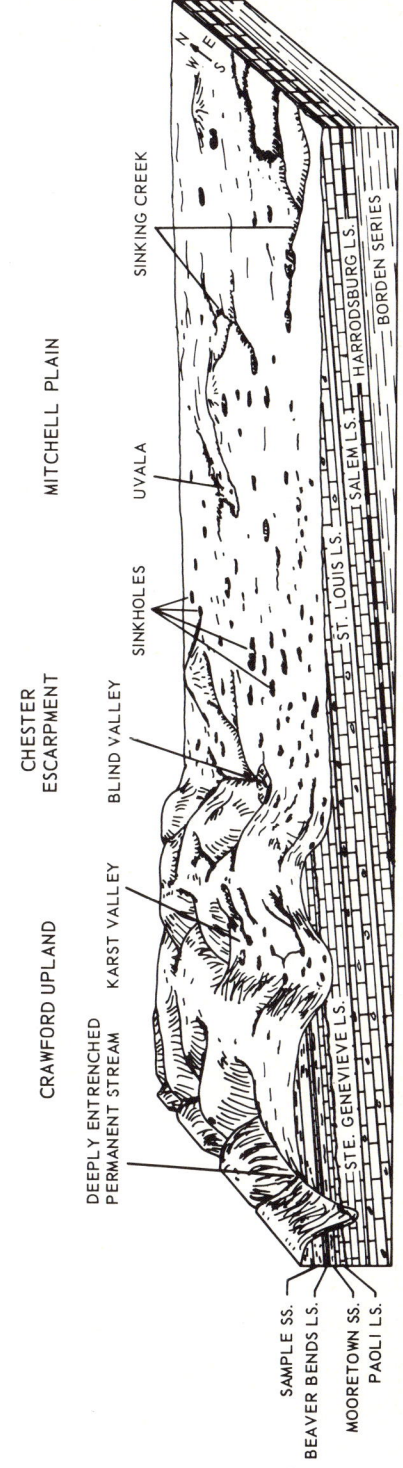

Fig. 2. Plan view of Chester Escarpment and Mitchell Plain in southern Indiana (from Powell, 1961; Wayne, 1950).

However, the carbonate rocks may become case-hardened and resistant to erosion in some places where the cover is removed or becomes thin due to surface erosion. Such resistant surfaces occur along some of the river valleys, such as the canyon of Rio Guajataca, Puerto Rico. Natural walls or ramparts formed as the river deepened its channel on one side of the rampart and the land surface was lowered by solution of the carbonate rocks in a sinkhole plain on the other side of the rampart. The most prominent rampart is on the west side of Rio Guajataca where the top of the rampart is 155 m above the river (Monroe, 1976, p. 39). The west side of the rampart slopes down to a sinkhole plain about 15—35 m below the top of the rampart. Groundwater in the sinkhole plain can move through the rampart to the river.

In addition to the ramparts, carbonate rocks that become indurated and case-hardened in the tops of hills among sinkholes may become a factor in determining whether the hills will have the shape of towers (tower or turm karst) or cones (cone or kegel karst). Other examples of the effects of case-hardening on differential erosion have been reported by Flint et al. (1953) in Okinawa, and Isphording (1974) in Yucatan.

In a study of solution effects upon bare limestone as related to differences in slope, Smith and Albritton (1941) found a gradation from shallow solution pits on gentle slopes through solution facets on moderate slopes, to solution furrows on the steep slopes which they designated as lapies. Apparently, there were no sinkholes and no solution in the subsurface.

In regions such as the Hueco Plateau in southeastern New Mexico, subsurface solution may occur where water without natural acids moves down from bare rock surfaces through cracks to the beds that are more soluble than carbonate rocks. At one locality in the Hueco Plateau, Rapp (1957) reported a system of parallel fractures as much as 2.4 km (1.5 mi.) long. The surface expressions of these fractures known as "dangerous cracks" is as much as 1.5 m (5 ft.) wide and 6 m (20 ft.) deep. Under these conditions typical sinkhole topography does not develop extensively as it does in the first stage of karstification in regions having some type of cover. For example, in the area of Carlsbad Caverns in southeastern New Mexico, much of the water responsible for solution that formed the caverns, apparently entered the groundwater circulation system through large open fractures and did not form typical sinkhole topography. East of the caverns, however, in the Pecos River Valley, where the soluble Paleozoic rocks were overlain by the less soluble Ogallala Formation of Late Tertiary age, solution began along escarpments as the Pecos River and its tributaries cut through the less soluble cover. As these escarpments retreated, sinkholes, solution shafts, and other karst features developed.

With the lowering of base level after the first stage of karstification in thick sequences of carbonate rocks, especially with consolidated formations, as the carbonate rocks of Paleozoic age in the U.S.A., the initial sinkhole pattern may not be inherited by the underlying carbonate rocks. That may be due, in part, to differences in the joint and fracture patterns with increased depth (Secor, 1971, p. 699). Also, as suggested by J.G. Newton (pers. commun.,

1972) modification of the solution system in the bedrock after lowering of base level might result in variations between original sinkhole patterns and what is inherited by underlying rocks. Examples of changes in sinkhole patterns in later stages of karstification can be seen in carbonate terranes of the Ozark Uplift in Missouri, the Nashville dome in central Tennessee, and the Cumberland Plateau in Alabama. In these regions the sinkhole pattern that developed in the initial karstification in the carbonate rocks of Mississippian age, as the overlying clastic rocks of Pennsylvanian age were removed, was not inherited by the underlying carbonates. In later stages of karstification, as base level was lowered, fewer sinkholes developed. Another example of differences in the sinkhole patterns is in two karst areas in southern Indiana where the Pennsylvanian clastic rocks have been removed by erosion exposing Mississippian carbonate and older Paleozoic rocks on the west flank of the Cincinnati arch (Stringfield et al., 1977, pp. 20, 47). As the scarps retreated, a sinkhole plain (Mitchell plain) formed on the Mississippian rocks where there are as many as 1000 sinkholes within 5.9 km^2 (1 mi.^{2}) in Orange County, Indiana (Thornbury, 1954, p. 320). Farther east, however, in the areas where the Mississippian rocks have been removed by erosion, the older Paleozoic rocks did not inherit the sinkhole patterns of the Mississippian rocks farther west. Where the older formations did not inherit the sinkhole patterns, solution of the rocks was chiefly on the surface, and conditions were not favorable for the development of secondary permeability in the subsurface.

In the Tertiary carbonate rocks of Florida, Puerto Rico and Jamaica, movement of water from the surface down through joints, fractures and other openings, and out to a discharge area initiated a circulation system that formed lateral solution channels and other openings below the water table. In these areas, the sinkhole pattern that developed in the initial stage of karstification was inherited by the underlying carbonate rocks as base level and the top of the zone of saturation were lowered.

There is general agreement that sinkholes, solution shafts, and other vertical openings, in the zone of aeration, increase the hydraulic permeability of carbonate rocks in a vertical direction. They form as the circulation system develops lateral solution channels and openings which enhances the potential for increased vertical movement of water. Some deep movement of water occurs in aquifers in which the present circulation system has renewed movement in permeable zones of paleokarst, as in the principal artesian aquifer system of Tertiary age (Floridan aquifer in Florida) in the southeastern U.S.A.

EFFECTS OF GEOLOGIC STRUCTURE

In compact carbonate rocks, such as some Paleozoic rocks, where there is no intergranular permeability, joints or fractures in the rocks are essential for the initiation of downward percolation of water. Also, joints or fractures and bedding planes may be essential in the initiation of lateral movement of water in the zone of saturation. Where there is intergranular permeability, as occurs in some carbonate rocks of Tertiary age, a circulation system may develop independent of joints and fractures.

If conditions of recharge and discharge are favorable, as described by Reeves (1932) and Cady (1936) in the Shenandoah Valley of Virginia, groundwater may move parallel to the dip from a recharge area on the crest of an anticline to a discharge area on the flank of an adjacent syncline. However, the movement of water in most carbonate rocks is not necessarily parallel to the dip or in the direction of the dip. Even under artesian conditions, as in the Tertiary artesian aquifer, which dips to the east in southeastern Georgia, the general direction of movement of the water is northeast toward a discharge area in the Atlantic Ocean northeast of Savannah.

Although the major geologic structures which extend to the land surface where erosion and solution exposed carbonate rocks, such as limestone of the Ocala uplift in Florida, have provided favorable conditions for karst, the hydraulic gradient and direction of the movement of the water may be lateral and therefore independent of the dip. For example, in the Polk County recharge area in central Florida (Stringfield, 1936, p. 148, pl. 12), on the south flank of the Ocala uplift, lateral movement of the water takes place in all directions from the center of the recharge area; the geologic structure has little or no effect on the direction of movement.

On breached anticlines and domes such as the Sequatchie anticline on the Cumberland plateau in Tennessee and Alabama (Milici, 1967; Hooks, 1969) and Burkes Garden, a dome in Virginia (Cooper, 1944), carbonate rocks were exposed to meteoric water as the clastic rocks on the crests of the structures were removed by erosion. After erosion removed a sufficient amount of clastic rocks to permit water to reach the underlying carbonate rocks, the upper part of the carbonates became saturated with meteoric water. Solution of the carbonate rocks began as a groundwater circulation system developed. Where the clastic cover surrounding the exposed, carbonate rocks contains relatively impervious beds, and the dip is relatively steep, the water in the carbonate rocks may not move out of the area except as surface overflow across the clastic rocks to a valley or surface stream. Such outlets from the carbonate area are lowered by erosion of the clastic rocks. Erosion of a breached dome may form a circular basin such as Burkes Garden. As the carbonate area is eroded by solution on breached domes, another outlet may develop at a lower level, leaving the former outlet as a water gap.

On a breached anticline, such as the Sequatchie anticline in Tennessee and Alabama, a good circulation system may not develop in the exposed carbonate rocks before a perennial stream, such as the Tennessee River, cuts the anticline and provides a discharge area. Solution of the carbonate rocks formed the Sequatchie Valley parallel to the longer axis of the anticline. The walls of the valley, capped by less soluble clastic rocks, retreated as solution removed the carbonate rocks. On both flanks of the anticline, however, the carbonate rocks dip below the circulation systems of the groundwater, and solution of the carbonate rocks is confined to the valley and edges of adjacent escarpments along the axis of the anticline. The Sequatchie River rises in the northern part of the valley in Tennessee where the erosion is removing the clastic cover overlying the carbonate rocks. The river follows the valley and becomes a tributary of the Tennessee River near Chattanooga, Tennessee, where it enters the karst valley from the east and flows southward many kilometers along the axis of the anticline into Alabama. In Alabama it eventually turns west through a valley which cuts across the clastic escarpment.

The general direction of the movement of both groundwater and surface water in the valley is parallel to the strike of the formations. At the northeast end of the anticline in Tennessee where the clastic cover of the Pennsylvanian sandstones, conglomerates, siltstones, and shales has not been removed completely from the underlying Mississippian carbonate rocks on the anticline, large coalesced sinkholes, or uvalas that are locally called "coves" are surrounded by escarpments of Pennsylvanian rocks (Milici, 1967, p. 183). Mississippian shales form the slopes of the coves. Underground solution and erosion will eventually cause the coves to connect with the Sequatchie valley. Similar conditions of karstification are present at the southern end of the anticline in northern Alabama. Fig. 3 represents a cross section of the valley at Blount Springs, Alabama.

Milici (1967, pp. 190—193) considers the inversion of topography from anticlinal hills and mountains to the anticlinal Sequatchie Valley began when the Pennsylvanian formations were breached and the underlying carbonate rocks were exposed along the crest of the anticline about 1.2—1.5 km (4000—5000 ft.) above the present valley floor. The valley was subsequently developed by headward erosion of drainage that was tributary to the Tennessee River and by underground solution of structurally elevated carbonate rocks where overlying formations were breached. The Tennessee River eroded headward across the Sequatchie anticline in Mesozoic time and initiated the drainage of the Sequatchie Valley (Milici, 1967, p. 183).

Other areas in which regional movement of water parallel to the strike of the geologic structures has been observed include carbonate rocks on the flanks of the Black Hills in South Dakota (Rahn and Gries, 1973, p. 17, pl. 3) and the Camp Branch anticline in Shelby County, Alabama.

Under some conditions a syncline may affect the circulation of the water and solution of the rocks; for example, in the Huntsville area in northern Alabama, where cavernous limestone of Paleozoic age overlies the Chattanooga Shale, LaMoreaux and Powell (1960, p. 370) found that, not only joint and bedding planes, but also local folds and differences in regional dip influenced the direction of movement of the groundwater. In that area where the Chattanooga Shale serves as a relatively impervious floor, general movement of the water is to the south and southeast. Under these conditions the solution of the carbonate rocks was most rapid in the trough of a syncline which is parallel to the regional structure (Stringfield, 1966, p. 133).

Faults may restrict the lateral movement of water, especially if water-bearing beds are faulted against relatively impervious beds. Conversely, some

326

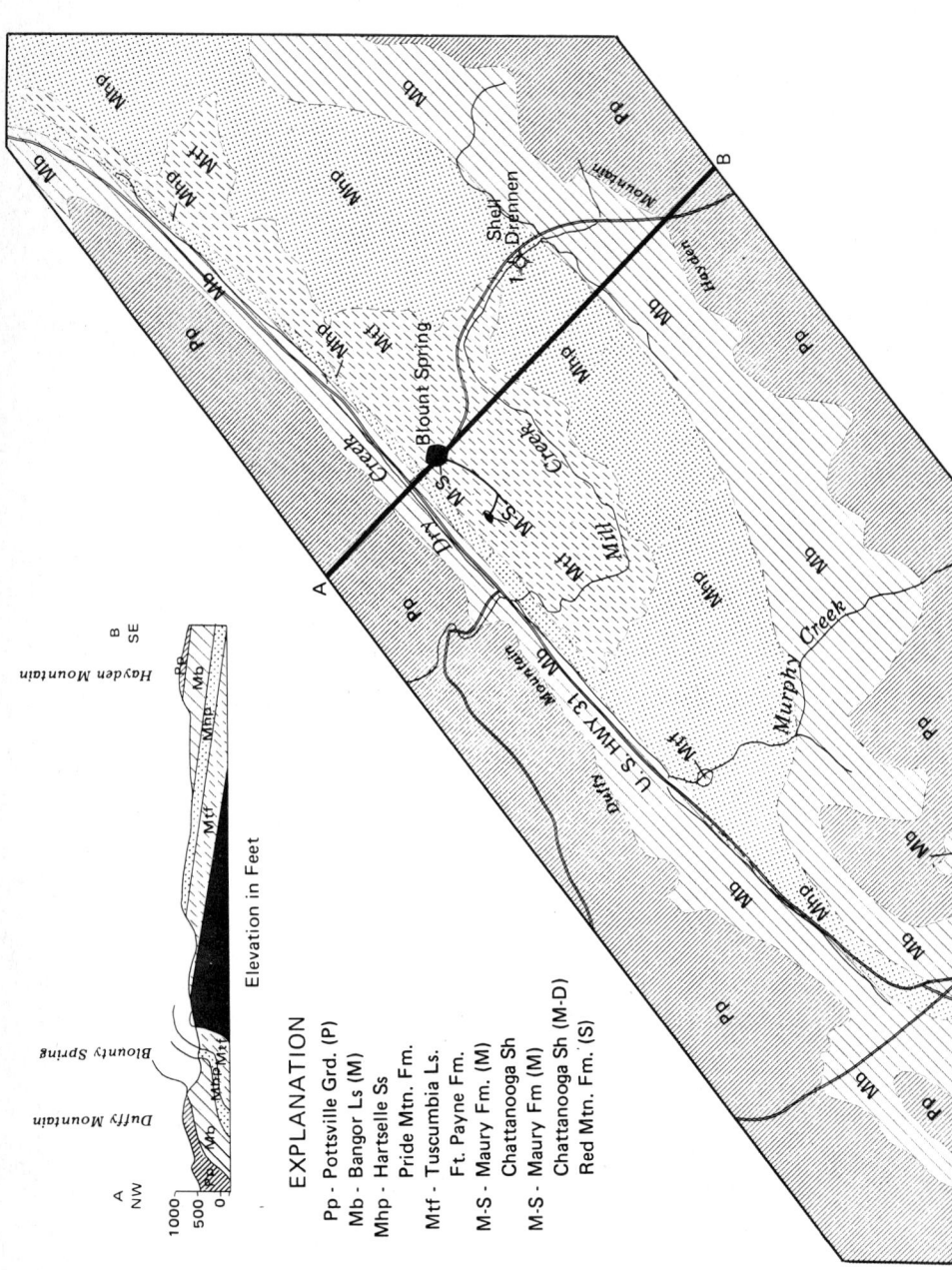

Fig. 3. Geologic map and cross-section of the Sequatchie anticline, Blount Springs, Alabama (from Hooks, 1969, p. 29).

fault zones may serve as avenues through which the water may move. As concluded by Livingston et al. (1936), and subsequent investigations, including Maclay and Small (1976, p. 21), faults in the Balcones fault zone in the San Antonio area, Texas, have had a pronounced effect on the circulation of water in the Edwards aquifer which is one of the most productive carbonate aquifers in the U.S.A. A large part of the recharge to the aquifer is due to downward movement of water along faults where surface streams flow into the area. Down-gradient in the discharge area of the aquifer, water moves laterally along fault zones to the land surface and form springs. Apparently, circulation of the water in the aquifer began after faulting had started and continued as the fault blocks of the Cretaceous Edwards Limestone and associated formations moved into their present position. The lateral circulation of the water probably began in a paleokarst zone in the aquifer. Down-gradient, the discharge from the aquifer system was, and still is, along fault zones. The early circulation system flushed out the saline water and increased the permeability in the productive parts of the aquifer. Although the paleokarst extends laterally into the saline water zone, the early circulation system did not have sufficient head to move farther down-gradient beyond the faults that serve as discharge areas.

KARST AQUIFERS

Karst aquifers, chiefly carbonate rocks, range in thickness from a few meters to more than 300 m. They range in areal extent from a few square kilometers to hundreds of square kilometers. Among the thickest and most extensive is the principal artesian aquifer system of Tertiary age which underlies all of Florida, southeastern Georgia, and adjacent parts of Alabama and South Carolina. In Florida it is known as the Floridan aquifer.

The water-bearing capacity of karst aquifers depends to a large extent on the enlargement of openings by solution. The least productive aquifers are those in which water in the zone of saturation is in openings that have not been enlarged by solution. The most productive aquifers are those in which the water in a thick zone of saturation occupies large interconnecting solution openings.

Karst aquifers may be placed in three groups according to water-bearing capacity as follows:

(1) Aquifers of low yield to wells.
(2) Aquifers in which the yield of water to springs and wells ranges from low to intermediate.
(3) Aquifers in which the yield of water to springs and wells ranges from intermediate to very large.

The water in aquifers of group *1* occurs chiefly in joints and fractures that have not been appreciably enlarged by solution. There has been little or no solution in the subsurface in the part of the aquifer occupied by the zone of saturation. The occurrence and movement of the water is comparable to that in igneous and metamorphic rocks in the Piedmont Province in the eastern

U.S.A. In general, the shape of the water table conforms to the topography of the land surface. The water table is highest in recharge areas on the hilltops and is nearer to the surface in discharge areas in the valleys.

Aquifers of group 2, with low to intermediate yields to wells and springs, are those in which water occurs in joints and fractures with some cavities and channels enlarged by solution. Typically, the zone of saturation is thin and the water is nonartesian. The yield of wells decreases with depth. Below the freshwater circulation system salty water may be present. There may be a large range in fluctuations of the water table except in discharge areas. Some wells may yield little or no water. The most productive wells are those that reach open solution channels. This type of aquifer is common in Paleozoic carbonate rocks, such as the Mississippian rocks in the Mammoth Cave region of Kentucky and in southern Indiana. Both areas are adjacent to solution escarpments of clastic formations which overlie the carbonate rocks (Fig. 1). The sinkhole plain, and the secondary permeability of the aquifer due to solution, developed as the solution escarpments retreated to their present position. The carbonate rocks dip gently on the flank of the Cincinnati arch, a large regional structure west of the Appalachian Mountains. The sinkhole plain is in the area of the youngest carbonate rocks exposed by retreating solution escarpments. Up-dip, where karstification has removed the younger rocks, older carbonate rocks, which did not inherit the sinkhole pattern that developed in the first stages of karstification, are exposed. These formations are less productive as aquifers because there has not been sufficient solution in the subsurface to form a circulation system and to develop secondary permeability. Where water occurs in open joints and fractures that extend to considerable depths with a thick zone of saturation, such as the London Chalk in England, the water may occur under both water-table and artesian conditions. The yield of wells is larger than that of similar aquifers in which the thickness of the zone of saturation is not as great.

Aquifers of group 3 are those in which the yields of wells and springs range from intermediate to very large. In this group are five of the most productive aquifers systems in the U.S.A.: (1) the principal artesian aquifer of Tertiary age (known as the Floridan in Florida) in the southeastern states; (2) a shallow nonartesian aquifer known as the Biscayne aquifer, chiefly of Quaternary age in southeastern Florida; (3) aquifers of Paleozoic age on the flanks of the Ozark Uplift in southern Missouri; (4) an aquifer system consisting of the Edwards Limestone and associated formations of Cretaceous age on the Edwards Plateau in the San Antonio area of Texas; and (5) the Roswell artesian aquifer of Paleozoic age in southeastern New Mexico. All these aquifer systems except the Biscayne, include buried paleokarst.

The first stages of karstification in these areas of productive aquifers were similar to those of the less productive aquifers in group 2. However, the present circulation system of the more productive aquifers reached buried paleokarst in which the secondary permeability was reactivated. In aquifers of this type the permeability may increase with depth. The water occurs under both artesian and nonartesian conditions. Movement of the water in some parts of the aquifer is through conduits, tubular openings, and other cavities similar to those of less productive aquifers in group 2 and also in the zone of aeration above the water table. Where the conduits and cavities are in the upper part of the zone of saturation, movement of water may be similar to that in the conduits and cavities in the upper part of the zone of saturation in group 2. In areas under water-table conditions where there is recharge through sinkholes and other openings, the water may become turbid in the zone of saturation after rains; the water table will then rise and the discharge of springs will increase. Where there are conduits and cavities above the water table, the recharge water moving downward from the surface may move a considerable distance laterally through conduits before reaching the zone of saturation. In both types of aquifers the water in the conduits in the upper part of the zone of saturation may be turbulent. However, flow through conduits and cavities under natural conditions below the upper part of the zone of saturation and under artesian conditions is laminar. In view of the fact that some aquifers, such as the principal artesian aquifer in southeastern U.S.A., have both types of "conduit flow", it is misleading to classify an aquifer as a "conduit type" where the water is turbulent in the conduits and as a diffused type where the water is laminar.

SUMMARY

Karst is a term used to describe the results of the processes of solution and leaching of limestone, dolomite, gypsum salt and other soluble rocks. Karst, a German word, is from the Indo-European term "kar", meaning rock, and the Slovenian word "kras" as used in the classic karst region in Yugoslavia where carbonate rocks are exposed.

Karst features may be divided into two groups as follows: (1) surficial features that do not extend far below the surface; and (2) deep karst features such as sinkholes, solution shafts, solution cavities and caverns. Important controlling factors responsible for development of karst and permeability in carbonate rocks include climate, the conditions under which the rocks are first exposed to meteoric water, solubility of the rocks, geologic structure, underground circulation of water and base level.

Movement of water from the surface of carbonate rocks down through joints, fractures, and other openings in the rocks and out to discharge areas initiates a circulation system that forms lateral solution channels and other openings below the water table.

Sinkholes, solution shafts and other vertical openings, which increase the hydraulic permeability of the carbonate rocks in a vertical direction, form in the zone of aeration as the circulation system develops lateral solution channels and openings. Deep movement of water occurs chiefly in aquifers in which the present circulation system has renewed movement in permeable zones of paleokarst, as, for example, in the principal artesian system of

Tertiary age (Floridan aquifer in Florida) in the southeastern U.S.A.

Where carbonate rocks are not horizontal, water may tend to move approximately parallel to the dip. However, the direction of the movement of water in carbonate rocks is not necessarily parallel to the dip or in the direction of the dip. On breached anticlines such as the Sequatchie anticline on the Cumberland Plateau in Tennessee and Alabama, the general direction of the movement of the water is parallel to the strike.

Faults may restrict the lateral movement of water, especially where water-bearing beds are faulted against relatively impervious beds. Conversely, some fault zones serve as avenues through which the water moves.

Karst aquifers may be placed in three groups according to yield. The aquifers with the largest yield include five of the most productive in the U.S.A. The water-bearing beds of all but one of these productive aquifers contain paleokarst in which the permeability has been reactivated and enlarged by the present circulation system.

REFERENCES

Back, W. and Hanshaw, B.B., 1965. Chemical geohydrology. In: V.T. Chow (Editor), Advances in Hydroscience, Vol. 3, Academic Press, New York, N.Y., pp. 49—109.

Back, W. and Hanshaw, B.B., 1970. Comparison of chemical hydrogeology of the carbonate peninsulas of Florida and Yucatan. J. Hydrol., 10: 330—368.

Briggs, R.P., 1966. The blanket sands of northern Puerto Rico. 3rd Caribbean Conf., Kingston, 1962. Jam. Geol. Surv., Publ. 95: 60—69.

Burdon, D.J. and Papakis, N., 1963. Handbook of karst hydrogeology with special reference to the carbonate aquifers of the Mediterranean region. Inst. Geol. Subsurface Res., Athens, U.N. Spec. Fund. Mimeogr., 276 pp.

Cady, R.C., 1936. Ground-water resources of the Shenandoah Valley, Virginia, with analysis by E.W. Lohr. Va. Geol. Surv., Bull. 45, 137 pp.

Carson III, R.G., 1964. Tolleys Cave, Virginia, Va. Dep. Conserv. Econ. Dev., Div. Miner. 10(1), 7 pp.

Chinese Academy of Geological Sciences, 1976. Karst of China. Inst. Hydrogeol. Eng. Geol., People's Publ. House, Shanghai.

Cooper, B.N., 1944. Geology and mineral resources of the Burkes Garden quadrangle, Virginia. Va. Geol. Surv., Bull. 60, 299 pp.

Corbel, A., 1959. Érosion en terrain cabaire — vitesse d'érosion et morphologie. Ann. Géogr., 366: 97—120.

Dilamarter, R.R. and Csallany, S.C. (Editors), 1977. Hydrologic Problems in Karst Regions. Western Kentucky University Press, Bowling Green, Ky., 481 pp.

Fish, J., 1972. Deep caves of the Americas. Can. Cover, 4(1): 39.

Flint, D.E., Corwin, G., Dingo, M.G., Fuller, W.P., MacNeil, F.S. and Saplis, R.A., 1953. Limestone walls of Okinawa. Geol. Soc. Am. Bull., 64: 247—1260.

Gilmer, F.W., 1818. On the geological formation of the Natural Bridge of Virginia. Am. Philos. Soc. Trans., 1: 187—192.

Gvozdetsky, N.A., 1972. The Problems of Karst Study and Practice. Nauka, Moscow, 340 pp.

Herak, M. and Stringfield, V.T. (Editors), 1972. Karst — Important Karst Regions of the Northern Hemisphere. Elsevier, Amsterdam, 551 pp.

Herrick, S.M. and LeGrand, H.E., 1964. Solution subsidence of a limestone terrane in southwest Georgia. Int. Assoc. Sci. Hydrol. Bull., 9(2): 25—36.

Hooks, W.G. (Editor), 1969. The Appalachian structural front in Alabama. Ala. Geol. Soc. Guideb., 7th Annu. Field Trip, 69 pp.

Howell, J.V. (Editor), 1957. Glossary of Geology and Related Sciences. Am. Geol. Inst., Washington, D.C., 325 pp.

Isphording, W.C., 1974. Weathering of Yucatan limestones; the genesis of terra rosas. In: A.E. Weidie (Editor), Field Seminar on Carbonate Rocks of the Yucatan peninsula. New Orleans Geol. Soc., New Orleans, La., pp. 78—93.

Jacobson, R.L. and Langmuir, D., 1970. The chemical history of some spring waters in carbonate rocks. Ground Water, 8(2): 5—10.

Jennings, J.N., 1971. Karst. Mass. Inst. Technol. Press, Cambridge, Mass., 252 pp.

LaMoreaux, P.E. and Powell, W.J., 1960. Stratigraphic and structural guides to the development of water wells and well fields in a limestone terrane. Int. Assoc. Sci. Hydrol. Bull., 52: 363—375 (reprinted by Ala. Geol. Surv., Repr. Ser. 6).

LaMoreaux, P.E., Raymond, D. and Joiner, T.J., 1970. Hydrology of limestone terranes — annotated bibliography of carbonate rocks. Ala. Geol. Surv. Bull., Part A, 94, 242 pp.

LaMoreaux, P.E., LeGrand, H.E., Stringfield, V.T. and Tolson, J.S., 1975. Progress of knowledge about hydrology of carbonate terranes. Ala. Geol. Surv., Bull., Part E, 94: 1—27.

Langmuir, D., 1971. The geochemistry of some carbonate ground waters in central Pennsylvania. Geochim. Cosmochim. Acta, 35: 1023—1047.

Livesay, A. and McGrain, P., 1962. Geology of Mammoth Cave, National Park Area. Univ. Ky., Spec. Publ. No. 7.

Livingston, P.P., Sayre, A.N. and White, W.N., 1936. Water resources of the Edwards limestone in the San Antonio area. Texas. U.S. Geol. Surv., Water-Supply Pap. 773-B, 55 pp.

Maclay, R.W. and Small, T.A., 1976. Progress report on geology of the Edwards aquifer, San Antonio area, Texas, and preliminary interpretation of borehole geophysical and laboratory data on carbonate rocks. U.S. Geol. Surv., Open-File Rep. 76-627, 65 pp.

McAllister, B., 1976. In: *The Washington Post*, February, 1976, p. C2.

Milici, R.C., 1967. The physiography of Setquatchie Valley and adjacent portions of the Cumberland Plateau, Tennessee. Southeast. Geol., 8(4): 179—193 (also Tenn. Dep. Conserv. Dev., Div. Geol., Rep. Invest. No. 22, 1968).

Monroe, W.H., 1966. Formation of tropical karst topography by limestone solution and reprecipitation. Caribb. J. Sci., 6(1—2): 1—7.

Monroe, W.H., 1968. The karst features of northern Puerto Rico. Natl. Speleol. Soc. Bull., 30(3).

Monroe, W.H., 1970. A glossary of karst terminology. U.S. Geol. Surv., Water-Supply Pap. 1899-K, 26 pp.

Monroe, W.H., 1974. Dendritic dry valleys in the cone karst of Puerto Rico. U.S. Geol. Surv., J. Res., 2(2): 159—163.

Monroe, W.H., 1976. The karst landforms of Puerto Rico. U.S. Geol. Surv., Prof. Pap. 899, 69 pp.

Moravec, G.I., 1974. Development of karren karst forms on the Newala Limestone in Cahaba Valley, Alabama. In: H.W. Raugh and E. Werner (Editors), Proceedings of the Fourth Conference on Karst Geology and Hydrology. W.Va. Geol. Econ. Surv., pp. 113—121.

Petrik, M. and Herak, M. (Editors), 1969. Krš Jugoslavije Carsus Iugosl., 622 pp.

Pohl, E.R., 1955. Vertical shafts in limestone caves. Natl. Speleol. Soc., Occas. Pap. 2, 24 pp.

Powell, R.L., 1961. Caves of Indiana. Ind. Indiana Geol. Surv., Circ. 8, 127 pp.

Rahn, P.H. and Gries, J.P., 1973. Large springs in the Black Hills, South Dakota and Wyoming. S.D. Geol. Surv., Rep. Invest. 107, 48 pp.

Rapp, J.R., 1957. Hueco Plateau area, Otero County, New Mexico. U.S. Geol. Surv., Opne-File Rep., 11 pp.

Rauch, H.W. and Werner, E. (Editors), Proceedings of the Fourth Conference on Karst Geology and Hydrology. W.Va. Geol. Econ. Surv., 185 pp.

Rauch, H.W. and White, W.B., 1970. Lithologic controls on the development of solution porosity in carbonate aquifers. Water Resour. Res., 2(4): 1175—1192.

Reeves, F., 1932. Thermal springs of Virginia. Va. Geol. Surv., Bull. 36, 185 pp.

Roglic, J., 1972. Historical review of morphologic concepts. In: M. Herak and V.T. Stringfield (Editors), Karst — Important Karst Regions of the Northern Hemisphere, Elsevier, Amsterdam, pp. 1—18.

Secor, Jr., D.T., 1971. Extension fracturing in rocks with internal pore pressure. Geol. Soc. Am., Abstr. Program, 3(7): 699 (abstract).

Shuster, E.T. and White, W.B., 1971. Seasonal fluctuations in its chemistry of limestone springs: a possible means for characterizing carbonate aquifers. J. Hydrol., 14: 93—128.

Silar, J., 1965. Development of tower karst of China and North Vietnam. Natl. Speleol. Soc. Bull., 27(2): 35—46.

Smith, Jr., J.F. and Albritton, Jr., C.C., 1941. Solution effects on limestone as a function of a slope. Geol. Soc. Am. Bull., 52: 61—78.

Stringfield, V.T., 1936. Artesian water in the Florida Peninsula. U.S. Geol. Surv., Water-Supply Pap. 773-C: 115—195.

Stringfield, V.T., 1966. Artesian water in Tertiary limestone in the southeastern States. U.S. Geol. Surv., Prof. Pap. 517, 226 pp.

Stringfield, V.T. and LeGrand, H.E., 1969a. Hydrology of carbonate rocks terranes — a review. J. Hydrol., 8(3): 349—411.

Stringfield, V.T. and LeGrand, H.E., 1969b. Relation of sea water to fresh water in carbonate rocks in coastal areas with special reference to Florida and Cephalonia (Kephallinia), Greece. J. Hydrol., 9(4): 387—404.

Stringfield, V.T. and LeGrand, H.E., 1971. Effects of karst features on circulation of water in carbonate rocks in coastal areas. J. Hydrol., 14: 139—157.

Stringfield, V.T. and LeGrand, H.E., 1974. Karst hydrology of Northern Yucatan Peninsula, Mexico. In: A.E. Weidie (Editor), Field Seminar on Water and Carbonate Rocks of the Yucatan Peninsula, Mexico, New Orleans Geol. Soc., New Orleans, La., p. 274 pp. 25—44 (also in: A.E. Weidie and W.C. Ward (Editors), Carbonate Rocks and Hydrogeology of the Yucatan Peninsula, New Orleans, Technological Soc., New Orleans, La., pp. 191—210.

Stringfield, V.T., LeGrand, H.E. and LaMoreaux, P.E., 1977. Development of karst and its effects on the permeability and circulation of water in carbonate rocks, with special reference to the southeastern States. Geol. Surv. Ala., Bull., Part G, 94, 68 pp.

Sweeting, M.M., 1972. Karst Landforms. Macmillan, London, 362 pp.

Thornbury, W.D., 1954. Principles of Geomorphology. Wiley, New York, N.Y., 618 pp.

Thrailkill, J., 1968. Chemical and lithologic factors in the excavation of limestone caves. Geol. Soc. Am. Bull., 79: 19—46.

Thrailkill, J., 1972. Carbonate chemistry of aquifer and stream water in Kentucky. J. Hydrol., 16: 93—104.

Tolson, J.S. and Doyle 4, F.P. (Editors), 1977. Karst Hydrology. University of Alabama Press, Huntsville, Ala., 578 pp.

UNESCO—FAO, 1972. Glossary and Multilingual Equivalents of Karst Terms. UNESCO, Paris, 72 pp.

Warren, W.M. and Moore, J.D., 1975. Annotated bibliography of carbonate rocks. Ala. Geol. Surv., Bull., Part E, 74: 31—168 (supplement to Part A, 1970).

Wayne, W.J., 1950. Description of the Indiana karst. Compass, 27(4): 215—223.

White, P.T., 1968. The Mekong, river of terror and hope. Natl. Geol. Mag., 134(6): 737—787.

Williams, P.W., 1970. Limestone morphology in Ireland. Ir. Geol. Stud., Belfast, 7: 105—124.

[2]

EVALUATION TECHNIQUES OF FRACTURED-ROCK HYDROLOGY

HARRY LEGRAND

331 Yadkin Dr., Raleigh, NC 27609 (U.S.A.)

(Accepted for publication February 6, 1979)

ABSTRACT

LeGrand, H., 1979. Evaluation techniques of fractured-rock hydrology. In: W. Back and D.A. Stephenson (Guest-Editors), Contemporary Hydrogeology — The George Burke Maxey Memorial Volume. J. Hydrol., 43: 333—346.

Areas in which fractured rocks are part of the hydrologic circulatory system are widespread. Certain aspects of the historical features of hydrology tend to influence fractured rocks, and the results of these influences in turn control certain aspects of applied hydrology. The dynamic changes in the fractured-rock system can be cast in spatial relations that are usefully studied in the field and in the context of their historical development. For example, circulating water in carbonate rocks causes enlargement of fractures by solution and the development of secondary permeability that further influences movement and storage of water; solutes moving in fractures of less soluble rocks control the degree of development of overlying insoluble residue, and to the degree present, this residue forms another type of subsurface water medium that must be considered with the fractured-rock system.

The treatment of fracture systems of consolidated rocks in a way similar to porous granular media, in a context of fractures analogous to pores — strictly a matter of scale — is not an adequate base for analyzing problems of fractured-rock hydrology. A key problem is the uneven distribution of permeability, a condition resulting from geologic and hydrologic processes since consolidation of the rock. Classifying the patterns of uneven distribution lead to appropriate models on which useful mean values and deviations can be described. Reconstruction of the hydrogeologic history of a fractured-rock system provides an elegant foundation for the evaluation of the existing hydrology.

INTRODUCTION

In the computer age, when hydrologic problems are attacked by mathematical techniques, a student might consider somewhat primitive a technique using inferences to evaluate the hydrology of fractured rocks. Yet, it is the purpose of this paper to show that an adequate appraisal of the hydrology of fractured rocks must depend on a continual mixture of data collection, observations and hydrogeologic inferences. More particularly, emphasis is placed on reconstructing the geologic and hydrologic history as an essential start to understanding and modeling the fractured-rock system.

The concentration of research on the hydrology of porous granular sands, clays and gravels has overshadowed research on the hydrology of fractured rocks; yet, fractured rocks are far more widespread in areal distribution near the land surface. This apparent disproportionate attention to granular materials is due partly to the fact that saturated sand deposits are the most common good aquifers. Moreover, research workers have found porous granular materials to be interesting and amenable to sophisticated study.

Can the advanced technology applicable to the hydrology of porous granular materials be applied to the hydrology of fractured rocks? An affirmative answer is partly correct, because the same hydrodynamic principles apply to both rock systems. Yet, an objective here is to show that treatment of fracture systems of consolidated rocks as porous media with fractures analogous to pores — strictly a matter of scale — is not an adequate base for analyzing problems of fractured-rock hydrology. An overview of the hydrology of fractured rocks follows.

Some problems that may arise in fractured-rock systems are difficulties of: (1) mapping hydrologic features except on large-scale maps; (2) interpolating and extrapolating hydrologic conditions; (3) determining: (a) the uneven distribution of permeability; and (b) the relation of surface sand and clay media with the underlying rock-fracture media; (4) predicting groundwater supply; (5) predicting waste management potential; and (6) determining water-level characteristics.

FRACTURED-ROCK SYSTEM

Igneous and metamorphic rocks and consolidated sedimentary deposits tend to be fractured to some extent where they occur within several hundred meters of the land surface. In these rocks the chief avenues for movement of water are the planar openings that are referred to interchangeably as joints or fractures.

The distribution and character of fractures can be viewed in several ways, according to: (a) texture of permeability; (b) changes with depth; and (c) degree of enlargement by solution. As regards texture of permeability, the fractures may be considered rather evenly and closely spaced, unevenly and sparsely distributed, or unevenly spaced and elongated. The fractures can be viewed according to their distribution with depth below the top of rock; some fractures extend to considerable depth, but most show a decrease in size and number with increasing depth. An additional distinction can be made on the basis of enlargement of fractures by solution; rocks of low solubility, such as shales and granites, have fractures that are not enlarged by solution, whereas carbonate rocks near land surface and within the groundwater circulation system may have cavernous openings that resulted from solution along fractures.

Consolidated rocks underlie at some depth all land areas, but those covered deeply by unconsolidated material may not be fractured, and they are of no concern to this discussion where they lie below the groundwater circulation system. Such conditions occur in the Atlantic and Gulf Coastal Plain, where hundreds and even thousands of meters of unconsolidated deposits lie over the bedrock. In most other regions the fractured rocks are: (1) exposed at the land surface; (2) covered by a layer of residual soil and decomposed rock; (3) covered by alluvium in some valleys; or (4) covered in former glaciated regions by some glacial deposits. The covered unconsolidated materials represent aquifers where moderately thick, and even where they are thin, they represent the upper part of the subsurface hydrologic system; such a system is complex because water entering as recharge or leaving as discharge passes somewhat massively through porous granular materials either after or before circulating through particular fractures in the rock.

DEVELOPMENT OF THE HYDROLOGIC SYSTEM

In contrast to the intrinsic and syngenetic permeability of sands, gravels and clays, the permeability of fractured rocks is epigenetic. Most consolidated rocks within a hundred meters of land surface contain one or more sets of systematic tension fractures that have developed or enlarged in the recent geologic past. Near the land surface the fractures tend to be larger or more pronounced than at deeper levels. Prior to the latest erosional history of an area, most of the present fractures were only hair-line cracks at considerable depth. As the erosional history is extended, the incipient fractures tend to be enlarged by solution or by other forces as erosion at the land surface removes overlying material. As overlying material is progressively removed by erosion, the rocks that were once deeply buried get closer to the land surface and, in the process, develop tension fractures and become part of the subsurface water circulation system.

The continual erosion through geologic time causes a concomitant lowering of stream valleys, of interstream areas, and of the water table. Thus, fractures which at slightly earlier geological time were below the water table ultimately are above the water table; as erosion progresses, rocks containing these fractures are destroyed by erosion, and these destroyed fractured rocks are replaced in relative space by fractured rocks that appear to have "risen with respect to the land surface" and are now above or slightly below the water table. The lowerings: (1) of a stream valley; (2) of its nearby interstream area; and (3) of the water table are not necessarily in unison and are not at a steady and uniform rate. These imbalances from place to place lead to a range of topographic features, to a range in thickness of soil cover, and to a range in water-table gradients. Thus, the imbalances in erosion lead to a variety of hydrogeologic settings that are continually and dynamically changing, and the hydrology everywhere within these settings is changing.

Some hydrologic features of fractured rocks that are not common to porous granular material follow:

(1) In contrast to flat-lying and undisturbed unconsolidated granular materials, a particular fractured rock or formation may have been folded and

faulted in many cases to the extent that only its exposed truncated end has any circulation of subsurface water; the covered parts of the formation dip into the ground with no means of circulation and discharge of water at great depths. Under such conditions, each type of rock may have its own fracture characteristics, and some hydrologic boundaries may coincide with rock boundaries in the shallow subsurface.

(2) Storage capacities of fractured rocks are generally very small. Although some karstified carbonate rocks have networks of large caverns, these caverns are generally above the water table and may not therefore be available for storage of water.

(3) No simple, clearcut statement can be made about the transmitting capacity of fractured rocks, but there may be limiting conditions not present in loose porous granular materials. For example, in both soluble carbonate rocks and non-soluble rocks there is generally a pronounced decrease in size and number of fractures with depth; thus, permeability decreases with depth and the base of the interconnecting openings may be indistinct. In the lateral view some interconnecting fractures are much smaller than others, and where fractures are sparse these smaller fractures represent constrictions in flow. Artesian aquifers are limited because of limitations on continuity of fractures and difficulty of movement of water through constrictions.

(4) Residual and transported soils and weathered materials irregularly cover rocks in most places. The covering material of porous granular material results in a complex two-media hydrologic system, no part of which should be neglected in evaluation.

(5) The smooth curvilinear water table in loose porous granular materials can be conceived in fractured rocks where large areas are considered; yet, locally in fractured rocks the water table is discontinuous where fractures are not present, and in many places the water table has a step-like pattern.

CONCEPTUAL MODEL OF A SELECTED FRACTURED-ROCK SETTING

The wide range of hydrologic conditions and the variety of hydrologic-related problems in fractured-rock regions poses a question of proper methods to study and to attack the problems of such regions. One approach is to identify a region or setting that has some distinctive characteristics and then develop a conceptual model to which selected specific information can be used to advantage. Following are concise statements that can form a basis for a conceptual model of the settings of the Piedmont and Blue Ridge Province of the southeastern U.S.A., underlain by igneous and metamorphic rocks.

Igneous and metamorphic rocks underlie the Piedmont and Blue Ridge region of the southeastern U.S.A. as shown in Fig. 1. The following general statements by LeGrand (1967) apply:

(1) The climate is temperate and humid; average annual rainfall is slightly more than 100 cm and is fairly evenly distributed throughout the year.

(2) The region is underlain by igneous and metamorphic rocks; the rocks range in chemical composition between that of granite (mainly silica and silicates of aluminum and potassium) and that of gabbro (chiefly silicates of aluminum, iron, magnesium and calcium).

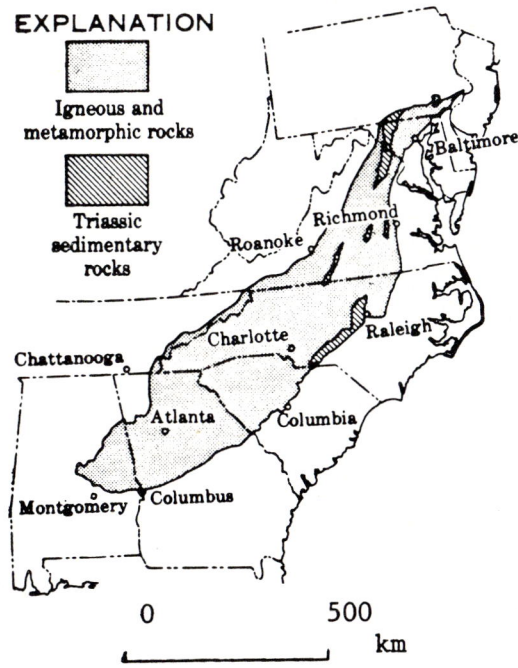

Fig. 1. Piedmont and Blue Ridge region of southeastern U.S.A. where fractured igneous and metamorphic rocks occur.

(3) A layer of residual weathered material lies on the fresh rock in most places; the thickness of the weathered material ranges from a feather edge to slightly more than 40 m (Fig. 2).

(4) Water occurs in two types of media: (1) clayey granular weathered material; and (2) underlying fractures and other linear openings in the bedrock.

(5) A close network of streams prevails, and in few places on an interstream area is a perennial stream more than 1 km away. A hill and dale topography occurs, commonly with gentle slopes.

(6) Toward each stream is a continuous flow of groundwater. Some of the outflowing groundwater is consumed as evapo-transpiration in valleys; the remainder discharges as small springs and as bank and channel seepage into the stream.

(7) Since all the perennial streams receive groundwater from adjacent interstream areas, streams are the linear sinks in the water table. This part of the

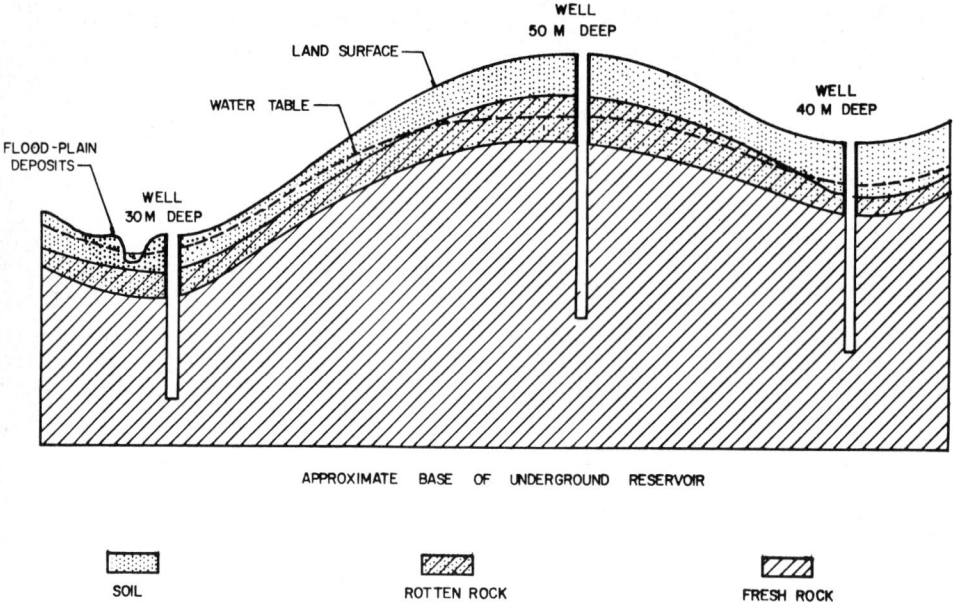

Fig. 2. Sketch showing a typical profile of underground conditions in the igneous and metamorphic rocks of the Piedmont and Blue Ridge Province of southeastern U.S.A.

water table is directly observable. The topography of the water table is similar to that of the land surface, but its relief is less. Thus, it is easy to construct synthetic water-table maps and to predetermine the general direction of the natural movement of groundwater.

(8) The path of natural movement of groundwater is relatively short and is almost invariably restricted to the zone underlying the gross topographic slope extending from the surface divide to the stream.

(9) From a point source of infiltration, water, or waste that might be with it, extends as a fan or expansive trail down-gradient toward the stream; its dispersal depends on the kind and degree of permeability, on the hydraulic gradient, and on the distance to the stream.

(10) Almost all recharge and discharge are through porous granular material (clayey soil or floodplain deposits), but some intermediate flow may be in bedrock openings.

(11) Bedrock fractures tend to decrease in size and number with depth, and in most places there is an insignificant storage and circulation of groundwater below a depth of 125 m.

(12) The saturated zone is not simple to define. Its top boundary is the water table, which lies in the clayey weathered material more often than not but which becomes discontinuous where it lies in fractured bedrock. The lower boundary is irregular and indistinct; it is represented by the base of the zone in which interconnecting fractures exist. The saturated zone is absent where

unfractured rocks crop out, but it is commonly 10—30 m thick. Water-yielding capacity within the zone ranges through several orders of magnitude; commonly it is less near the base of the saturated zone than near the top.

(13) The water table is near land surface in valleys and as much as 8—20 m below land surface beneath hills. The range of seasonal fluctuation of the water is as little as 1 m in valleys and as much as 4 m beneath hills.

(14) Many fractures are enlarged by the action of solution, especially in gneisses and schists containing silicates of calcium. Many of these enlarged fractures underlie "draws" or linear sags in the surface topography. These draws, representing zones of relatively high permeability in the bedrock, are sites for the best-producing wells.

(15) The yields of wells range from less than 4 l/min. to as much as 900 l/min.; the sustained yield of most wells is less than 400 but more than 20 l/min. The cone of pumping depression of a domestic well in one hectare does not generally extend to the cone of a pumping well a few hundred meters away.

(16) The yield of a well cannot be predetermined, but experience developed by using past records of yield from various topographic and rock conditions allows useful predictions based on degrees of probability.

(17) Two distinctive classes of groundwater are present. The first includes soft, slightly acidic water low in dissolved mineral constituents; water of this type comes from light-colored rocks resembling granite in composition, and includes granite, granite gneiss, mica schist, slate and rhyolite flows and tuffs. The second includes a hard slightly-alkaline water relatively high in dissolved solids; water of this type comes from dark rocks, such as diorite, gabbro, hornblende gneiss, and andesite flows and tuffs.

Although the list of generalizations can be extended, it is adequate to show that prior work and knowledge, properly abstracted, help to solve many hydrological problems of fractured rocks. If generalizations are to be useful they must be separated from masses of unwieldy data. To be superimposed on this background of knowledge are selected specific data and observations that may be readily available and maps of the pertinent area showing the geology, soils, topography and the water table. The data and observational features, coupled with proper inferences, should more closely delimit the possible answers.

APPROACHES TO EVALUATION

A fractured-rock system can be viewed as a porous medium with fractures analogous to pores — strictly of scale (Bredehoeft, 1971). This is a tack used with much good logic by hydrologists in considering the subject of aquifer performance. The determination of the behavior of aquifers or hydrologic systems is the main goal and end-product for many hydrologists. The system approach provides a good foundation for most problems of the hydrology of porous granular materials because interpolations and averaging of values are in general very useful. Yet, for problems in fractured rocks, an indiscriminate averaging of values may lead to errors in specific cases, and interpolations and predictions are not simple. Many small local actions appear to be only slightly related to the larger system. For example, a statement that the potential groundwater yield in a river basin in the Piedmont region underlain by fractured rocks is, say, 10^6 l day^{-1} km^{-2} cannot be meaningful to the present and future domestic well owners in the basin; these well owners may have some problems, but none can be related specifically to the "average yield" in the basin as a whole. An adequate domestic supply is commonly available, but the permeability is low and erratically distributed to the extent that overdevelopment of groundwater throughout a river basin in the region is unlikely. Thus, some local hydrologic assets and some real problems are not within the circumscription of the "whole-system" approach.

In highly-permeable porous granular materials the approach of considering an entire aquifer or large river basin as a unit is proper. However, in fractured rocks of low overall permeability this approach may get us nowhere. There may be a multitude of small, individual systems instead of a single big system that characterize the region. We may be unable to model the region without first modelling a small unit, and then we can show that there is good recurrence patterns of hydrologic conditions. In many instances there is an arterial distribution of permeability. Groundwater divides separate distinct but similar hydrologic units. Fig. 3 is a part of 7.5′ quadrangle in the Piedmont of North Carolina underlain by fractured igneous and metamorphic rocks; such a small drainage basin can be used as a model for other areas of the region. Conditions in each basin that has a perennial stream are similar to those in other basins of the entire Piedmont and Blue Ridge region, as described earlier (see Fig. 2). Thus, there are thousands of small but similar systems that do not need to be studied in detail separately because the similar recurrent hydrogeologic patterns provide excellent background information. The background information (precised in the conceptual model stated earlier in the paper) can be enmeshed with a few key bits of information about a site or area to provide the proper base for judgement.

The problems of scale are realistic. Conditions change locally within short distances, influencing the development of groundwater supplies, the management of wastes, and other concerns of man. The scale of mapping to convey information to others is important. In Fig. 3 the scale of mapping of useful hydrogeologic features is quite acceptable for the Piedmont and Blue Ridge regions. On a smaller scale map, covering an area as large as a county, the same features cannot be presented very well.

Some local conditions of great importance to man's use in fractured-rock regions may not be evident where an approach of evaluation of a large area is taken. For example, in a drainage basin study, hydrologists commonly take a water-budget approach in which the total annual precipitation is equated against the sum of the groundwater storage. In this approach, hydrologic values of an entire drainage basin above a position in a stream are averaged, and a good overview of the large-scale hydrology can be made. Yet, this

Fig. 3. A part of the Rural Hall topographic map in North Carolina showing the Ogburn Branch drainage basin. Both surface and subsurface hydrologic features in this area are similar to those throughout the entire Piedmont and Blue Ridge Province.

"averaging out at the gage" should be supplemented by determining the range of local hydrologic conditions, where the range is considerable. In regions underlain by fractured rocks the range of hydrologic conditions is likely to be great, chiefly because of the uneven distribution of permeability.

Analytical equations to obtain values of transmissivity, storage coefficient, and other hydraulic properties are widely used for sands, clays, and gravels, and such equations have some limited use with regard to fractured rocks where input data are reliable. The validity of a mathematical model should be carefully checked for each case to determine if the model can be usefully applied to a particular fractured-rock setting. A matter to be considered is that of scale, and a question to be asked is whether an averaging of permeabilities is valid for a solution to a particular problem. Averaging of permeabilities will oversimplify many problems. The composite two-media setting with an intergranular type of permeability in soil of varying thickness and with fracture permeability in the underlying rock is difficult to model mathematically. Moreover, different permeability regimes within the rock need consideration. The low or negligible permeability of the matrix rock represents one regime, and the fracture system represents another; another classification of regimes would bring out the distinction in many cases of relatively high permeability near the top of rock and of low permeability in a zone below. The distinction between the zone of high permeability and the underlying zone of low permeability is generally arbitrary, or course; yet, the lower zone commonly decreases exponentially in permeability with depth, and an arbitrary depth to the base of the zone must be considered in any case. The complex regimes of permeability place limitations on the usefulness of analytical equations.

The major approach taken here emphasizes inferential hydrogeology and relies on conceptual models based on best use of principles and of past knowledge with continual upgrading of inferences as new information becomes available. Helpful inferences can be developed from the fact that the variations of the fracture system with time have counterparts in space so that time and space features can be viewed somewhat mutually and interchangeably.

Stages in the development of karst topography were described by Grund (1914) and applied to the development in carbonate rocks by LeGrand and Stringfield (1973). In karst regions, stages of youth, maturity, and old age can be readily identified. This identification is not merely academic; an understanding of the stages of karst development provides much information about the distribution of permeability. Youthful, mature, and old age features can develop simultaneously, but at different places in the same carbonate-rock area. The progressive karst development leads toward large solution openings during the mature stage that results in a high permeability; the high permeability causes the water table to be depressed and results in many large solution cavities lying above the saturated zone and not necessarily available for storage of groundwater. Very useful conceptual models of the hydrology can be formulated for each stage of karst development and for its counterpart in space. A simplified model of the mature stage is shown in Fig. 4.

Other types of fractured rocks may have less distinctive space and time correlations in hydrologic development, but in all cases there are developments in topographic stages that related in some cases to development of permeability and that in all cases relate to recharge and discharge features of the system. If we can pick out the stages in erosion and the delicate imbalances in erosion, and if we can identify the stages in development of the landscape on fractured rocks, we have at our finger tips more hydrologic information than we realize. Reconstruction of the **hydrogeologic** history of a fractured-rock

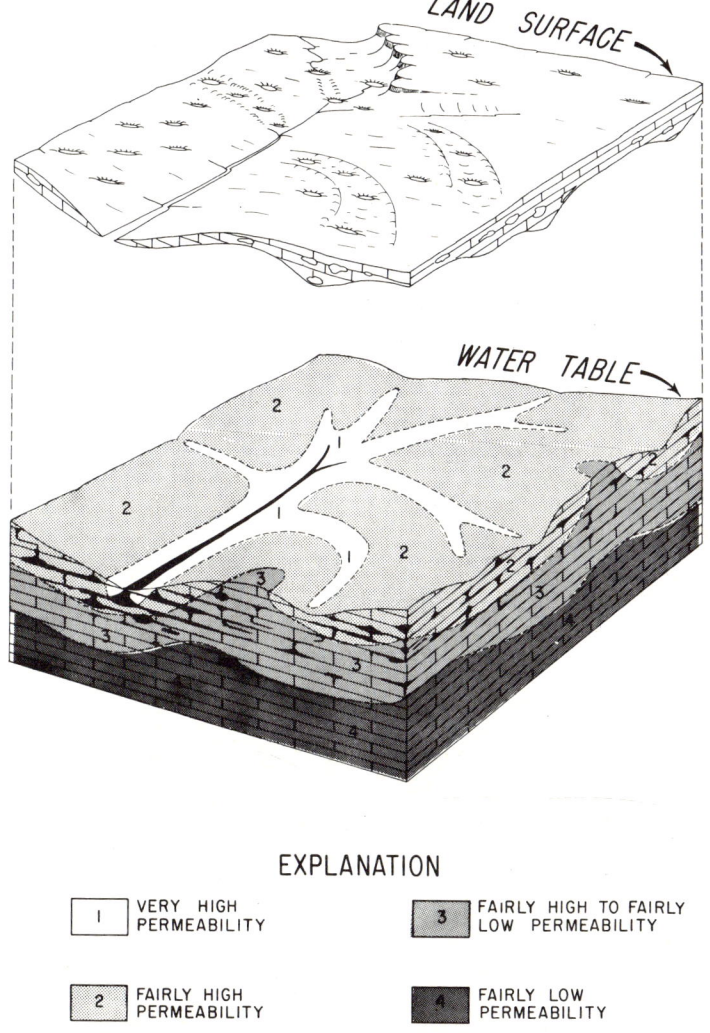

Fig. 4. Sketch for mature karst features showing idealized fully developed karst type aquifer. Large conduit, or arterial, permeability is shown in zone *1* and may occur also in other zones.

area is a useful and inexpensive method of evaluating the hydrology. This approach of developing inferences with emphasis on paleohydrology cannot be formularized into mathematical equations.

If we do not use these space and time differences to good advantage we are handicapped greatly by our inability to extrapolate and interpolate simply in the fracture—rock system. Simple arithmetic spatial extrapolations are not necessarily valid, as is generally the case in loose porous granular material. For example, the yields of five wells in profile down a topographic slope and spaced 35 m apart in fractured rocks of the Piedmont region of southeastern U.S.A. might be 60, 15, 100 and 150 l/min., respectively. Extrapolations based on a study of recurrence patterns of specific conditions and applied in a framework of probability may be reasonable; the recurrence patterns may be derived from a statistical study of spatial data, or they may be derived from knowledge of time changes in hydrogeologic conditions that can be transferred and applied to spatial patterns.

If we cannot interpolate and extrapolate simply in fractured rock systems, we face serious problems of prediction. There is an element of indeterminacy to the extent that we cannot be sure of the position and extent of each fracture, or each solution cavity in the karst area. In the context used here the principle of indeterminacy refers to (Leopold and Langbein, 1963):

> "those situations in which the applicable physical laws may be satisfied by a large number of combinations of values of interdependent variables. As a result, a number of individual cases will differ among themselves, although their average is reproducible in different samples. Any individual case, then, cannot be forecast or specified except in a statistical sense. The result of an individual case is indeterminate."

The principle of indeterminacy must be reckoned with in a systematic evaluation of fractured-rock hydrology, but the uncertainties need not be discouraging. Leopold and Langbein point out that as more is known about the processes operating and as more is learned about the factors involved, the range of uncertainty will decrease, but it never will be entirely removed. The need for reliance on hydrogeologic inferences is obvious.

An overview of the distribution of permeability of a fractured-rock setting needs to be made early. Whereas the relatively low gross permeability of most igneous and metamorphic rocks provides hydrologic patterns typified by the Piedmont region, the extremely high permeability of some karstified carbonate-rock settings provide other hydrologic patterns to be considered. Very highly cavernous and permeable unsaturated zones tend to keep the water table depressed below land surface in many karst regions, a condition that leads to a low density of perennial streams. The uneven distribution of permeability beneath surface karst streams causes them to lose or gain much water in short distances, depending on the position of the water table with reference to stream level. Other striking characteristics of some karst regions that are related to fracture permeability include: large springs rather than small springs and diffuse seepage, lack of concordance of topographic divides with water-table divides, and a shifting of groundwater divides away from major trunk streams during dry seasons (which leads to a temporary enlargement of some drainage basins and movement of groundwater as underflow from one valley to another beneath some upland areas). These examples show that there are patterns that can be identified when one has an understanding of the hydrogeologic framework and an understanding of the principles of karstification.

CONCLUSIONS

Some methods of study of the hydrology of porous granular materials can be used in the study of the hydrology of fractured rocks. Yet, the fracture type of permeability is generally somewhat erratic in distribution; since most studies depend in some way on permeability distribution, many predictions concerning the hydrogeology of fractured rocks must be made with guarded qualifications. Other characteristics of some fractured rocks that are not common in loose granular materials are: indistinct base of an aquifer or of the freshwater circulation system, tilted or compex structural setting, and soils on parts of rock which provide a complex two-media hydrologic system.

An evaluation of the hydrology of fractured rocks can be made in several ways. An effective way is to develop a conceptual model that can be upgraded progressively. This approach requires an understanding of the development of the permeability in the system. In contrast to the inherent permeability of sands, clays and gravels, the permeability of fractured rocks is epigenetic and has developed with the specific erosional history of each area. Many good inferences about the hydrogeology of a specific area can be made by reconstructing the erosional history and by relating stages of development with land surface features.

A conceptual model is built on many aspects of a hydrogeological overview, including a determination of: (a) what features are similar to those of other settings already studied; (b) degree of variance of features; (c) various physical boundary conditions; (d) base of the system; (e) effects of topography; (f) effects of climate; (g) dependence on porous granular materials such as soil on slopes, and alluvium in valleys; (h) geologic and hydrologic history; (i) significance of surface expression in evaluation; (j) value of interpolation and extrapolation; and (k) characteristics of water-table behavior. The aspects above are not of equal value, and efforts to get useful information about each need not be great.

An understanding of the hydrologic system of fractured rocks leads to conceptual models that are fundamental to evaluation. The major approach taken here emphasizes inferential hydrogeology and relies on conceptual models based on best use of principles and past knowledge with continual upgrading in inferences as new information becomes available.

REFERENCES

Bredehoeft, J.D., 1971. Analysis of flow in fracture systems — a porous medium model. Annu. Meet. 1971, Geol. Soc. Am., Abstr. Progr., 3(7): 512.
Grund, A., 1914. Der geographische Zyklus in Karst. Z. Ges. Erdkd. Berlin, pp. 621—640.
LeGrand, H.E., 1967. Ground Water of the Piedmont and Blue Ridge Provinces in the Southeastern States. U.S. Geol. Surv., Circ. 538, 11 pp.
LeGrand, H.E. and Stringfield, V.T., 1973. Karst hydrology — a review. J. Hydrol., 20: 97)120.
Leopold, L.B. and Langbein, W.B., 1963. Association and indeterminacy in geomorphology. In: Fabric in Geology, Addison—Wesley, Reading, Mass., pp. 184—192.

[2]

SECONDARY PERMEABILITY AS A POSSIBLE FACTOR IN THE ORIGIN OF DEBRIS AVALANCHES ASSOCIATED WITH HEAVY RAINFALL

A.G. EVERETT

Everett and Associates, 416 Hungerford Drive, Rockville, MD 20850 (U.S.A.)

(Accepted for publication April 25, 1979)

ABSTRACT

Everett, A.G., 1979. Secondary permeability as a possible factor in the origin of debris avalanches associated with heavy rainfall. In: W. Back and D.A. Sephenson (Guest-Editors), Contemporary Hydrogeology — The George Burke Maxey Memorial Volume. J. Hydrol., 43: 347—354.

Throughout much of the Appalachian region, heavy rainfall leads not only to flooding but also to extensive erosion and landsliding. Such a heavy rainfall occurred on August 18, 1972, in southwestern West Virginia with the heaviest rainfall and resultant damage centered over the Gilbert and Bens Creek drainages. Landslides, in the form of debris avalanches, were the principal form of mass movement during the storm. A number of these landslides appear to have resulted from soil zone saturation associated with temporary spring flows issuing from bedrock to the soil mantle where joints in highly fractured shales and coal beds intersected underlying, relatively unjointed, sandstones. Thus, it appears that, in this case, increased pore-water pressures as an effect of concentration of flow through secondary permeability played a substantial role in the localization and formation of slide masses. How widespread this phenomenon may be is not known.

INTRODUCTION

On August 18, 1972, a storm system over the central Appalachian states produced heavy rains that centered in the drainage basins of Gilbert and Bens Creek in Mingo County, West Virginia. Extensive flooding produced by this rain was accompanied by landsliding; locally, property damage was severe. That the rain was regionally widespread is indicated by the fact that increased storm flows were measured along major streams over most of southern West Virginia starting on August 19 (Fig.1), but the heaviest downpours were centered in the Mingo County area during two periods on the 18th. The first brought an estimated 18 cm of rain in a 2- to 3-hr. storm in the morning, and the second added perhaps more than seven additional centimeters in the afternoon. No formal or informal meteorological data are known to have been kept during this rain in the Mingo County area, hence the figures cited are estimates made by local residents on the basis of observations of buckets

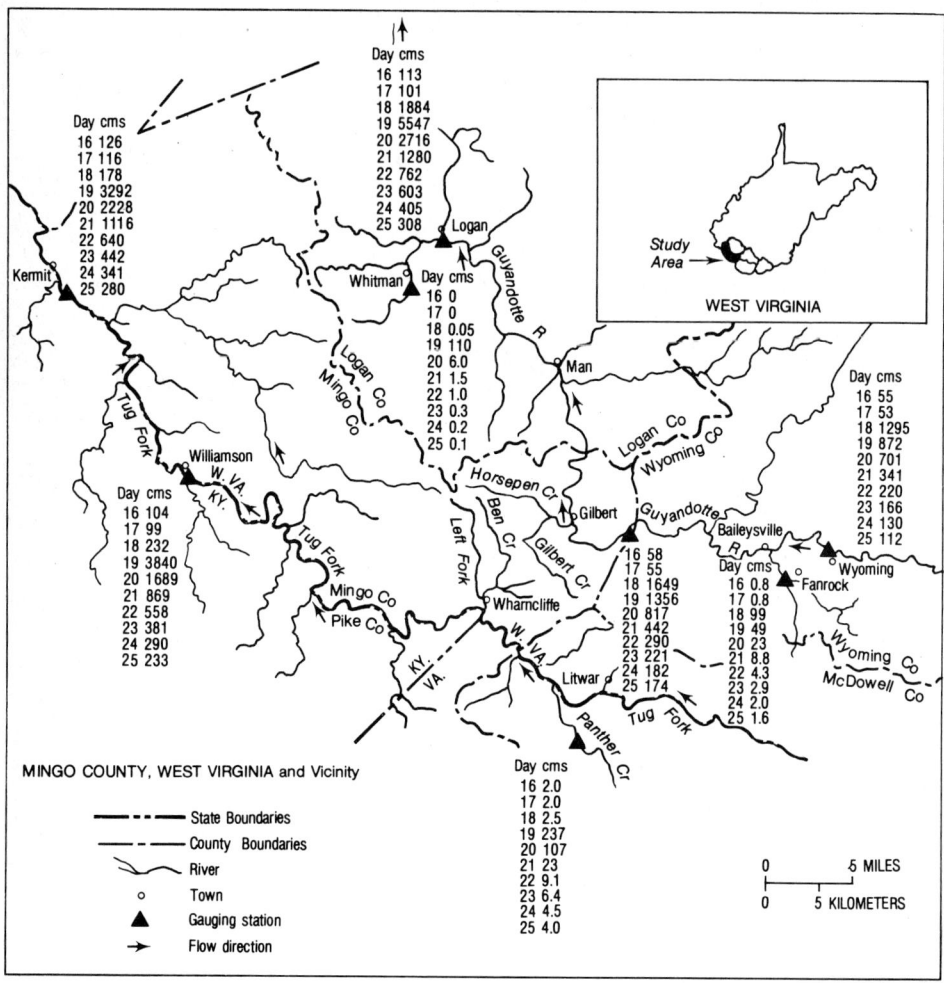

Fig.1. Discharge data, in cubic meters per second, for the Guyandotte River Basin, August 16—25, 1972.

or other containers being filled to overflowing (*The Intelligencer*, 1972). During the field investigation on which this report is based, a Gilbert resident stated that his swimming pool, on which he maintained a 25-cm freeboard, overflowed during the afternoon rain.

In the four-County area comprised of Mingo, Logan, McDowell and Wyoming Counties, flood and landslide damage caused over 70 dwellings and house trailers to be condemned and over 300 automobile bodies to be removed during the post-flood cleanup. The four Counties were designated as a Federal disaster area and seven million U.S. dollars were allocated for disaster relief.

GENERAL DESCRIPTION OF THE AREA

Bedrock in this region consists of coal-bearing formations of Pennsylvanian age, with the Kanawha Formation, the uppermost formation of the Pottsville Group, composing most of the hills; the tops of the highest hills are capped by the basal units of the Alleghany Formation. The rocks are an interbedded cyclothemic sequence of lithic, feldspathic sandstones, silty shales, thin and generally discontinuous limestones, underclays, and coal beds. Hennen (1915) described a number of partial stratigraphic sections throughout Mingo and Logan Counties.

The rocks tend to weather to a soil that becomes heavily vegetated. Soil profiles exposed by erosion during the August 18 storm range from a few centimeters to about 1.8 m in depth. The soils commonly lack zonation, containing rock fragments throughout, but there is a high organic content in the upper few centimeters of forested soils.

Hill slopes in the Gilbert Creek valley range from 17° to 36°, averaging about 27°; in Bens Creek valley, a geomorphically "younger" valley in form, slopes range from 21° to 36°, averaging about 31°. No asymmetry of slopes is evident.

Valley floors are generally narrow, ranging from 30 to 100 m in width, and have gradients usually of less than 10 m per km and rarely over 38 m per km (Table I).

TABLE I

Stream and drainage basin data (modified from Hennen, 1915)

Name	Location	Total distance (km)	Gradient (m/km)	Area of drainage basin (km^2)
Gilbert Creek	Mudlick Branch to Horsepen	7.9	11.2	76.13
Gilbert Creek	Horsepen to mouth	3.9	5.5	
Horsepen Creek	Smith Branch to mouth	7.1	10.4	40.00
Bens Creek	Walnut Hollow to mouth	7.3	10.58	58.30
Left Fork of Bens Creek	Lefthand Fork to mouth	7.4	14.4	

The drainage basins of Gilbert and Bens Creek are 76.13 and 58.30 km^2, respectively (Hennen, 1915, pp. 21 and 22). Within the Gilbert Creek drainage basin, the Horsepen Creek basin comprises 40 km^2.

Strip mining of coal is the principal industry and source of employment in the region; the strip mines are located along the upper slopes of the hills and residences and commercial establishments are concentrated along narrow valley bottoms. The area has been mined by deep mines and by strip pits since the latter part of the nineteenth century (Hennen, 1915).

FLOODING AND EROSION RESULTING FROM HEAVY RAINFALL

Personnel of the U.S. Geological Survey (USGS) estimated that the flows on August 18 reached peaks of 915 m^3/s on Horsepen Creek and 1372 m^3/s on Gilbert Creek at sites near the town of Gilbert (G.S. Runner, pers. commun., 1972). During and immediately after the storm, the creeks carried large sediment loads derived from hillslope erosion and also from landslides. Debris, including a number of old auto bodies, clogged the stream channels and forced water over the banks onto the narrow floodplains. Small wooden bridges, for both foot traffic and automobile access to driveways, were swept away and added to the debris load, isolating many residences from the main valley-bottom road. These bridges did not seem to have survived to serve as barriers to flow along either Gilbert or Bens Creek, but rather they became part of the debris dams that formed in the channels.

Many of the driveway access bridges were replaced within weeks by the U.S. Army Corps of Engineers with bridges built of four to five 40-cm I-beams, about 8 m in length, set on timber cribbing and covered by 5 cm thick planks. The Corps' design clearance for the bridges was specified to be a minimum of 1.54 m (5 ft.) above the stream channels. In the upper reaches of creeks, stream channels are sufficiently entrenched to allow construction with the specified clearance. Toward the mouths of such streams as Gilbert and Horsepen, however, stream banks are low and the bridges are only 0.67—1 m above the channel. Such stout bridges can be expected to clog with, and retain, debris during future floods; this will force the flood waters out of the channel and onto the adjacent narrow floodplains where the majority of the residential and commercial structures are located. These bridges will increase the probability of property damage in future floods.

The August floods not only carried a great deal of sediment out of the drainages but also deposited a great deal on valley floors where numerous deltas were formed by high-gradient side-channel tributaries. Gilbert Creek formed a point bar along the bank of the Guyandotte River, with much of that sediment load still in evidence three weeks after the peak flood stage of the Guyandotte River on August 19th.

Rillwork was common on exposed soil banks, mine waste piles and bare ground. Overbank deposits from strip-mine operations showed evidence of sediment erosion, but relatively few such waste deposits were involved in the regionally widespread landsliding. West Virginia strip-mining regulations limit strip mining to slopes not in excess of 33°. During the 1972 field work, no active strip mining was seen on slopes in excess of 28°, but overbank deposition of mine spoil locally raised some slopes to 35°.

Debris avalanches associated with flooding

Field study in 1972 showed that "chutes," or landslide scars, were common in the Gilbert and Bens Creek areas. Some of these chutes had been freshly formed in association with the August 18 rainfall whereas others were partially revegetated scars from previous periods of landsliding. These chutes were formed by debris avalanches, a type of slide characteristically occurring (Hack and Goodlett, 1960, p.44):

"in humid climates . . . accompanied or preceded by heavy rains."

Terzaghi (1962, p.105) observed that:

"exposure to rain or melting snow belongs to the normal existence of a slope. Hence, if a slope is old, heavy rainstorms or rapidly melting snow can hardly be the sole cause of a slope failure, because it is most unlikely that they are without any precedent in the history of a slope. They can only be considered contributing factors."

In the Appalachians, however, the close association of heavy rainfall and landsliding is well documented as a major mass movement erosional mechanism and a source of sediment for streams (Hack and Goodlett, 1960; Williams and Guy, 1973; Lessing et al., 1976). Similar relationships of heavy rainfall and widespread landsliding have been documented for other areas by Campbell (1975) and by Nilsen et al. (1976), among others.

Work currently underway by the USGS on slope stability in the Appalachian Plateau indicates that from 50 to 90% of the area has active and older slides with the bulk being earth slides and slump; debris avalanches are numerous in some areas (W.E. Davies, pers. commun., 1979).

Factors related to susceptibility of the West Virginia region to landsliding have been given by Lessing et al. (1976, pp. 23 and 31). The slope angles for 81% of the landslides that they reviewed were between 15 and 35%; on that basis alone, they estimated that:

". . . 2/3 of the State has slopes that are within the range of slide-prone areas."

They found a clear correlation of slide susceptibility with specific stratigraphic, hence lithologic, units, but a low correlation with mine waste deposits. About 85% of the recent slides that they studied were in near proximity to older slides.

In the Gilbert and Bens Creek area, only 5—10% of the landslides were in association with present or past strip-mine activity, whereas about 50% were associated with road building and other construction activity that increased flat areas along valley floors and walls by cutting away the lateral support from hillslopes. The remainder of the slides were on hillslopes presently undisturbed by man (although undoubtedly many had been cut over by logging one or more times in the last century). The latter slides commonly occurred on heavily forested hillslopes with thick undercovers of grasses or brush. Thus, it appears that the binding action of roots did not inhibit slides but rather served to bind the uppermost part of the slide mass into a reasonably coherent layer that was carried passively during sliding. Deformation of the lower part of slide masses was indicated at their fronts by incoherent fluidized soil bulges that commonly showed transverse fissures and small rivulets.

The slide areas were examined about three weeks after the events of August 18. A light rain had fallen for several days prior to, as well as during, the first two days of the field examination. During that time, there was a slight flow of water from many of the highly fractured coal beds associated with slide scars. Discussions with several residents of the area revealed that springs in the region commonly issued from coal beds overlying sandstone beds.

All of the landslides examined involved the soil horizons, most penetrating only a few centimeters into decomposed rock. No fresh or slightly weathered rock was involved in any of the landslides; instead such rock, mantled only by a very thin soil cover, served as a base of the chutes. Slides developed along surfaces at the base of the root zone of the grasses and underbrush, which served only to keep the uppermost soil material intact. Large trees usually were uprooted along with other vegetation. Slides ranged in depth from more than 0.3 m up to about 2 m, penetrating to the base of the development of soil and of highly decomposed rock.

The apparent role of secondary permeability in the formation of debris avalanches

The mechanisms usually described as being involved in landslides are those described by Terzaghi (1962), namely, a

"... decrease in shearing resistance along potential surfaces of slides"

with a rise in the piezometric surface. The role of specific lithologic units involved and the nature of their effective permeability, however, has received rather scant attention. Some workers have noted the relationship between lithologic associations and landslide phenomena in the Appalachians. Hack and Goodlett (1960) noted that erosional phenomena in a Virginia Appalachian region to the southeast were concentrated in the Hampshire Formation, a sequence of alternating shale and sandstone that bears a gross resemblance to the Kanawha Formation of West Virginia.

Terzaghi and Peck (1967, p.427) noted the regionally widespread Appalachian phenomenon of slides in silty shales:

"in the Alleghany region of West Virginia, southern Pennsylvania, and eastern Ohio..."

Stringfield and Smith (1956, p. 17), in their report on landslides, associated with heavy rains in the Petersburg area of West Virginia commented:

"it seems quite likely that the rainfall on the east side of the mountain first saturated the soil. Then as rainfall continued, water moved ... in and beneath the soil on top of the relatively impermeable sandstone [Tuscarora "quartzite"], and emerged at the heads of the hollows where the sandstone dipped beneath shale, thoroughly wetting the slide surface."

A phenomenon somewhat similar to the one described above seems to have been involved in many of the landslides in the Gilbert and Bens Creek areas. On slopes not recently modified by man, as well as on some slopes for which lateral support was removed in the course of construction activity, it appears that rainfall first saturated the soil cover with water which then penetrated the joints that form the secondary permeability of the highly fractured shales and coal beds. The interbedded sandstones, being dominantly lithic arenite composed of numerous micaceous metamorphic rock fragments and mica flakes as part of both the matrix and grain components, have joints with a wider spacing, averaging about a factor-of-ten fewer joints per unit of distance, than do the overlying shales and coals. It thus appears that a heavy flow of water moved rapidly downward through the soil zone into the shale and coal beds. When the water flow intersected a sandstone bed with its lessened secondary permeability, it moved laterally toward a hillslope where it debouched into the mantling regolith.

Terzaghi and Peck (1967, p.428) describe the mechanism for landsliding in Appalachian silty shales as follows:

> "the slides are preceded by a sudden, but temporary and local, increase of the pore-water pressure in the zone of sliding. The shale within this zone consists of macroscopic fragments which are in the process of progressive deterioration. Because of the slope the accumulation of fragments is acted on by shearing forces and the joints between the fragments open. During wet spells the open spaces are filled with water. As soon as the deterioration reaches a critical stage, which occurs in different places at different times, the fragments break down during a wet spell under the combined influence of the overburden and seepage pressures. The weight of the overburden is temporarily transferred to the water, whereupon the effective pressure and the corresponding shearing resistance along a potential surface of sliding are reduced and a slide occurs. It may stop rather abruptly, because the excess porewater pressure dissipates rapidly on account of the relatively high permeability of the accumulation of shale fragments."

This mechanism may have been operative in the Mingo County area during the 1972 rainstorms, but it should be noted that no slides were observed to have been formed by obvious local floodwater loading of slopes, such as by breaching of the berms of silt retention ponds that are located along strip benches high on valley walls. Erosion associated with such suddenly-released volumes of water from breaches in the pond berms was in the form of channels cut through the soil profile to bedrock. The slide mechanism seems to be a rate-related phenomenon in which the pore-water pressure increased rapidly but only to the extent that fractures filled without forming a mass consisting more of water than of soil and rock debris.

CONCLUSIONS

Landslides and flooding are a common result of heavy rainfall in the Appalachian region. During the Gilbert and Bens Creek floods of August, 1972, there was evidence that water from very heavy rainfall moved downward through extensively fractured silty shales and coal beds to the intersection of joints and the surface of the relatively less jointed sandstone beds. Flow was then lateral to the soil mantle of hillslopes, where it increased pore pressures, triggering debris avalanches. How widespread this postulated mechanism involving secondary permeability may be is not known. The extensive occurrence of debris avalanches in the Appalachian region suggests that the mechanism may be operative elsewhere in lithologies that have vertical differences in secondary permeability.

ACKNOWLEDGMENTS

Field work for this report was done in September, 1972, while the writer was an employee of the U.S. Environmental Protection Agency, but the observations and opinions expressed herein are solely his and do not reflect any position or opinion of that Agency. C. Sullivan assisted in the project by compilation of data and by drafting; J.J. Anderson materially improved the manuscript with his suggestions. The errors, idiosyncrasies, and opinions are the sole responsibility of the writer.

REFERENCES

Campbell, R.H., 1975. Soil slips, debris flows, and rainstorms in the Santa Monica Mountains and vicinity, southern California. U.S. Geol. Surv., Prof. Pap. 851, 51 pp.
Hack, J.T. and Goodlett, J.C., 1960. Geomorphology and forest ecology of a mountain region in the central Appalachians. U.S. Geol. Surv., Prof. Pap. 347, 66 pp.
Hennen, R.V., 1915. Logan and Mingo Counties — Detailed County Geologic Reports. W. Va. Geol. Econ. Surv., 775 pp.
Lessing, P., Kulander, B.R., Wilson, B.D., Dean, S.L. and Woodring, S.M., 1976. West Virginia landslides and slideprone areas. W. Va. Geol. Surv. Environ. Geol. Bull. 15, 64 pp.
Nilsen, T.H., Taylor, F.A. and Dean, R.M., 1976. Natural conditions that control landsliding in the San Francisco Bay region — An analysis based on data from the 1968—69 and 1972—73 rainy seasons. U.S. Geol. Surv., Bull. 1424, 35 pp.
Stringfield, V.T. and Smith, R.C., 1956. Relation of geology to drainage, floods, and landslides in the Petersburg area, West Virginia. W. Va. Geol. Econ. Surv., Rep. Invest. 13, 19 pp.
Terzaghi, K., 1962. Mechanism of landslides. In: S. Paige (Editor), Application of Geology to Engineering Practice, The Geological Society of America, New York, N.Y., pp. 83—123.
Terzaghi, K. and Peck, R.B., 1967. Soil Mechanics in Engineering Practice. Wiley, New York, N.Y., 729 pp.
The Intelligencer, 1972. Newspaper of Wheeling, W. Va., August 29, 1972.
Williams, G.P. and Guy, H.P., 1973. Erosional and depositional aspects of hurricane Camille in Virginia, 1969. U.S. Geol. Surv., Prof. Pap. 804, 80 pp.

[2]

SEASONAL CHEMICAL AND ISOTOPIC VARIATIONS OF SOIL CO_2 AT TROUT CREEK, ONTARIO

E.J. REARDON[1], G.B. ALLISON[2] and P. FRITZ[1]

[1] *Department of Earth Sciences, University of Waterloo, Waterloo, Ont. (Canada)*
[2] *CSIRO, Division of Soils, Glen Osmond, S.A. (Australia)*

(Accepted for publication May 28, 1979)

ABSTRACT

Reardon, E.J., Allison, G.B. and Fritz, P., 1979. Seasonal chemical and isotopic variations of soil CO_2 at Trout Creek, Ontario. In: W. Back and D.A. Stephenson (Guest-Editors), Contemporary Hydrogeology — The George Burke Maxey Memorial Volume. J. Hydrol., 43: 355—371.

The partial pressure of carbon dioxide in soil gas and its ^{13}C isotopic composition were monitored in a sandy calcareous soil during 1977—1978. These measurements were coupled with chemical and isotopic analyses of groundwater from a water-table piezometer. The ^{13}C isotopic composition of the groundwater at the water table is in apparent equilibrium with the soil zone CO_2 gas phase. The P_{CO_2} is primarily controlled by root respiration, diffusive loss to the atmosphere and uptake by the aqueous phase. Steep P_{CO_2} gradients towards the water table are evident during periods of low CO_2 production in the root zone. By early spring, diffusive loss of CO_2 along these gradients extends to depths of 7 m or more. During the growing season this gradient is reversed, resulting in CO_2 buildup below the root zone. Below a depth of 7 m, soil CO_2 content showed little seasonal change.

Leaching of soil carbonate extends to depths of 2 m or more in the study area. A rapid neutralization of slightly acid soil water by soil carbonate as the water is displaced across the leached—unleached zone boundary due to recharge events is believed to account for the deposition of iron and manganese oxides at this interface.

Saturation of soil water with respect to calcite is attained in the unsaturated zone and a state of supersaturation is reached by the time the soil water reaches the water table. Two mechanisms may account for this supersaturation: (1) slow dissolution of dolomite by infiltrating soil water after calcite saturation is reached; and (2) CO_2 degassing of soil water in response to CO_2 losses from the soil profile during fall and winter.

INTRODUCTION

Seasonal variations in CO_2 pressures in the unsaturated zone are well established (Lundegardh, 1927; Witkamp, 1969; De Jong and Schappert, 1972; Garrett and Cox, 1973; Edwards, 1975; Albertsen, 1977). The principal mechanisms by which CO_2 is produced in forested soils are root respiration, oxidation of organic material and microbial respiration. Generally, root respiration is the major contributor of CO_2 to soil air in forested areas (Lundegardh, 1927; Nilouskaya et al., 1970; Edwards, 1975). Most studies of CO_2 transport in the unsaturated zone have focused on the upward flux of CO_2 to the atmosphere. This study was undertaken to examine CO_2 transport below the highly productive root zone and to study the carbonate chemical and isotopic evolution of the gas—water system in the unsaturated zone. An understanding of the CO_2—water equilibrium in soils may be important for the interpretation of groundwater ^{14}C data because "age correction" procedures are based on assumptions regarding the initial P_{CO_2} of infiltrating groundwater and its ^{13}C isotopic composition.

STUDY AREA

Trout Creek, a small stream within the Big Creek Drainage Basin, is located approximately 5 km SE of Delhi, Ontario (Fig.1). Devonian bedrock in the area does not outcrop in the drainage basin due to the extensive overlying deposits of Pleistocene tills, and deltaic and lacustrine sands and gravels. The uppermost unit, forming the Norfolk Sand Plains, is a homogeneous aeolian-reworked clean medium-grained sand containing dominantly quartz and feldspar and approximately 15 wt.% calcite and 5 wt.% dolomite. This unit varies from zero to tens of meters in thickness over the study area.

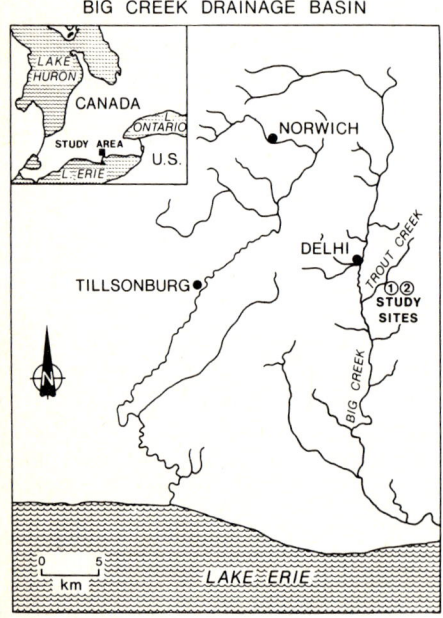

Fig.1. Location of study area. Study sites *1* and *2*, refer to sites developed in 1977 and 1978, respectively.

Detailed hydrogeologic studies of the Big Creek Drainage Basin can be found in Yakutchik and Lammers (1970), and Novakovic and Farvolden (1974). A study by Wallis (1975) examined the variation of dissolved organic carbon in both the saturated and unsaturated zone in the study area.

CLIMATE AND VEGETATION

Norfolk County of southern Ontario incorporates the Big Creek Drainage Basin. This area experiences a temperate climate with average annual precipitation of 95 cm and mean temperatures for January and July of −6 and +20°C, respectively. Extensive tracts of land are devoted exclusively to tobacco farming. The study site at Trout Creek is one of the few uncultivated areas in the region. Vegetation consists of mixed-conifer deciduous forests with red pine, spruce and maple predominating.

SAMPLING METHODS AND ANALYTICAL TECHNIQUES

During early spring, 1977, two holes were hand-augured to the water table at study site *1* (Fig.2) and CO_2 sampling devices were installed for regular monitoring of CO_2 pressures in the unsaturated zone. A sampling device consisted of an inverted Eppendorf pipette tip attached to 2 mm I.D. Tygon® tubing at the tip and wrapped with 80 mesh nylon screening at the other end. The devices were lowered to various depths and the hole backfilled with the original soil material. Care was taken not to damage the Tygon® tubing which terminated at the soil surface and to ensure that no humic material accidentally fell into the hole. At the surface, Eppendorf pipette tips were attached to the Tygon® tubing and capped with rubber septa suitable for gas sampling using 50-ml syringes.

In addition to the two augured holes for installation of CO_2 sampling devices, three 4-mm I.D. steel tubes were driven to specific depths and capped with rubber septa. These were sampled on occasion for comparison of sampling methods. No differences in measured P_{CO_2} between these two sampling devices were noted during the study. The locations of all CO_2 sampling devices at this site are shown in Fig.2; all were installed within a 10-m² area. Total volume of the deepest sampling device was approximately 15 ml, thus, two separate 15-ml volumes were extracted and discarded to ensure flushing of the system of any atmospheric component before sampling. Generally a 10—20-ml aliquot was extracted from each sampling device for CO_2 analysis. When ^{13}C determinations were to be made, additional 50-ml aliquots were extracted after the complete profile had been sampled for CO_2. Analyses for CO_2 were performed on a Hamilton—Fisher® gas partitioner as soon after sampling as was feasible and generally within 6—24 hr. In the early part of the study plastic syringes were used and diffusive loss of CO_2 was sometimes noted by comparison with P_{CO_2}'s determined on large-volume glass syringes collected for ^{13}C analyses, but this loss rarely amounted to more than 10% and was corrected for. After August, 1977, glass syringes were used exclusively for CO_2 gas sampling.

A piezometer, driven 5.8 m to the water table and having a slotted intake zone 50 cm below the water table, was installed in April, 1977. This piezometer allowed routine monitoring of water levels, and groundwater sampling for $\delta^{13}C$ and chemical analyses. Measurement of pH, temperature and alkalinity were performed on site. Water samples were filtered in the field through 0.45-μm Millipore® filters with both acidified and non-acidified samples collected for analyses of major and minor constituents and separate samples for ^{13}C analyses.

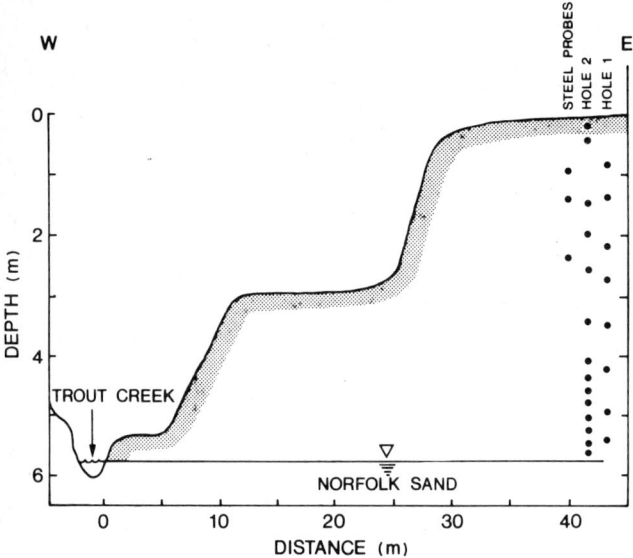

Fig.2. Cross-section of study site 1 with locations of CO_2 sampling devices.

Towards the end of May 1977, a separate hole was drilled and cased with aluminum tubing for soil-moisture measurements using an NEA® model A3-16 neutron probe. Periodic measurements made during spring and early summer of 1977 and 1978 allowed calculation of volumetric moisture content to 4 m.

^{13}C analyses of groundwater samples and soil gas were made at the University of Waterloo using a VG Micromass® 602D mass spectrometer. ^{13}C contents are recorded in the conventional δ (‰) notation and refer to the PDB standard.

RESULTS

Precipitation and water-table levels from May to November, 1977, are shown in Fig.3. Precipitation data are from the Tobacco Research facility 5 km from the study area at Delhi, Ontario. These precipitation records were compared with data from the Simcoe Meteorological Station located 30 km from the study area. Although significant daily variations in the amounts of precipitation existed between these two stations, the total amounts were quite similar on a weekly and seasonal basis. Aside from the apparent water-table rise only one week after the piezometer was installed, there were no water-table responses to rainfall events, except after 77/9/15.

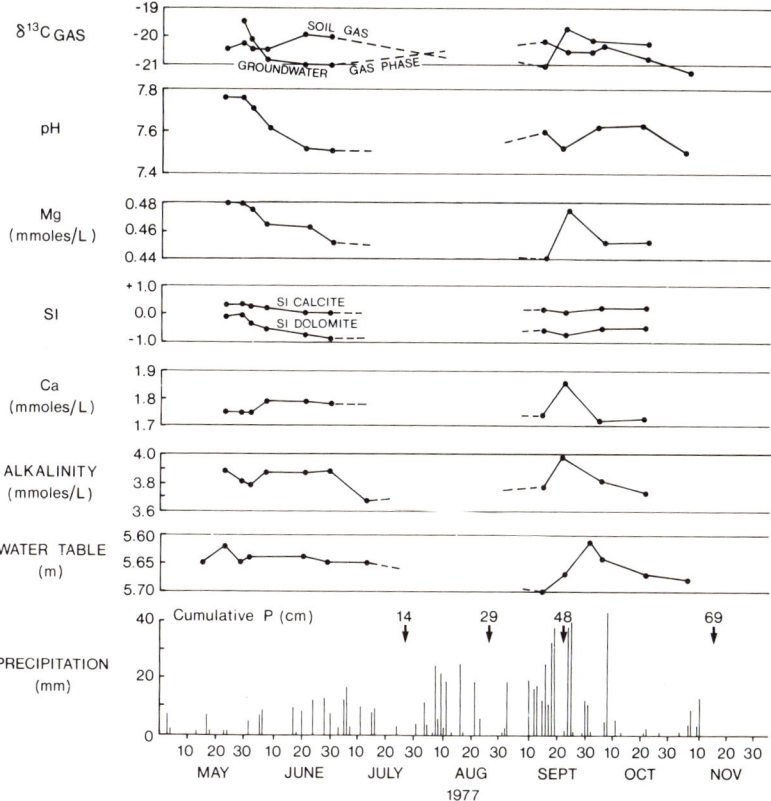

Fig.3. Temporal variations of some chemical and physical properties of study site *1*. Precipitation data are from the Delhi Tobacco Research Center, 5 km from the study area. $\delta^{13}C$ (soil gas) are average values determined from Fig.9 and $\delta^{13}C$ (groundwater gas phase) are computed from the ^{13}C isotopic composition of the groundwater using eq.2 (see text).

Neutron moisture meter (NMM) measurements were made periodically during the spring and early summer of 1977 and 1978 (Fig.4). The NMM was field-calibrated by taking soil samples for water content determinations during installation of the aluminum access tube. The upper 2 m of the soil profile is not representative of the soil at depth due to the presence of humus material, root systems, Fe-oxyhydroxide coatings on grains and the absence of carbonate material. Below this depth, the soil is very uniform with a 20 wt.% carbonate content and a total porosity of 0.38.

The 1978 NMM data showed that volumetric water contents fell below 0.1 in the upper 2 m by late June. Volumetric water contents below 2 m were uniform at this time near 0.1. A similar moisture content was evident for the soil in June, 1977. This value is assumed to be representative for field capacity conditions of the soil in later discussions. The implication of field capacity is that the unsaturated hydraulic conductivity of the soil is so low that rates of water movement are negligible below this water content.

Soil material from one drill hole (cased during drilling) was analyzed for total carbonate content by weight loss after ignition at 950°C following heating to 550°C, to combust organic material. Uncertainty in carbonate content by this method, due to a water weight loss contribution from clays

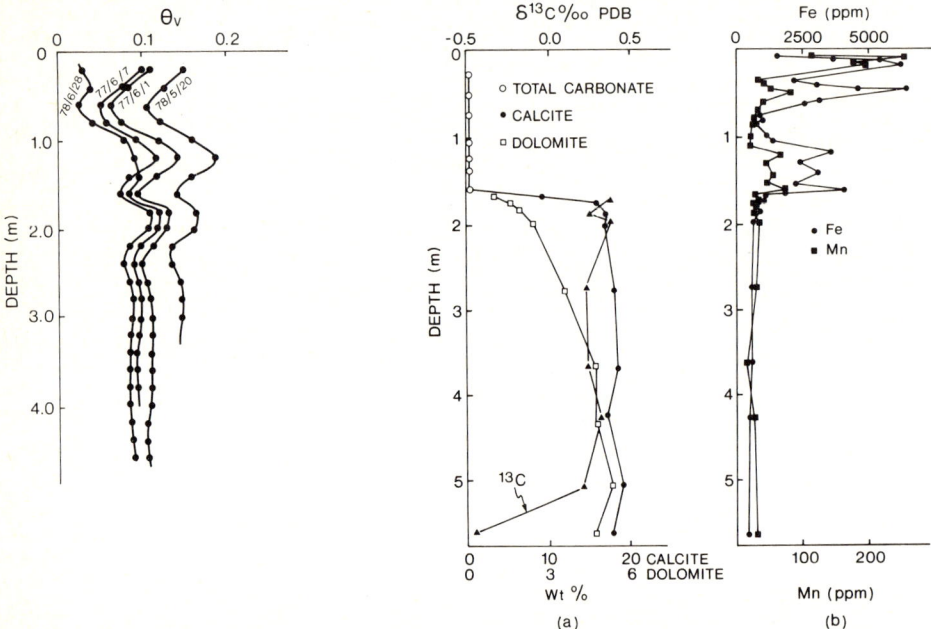

Fig.4. Volumetric soil-moisture content at study site *1* on various dates by neutron moisture meter (NMM). Similarities in the variation of moisture content with depth in the upper 2 m of the soil among sampling dates are believed due to the varying water retention characteristics of the soil related to mineralogic variations such as zones of Fe-oxide build-up and removal, calcite leaching, etc.

Fig.5. a. Weight loss on ignition at 950°C after heating to 550°C for soil material from study site *1* (expressed as wt.% CaCO$_3$) (*circles*). Above 0.5% weight loss, calcite (*closed circles*) was differentiated from dolomite (*boxes*) by Ca and Mg analyses of acid extracts. Included in the diagram are $\delta^{13}C$ values for soil carbonate material, shown as *triangles*.
b. Hydroxylamine hydrochloride leachable Fe and Mn from soil material with depth. Results expressed as ppm of dry weight of soil.

and hydrous silicate material upon heating, was determined to be less than 0.3%. Percent dolomite of total carbonate was calculated from Ca and Mg analyses of concentrated HCl solutions reacted with soil material for 48 hr. The results are recorded in Fig.5a. The depth of carbonate removal by weathering was found to be to 1.7 m in the cased hole while in the other two holes depths of 1.5 and 1.8 m were found. ^{13}C isotope compositions for the soil

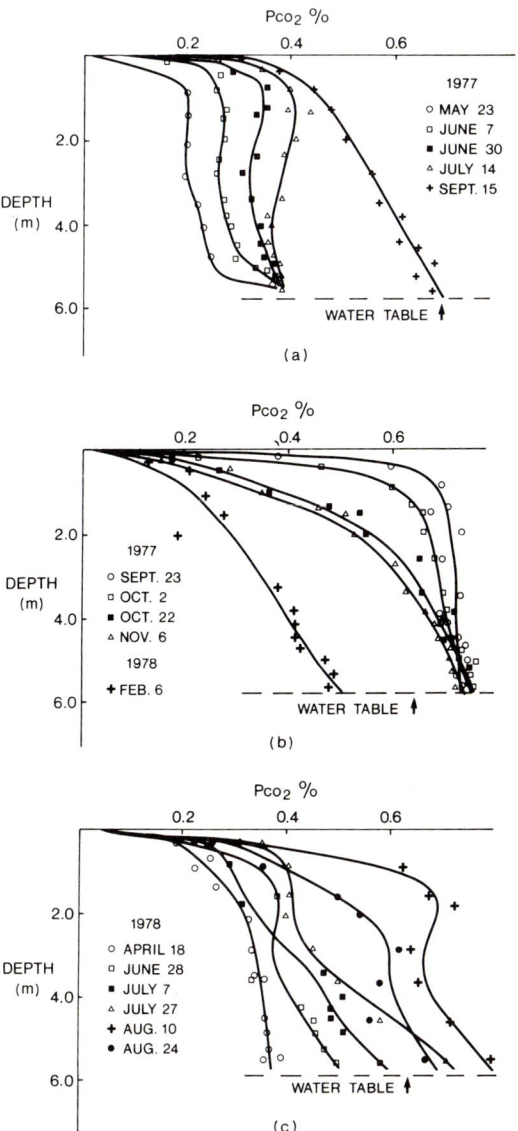

Fig.6. P_{CO_2} vs. depth for study site 1 for: (a) 77/5/23—77/9/15; (b) 77/9/23—78/2/6; and (c) 78/4/18—78/8/24. Curves have been drawn in by hand and are meant only to simplify the display of changes in the P_{CO_2} profile with time.

carbonate material are also displayed in Fig.5a. An average $\delta^{13}C$ of +0.27 ± 0.12‰ relative to PDB for two standard deviations is obtained for seven samples taken above the water table. A shift towards a slightly lower $\delta^{13}C$ (−0.5‰) for the soil carbonate at the water table is suggestive of the presence of minor amounts of secondary calcite which would have an isotopic composition of approximately −10‰, given the observed groundwater chemical and isotopic composition.

All soil samples were analysed for leachable Fe- and Mn-oxides by a 24-hr. reaction with 0.1-m hydroxylamine hydrochloride in 0.2-m potassium tetroxalate to buffer the pH near 2. Results for the soil material with depth are shown in Fig.5b.

Pertinent groundwater chemical and isotopic variations are recorded in Fig.3. The piezometer installed for water-table sampling had a slotted intake zone of approximately 50 cm. Samples collected over such a length must incorporate waters representing several recharge events. Thus it is unrealistic to expect variations in water-table chemistry to correlate with chemical changes in the unsaturated zone except in the case of major recharge events.

Included in Fig.3 is the $\delta^{13}C$ of the CO_2 gas phase in equilibrium with the groundwater samples computed using ISOTOP (Reardon and Fritz, 1978), a computer subroutine linked to WATEQF (Plummer et al., 1976). These values are computed using pH, $\delta^{13}C$ of the water and the concentrations of all dissolved carbonate species and appropriate isotope fractionation factors between the different carbonate species and CO_2. Agreement is within 1‰ when compared to the average $\delta^{13}C$ of the soil CO_2 gas phase.

Soil CO_2 pressure profiles for 1977 and 1978 are displayed in Fig.6 and $\delta^{13}C$ values for the CO_2 gas phase are plotted vs. depth for various sampling times in Fig.9.

DISCUSSION OF RESULTS

Soil CO_2 gas sampling began in May, 1977. The CO_2 content of the soil profile at that time showed little variation with depth (averaging 0.2% or $10^{-2.7}$ atm.) except for an expected sharp diffusion gradient near the soil—atmosphere interface and another gradient near the water table (see Fig.6). This latter gradient suggests that groundwater will exsolve CO_2 into the soil profile when soil zone P_{CO_2} is lower than that of groundwater. Groundwater P_{CO_2} was uniform during 1977 and averaged 0.4% ± 0.1 as calculated by WATEQF.

The May 23 profile is thought to be a result of very low production of CO_2 in the root zone as plants had not started significant growth at this time. The CO_2 concentrations in the unsaturated zone are low due to the effects of infiltration of snow melt dissolving CO_2 (enhanced by carbonate dissolution) and diffusive loss to the soil—atmosphere interface. Both these processes remove much of the soil zone CO_2 remaining from the previous autumn.

During the period May 23—July 14, CO_2 appears to have exsolved from the water table (evidenced by the P_{CO_2} gradients in the soil) and moved upwards by diffusion. At the same time, root zone activity built up high P_{CO_2} in the top 2 m and CO_2 moved down the profile via diffusion and the recharge flux of water, which is quite small at this time. Evidence for two sources of CO_2 in the soil profile is also given by the fact that there is a minimum in the P_{CO_2} profile at 3—4 m throughout this period.

Other mechanisms of CO_2 buildup below the 2 m depth may be respiration from deeper root systems and/or lateral mass transport of a CO_2-rich regime of soil air located in a westerly direction from study site 1 where the highly productive root zone lies progressively closer to the water table (see Fig.2). CO_2 measurements in soil near the banks of Trout Creek during the summer of 1978 revealed CO_2 concentrations as high as 10%. CO_2 concentrations of this magnitude would be expected where a shallow depth to the water table would act as a boundary for the buildup of CO_2 produced by root respiration. Diurnal fluctuations in barometric pressure could conceivably transport this CO_2-rich soil air laterally into the soil along zones of high permeability. The steep-sided banks of Trout Creek are likely avenues for entry of atmospheric gas to equalize increasing barometric pressures since the ratio of the horizontal to vertical hydraulic conductivity is likely to be quite high. Novakovic and Farvolden (1974) reported a value of 3:1 for similar sands in the study area.

A barometric pressure increase then would breathe in a CO_2-rich soil air laterally into the soil where the processes of dispersion and diffusion would tend to reduce its CO_2 concentration. A subsequent barometric pressure decrease would thus breathe out soil air with a lower CO_2 concentration. This mechanism of mass transport coupled with dispersive and diffusive processes may explain the P_{CO_2} increases at depth observed in 1978 at study site 1 (see Fig.6c). Through spring and summer, P_{CO_2} increased throughout the soil profile in spite of a strong gradient away from the water table. Unfortunately, we have insufficient data to appraise the importance of this factor to CO_2 transport at the study site.

Regular monitoring of P_{CO_2} was interrupted in early July, 1977. The next sampling period was not until September 15. A dramatic change in the profile is evident. The average P_{CO_2} of the profile nearly doubled and there is a strong gradient away from the water table. The profile on September 23 depicts a significant increase in CO_2 compared to September 15. This may be due to an increase in soil microbial activity related to a corresponding increase of soil moisture and surface derived organic nutrients in the soil accompanying the September rains. The relationship of soil moisture to microbial activity has been discussed by Matthess (1973).

The subsequent evolution of soil P_{CO_2} profiles after leaf fall from October to February, 1978, depict an ever-steepening gradient towards the water table caused by the continuing loss of CO_2 from the soil by diffusion and reduction of root respiration. The P_{CO_2} profile in April, 1978, shows no diffusion gradient towards the water table as was observed in the spring of 1977 probably because measured P_{CO_2} at the water table (0.38%) is very near average groundwater P_{CO_2} (0.40%) as calculated by WATEQF.

The ^{13}C isotopic composition of soil CO_2 (see Figs. 3 and 9) shows no substantial seasonal or depth variations. $\delta^{13}C$ values range from −19 to −22°/00 characteristic for soils in which root respiration is a dominant component.

GROUNDWATER CHEMICAL AND ISOTOPIC VARIATIONS

Groundwater recharge in the study area occurs principally during periodic winter thaws and in spring after snow melt. By late June, the soil establishs a moisture deficit (see Fig.4), in the root zone and below this depth volumetric water contents are near 0.1. Some recharge may occur during the fall as observed in 1977 due to the combined effects of increased precipitation and reduction of evapotranspiration. At both these times, P_{CO_2} profiles were either uniform or displayed a gradient away from the water table and soil water will continually encounter the same or greater CO_2 pressure as it moves through the profile.

During the summer, recharge is low because rainfall serves mostly to make up the soil moisture deficit established since the previous rainfall event. Only the unusually large quantities of precipitation in August and September of 1977 served to wet up the soil profile and the first water-table response to precipitation events occurred on September 23. During the preceding period, the soil moisture deep in the profile (>2 m) would be expected to be relatively constant, close to a volumetric water content of 0.1. Because it can be shown that under these conditions, considerably more carbonate is held in solution than is in the gas phase (Fig.7), soil moisture serves to buffer soil-

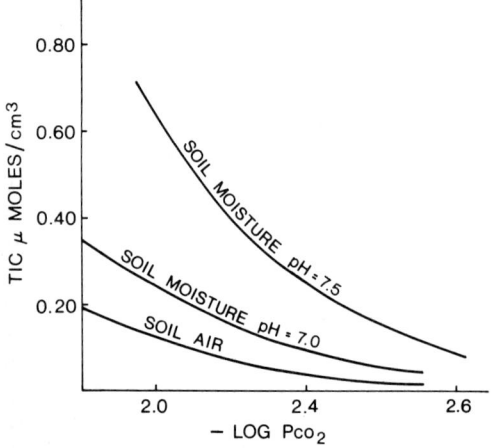

Fig.7. Total inorganic carbon expressed in micromoles per cm³ of soil vs. − log P_{CO_2} at 10°C plotted for soil air and for soil moisture at pH 7.0 and 7.5. A volumetric air content of 28% and water content of 0.1 were used in the construction of this diagram.

gas CO_2 increases during this period through the reaction:

$$CO_2 + H_2O + CaCO_3 \rightarrow Ca^{2+} + 2HCO_3^- \tag{1}$$

Aside from a slight increase in Ca, Mg and alkalinity which correlated with a water-table response in September, 1977, groundwater chemistry showed little seasonal variation. This is to be expected since the intake depth of the piezometer ($\sim$ 50 cm) must incorporate waters of different recharge events. The solubility of carbonates increases with P_{CO_2} and thus increases in Ca, Mg and alkalinity observed in September suggest that soil water equilibrated with the high summer P_{CO_2} was displaced to the water table in response to substantial rainfall events during this period.

Saturation index [$\log(IAP/K_{sp})$] calculations using WATEQF showed that the groundwaters are slightly supersaturated with respect to calcite (0.05— 0.3 units) and undersaturated with respect to dolomite (-0.3 to -0.7).

Under negligible soil drainage rates (10 vol.% water content) we calculate that the soil profile would contain 57 cm of water. The total yearly precipitation in the study area averages 95 cm and only a portion of this would actually infiltrate below the zone of evapotranspiration. The residence time of a particle of water within the unsaturated zone, then, could be of the order of a year. Soil water may thus experience a large variation in P_{CO_2} conditions from the time it infiltrates into the soil horizon to the time it recharges to the groundwater. A state of supersaturation with respect to calcite should thus be achieved in soil water during periods of CO_2 loss from the soil.

The $\delta^{13}C$'s for the CO_2 gas phase that would be in equilibrium with the groundwater calculated using ISOTOP (Reardon and Fritz, 1978) are shown in Fig.3. These values have been calculated using the expression:

$$\delta^{13}C_{CO_2} = \left(C_T \delta^{13}C_w - \sum_{i=1}^{n} m_i (K_i - 1) \times 1000 \right) \Big/ \sum_{i=1}^{n} m_i K_i \tag{2}$$

where $\delta^{13}C_w$ is the isotopic composition of the groundwater; m_i and K_i refer to the molal concentrations of the various carbonic, bicarbonate and carbonate species, and their associated fractionation factors with CO_2 (g) and C_T, total inorganic carbon. $\delta^{13}C_{CO_2}$ values for the groundwater samples are generally within 1 ‰ of the average measured $\delta^{13}C$ of the soil CO_2 for each sampling date. This evidence lends credence to the suggestion made by some workers (Langmuir, 1971; Rightmire and Hanshaw, 1973; Deines et al., 1974; Fritz et al., 1978) that dissolution of carbonates in the unsaturated zone proceeds by open-system processes, that is, the rock component of the infiltrating waters is exchanged and equilibrated with the CO_2 of the soil gas reservoir. However, it may also be argued that a 50/50 contribution to the dissolved inorganic carbon (DIC) from rock carbonate (+0.3 ‰) and soil gas (-21 ‰) would also result in a similar ^{13}C

isotopic composition for the groundwater, assuming no isotopic exchange of the DIC with the soil CO_2 gas phase. A 50/50 contribution to DIC from rock carbonate and soil CO_2 would be expected from the stoichiometry of reaction (1) provided $H_2CO_3^0$ is the dominant acid responsible for the dissolution and HCO_3^- is the dominant carbonate species after reaction. Since alkalinities are nearly double Ca + Mg and pH's are near 7.5, these conditions appear to be met in the study area. In order to document that C isotopic equilibrium does occur in the unsaturated zone, ^{14}C measurements of soil gas would have to be compared with values for water-table groundwater samples. This work is presently in progress.

CHEMICAL EVOLUTION OF INFILTRATING SOIL WATER

Without a knowledge of the chemical composition of soil water with depth, a detailed description of the chemical evolution of infiltrating soil water is impossible but some basic aspects of this process can be deduced from the available data. First, the depth of carbonate leaching, although highly variable at the study site (Cope, 1977), generally ranges from 1.5 to 2.0 m. Second, the calcite content of the soil profile increases with depth to a uniform content over very short distances ($\sim$ 10—20 cm) whereas dolomite content increases more gradually with depth. Third, groundwater samples from the water-table piezometer were invariably found to be slightly supersaturated with respect to calcite and undersaturated with respect to dolomite. Attainment of calcite saturation then must occur within the soil profile. The bulk of dissolution of calcite must also occur over short distances, otherwise, soil calcite content would increase more gradually with depth. The gradual increase in dolomite content with depth is consistent with the state of undersaturation observed for groundwater samples and the slow kinetics of dolomite dissolution.

Once calcite saturation is attained within the soil profile, continued dissolution of dolomite should result in a state of calcite supersaturation. We have found calcite saturation indices for groundwaters to be generally above 0.1 log units.

A large proportion of the annual groundwater recharge in the study area occurs during periodic winter thaws and in the spring as a result of infiltration of snow melt. It is also expected that a medium-grained sand could drain to a moisture content close to field capacity conditions relatively quickly. It appears reasonable then that the bulk of calcite dissolved by infiltrating soil water occurs not only over short distances but over short periods of time.

Plummer et al. (1979) have shown that the kinetics of calcite dissolution are highly dependent on pH and $H_2CO_3^0$ concentrations. They have found that rates of calcite dissolution increase sharply for pH's less than 5. The results of Plummer et al. (1979) are not directly applicable to reaction rates of calcite dissolution in the unsaturated zone at low pH's of soil moisture since their study involved well-mixed calcite—water systems. In an unsaturated porous media the limiting factor governing reaction rates at low pH may be diffusion of CO_2 from the gas phase into the water phase to form carbonic acid or the subsequent diffusion of H_2CO_3 and H^+ through water films to the calcite—water interface.

In the upper 2 m of carbonate leached soil, soil-water pH, due to a variety of acid producing reactions and in the absence of an effective buffering agent, such as calcite, may be relatively low. These acid producing reactions include the oxidation and hydrolysis of ferrous iron liberated from the weathering of silicates and the dissociation of carbonic and organic acids. As infiltration of snow melt or precipitation displaces the soil water within the upper 2 m of soil horizon into contact with carbonate containing soil, neutralization of this acidity by calcite dissolution should result.

It is well documented that the solubility of ferric oxyhydroxides decreases and the kinetics of ferrous iron oxidation increases with increasing pH (Stumm and Morgan, 1970). The same is true for Mn, and thus one would expect precipitation of Fe- and Mn-oxide phases at the point of soil-water neutralization. The sharp drop in both the hydroxylamine hydrochloride leachable Mn and Fe from the soil cores (see Fig.5b) coincides with the increase in soil carbonate. These findings emphasize the importance of the depth of leaching to the deposition of these phases in the soil profile. In fact, the variation in the amount of Fe- and Mn-oxyhydroxides immediately above the leached—unleached zone boundary of calcareous soil may well be correlated with the rate of downward movement of this interface with time which would in turn be a function of the historic soil P_{CO_2}, temperature, and soil-water acidity relationships.

Soil water moving across the calcite leached—unleached zone boundary must act as a sink for soil CO_2 gas during infiltration events because CO_2 is used up as indicated by reaction (1). If the calcite leached—unleached zone boundary lies below the zone of root respiration, then this reaction could conceivably induce diffusion of CO_2 to the site of reaction. If the diffusion of CO_2 is slow compared to the rate of calcite dissolution, then diffusion gradients towards the calcite leached—unleached zone boundary would be created. Our data unfortunately are not sufficiently detailed either spatially or temporally to offer evidence for or against the importance of this effect to CO_2 transport in the unsaturated zone.

RESULTS OF GAS SAMPLING AT STUDY SITE 2

After it was noted that P_{CO_2} increased during the summer of 1977 by a factor of 4 to depths of more than 5 m at study site 1 it was decided to install gas sampling devices at site 2 which has a greater depth to the water table where the boundary effect of a water table to CO_2 buildup would not be present. Site 2 is located ~ 200 m to the east of study site 1 (see Fig.1). Twenty-five sampling devices distributed over 12-m depth were installed to allow sampling of soil gas for P_{CO_2} and ^{13}C determinations. Figs. 8 and 9 show a summary of the results. During spring and summer of 1978, soil CO_2 content increased to a depth of 7 m but remained uniform below this depth throughout the sampling period. As compared to the shallower site (Fig.6) the greater depth prevented the water table from acting as a boundary to the buildup of soil CO_2. $\delta^{13}C$ values plotted in Fig.9 show no variation with depth throughout the sampling period. There is, however, a decrease of $2^0/_{00}$ from 78/5/24 to 78/6/28. We have no adequate explanation for this decrease. It may be argued that soil gas would be expected to be heavier in spring since exsolution of isotopically lighter CO_2 from soil moisture is occurring in response to diffusive loss of CO_2. Over time, this process would enrich the soil atmosphere in the heavier isotope as the soil-moisture CO_2 reservoir became depleted. Sometime between 78/5/24 and 78/6/28 the soil P_{CO_2} profile changed from one characterized by upward diffusive loss of CO_2 to one of downward diffusive gain of CO_2 (see Fig.8) from the zone of root respiration. The mass flux of root zone CO_2 would then enrich the soil atmosphere again in the lighter isotope. The argument against this is that one would expect $\delta^{13}C$ variations with depth over time which are not observed in the soil profile.

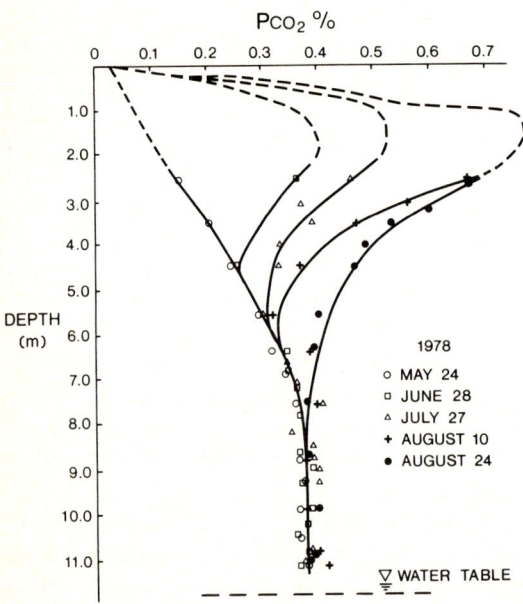

Fig.8. P_{CO_2} vs. depth for study site 2 during spring and summer of 1978. *Dashed* portions of the curves are extrapolated to atmospheric P_{CO_2} (0.03%).

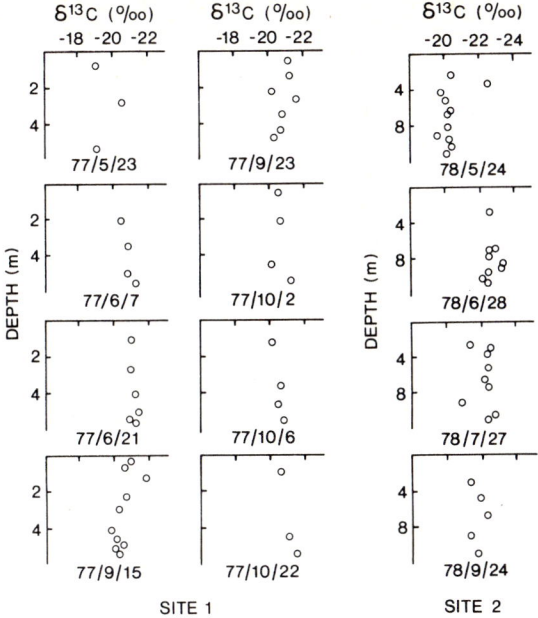

Fig.9. $\delta^{13}C$ of soil CO_2 vs. depth for study sites *1* and *2* at selected sampling dates when samples from three or more depths were analysed.

Diffusive flux of CO_2 from the root zone appears adequate to account for the increases of CO_2 below the 3-m depth from 78/6/28 to 78/8/24. Millington (1959) and Lai et al. (1976) have suggested an equation to calculate the diffusion coefficient of CO_2 in dry porous soils:

$$D/D_0 = S^{4/3} \qquad (3)$$

where D/D_0 is the ratio of CO_2 diffusion coefficient in soil to that in air, and S is the air-filled porosity. Since diffusion coefficients for wet porous media are generally lower, CO_2 flux calculations using coefficients determined from eq.3 should yield maximum estimates. An instantaneous CO_2 flux can be calculated using Fick's first law of diffusion: $q = D(dc/dx)$ where q is the flux in g cm^{-2} s^{-1} and D has units of s cm^{-1}. Assuming a volumetric water content of 0.1 below a depth of 3 m from June 28 to August 24 and using an air-filled porosity of 0.28 (see earlier) and integrating instantaneous CO_2 fluxes calculated for each sampling date over this time interval, we calculate a downward net CO_2 transfer across the 3-m depth of $7.6 \cdot 10^{-4}$ g cm^{-2}. The actual CO_2 addition to the soil air below the 3-m depth is calculated to be $2.8 \cdot 10^{-4}$ g cm^{-2}. This value was determined from the difference in the

June 28 and August 24 P_{CO_2} profiles using an air-filled porosity of 0.28. An additional sink of CO_2 in moist soils, however, is CO_2 dissolution in soil moisture in response to increases in P_{CO_2}. At a specific equilibrium CO_2 pressure, total dissolved inorganic carbon is defined solely by pH (see Fig.7). Assuming a soil moisture pH equal to that measured for groundwater, we calculate a net increase of DIC in soil moisture assuming field capacity conditions from June 28 to August 24 of $16.0 \cdot 10^{-4}$ g cm^{-2}. It may be argued that in this calcareous soil an increase in soil P_{CO_2} should cause dissolution of calcite so that a portion of the predicted increase in DIC should come from this source. Assuming that calcite is just at saturation below 3 m on June 28 and August 24, and assuming no significant pH change for the soil solution, the DIC contribution from soil carbonate would be approximately half that from soil CO_2. Thus $10.8 \cdot 10^{-4}$ g cm^{-2} represents a minimum estimate for the DIC increase below the 3-m depth for soil air and soil moisture. This value is very close to the calculated amount transferred by diffusion $7.6 \cdot 10^{-4}$ g cm^{-2}. Although calculations of this type are subject to considerable uncertainty without soil-water chemical data it does show the importance of soil moisture to any calculation of CO_2 budgets in soil profiles. In this regard infiltration and groundwater recharge may be important mechanisms for CO_2 redistribution or losses from soil profiles.

CONCLUSIONS

The following conclusions are consistent with the data presented in this paper:

(1) Soil P_{CO_2} shows expected strong seasonal variations and the degree to which CO_2 production during the growing season can increase soil CO_2 content depends on the depth to the water table.

(2) $\delta^{13}C$ of soil CO_2 in the study area shows no significant seasonal variations.

(3) The ^{13}C isotopic composition of water-table groundwater samples are consistent with the isotopic equilibration of soil-water with soil-gas CO_2 before recharge. However, since a 50/50 contribution to the DIC from soil gas and soil carbonate could also account for the observed ^{13}C isotopic composition of groundwater assuming no isotopic exchange, it cannot be concluded that isotopic equilibration has in fact occurred. A comparison of ^{14}C measurements for soil-gas and groundwater samples from the water table may answer this problem.

(4) Soil water may attain supersaturation with respect to calcite in soils containing dolomite in addition to calcite. This is caused by due to the slow dissolution of dolomite after calcite saturation is reached or by degassing of soil water in response to seasonal losses of CO_2.

(5) In calcareous soils with developed carbonate leached zones, Fe- and Mn-oxyhydroxides may be precipitated at the leached—unleached zone boundary as mildly acid soil water displaced across this boundary is neutralized by reaction with soil carbonates.

ACKNOWLEDGEMENTS

The authors express their appreciation to J.A. Cherry and S. Feenstra for their valuable comments on the manuscript and to W. Cameron for assistance in the field.

REFERENCES

Albertsen, M., 1977. Und Felduntersuchungen zum Gasaustausch zweischen Grundwasser und Atmosphäre über natürlichen und verunreinigten Grundwasser. Ph.D. Thesis, University of Kiel, Kiel (unpublished).
Cope, G., 1978. Variations in the depth of carbonate leaching in sands at Trout Creek, Ontario. B.Sc. Thesis, University of Waterloo, Waterloo, Ont. (unpublished).
Deines, P., Langmuir, D. and Harmon, R.S., 1974. Stable carbon isotopes to indicate the presence or absence of a gas phase in the evolution of carbonate groundwaters. Geochim. Cosmochim. Acta, 38: 1147—1164.
De Jong, E. and Schappert, H.J.V., 1972. Calculation of soil respiration and activity from CO_2 profiles in the soil. Soil Sci., 113: 328—333.
Edwards, N.T., 1975. Effects of temperature and moisture on carbon dioxide evolution in a mixed deciduous forest floor. Soil Sci. Soc. Am. Proc., 39: 361—365.
Fritz, P., Reardon, E.J., Barker, J., Brown, R.M., Cherry, J.A., Killey, R.W.D. and MacNaughton, D., 1978. The carbon isotope geochemistry of a small groundwater system in Northeastern Ontario. Water Resour. Res., 14: 1059—1067.
Garret, H.E. and Cox, G.S., 1973. Carbon dioxide evolution from the floor of an oak—hickory forest. Soil Sci. Soc. Am. Proc., 37: 641—644.
Lai, Sung-Ho, Tiedja, J.M. and Erickson, A.E., 1976. In situ measurements of gas diffusion coefficients in soils. Soil Sci. Soc. Am. Proc., 40: 3—6.
Langmuir, D., 1971. The geochemistry of some carbonate groundwaters in central Pennsylvania. Geochim. Cosmochim. Acta, 35: 1023—1045.
Lundegardh, H., 1927. Carbon dioxide evolution of soil and crop growth. Soil Sci., 23: 417—453.
Matthess, G., 1973. Die Beschaffenheit des Grundwassers. In: Lehrbuch der Hydrogeologie, Vol. 2, Borntraeger, Berlin, 324 pp.
Millington, R.J., 1959. Gas diffusion in porous media. Science, 130: 100—102.
Nilouskaya, N.T., Kovalenko, V.K. and Laptev, V.V., 1970. Uptake and liberation of carbon dioxide by plants and micro-organisms under artificial environmental conditions. J. Plant Physiol., 17: 567—572.
Novakovic, B. and Farvolden, R.N., 1974. Investigations of groundwater from systems in Big Creek and Big Otter drainage basins, Ontario. Can. J. Earth Sci., 11: 964—975.
Plummer, L.N., Jones, B.F. and Truesdell, A.H., 1976. WATEQF. U.S. Geol. Surv., Water Res. Invest., No. 76-13, 61 pp.
Plummer, L.N., Wigley, T.M.C. and Parkhust, D.L., 1979. A critical review of calcite dissolution kinetics. Am. J. Sci. (in press).
Reardon, E.J. and Fritz, P., 1978. Computer modelling of groundwater ^{13}C and ^{14}C isotope compositions. J. Hydrol., 36: 201—224.
Rightmire, C.T. and Hanshaw, B.B., 1973. Relationship between the carbon isotope composition of soil CO_2 and dissolved carbonate species in groundwater. Water Resour. Res., 9: 958—967.
Stumm, W. and Morgan, J.J., 1970. Aquatic Chemistry. Wiley—Interscience, New York, N.Y., 583 pp.
Wallis, P., 1975. Dissolved organic carbon in groundwater and streams. M.Sc. Thesis, University of Waterloo, Waterloo, Ont. (unpublished).
Witkamp, M., 1969. Cycles of temperatures and carbon dioxide evolution from litter and soil. Ecology, 50: 922—924.
Yakutchik, T.J. and Lammers, W., 1970. Water resources of the Big Creek drainage basin. Ont. Water Resour. Comm., Water Resour. Rep. 2, 172 pp.

[2]

ARSENIC SPECIES AS AN INDICATOR OF REDOX CONDITIONS IN GROUNDWATER

J.A. CHERRY[1], A.U. SHAIKH[2], D.E. TALLMAN[2] and R.V. NICHOLSON[1]

[1] *Department of Earth Sciences, University of Waterloo, Waterloo, Ont. N2L 3G1 (Canada)*
[2] *Department of Chemistry, North Dakota State University, Fargo, ND 58102 (U.S.A.)*

(Accepted for publication May 28, 1979)

ABSTRACT

Cherry, J.A., Shaikh, A.U., Tallman, D.E. and Nicholson, R.V., 1979. Arsenic species as an indicator of redox conditions in groundwater. In: W. Back and D.A. Stephenson (Guest-Editors), Contemporary Hydrogeology — The George Burke Maxey Memorial Volume. J. Hydrol., 43: 373—392.

Although the thermodynamically based concept of oxidation—reduction potential has for many decades been an accepted tool for interpretation of the chemistry of hydrochemical systems, attempts at measurement of actual redox levels in natural waters have been fraught with difficulty. Existing methods of measurement involve use of potential-sensing inert metal electrodes or analytical determination of redox-indicator species such as dissolved O_2 or Fe^{2+} or redox couples such as SO_4^{2-}—HS^- and HCO_3^-—CH_4. As a result of recent advances in analytical methods, it is now possible to determine the concentrations of both As(III) and As(V) at sufficiently low levels so that the apparent redox condition, as pE or Eh, can be computed from measured concentrations of As(III) and As(V) species. The arsenic pE or Eh domain obtained using published thermodynamic data for As species and the assumption of redox equilibrium, provides a basis for obtaining an indication of redox levels within the central portion of the redox field for natural waters. The redox domain for the As couple is largest at high total dissolved As concentrations, but even at concentrations as low as 1—10 µg/l the domain has significant extent.

Oxidation and reduction of As(III) and As(V) in laboratory trials with redox agents common to natural waters, such as O_2, H_2S and Fe, suggests that oxidation or reduction of As species in natural waters occurs at rates sufficiently slow to enable water samples to be collected, transported and analysed before excessive change in species distribution takes place, but rapid enough for As species to adjust to the dominant redox condition of the water if periods of years or longer are available for equilibration. Because of the long equilibration time and the position of the pE—pH domain for the As couple, groundwater is best suited for use of As as a redox indicator.

INTRODUCTION

Based on equilibrium thermodynamic considerations the oxidation—reduction status of aqueous solutions is described by the parameters pE, which is a measure of the electron activity of the solution in a manner that is analogous to pH, or Eh, which is the thermodynamic redox potential. Eh is expressed in terms of volts and pE, represents the negative logarithm of the electron activity. These two parameters serve the same purpose and are related by the expression:

$$pE = (F/2.3RT)Eh \tag{1}$$

where F is Faraday's constant; R is the international gas constant; and T is temperature in kelvin. In the literature of the geosciences prior to the 1970's, Eh was used almost exclusively to express the redox condition of natural waters. More recently pE, largely for computational and conceptual convenience, has come into common use. Natural geochemical systems are commonly interpreted within the framework of equilibrium thermodynamics with pE or Eh and pH as master variables. This approach is exemplified in the texts by Garrels and Christ (1965), and Krauskopf (1967).

There is little doubt that during the 1950's and 1960's the use of equilibrium thermodynamic concepts with emphasis on the use of Eh and pH as master variables fostered considerable progress in the understanding of the geochemistry of natural waters. A severe limiting factor, however, has been the difficulty of acquiring meaningful measurements of the redox conditions of natural waters. The purpose of this paper is to draw attention to the potential for use of naturally occurring concentrations of dissolved species of As in the +3 (III) and +5 (V) valence states as an indicator of the redox status of aqueous systems, with emphasis on applicability in groundwater studies. This work has been prompted by recent advances in the capability for analysis of As(III) and As(V) in water at concentrations lower than 1.0 ng/l (Foreback, 1973; Shaikh and Tallman, 1978).

MEASUREMENT OF REDOX CONDITIONS

For measurement purposes the Eh of natural waters is commonly defined as the potential (volts) developed at an inert metallic electrode expressed relative to the normal hydrogen electrode of zero potential. Platinum is normally used as the inert electrode. A comprehensive review of the use of electrodes in geochemical redox studies is provided by Langmuir (1970). Germanov et al. (1959), and Back and Barnes (1961) have described measurement techniques based on the electrode approach. For electrode measurements to yield meaningful data the following conditions are required: (1) the important multivalent species in solution must be electroactive, that is, they must undergo voltage-generating reactions with the metallic electrode; (2) the electrode surface must be inert, that is, it must be free of "poisoning". due to reactions with dissolved oxygen, iron hydroxide or sulfide that may cause electrode coatings; and (3) reactions must be thermodynamically reversible. For the data to have exact meaning in the thermodynamic sense the redox species in the solution must be at equilibrium. This is not a requirement, however, that must be met in order for redox measurements on natural waters to be useful. Providing that the redox measurements are reproducible and reflect important relations with major redox species in the solution, valuable insight can be gained even when measured values are different than predicted, using equilibrium models. The lack of equilibrium and the need for additional information or more sophisticated theory are then made clear (Stumm and Morgan, 1970). The conditions governing electrode measurements of redox potential in natural waters commonly do not meet the three requirements indicated above. This is reflected in severely drifting electrode potential readings as a function of time or as readings that arise from potentials generated by minor redox species. Spurious and uninterpretable data are commonly acquired. Whitfield (1974), in a review of the capability of the electrode technique for measurement of Eh, indicates that its usefulness is restricted to a region that is bounded by the Pt-oxide and Pt-sulfide stability fields, which at a pH of 7 is the Eh interval of approximately 0—400 mV.

A second approach for measurement of the redox status of natural waters involves analysis of water samples for concentrations of two or more dissolved species containing the same element in different oxidation states. The concentrations are converted to activities using the Debije-Hückel or mean-salt relations and the pE or Eh values are obtained as indicated below.

The equilibrium relation for the two dissolved species, expressed as an electrochemical half-cell, is:

(reduced species) $\rightleftharpoons$ (oxidized species) + ne^-

with the redox potential (volts) expressed as:

$$Eh = E^0 + \frac{2.3RT}{nF} \log \left[\frac{\text{(activity of oxidized species)}}{\text{(activity of reduced species)}}\right] \qquad (2)$$

where: E^0 is the standard cell potential; R is the universal gas constant; n is the number of moles of electrons per mole of oxidant; T is temperature in kelvin; and F is Faraday's constant.

Or, the half-cell reaction can be expressed as an equilibrium constant from the law of mass action:

$$K_{eq} = \frac{[\text{oxid}][e^-]}{[\text{reduced}]} \qquad (3)$$

where the brackets denote activities. From this, the expression for pE follows:

$$pE = p[\text{reduced}] - p[\text{oxid}] + pK \qquad (4)$$

where p denotes the negative logarithm, base 10, of the quantity.

The quantities Eh and pE are related by eq. 1. With these relations pE or Eh values can be calculated from the results of analytical determination of redox pairs such as SO_4^{2-} and HS^- or S^{2-}, HCO_3^- or CO_2 and CH_4, NO_3^- and NH_4^+, Fe^{3+} and Fe^{2+}, or Mn(IV) and Mn(II). For a system at equilibrium, computed redox levels from each of these analytical pairs have the same value. Examples of the use of these redox couples in studies of the redox levels of natural waters are provided by Thorstenson (1970), Berner (1971) and Carpenter and Stoufer (1977). This approach to studies of pE is described by Stumm and Morgan (1970).

Although conceptually the redox-pair approach to the acquisition of pE or Eh data is attractive, its applicability to natural waters has been severely restricted because of two factors: (1) many of the important redox reactions (those involving N, S, or C) are biologically mediated, with irreversibility as a characteristic feature; and (2) in the pE—pH domain of natural waters, the concentration of one of the species in a redox pair is commonly below the limits of analytical detection, for example S^{2-} or HS^- in the SO_4^{2-}—HS^- pair, or Fe^{3+} in the Fe^{3+}—Fe^{2+} pair.

As a supplement to electrode measurements of redox potential or to the above-mentioned redox pairs, we propose the use of As pairs that arise from the presence of As(III) and As(V) in natural waters. The remainder of this paper is directed towards a review of the thermodynamic properties of these As species in the pE—pH domain typical of natural waters and presentation of the results of preliminary laboratory studies of some of the kinetic properties of arsenic behaviour in aqueous solutions.

ARSENIC AS A REDOX INDICATOR

The free energies of formation for As species occurring in aqueous solutions at 25°C are listed in Table I. Half-cell reactions for these species are indicated in Table II. From this information the pE—pH diagram shown in Fig. 1 was prepared. Each line separating the domains of two species represents points of equal activities for the two species. From each line towards the interior of a domain the proportion of the species indicated for the domain increases, while the total activity of dissolved As remains constant. Fig. 1 indicates that

TABLE I

Thermodynamic data for the relevant species

Species	State	ΔG_f^0 (kcal./mol)	Reference*	Species	State	ΔG_f^0 (kcal./mol)	Reference*
$H_3AsO_4^0$	aq	−184.0	(1)	As_2O_5	c	−186.9	(1)
$H_2AsO_4^-$	aq	−181.0	(1)	CO_3^{2-}	aq	−126.17	(2)
$HAsO_4^{2-}$	aq	−171.5	(1)	HCO_3^-	aq	−140.26	(2)
AsO_4^{3-}	aq	−155.4	(1)	$H_2CO_3^0$	aq	−148.94	(2)
$H_3AsO_3^0$	aq	−154.4	(1)	CH_4	g	−12.14	(3)
$H_2AsO_3^-$	aq	−141.8	(1)	Fe^{2+}	aq	−18.85	(2)
$HAsO_3^{2-}$	aq	−125.3	(1)	$Fe(OH)_3$(ppt)	amorph	−166.0	(2)
$HAsS_2^0$	aq	−11.61	(1)	$FeCO_3$(siderite)	c	−159.35	(2)
AsS_2^-	aq	−6.56	(1)	FeS_2(pyrite)	c	−39.9	(2)
AsS (realgar)	c	−16.81	(1)	NO_3^-	aq	−26.43	(3)
As_2S_3(orpiment)	c	−40.25	(1)	NH_4^+	aq	−19.0	(3)
As(metal)	c	0	(1)	H_2S^0	aq	−6.54	(2)
AsH_3^0	aq	23.8	(1)	HS^-	aq	−3.01	(2)
AsH_3	g	16.5	(1)	SO_4^{2-}	aq	−177	(2)
As_2O_3	c	−140.8	(1)	S	c	−0	(3)

*References: 1 = Ferguson and Gavis (1972); 2 = Hem (1977); 3 = Garrels and Christ (1965).

TABLE II

Chemical reactions and the corresponding thermodynamic equations for the arsenic–sulfur–water system

Chemical reaction	Thermodynamic equation
$H_3AsO_4^0 \rightleftharpoons H_2AsO_4^- + H^+$	$pH = 2.2$
$H_2AsO_4^- \rightleftharpoons HAsO_4^{2-} + H^+$	$pH = 6.96$
$HAsO_4^{2-} \rightleftharpoons AsO_4^{3-} + H^+$	$pH = 11.5$
$H_3AsO_3^0 \rightleftharpoons H_2AsO_3^- + H^+$	$pH = 9.2$
$H_2AsO_3^- \rightleftharpoons HAsO_3^{2-} + H^+$	$pH = 12.1$
$HAsS_2^0 \rightleftharpoons AsS_2^- + H^+$	$pH = 3.7$
$H_3AsO_3^0 + H_2O \rightleftharpoons H_3AsO_4^0 + 2H^+ + 2e^-$	$pH + pE = 9.9$
$H_3AsO_3^0 + H_2O \rightleftharpoons H_2AsO_4^- + 3H^+ + 2e^-$	$pH + \frac{2}{3}pE = 7.35$
$H_3AsO_3^0 + H_2O \rightleftharpoons HAsO_4^{2-} + 4H^+ + 2e^-$	$pH + \frac{1}{2}pE = 7.25$
$H_2AsO_3^- + H_2O \rightleftharpoons HAsO_4^{2-} + 3H^+ + 2e^-$	$pH + \frac{2}{3}pE = 6.6$
$As(c) + 3H_2O \rightleftharpoons H_3AsO_3^0 + 3H^+ + 3e^-$	$pH + pE = 1.0$
$As(c) + 3H_2O \rightleftharpoons H_2AsO_3^- + 4H^+ + 3e^-$	$pH + \frac{3}{4}pE = 3.1$
$As(c) + 3H_2O \rightleftharpoons HAsO_3^{2-} + 5H^+ + 3e^-$	$pH + \frac{3}{5}pE = 4.9$
$H_2S^0 + 4H_2O \rightleftharpoons SO_4^{2-} + 10H^+ + 8e^-$	$pH + \frac{4}{5}pE = 4.1$
$HS^- + 4H_2O \rightleftharpoons SO_4^{2-} + 9H^+ + 8e^-$	$pH + \frac{8}{9}pE = 3.8$
$H_2S^0 \rightleftharpoons S(c) + 2H^+ + 2e^-$	$pH + pE - \frac{1}{2}pH_2S = 2.4$
$S(c) + 4H_2O \rightleftharpoons SO_4^{2-} + 8H^+ + 6e^-$	$pH + \frac{3}{4}pE + \frac{1}{8}pSO_4^{2-} = 4.56$
$As(c) + H_2S^0 \rightleftharpoons AsS(c) + 2H^+ + 2e^-$	$pH + pE - \frac{1}{2}pH_2S = -3.76$
$2As(c) + 3H_2S^0 \rightleftharpoons As_2S_3 + 6H^+ + 6e^-$	$pH + pE - \frac{1}{2}pH_2S = -2.5$
$2AsS(c) + H_2S^0 \rightleftharpoons As_2S_3(c) + 2H^+ + 2e^-$	$pH + pE - \frac{1}{2}pH_2S = -0.55$
$AsS(c) + H_2S^0 \rightleftharpoons AsS_2^- + 2H^+ + e^-$	$pH + \frac{1}{2}pE + \frac{1}{2}pAsS_2^- - \frac{1}{2}pH_2S = 6.15$
$AsS_3(c) + H_2S^0 \rightleftharpoons 2AsS_2^- + 2H^+$	$pH + pAsS_2^- - \frac{1}{2}pH_2S = 12.3$

for the pE—pH realm of natural waters, As^0, $H_3AsO_3^0$, $H_2AsO_4^-$ and $HAsO_4^{2-}$ with oxidation states 0, III, V and V, respectively, are the important As species. If the water contains species of dissolved sulfur, AsS_2^- can be important species of As(III) at low redox levels.

In the lowest part of the pE—pH domain in which H_2O is stable, concentrations of As(III) in solution are limited to microgram per litre levels by the solubility of the As-sulfide minerals orpiment and realgar. There is evidence to indicate, however, that Fe(II) would limit the sulfide activity and hence As sulfide would not reach saturation (Kanamori, 1965). The location of this pE—pH zone is shown in Fig. 2. In all other pE—pH zones, however, concentrations of dissolved As are not limited by the solubility of As compounds occurring in nature. The presence of As in water is therefore primarily dependent on the availability of As in the geologic materials in contact with the water or on input from sources of contamination. In some situations concentrations of dissolved As may be limited by adsorption. Levels of total dissolved As in natural waters are generally very low, but nevertheless appreciable relative to the detection limits of modern analytical methods.

Modern analytical methods provide for species specific As detection to

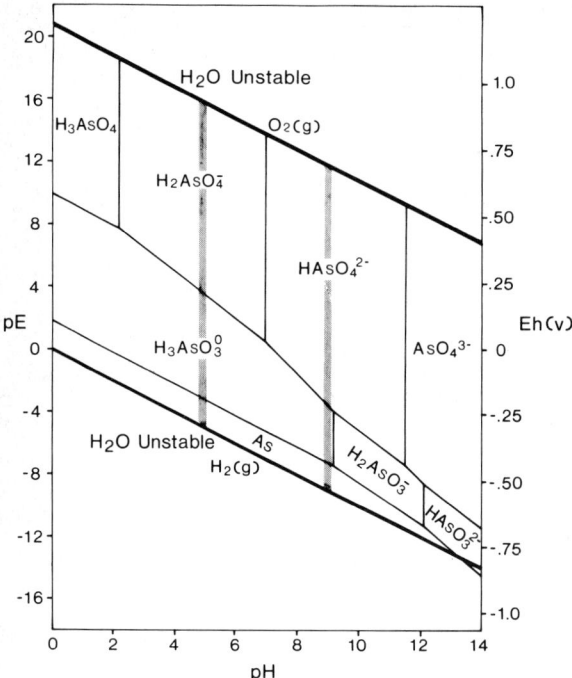

Fig. 1. pE—pH diagram for the As—H$_2$O system at 25°C. Total dissolved As species is set at $10^{-6.176}$ mol/l (50 µg/l). The *area* within the *vertical bars* represents the common pE—pH domains for natural waters.

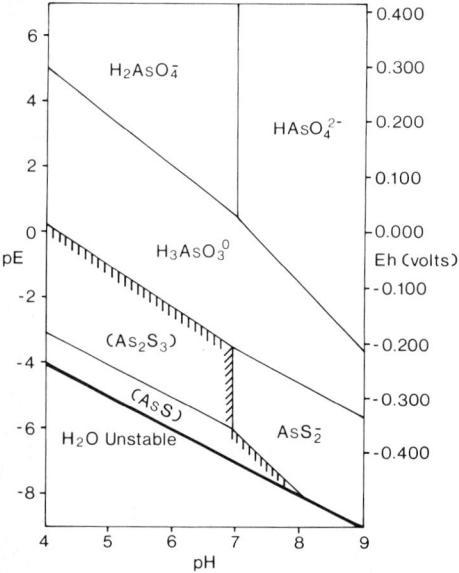

Fig. 2. pE—pH diagram for the As—S—H$_2$O system at 25°C. Total dissolved As species is set at $10^{-7.176}$ mol/l (5 µg/l) and S species is set at 10^{-3} mol/l (32 mg/l). The *area* within the *hatched lines* denotes that the solid phases are predominant (total dissolved As species are present at less than $10^{-7.176}$ mol/l (5 µg/l).

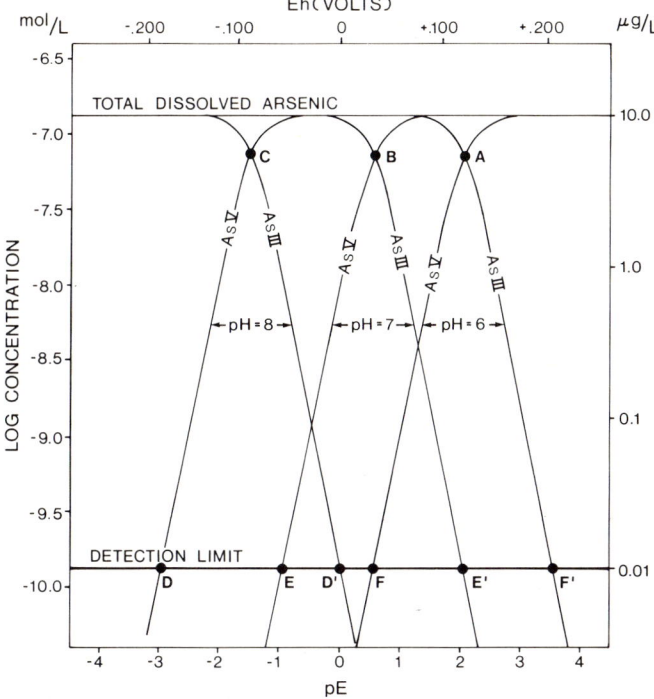

Fig. 3. Log(concentration) vs. pE diagram shows the variation of concentration of both As(III) and As(V) species with pE at pH values of 6, 7 and 8. The lettered points refer to those on Fig. 4.

1 ng by flameless atomic absorption spectrometry (Shaikh and Tallman, 1978) and to 0.05 ng by use of a d.c. discharge source and monochrometer—detector system (Foreback, 1973). For 100 ml water samples these methods yield detection limits of 10^{-2} and $5 \cdot 10^{-4}$ μg/l respectively. With either of these analytical capabilities, it is possible with waters of low total As concentration, to obtain detectable concentrations for two of the As species in pE—pH domains that are often of considerable interest.

Fig. 3 shows log concentration—pE diagrams for pH conditions of 6, 7 and 8, and a total As concentration of 10 μg/l. The detection limit of 10^{-2} μg/l, noted above for flameless atomic absorption, is represented by the low horizontal line on the graph. The upper horizontal line represents the total concentration of dissolved As, which in this example is arbitrarily specified at 10 μg/l. This is considerably below the limit specified for drinking water (50 μg/l) and therefore represents a level that is not unusually large for groundwater in many areas. The pE (or Eh) range within which both As species can be detected for each specified pH condition is indicated. The graphs were obtained from the half-cell reactions:

$$H_3AsO_3^0 + H_2O \rightleftharpoons H_2AsO_4^- + 3H^+ + 2e^- \tag{5}$$

$$H_3AsO_3^0 + H_2O \rightleftharpoons HAsO_4^{2-} + 4H^+ + 2e^- \tag{6}$$

expressed in the form:

$$pE = 11 - \tfrac{3}{2} pH - \tfrac{1}{2} \log \frac{[H_3AsO_3^0]}{[HAsO_4^{2-}]} \tag{7}$$

$$pE = 14.5 - 2pH - \tfrac{1}{2} \log \frac{[H_3AsO_3^0]}{[HAsO_4^{2-}]} \tag{8}$$

For waters within the pE intervals noted on the graphs in Fig. 3, analysis of the two As species and measurement of the in situ pH provides for calculation of the redox condition, expressed as pE, from eqs. 7 or 8, or as Eh, from eq. 2.

The width of the pE range for detection of both species, which we will refer to as the *redox window* for As, depends on the As detection limit and the total concentration of As in solution as indicated on Fig. 3. The width of the redox window increases with increasing total As values and lower limits of detection. The position of the redox window for As and its width for total As concentrations of 1, 10 and 100 μg/l are shown in Fig. 4. As a further explanation of Figs. 3 and 4, point A labelled on these two diagrams represents

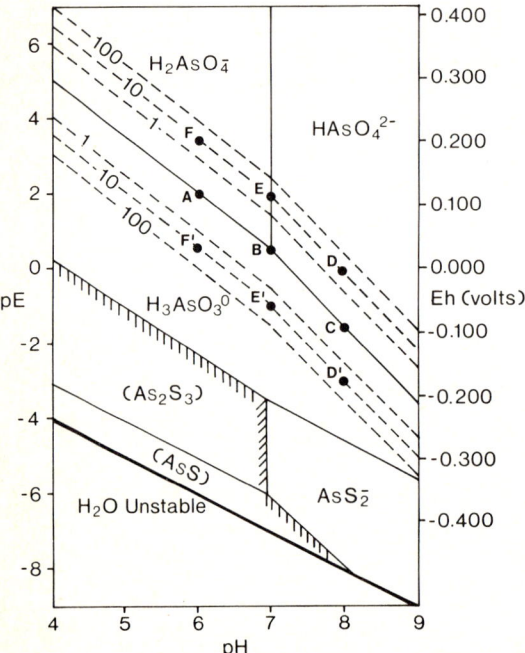

Fig. 4. The pE—pH diagram is the same as that in Fig. 2. The *numbers* on the *dashed lines* refer to total dissolved concentrations of As in μg/l. The sets of parallel lines form a *redox window* in which As may be used as a redox indicator.

a single redox condition displayed differently on each diagram. From Fig. 3 it is evident that point A on the species-domain boundary line on Fig. 4 represents a condition of equal concentration of the two species $H_2AsO_4^-$ and $H_3AsO_3^0$, at pH = 6. Points B and C represent equi-concentration conditions for pH of 7 and 8, respectively. On Fig. 3, the points D, D', E, E', F and F' are at the arbitrarily specified "detection limit" for As(III) and As(V), and correspond to those points in Fig. 4 on the pE—pH diagram. It follows that the two dashed lines representing total As concentrations of 10 μg/l on Fig. 4 depict a continuum of intersection points for pH values between 4 and 9. The domain between the upper and lower 10 μg/l lines is the As redox window for the specified detection limit for total dissolved As. The other two pairs of dashed lines represent the redox windows at the same detection limit (0.01 μg/l) for total As concentrations of 1 and 100 μg/l. The width of the redox window for each total concentration can be expanded considerably if a much lower detection limit is achieved, such as is feasible if larger water samples are collected for analysis by the method of Shaikh and Tallman (1978) or if an analytical technique with a lower detection limit, such as that described by Foreback (1973) is used.

For illustrative purposes in these diagrams, a detection limit of 0.01 μg/l is used because, in conjunction with the methods described by Shaikh and Tallman (1978), this limit can be obtained routinely using instrumentation that is available in many water quality laboratories.

The presence or absence of one of the As species, As(III) or As(V), species can be used as a qualitative indication of redox levels. If only As(V) species are detected in the water, the redox level is above the As window. In this case, at pH > 7, $HAsO_4^{2-}$ is the dominant species of As(V) and at pH < 7, $H_2AsO_4^-$ is the dominant As(V) species. At redox levels below the redox window, only As(III) species are detectable with $H_3AsO_3^0$ being the dominant species under moderately reducing conditions and AsS_2^- being important at very low redox levels.

The redox windows for other dissolved species that commonly occur in natural waters are displayed in Fig. 5, along with the redox domain in which the electrode potential method can in some situations yield useful results. The detection limits used for calculation of these redox windows are listed in Table III. These correspond to the Winkler method limit for dissolved oxygen and common laboratory limits for the N and C species. The quoted ferrous iron, sulfide and sulfate detection limits may be achieved using methods described by Lee and Stumm (1960), Butler and Locke (1976) and Kloster and King (1977). Figs. 4 and 5 indicate that, with the exception of the Fe pair, the As window is quite large relative to the redox windows for the other pairs. The As window is well above the windows of SO_4^{2-}—HS^- and HCO_3^-—CH_4 and far below the window for dissolved oxygen. The As window partially overlaps with the Fe window and occurs within the redox domain in which electrode measurements can yield useful redox values in some situations.

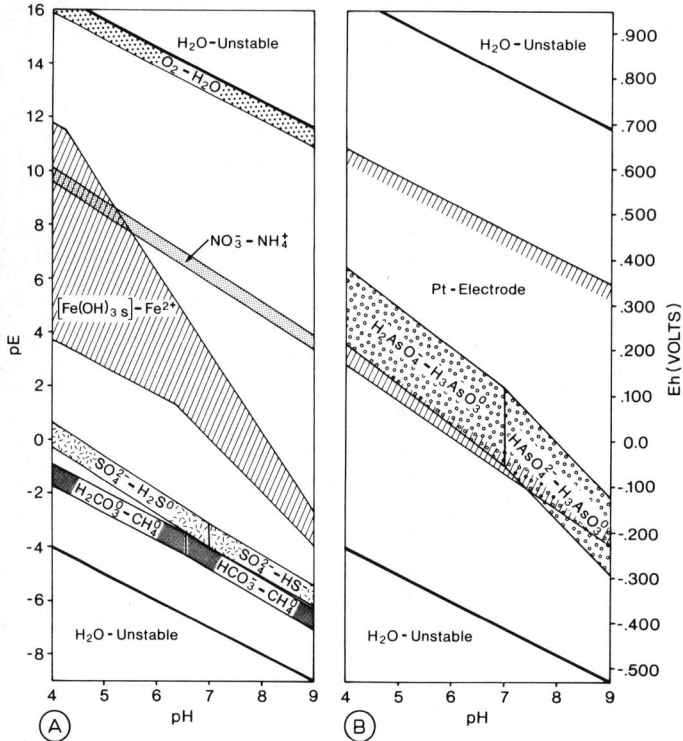

Fig. 5. A. Redox windows of various couples. The assumed total concentrations and detection limits for the species may be seen in Table III. The Fe^{2+} is assumed to be in equilibrium with the amorphous oxide phase ($Fe(OH)_3$).
B. Redox domain in which the Pt electrode may be useful and the As window as described in the text.

TABLE III

Redox couples given with the analytical detection limits and reasonable total concentrations for a groundwater environment

Couple	Species	Detection limit		Total concentration	
		(mg/l)	(µM)	(mg/l)	(mM)
H_2O—O_2	$O_2(aq)$	0.05	3.1	12	0.75
NH_4^+—NO_3^-	NH_4^+	0.01	0.7	5	0.34
	NO_3^-	0.01	0.7		
Fe^{2+}—$Fe(OH)_3$	Fe^{2+}	0.005	0.018	no limit	
$H_2S^0(HS^-)$—SO_4^{2-}	$H_2S(HS^-)$	0.005	0.16	25	0.78 (as S)
CH_4—HCO_3^-	CH_4	0.01	0.83	25	2.1 (as C)
	HCO_3^-	0.01	0.83		

Use of the Fe^{3+}—Fe^{2+} couple for estimation of redox levels is fraught with uncertainties because only one of the two Fe species is normally present in a sufficiently high concentration for analytical measurement. The other is obtained by calculation from measured pH values, with the assumption that it is in equilibrium with a specified solid phase. For example, Hem and Cropper (1959), Back and Barnes (1965), Back and Hanshaw (1965), and Hem (1977) described the use of measurements of total dissolved Fe and pH to obtain pE or Eh estimates in the part of the redox domain in which ferric hydroxide [$Fe(OH)_3$)] is in equilibrium with the predominantly ferrous solute species in solution. In this domain total dissolved Fe is almost entirely Fe^{2+} and as an approximation is set equal to Fe^{2+}. The Fe^{3+} activity is specified from the assumption that this species is in equilibrium with $Fe(OH)_3$. Ferric hydroxide as a solid phase exists in various forms, from meta-stable amorphous freshly precipitated forms through poorly crystallized material to the crystalline form. The Gibbs free-energy values for these forms vary considerably, and since the computed redox equilibria of the dissolved species are dependent on the free energy of the solid phase that is assumed to exert the controlling influence on the Fe species, there is considerable uncertainty associated with the value of the pE or Eh computed from measured Fe concentrations. More detailed discussions of the pE—pH relations for Fe solid phases are presented by Hem (1965, 1977). Fig. 5, following the approach of Hem (1977) indicates that the pE—pH domain within which ferric hydroxide is in equilibrium with predominantly ferrous solute species has a pE width, at pH 7, of 3.6—0.0, or as Eh, of 0.210—0.0 V. Below this redox window, the dissolved Fe concentration is controlled by the solubility of $FeCO_3$ (siderite) and FeS_2 (pyrite).

The position of the redox window for As is of particular interest in groundwater studies because in many groundwater flow systems, most of the water has progressed sufficiently far along the redox pathway to be devoid of dissolved oxygen and yet not sufficiently far to exhibit detectable concentrations of H_2S, HS^-, or CH_4. In much of this domain As species may serve as a redox-level indicator for comparison to or as an alternative to redox values obtained from dissolved Fe or electrode measurements.

For As to be most useful as a redox indicator it must occur in the water at concentrations that yield a useful width for the redox window. The "maximum contaminant level" set by the USEPA (1975) for As in drinking water is 50 µg/l. From Fig. 5 it is apparent that total As concentrations well below this limit are sufficiently large to produce a redox window of significant width. In a survey of As concentrations in rivers and lakes quoted by Ferguson and Gavis (1972), it was found that of 727 samples analysed, about 20% contained As at levels above 10 µg/l. Ferguson and Anderson (1974) also noted that As is the inorganic constituent that more commonly than any other constituent exceeds the maximum permissable limit for drinking water supplies. Because of the position of its redox window, As should generally be better suited as a redox indicator for use in groundwater than for surface water. However, we know of no published general surveys of As concentrations in groundwater and therefore we cannot at present specify in which hydrogeologic regions the technique can be expected to have the greatest potential for application.

EXPERIMENTAL STUDIES

For As species to be useful as an indicator of redox levels of aqueous systems, the redox reactions that control the species distribution must proceed at rates that are rapid relative to the rates of the geochemical or biogeochemical processes that control the redox level (i.e., electron activity) of the solution. Since As normally occurs in only very small concentrations in natural waters, its speciation involves transfer of only a small number of electrons relative to the transfers that control the dominant redox pairs in the system. Therefore, if the distribution of As species attains equilibrium rapidly, the distribution will be an index of the redox level of the water.

A second major requirement for As species behaviour must also be met. The rate of oxidation of As(III) to As(V) species must be sufficiently slow to allow for sample storage, preparation and analysis of the species. To provide at least a semi-quantitative basis for appraisal of the rates of As oxidation and reduction in aqueous solutions, the laboratory experiments described below were performed.

The apparatus and methodology used for As speciation analysis in this study are described by Shaikh and Tallman (1978). Distilled deionized water was used to prepare all solutions. All reagents were of analytical grade and were used without further purification. Ultrahigh purity hydrogen sulfide, oxygen, and hydrogen gases were used in those experiments involving gaseous oxidants and reductants. No detectable amount of As was found in any of the reagents employed to oxidize As(III) or reduce As(V) species, or in solutions used to buffer the experimental solutions. Sodium borohydride (Alfa Chemicals, Inc.) used to generate arsine in the analysis procedure was found to contain a small amount of As and, hence, all data were blank corrected. All reagents were stored and reactions conducted in Pyrex® glass containers so as to minimize the loss of As (Pierce and Brown, 1976).

To assess the effect of pH on the stability of the As(III)/As(V) ratio a series of solutions of pH between 2 and 10.5 were prepared in deoxygenated buffer solutions. Deaeration was attained by passage of ultrahigh-purity N_2 through the solutions for 1.5 hr. 1.0 ml each of 5 ppm As(III) and 5 ppm As(V) were taken and the volume was made up to 100 ml with appropriate deoxygenated buffer. These solutions, each of which contained 50 ppb of As(III) and As(V), were sealed and analysed periodically for the two As species (Table IV).

The influence of various redox agents on the As(III)/As(V) ratio was examined by preparing solutions to contain the appropriate amount of oxidizing or reducing agent in deoxygenated water. Each solution was then made 50 ppb each in As(III) and As(V) as described above. A "control" solution was prepared in deoxygenated distilled water containing the same

TABLE IV

Stability of As(III)/As(V) ratio as a function of time

Solution	pH	1–2 days		12–13 days		19–20 days		71–78 days	
		[As(III)] (mol%)	% change in total As	[As(III)] (mol%)	% change in total As	[As(III)] (mol%)	% change in total As	[As(III)] (mol%)	% change in total As
HCl + NaOH, $\mu = 0.1$	2	50.0	0	53.4	+0.4	55.1	+0.2	48.9	−8.0
Acetate + acetic acid, $\mu = 0.1$	4	51.6	−2.0	53.1	−1.0	54.1	0	48.6	−12.4
Acetate + acetic acid, $\mu = 0.1$	6	50.1	−1.4	51.1	−1.8	52.3	−0.6	47.8	−9.6
Standard phosphate, $\mu = 0.05$	7	49.6	0	50.6	−2.8	52.2	−1.2	45.2	−10.2
Phosphate + NaOH, $\mu = 0.05$	8.5	50.5	+1.4	52.4	0	52.3	+1.0	44.6	−9.0
Standard phosphate buffer, $\mu = 0.05$	10	51.4	0	52.2	−2.4	51.8	+0.4	43.4	−11.6

Reactions were carried out with different pH media as described above. Each solution contains 50 ppb each in As(III) and As(V).

amounts of As(III) and As(V) but no oxidizing or reducing agents. When molecular oxygen or hydrogen was used as the redox reagent, distilled water was saturated with either ultrapure O_2 or H_2 for 3—4 hr. and these solutions were used immediately to prepare the As(III)/As(V) reaction solutions. Concentrations of O_2 and H_2 in solution were estimated from standard solubility values. When H_2S was employed as a reducing agent, distilled water was first saturated with H_2S for 1 hr., then diluted by a factor of 10 with deoxygenated distilled water and was then used to prepare the As(III)/As(V) reaction solution. From solubility values and pH of the solution, the concentration of total H_2S was calculated.

For the pH dependence studies of the oxidizing agent Fe^{3+}, HCl—NaOH was used for pH adjustment at pH 2, and acetate buffer at pH 5. The phosphate buffer was found to form a precipitate with Fe^{3+} at pH 7 and, therefore, pH 7 studies involving Fe^{3+} were conducted in the absence of buffer. For H_2S, pH was adjusted to 4, 5 or 7 either by adding HCl or NaOH. These solutions were used to prepare reaction solutions, 50 ppb in each As(III) and As(V) as described above. Each of these solutions was analyzed periodically for As(III) and As(V).

RESULTS

The results of analyses of a series of solutions containing As(III) and As(V) in the absence of added redox agents at various pH levels between 2 and 10.5 analyzed over a period of 2.5 months are summarized in Table IV. The ratio remains essentially unchanged within the precision of our measurement for a period of approximately three weeks. However, slow oxidation of As(III) may be occurring, with approximately 5—7% As(III) converted to As(V) after 2.5 months, possibly through air oxidation. The total As also remained constant over a period of three weeks. After 2.5 months, a 10% loss in total As was observed. It is possible that the slow increase in the As(V)/As(III) ratio attributed to oxidation may in fact reflect a preferential loss of As(III) from solution. It is interesting to note that the decrease in total As does not exhibit a significant pH dependence over the range investigated. The mechanism of loss of As from laboratory solutions is in need of further investigation, but in any event, the results of this study show that at pH's between 2 and 10.5 the As(III)/As(V) ratio remains relatively constant for up to three weeks without appreciable loss of As from solution. Repetitive analyses of filtered acid-preserved groundwater samples showed similar time dependent changes in total As and species distribution. Thus it should be possible to sample groundwater or other natural environments, return the sample preserved with acid to the laboratory and carry out the analysis without fear of perturbing the As(III)/As(V) ratio, providing the entire operation is carried out within a week or two.

From the results described above, it would appear that the As(III)/As(V) ratio under the influence of a wide range of pH is quite stable relative to the

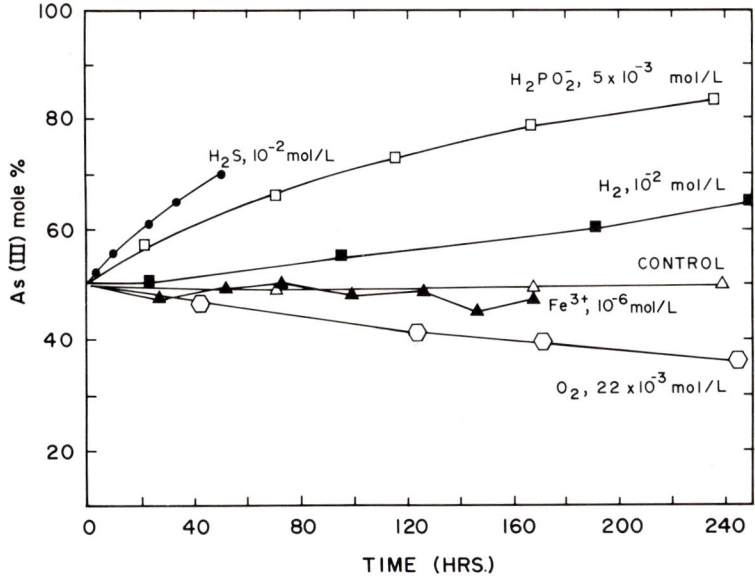

Fig. 6. Effect of various redox agents on As(III) and As(V). Each solution initially containing 50 ppb As(III) and 50 ppb As(V) at pH 7.0.

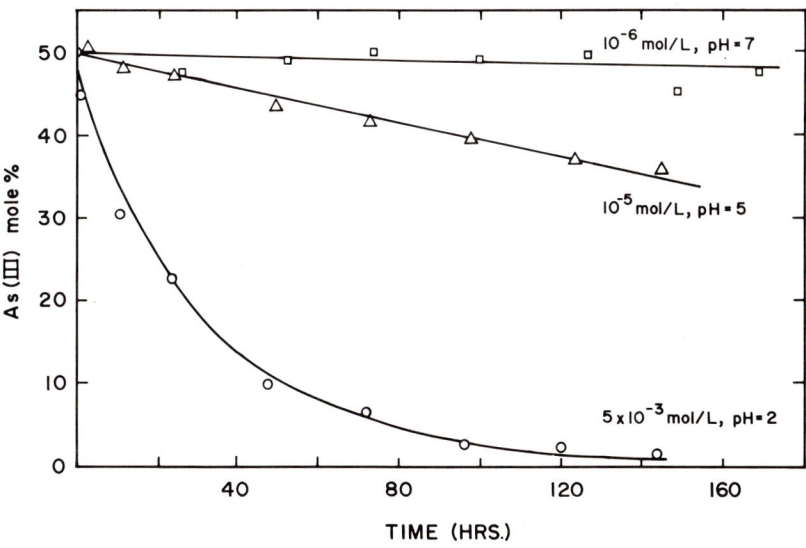

Fig. 7. pH dependence of Fe^{3+} oxidation of As(III). Each solution initially containing 50 ppb As(III) and 50 ppb As(V).

time scale of sample storage and preparation for analysis. The next question that arises is whether or not the ratio changes rapidly enough under the influence of redox agents so as to be a useful sensor of redox levels in natural aqueous environments. The results of the oxidation and reduction rate experiments must be appraised in this context.

The results of the oxidation and reduction rate experiments using H_2S, $H_2PO_2^-$, H_2, Fe^{3+} and O_2 are shown in Fig. 6, which indicates the As(III)/As(V) ratio changes appreciably over a time period of days as opposed to seconds or minutes and in the case of Fe^{3+} over much larger time periods. The rates of oxidation of As(III) by Fe^{3+} and the reduction of As(V) by H_2S were found to be pH dependent. Over a period of one week at pH 7, the rate of Fe^{3+} oxidation of As(III) is barely discernable, at pH 5 the change is appreciable, and at pH 2 almost all of the As(III) is oxidized to As(V) (Fig. 7). This behavior is most likely due to a decrease, at higher pH, of ferric ions as a result of the low solubility of ferric oxy-hydroxide phase. Differentiating the concentration and pH effects would require further study. Maximum Fe^{3+} concentrations were desired in order to achieve readily distinguishable trends within a reasonable experimental time period.

Fig. 8 shows a similar behaviour for the reduction of As(V) by H_2S. In this case the redox behaviour of As species can be accounted for if free molecular H_2S is the only dissolved S species capable of reducing As(V). In the pH range used in these experiments, total dissolved H_2S consists of $H_2S_{(aq)}$ and HS^-. At pH 7 (25°C and 1 bar), calculations based on a pK value of 7.00 (Latimer, 1952) indicate that 50% of the total dissolved sulphide is $H_2S_{(aq)}$. The ratio of H_2S/HS^- increases ten-fold for every one unit decrease of pH.

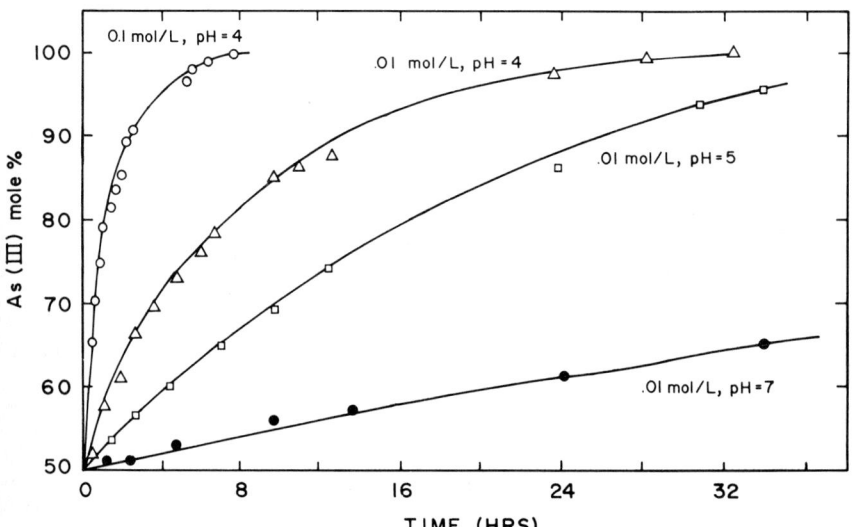

Fig. 8. pH dependence of H_2S reduction of As(V). Each solution initially containing 50 ppb As(III) and 50 ppb As(V). Total H_2S concentration is specified.

DISCUSSION

The results of the experimental studies indicate that the As(III)/As(V) ratio is quite stable under a wide range of pH conditions in the absence of known redox agents. Samples of natural waters can be kept in storage under normal conditions for many days without undergoing changes in the ratio. Fig. 6 indicates that even when water contains saturation levels of dissolved O_2, the rate of As(III)/As(V) change is slow on a laboratory time scale. These results indicate that it should be a feasible task to acquire water samples from the field, transport them to the laboratory, and store them for many days in tight Pyrex® glass containers without changes in the As species ratio from in situ values.

The results of the oxidation and reduction rate experiments suggest that if long periods of time are available, the As ratio in water containing redox agents such as H_2S, $H_2PO_2^-$, Fe^{3+}, H_2 and O_2 will adjust to a value indicative of the redox level determined by these species. The concentrations of the redox agents used in the experiments were much larger than those normally found in natural waters. The concentrations were chosen with a view to observing changes in the As(III)/As(V) on a time scale appropriate for laboratory work. The variation of the initial rates of reaction for the oxidizing and reducing agents indicated in Fig. 6 indicates first-order dependence on the concentration of the redox agent. At the concentrations of H_2S, $H_2PO_2^-$, H_2, Fe^{3+}, Fe^{2+} and O_2 that occur in natural water, oxidation (or reduction) of As(III) (or As(V)) could require many months or years rather than days or weeks observed in laboratory experiments at much higher concentrations.

Of the various environments in which natural waters exist, it is reasonable to expect that it would be the groundwater environment for which there would be the best possibility of using As species for sensing redox levels. Except in shallow zones near the water table or in flow regimes in highly permeable fractured or jointed rocks, groundwater is normally quite old. In many areas groundwater travels at rates less than a few metres per week. It commonly has many years, decades, or longer to adjust to the geochemical or biogeochemical influences of the subsurface geological domain in which it resides in a condition of slow movement. In a particular groundwater zone the As in solution would represent the accumulation of As acquired from the soil and geological strata through which the groundwater has passed enroute to its present location. Provided that the groundwater is moving slowly along its flow paths, it is expected that the ratio of As(III)/As(V) would adjust to the electron activity of the solution, which would be dominated by redox pairs such as NO_3^-—NH_4^+, Fe(III)—Fe^{2+}, Mn(IV)—Mn^{2+}, SO_4^{2-}—HS^-, or H_2S, HCO_3^- (or H_2CO_3)—CH_4.

SUMMARY OF CONCLUSIONS

Recent advances in analytical methods now make it possible to determine

As(III) and As(V) in natural waters at the nanogram per litre concentration level. It is therefore possible in waters that contain total dissolved As at the microgram per litre level to measure detectable concentrations of both As(III) and As(V) in a redox domain that occupies the central portion of the redox stability field for water at surface or normal groundwater temperatures and pressures. Detection of both species of As provides for calculation of the redox level of the water.

Laboratory studies of the rates of oxidation of As(III) and reduction of As(V) in aqueous solution with redox agents common to natural waters indicate that the oxidation rate upon exposure to oxygen is slow enough to expect sample collection and preparation for analysis to be accomplished prior to significant change in the As(V)/As(III) ratio. Reduction of As(V) in the presence of H_2S and oxidation of As(III) in the presence of Fe^{3+} were observed to occur at slow but significant rates over time periods of tens of hours. The concentrations of As and redox agents used in the experiments were large relative to concentrations typical of natural waters. In natural waters rates of oxidation or reduction would therefore be expected to be much slower than the laboratory rates but nevertheless significant over long time periods. From this it is concluded that use of the As(III) and (V) redox pair as an indicator of redox level would be most appropriate for groundwater rather than most surface waters because groundwater commonly has slow flow rates and residence times of many decades, or longer.

Like all of the other redox pairs that have been used by various investigators as indicators of redox levels and like the electrode approach, use of the As species pair as a redox indicator will undoubtedly have its limitations. We conclude, however, that the arsenic method is sufficiently promising to warrant its application and evaluation in a variety of groundwater environments.

In the calculation of redox conditions directly from concentrations of the As species determined by chemical analysis, it is assumed that the concentrations of the dissolved As represent the free ionic concentrations, such as $H_2AsO_4^-$, $HAsO_4^{2-}$ and $H_3AsO_3^0$. This, however, is not precise, since these activities should be corrected for the occurrence of As included in ion pairs and complexes in the water. Preliminary estimates using data for the analogous phosphate species indicate that at low concentration of As encountered in natural water systems, the extent of ion-pair or complex formation is likely to be negligible. Studies are currently under way to identify the major As complexes and their stability constants. As well, field studies are in progress in several hydrogeologic settings for delineation of dissolved As species and other more conventional redox indicators. It may be appropriate in future studies to direct some attention towards Se species as well as As species. The redox window for Se is in a higher pE range than As, but is well within the range of interest for many groundwaters. Analytical methods for Se species are described by Shomoisihi and Toei (1978).

REFERENCES

Back, W. and Barnes, I., 1961. Equipment for field measurement of electrochemical potentials. U.S. Geol. Surv., Water-Supply Pap. 424-C, pp. 366—368.

Back, W. and Barnes, I., 1965. Relation of electrochemical potentials and iron content to groundwater flow patterns. U.S. Geol. Surv., Prof. Pap. 498-C, 16 pp.

Back, W. and Hanshaw, B., 1965. Chemical geohydrology. In: V.T. Chow (Editor), Advances in Hydroscience, Vol. 1, Academic Press, New York, N.Y., pp. 49—109.

Berner, R.A., 1971. Principles of Chemical Sedimentology. McGraw-Hill, New York, N.Y., 240 pp.

Butler, J.W. and Locke, D.N., 1976. Photometric end point detection of the Ba-thorin titration of sulphates. Brinkman Instruments Inc., New York, N.Y., 7 pp.

Carpenter, A.B. and Stoufer, R.N., 1977. Influence of organic carbon and micro-organisms on iron and sulphide concentrations in groundwater. Mo. Water Resour. Res. Cent., N.T.I.S. Publ., PB-270 640, 85 pp.

Ferguson, J.F. and Anderson, M.A., 1974. Chemical forms of arsenic in water supplies and their removal. In: A.J. Rubin (Editor), Chemistry of Water Supply, Treatment and Distribution. Ann Arbor Science Publishers, Ann Arbor, Mich., pp. 137—158.

Ferguson, J.F. and Gavis, J., 1972. A review of the arsenic cycle in natural waters. Water Res., 6: 1259—1274.

Foreback, C.C., 1973. Some studies on the detection and determination of mercury, arsenic and antimony in gas discharges. Ph.D. Dissertation, University Microfilms International, Ann Arbor, Mich., 115 pp.

Garrels, R.M. and Christ, C.L., 1965. Solutions, Minerals and Equilibria. Freeman, Cooper and Co., San Francisco, Calif., 450 pp.

Germanov, A.I., Volkov, C.A., Lisitsin, A.K. and Serebrennikov, V.S., 1959. Investigation of the oxidation—reduction potential of groundwaters. Geochemistry (U.S.S.R.), 3: 322—329 (translated from the Russian).

Hem, J.D., 1965. Equilibrium chemistry of iron in groundwater. In: S.D. Faust and J.V. Hunter (Editors), Principles and Applications of Water Chemistry. Wiley, New York, N.Y., pp. 625—643.

Hem, J.D., 1977. Reactions of metal ions at surfaces of hydrous iron oxide. Geochim. Cosmochim. Acta, 41: 527—538.

Hem, J.D. and Cropper, W.H., 1959. Survey of ferrous—ferric chemical equilibria and redox potentials. U.S. Geol. Surv., Water-Supply Pap. 1459-A, 29 pp.

Kanamori, S., 1965. Geochemical study of arsenic in natural waters, III. The significance of ferric hydroxide precipitate in stratification and sedimentation. J. Earth Sci., Nagoya Univ., 13: 46—57.

Kloster, M.B. and King, M.P., 1977. The determination of sulfide with DPD. J. Am. Water Works Assoc., Oct. 1977: 544—546.

Krauskopf, K.B., 1967. Introduction to Geochemistry. McGraw-Hill, New York, N.Y., 721 pp.

Langmuir, D., 1970. Eh—pH determinations. In: R.E. Carver (Editor), Methods of Sedimentary Petrology. Wiley—Interscience, New York, N.Y., pp. 597—635.

Langmuir, D. and Whittemore, D.O., 1971. Variations in the stability of precipitated ferric oxyhydroxides. In: Non Equilibrium Systems in Natural Water Chemistry. Adv. Chem. Ser., 106: 209—234.

Latimer, W.M., 1952. The Oxidation States of the Elements and their Potentials in Aqueous Solutions. Prentice-Hall, Englewood Cliffs, N.J., 392 pp., 2nd ed.

Lee, G.F. and Stumm, W., 1960. Determination of ferrous iron in the presence of ferric iron with Batho-phenanthroline. J. Am. Water Works Assoc., 52: 1567—1574.

Pierce, F.D. and Brown, H.R., 1976. Inorganic interference study of automated arsenic and selenium determination with atomic absorption spectroscopy. Anal. Chem., 48: 693—695.

Shaikh, A.V. and Tallman, D.E., 1978. Species-specific analysis for nanogram quantities of arsenic in natural waters by arsine generation followed by graphite furance atomic absorption spectrometry. Anal. Chim. Acta, 98: 251—259.

Shomoisihi, Y. and Toei, K., 1978. The gas chromatographic determination of Se(IV) and total Se in natural water with 1,2-diamino-3,5-dibromo benzene. Anal. Chim. Acta, 100: 65.

Stumm, W. and Morgan, J.J., 1970. Aquatic Chemistry. Wiley—Interscience, New York, N.Y., 583 pp.

Thorstenson, D.C., 1970. Equilibrium distribution of small organic molecules in natural waters. Geochim. Cosmochim. Acta, 34: 745—770.

USEPA (U.S. Environmental Protection Agency), 1975. National interim primary drinking water regulations. U.S. Environ. Prot. Agency, Fed. Regist., 40, No. 248.

Whitfield, M., 1974. Thermodynamic limitations on the use of the platinum electrode in Eh measurements. Limnol. Oceanogr., 19: 857—865.

[2]

MODERN MARINE SEDIMENTS AS A NATURAL ANALOG TO THE CHEMICALLY STRESSED ENVIRONMENT OF A LANDFILL

MARY JO BAEDECKER and WILLIAM BACK

U.S. Geological Survey, National Center 432, Reston, VA 22092 (U.S.A.)

(Accepted for publication May 10, 1979)

ABSTRACT

Baedecker, M.J. and Back, W., 1979. Modern marine sediments as a natural analog to the chemically stressed environment of a landfill. In: W. Back and D.A. Stephenson (Guest-Editors), Contemporary Hydrogeology — The George Burke Maxey Memorial Volume. J. Hydrol., 43: 393—414.

Chemical reactions that occur in landfills are analogous to those reactions that occur in marine sediments. Lateral zonation of C, N, S, O, H, Fe and Mn species in landfills is similar to the vertical zonation of these species in marine sediments and results from the following reaction sequence: (1) oxidation of C, N and S species in the presence of dissolved free oxygen to HCO_3^-, NO_3^- and SO_4^{2-}; (2) after consumption of molecular oxygen, then NO_3^- is reduced, and Fe and Mn are solubilized; (3) SO_4^{2-} is reduced to sulfide; and (4) organic compounds become the source of oxygen, and CH_4 and NH_4^+ are formed as fermentation products. In a landfill in Delaware the oxidation potential increases downgradient and the redox zones in the reducing plume are characterized by: CH_4, NH_4^+, Fe^{2+}, Mn^{2+}, HCO_3^- and NO_3^-. Lack of SO_4^{2-} at that landfill eliminates the sulfide zone. Although it has not been observed at landfills, mineral alteration should result in precipitation of pyrite and/or siderite downgradient. Controls on the pH of leachate are the relative rates of production of HCO_3^-, NH_4^+ and CH_4. Production of methane by fermentation at landfills results in ^{13}C isotope fractionation and the accumulation of isotopically heavy ΣCO_2 (+10 to +18$^o/_{oo}$ PDB). Isotope measurements may be useful to determine the extent of CO_2 reduction in landfills and extent of dilution downgradient. The boundaries of reaction zones in stressed aquifers are determined by head distribution and flow velocity. Thus, if the groundwater flow is rapid relative to reaction rates, redox zones will develop downgradient. Where groundwater flow velocities are low the zones will overlap to the extent that they may be indeterminate.

INTRODUCTION

Problems associated with the movement of leachate generated by waste disposal are increasing. It has become clear that in order to manage disposal sites basic questions must be answered such as, should leachate be contained, allowed to purify, or to be treated; what are the reaction rates involved in the breakdown of constituents; and how can the increasing need for disposal be met without altering significantly the quality of potable water. To answer these questions a basic understanding of geochemical reactions that occur in these environments is necessary.

Significant contributions to the understanding of groundwater flow systems are being made through studies of low-temperature aqueous geochemistry, by adapting principles from other disciplines, and by comparative studies from other segments of the hydrochemical cycle. This paper applies principles of organic geochemistry and marine geochemistry to provide additional understanding of a chemically stressed groundwater system. Also a discussion of fractionation of isotopes associated with organic reactions and generation of gases demonstrates their role in groundwater chemistry. The analogy between interstitial waters of marine sediments and stressed groundwater systems is based primarily on the types of oxidation—reduction reactions that occur within the various redox zones during decomposition of organic compounds.

Concern with the chemical character of leachate and its affect on groundwater quality has resulted in numerous studies on monitoring landfills and identification of reactions involving major cations and anions (Hughes et al., 1976; Kunkle and Shade, 1976), heavy metals (Suarez and Langmuir, 1976; Griffin et al., 1977), organic constituents (Dunlap et al., 1976; Chian and DeWalle, 1977), redox zonation (Apgar and Langmuir, 1971; Matthess, 1972), and isotopes (Games and Hayes, 1976; Fritz et al., 1976). The purpose of this paper is: (1) to discuss the reactions and processes that mobilize and attenuate chemical constituents in stressed hydrologic systems with emphasis on those parameters which are most responsive to chemical reactions and hydrogeologic changes, such as pH, Eh, concentrations of Fe, Mn, C-, N- and S-containing species; (2) to discuss the significance of isotope fractionation resulting from production of gases by biochemical processes; and (3) to compare reactions in chemically stressed groundwater systems with those reactions occurring in marine sediments.

The composition and behavior of leachate in a hydrologic system are controlled by biological, geochemical and psysical processes. Physical processes include dilution and filtration, that are influenced in part by groundwater flow (Langmuir, 1972). Biogeochemical reactions include oxidation—reduction, solution—precipitation, ion exchange, adsorption, degradation and isotopic fractionation. In many cases different reactions are interrelated, such as the dependence of chemical mobilization of Fe and Mn on the biological decomposition of organic material. Many of these reactions are attenuation processes that result in reduction of solute concentration and an increase in pH; others such as solution of feldspars, carbonates and clays result in an increase of ions in solution.

Although organic material can decompose by non-biological chemical reactions, the biologically controlled reactions are generally faster and more likely to dominate in the near-surface environments. Evidence of this includes the large amounts of gases formed in leachates as end-products of the decomposition of organic material. Many organic compounds formed as intermadiates during processes of decomposition are sparingly soluble in water and are attenuated by sorption on minerals, especially the clays. This attenuation may be either by adsorption or ion exchange. Weiss (1969) demonstrated that organic ions can occupy exchange sites on clays and that polar organic molecules are adsorbed on clay surfaces.

Although organic constituents are much more difficult to determine quantitatively than most inorganic constituents, large numbers of organic compounds in water have been identified by the advances in liquid chromatography and gas chromatography coupled with mass spectrometry (e.g., see Robez, 1973). Disposal sites are favored environments for the study of aqueous organic reactions because the concentrations of compounds are sufficiently high to permit identification. Although improved methods for extraction of organic compounds from an aqueous phase include organic solvent partitioning, macroreticular resins (Leenheer and Huffman, 1976) and membrane ultrafiltration and gel permeation chromatography (Chian, 1977) problems of concentration and quantitative separation still are major obstacles in analyses [see Giger and Roberts (1978) for a recent review].

DEGRADATION OF ORGANIC MATTER

Generation of gases

High concentrations (up to several parts per thousand) of organic carbon in leachate results from decomposition of organic-rich materials such as domestic waste, sewage sludge, stumps, trees and discarded chemicals. In addition, when the bacterial population declines, the C, N and P originally removed from the leachate as nutrients for cell synthesis are returned to the system as organic bio-polymers. The number of possible compounds appears limitless; however, bacterial decomposition through well-defined pathways results in the formation of certain compounds of lower molecular weight which may be useful indicators for characterizing leachate. These include alcohols, amino acids and carboxylic acids; the former are infrequently reported because of the difficulty of analysis in water. Numerous studies have reported volatile acids, with concentrations up to 470 mg/l in landfills (Baedecker and Back, 1979) and 6,900 mg/l in oil-field brines (Willey et al., 1975). Chian and DeWalle (1977) characterized soluble organic matter in leachate and identified free volatile fatty acids as the most abundant fraction with a fulvic-like material being the next largest.

Generation of large amounts of organic acids should lower the pH. However, this does not always occur because many other reactions affect the hydrogen ion concentration in reducing environments. Organic acid anions in solution contribute to total alkalinity measurements and in some cases may control the alkalinity. Organic compounds such as fatty acids, amino acids and carbohydrates degrade in reducing environments by anaerobic decay using nitrate, sulfate or organic compounds as a source of oxygen in the absence of molecular oxygen. The end-products of complete dissimilation by bacteria are CO_2, CH_4, NH_3, H_2S, and H_2. Hydrogen gas is seldom detected in the breakdown of organic matter, especially if fatty acids are produced. In fact, the biochemical evolution of H_2 is unfavorable under standard conditions unless the partial pressure of H_2 is reduced by continuous H_2 removal (Wolin, 1974). Any H_2 gas formed will be utilized immediately by CH_4 producers (Toerien and Hattingh, 1969).

Although the conditions of formation are restrictive, CH_4 is generated in many environments such as anaerobic sewage digestors, waste disposal sites, poorly drained swamps, paddy soils, anoxic freshwater lake bottoms and marine sediments. Laboratory studies of CH_4 formation in soils which have been kept submerged show that an increase of CH_4 coincides with a decrease of organic acids, particularly acetic acid (Takai, 1970). The principal mechanisms of CH_4 formation in anaerobic environments are the degradation of acetic acid or the hydrogenation of CO_2:

$$\overset{\star}{C}H_3 COOH \rightarrow \overset{\star}{C}H_4 + CO_2 \qquad (1)$$

$$CO_2 + 4H_2 \rightarrow CH_4 + 2H_2O \qquad (2)$$

^{14}C labelling experiments have shown that 70% of the CH_4 formed in mixed-culture sewage sludges is from the methyl group of acetic acid (Smith and Mah, 1966; Jeris and McCarty, 1965). Although the reaction mechanism is not well understood CO_2, can be reduced by other anaerobes utilizing H_2 gas, fatty acids or alcohols as hydrogen donors (Takai, 1970):

$$4H_2 + 2CO_2 \rightarrow CH_3COOH + 2H_2O \qquad (3)$$

which provides another mechanism for hydrogen removal in anaerobic environments.

In addition to the generation of gases, the exchange of gases can modify the chemical composition of landfill leachate. Where gas exchange with oxygen of the atmosphere can occur it favors aerobic reactions, which increase the extent and rate of degradation. However, Apgar and Langmuir (1971) showed that in humid environments anaerobic leachate can move through ~12 m (40 ft.) of soil with little effective renovation. Within the saturated zone, restricted gas exchange favors anaerobic reactions.

pH controls

Controls on the pH of landfill leachate are not well understood. The main control in natural waters, the CO_2–HCO_3^-–CO_3^{2-} system, is modified by other reactions in environments where large amounts of organic matter are decomposing. It is surprising that many landfills in different environments have H^+ concentrations in the $10^{-6.5}$–10^{-7} m range. Field pH measurements have been reported within this range for landfills emplaced in a variety of terranes including sand, glacial till, carbonate and glauconitic clay. This narrow range of pH occurs regardless of whether the upgradient water has either a higher or lower pH. Apparently, reactions occurring in an organic-rich environment significantly increase the concentration of solutes to generate a highly buffered solution which is independent of the H^+ concentration of the less buffered upgradient water. These values from landfills are similar to the pH range of 6.5—6.7 observed in flooded paddy soils where CH_4 was forming (Takai, 1970). In extreme situations the pH can be controlled by the disposal of certain constituents such as caustic soda, ash and mine tailings. However, for normal disposal of domestic and industrial waste, reactions involving the C, N, S and Fe systems control the pH.

The reactions most important in controlling pH of leachate assuming minimal dilution are: (1) degradation of organic material producing CO_2 and smaller amounts of NH_3, which form the HCO_3^-, H^+ and NH_4^+ ions; (2) reduction of CO_2 to CH_4; (3) reduction of SO_4^{2-} to H_2S and additional CO_2; (4) reduction of NO_3^- to N_2 or NH_4^+ and additional CO_2; (5) reduction of $Fe(OH)_3$ to Fe^{2+}; (6) silicate hydrolysis of feldspars; and (7) H^+ exchange on clays. The first three are generally the most important reactions because of the high concentrations of these species. For example, the highest concentrations of HCO_3^- (including organic acid anions), NH_4^+ and CH_4 found in a Delaware landfill were 73, 34 and 1.4 mmol/l respectively (Baedecker and Back, 1979). Although in this particular landfill SO_4^{2-} concentration is low, reactions that generate H_2S and CO_2 are significant in landfills with high dissolved SO_4^{2-}. The similarity of pH's in landfills is due in part to the production of NH_4^+ which provides a sink for H^+ and prevents the formation of acid water.

Although many of the above reactions occurring in landfills are similar to those in marine sediments, typical pH's for the latter are in the 7—8 range while landfills are generally in the 6.5—7 range. The pH in landfills is lower primarily because of generation of large amounts of CO_2 from waste materials which forms carbonic acid as an intermediate product and dissociates to H^+ and HCO_3^-. Whereas, in marine sediments buffering by dissolution of calcareous material provides a major source of CO_3^{2-} which consumes H^+ to raise the pH. In some marine sediments where sulfate reduction is a dominant reaction the pH may be close to that of landfills because of the generation of H^+:

$$SO_4^{2-} + 2CH_2O \rightarrow HS^- + 2HCO_3^- + H^+$$

Enrichment of $\delta^{13}C$ in dissolved CO_2

During diagenesis biogenic CO_2 is generally formed with an isotopic composition about the same as the original organic matter. However, the formation of CH_4 by methanogenic bacteria during diagenesis results in isotopically-light CH_4 and isotopically-heavy dissolved CO_2. Heavy CO_2 has been

TABLE I

Sequence of oxidation—reduction reactions

Aerobic respiration:

$C_6H_{12}O_6$ + 6O_2 → 6CO_2 + 6H_2O
glucose

$CH_3CH(NH_2)COOH$ + O_2 → CH_3COOH + NH_3 + CO_2
alanine

Nitrification:

NH_4^+ + 1½O_2 → 2H^+ + H_2O + NO_2^-

NO_2^- + ½O_2 → NO_3^-

Nitrate reduction:

6NO_3^- + 5CH_3OH → 5CO_2 + 7H_2O + 6OH^- + 3N_2

Mn reduction:

2MnO_2 + 4H^+ + CH_2O → 2Mn^{2+} + 3H_2O + CO_2

Fe reduction:

4$Fe(OH)_3$ + 8H^+ + CH_2O → 4Fe^{2+} + 11H_2O + CO_2

Sulfate reduction:

SO_4^{2-} + H^+ + 2CH_2O → HS^- + 2CO_2 + 2H_2O

Fermentation:

$C_6H_{12}O_6$ + 2H_2O → 2CH_3COOH + 2CO_2 + 4H_2
glucose

$CH_3CH(NH_2)COOH$ + 2H_2O → NH_3 + CH_3COOH + CO_2 + 2H_2
alanine

CO_2 + 4H_2 → CH_4 + 2H_2O

---

found in natural gases (Wasserburg et al., 1963), interstitial water from a reducing fjord (Nissenbaum et al., 1972), sewage sludge, and landfill gases. The isotopic fractionation factor, $\alpha^{(*)}$ between ΣCO_2 (HCO_3^-, CO_3^{2-}, H_2CO_3, $CO_{2\,aq}$) and CH_4 for the above studies ranges from 1.054 to 1.076. The variation of fractionation factors probably results from the particular microbial community that develops in response to available nutrients (Games and Hayes, 1976) and is further complicated by the occurrence of two possible processes for CH_4 formation — by CO_2 reduction, if a hydrogen source is available or by acetate decomposition (Table I).

Another factor that may affect isotopic composition is the type of refuse that decomposes. However, a surprising uniformity exists in reported $\delta^{13}C$ values for ΣCO_2 in landfills. Games and Hayes (1976) report values of

---

*α (CO_2/CH_4) = ($\delta^{13}CO_2 - \delta^{13}CH_4$)/($\delta^{13}CH_4$ + 1000) + 1
($\delta^{13}C$ measurements relative to PDB standard.)

+16.1, +16.6 and +20.0‰ for three separate landfills and +17.0‰ for a laboratory refuse reactor that reached steady state in isotopic content between CO_2 and CH_4. These landfills differed in the type of refuse deposited and length of time since burial.

$\delta^{13}C$ measurements for ΣCO_2 from three locations at a Delaware landfill were +10.3, +15.3 and +18.4‰. The lightest value (most enriched in ^{12}C) was from the newest part of the landfill while the heaviest value was from the oldest area of the landfill (Baedecker and Back, 1979). An explanation for the heavy values in the older part is that although CH_4 is still forming, relatively little CO_2, is being generated and, therefore, a steady state with maximum isotopic fractionation exists between the two species. Two possible explanations for the isotopic values in the newer area are: (1) a smaller percentage of the ΣCO_2 has been reduced to CH_4 than in the older part and thus the $\delta^{13}C$ value for ΣCO_2 is not as heavy (+10.3‰); and (2) CO_2 is still being generated in large quantities which dilute the effects of fractionation and causes the isotopic value to be lighter. The latter explanation is favored on the basis of HCO_3^- concentrations. HCO_3^- concentrations for the newer part are significantly higher (35 mequiv.) than for the older portion of the landfill (20 mequiv./l), while the pH at both sites was 6.5. The lower HCO_3^- value at the older site supports the concept of less CO_2 being generated.

A Rayleigh plot similar to that used by Nissenbaum et al. (1972) in a reducing fjord, and Claypool and Kaplan (1974) in reducing deep-sea sediments can be used to graph the change in isotopic composition of CO_2 against the percent of CO_2 that has been reduced. The following equation is plotted in Fig.1:

$$\delta^{13}CO_2 = [\delta^{13}CO_2 + 1000(1 - x/100)^{1/\alpha - 1}] - 1000$$

where $x/100$ is the percent CO_2 reduced, α the fractionation factor (1.070, Games and Hayes, 1976), and $\delta^{13}CO_2$ is the initial isotopic composition of the organic matter. The initial $\delta^{13}C$ value used, $-25.5‰$, is the average for several nearby wells, not affected by landfill leachate, in which all natural organic matter has been oxidized. If we assume that the organic sources in the landfill have isotopic C values similar to terrigenous C ($-25‰$) the plot indicates (Fig.1) that 42—47% of the original CO_2 formed has been reduced to CH_4. However, this may not be a valid assumption owing to various C sources in the refuse. If we assume a range of initial $\delta^{13}C$ values of $-18‰$ for concrete and $-38‰$ for commercial CO_2 (T. Coplen, pers. commun., 1979) between 34 and 56% of the original CO_2 has been reduced. This interpretation also assumes that biological CH_4 production is solely from CO_2 and that CO_2 is neither added nor removed by other processes. Although these conditions cannot be met, compared to the work of others it appears that maximum fractionation has been achieved, at least on the older portion of the landfill. At this site the high rate of methane fermentation prevents isotopic dilution and ΣCO_2 remains at a constant isotopic composition.

Isotopic values for ΣCO_2 in anaerobic water downgradient from the Delaware landfill range from +7.0 to $-18.8^o/_{oo}$. The main control on this wide range in values is dilution caused by mixing of the leachate with groundwater which migrates in response to heavy pumping downgradient from the landfill. It is suggested that C isotopic measurements are useful in predicting the extent of contaminant movement and the progress of reactions in stressed environments.

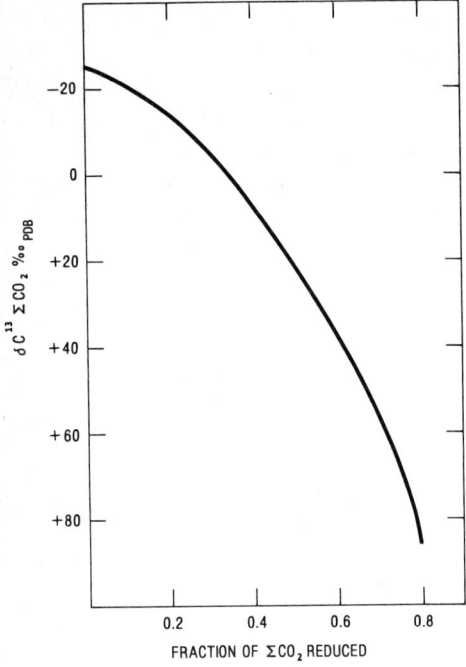

Fig.1. C-isotope fractionation of residual ΣCO_2 in closed-system single-stage Rayleigh distallation process.

MINERAL SOLUBILITY

As is well documented in the literature, controls on solution and precipitation of calcareous minerals are temperature, pH, ionic strength and concentration of Ca, Mg and HCO_3^- (see, e.g., Hanshaw and Back, 1979). However, effects of organic degradation on carbonate geochemistry have been less studied. Reactions which appear to be of primary influence are generation of gases (CO_2 and CH_4), formation of organic acids as intermediate products and formation of organic complexes of Ca. Although CO_2 is one of the final products in essentially all degradation reactions, its relative significance on mineral solubility is determined largely by particular reaction paths and reaction progress. In general, any overall series of reactions that increase H$^+$ concentration will cause solution of calcareous minerals and conversely any series of reactions that cause a pH increase can lead to precipitation of calcite.

It is equally valid that an increase in partial pressure of CO_2 in contact with minerals will decrease pH and permit solution of calcite; conversely a decrease in CO_2 by outgassing or some other process will raise the pH and favor supersaturation with respect to calcareous minerals. However, depending on other controls on pH and occurrence of mineral reactions much of the biogenetically generated CO_2 can be converted to HCO_3^- which can attain concentrations high enough to cause supersaturation.

Another source of HCO_3^- in sediments results from oxidation of organic material, including CH_4, by reduction of sulfate. This reaction occurs extensively on tidal flats and in the oceans and contributes HCO_3^- to form calcareous crusts. Although generation of HCO_3^- from CH_4 by sulfate reduction occurs at landfills the possibility of concomitant precipitation of calcite has not yet been investigated. However, even in the Delaware landfill where sulfate reduction is not a source of HCO_3^- because of the low concentration of SO_4^{2-} the saturation index, SI$^{(*)}$, for calcite was 0.67 and 0.54. These values indicate supersaturation and in some environments calcite would precipitate, thereby establishing equilibrium concentration of Ca^{2+} and HCO_3^-. If these SI values are typical of landfills, perhaps factors other than calcite solubility are controls on Ca and HCO_3^- concentrations.

Calcite precipitation may be prevented in a landfill due to formation of calcium soaps or salts of fatty acids similar to those formed in marine sediments during alkaline putrefication (Berner, 1971). The reactions are:

$$NH_3 + \underset{\text{fatty acid}}{RCOOH} \rightarrow NH_4^+ + RCOO^- \quad \text{and} \quad Ca^{2+} + 2RCOO^- \rightarrow Ca(RCOO)_2$$

where R = hydrocarbon chain. In the presence of excess free fatty acids Ca ions can form Ca-soaps which are more stable than calcite, thereby making Ca unavailable for calcite precipitation. On the other hand, when fatty acids are used up Ca-soap becomes unstable and can form calcite and hydrocarbons. Other factors that may retard calcite precipitation are the presence of hydrated Mg^{2+} ions which compete for sites on the carbonate crystals (Pytkowicz, 1965) and, more likely, in landfills, the absorption of polar organic compounds on $CaCO_3$ which blocks nucleation sites in marine water thus preventing precipitation (Berner, 1971). In calculating saturation indeces for landfills it is critical that calcium complexation is considered and that alkalinity reflects only HCO_3^- and not anions of organic acids, otherwise

---

*SI = log [(iron activity product)/(equilibrium constant)].

the SI will indicate erroneously high supersaturation for calcite. This was demonstrated in oil-field brines by Willey et al. (1975), and Carothers and Kharaka (1978).

High concentrations of cations (Ca, Mg, Na and K) in refuse originate in part from the trash, however, the distribution of cations suggests mineral dissolution is also important. Aluminosilicate minerals are generally in contact with leachate where cover material has been added and at the periphery of the pit. However, mineralogic and geochemical studies have not yet been undertaken at landfills to identify which reactions may release cations. It is well established that silicate and aluminosilicate minerals are dissolved in water with H^+ ions by such reactions as:

$$Mg_2SiO_4 + 4H^+ \rightarrow 2Mg^{2+} + H_4SiO_4$$

or by the weathering of silicates by hydrolysis where the water molecule itself is the source of the hydrogen ion made available by presence of CO_2 as follows:

$$feldspar + H_2O + CO_2 \rightarrow clay + cations + HCO_3^- + SiO_2$$

It would follow that organic degradation reactions from landfills which generate either H^+ or CO_2 will facilitate weathering of minerals. Experiments by Huang and Keller (1970) indicate that solubility of minerals is higher in organic acids than in natural water and probably results from the formation of cation—acid complexes. A recent review on mineralogy and chemistry of lake sediments (Jones and Bowser, 1979) includes a thorough discussion of processes of mobilization and fate of the major cations. Applications of results of these studies to cores of aquifer material contaminated by leachate may be fruitful in identifying the sources of major cations.

In addition to dissolving feldspars, leachate and its products of decomposition can alter clays. Functional groups of organic molecules may coordinate to surfaces and edges of minerals; certain clays can accomodate organic molecules in interlattice positions and retard degradation of the organic compounds (Degens and Mopper, 1976). They also demonstrated in the laboratory that clays may act as catalysts, for example, heterocyclic nitrogen compounds were synthesized from CO_2 and NH_3 in presence of an aqueous slurry of kaolinite. If reactions of this type occur at landfills with caustic wastes, it would provide another carbon and nitrogen sink, and possibly result in the synthesis of more toxic compounds.

Another clay—leachate reaction is sorption of leachate-generated ammonium ion (NH_4^+) onto clays in exchange for cations which could be a significant source of Ca, Mg, Na and K in some environments (Baedecker and Back, 1979). Ion exchange can also affect pH of the leachate in those situations where H^+ is exchanged for K^+ on clays.

OXIDATION—REDUCTION REACTIONS

Since the first measurements of oxidation potentials in groundwater, downgradient zonation of redox potentials and their significance in natural water have been recognized. Oxidation—reduction potentials, whether actually measured with a Pt electrode, calculated from various redox couples, or used conceptually as a qualitative model provide a basis for interpretation of controlling reactions in marine sediments and in natural and contaminated groundwater regimes. Cherry et al. (1979) provide a readily available discussion of the theoretical background for use of redox potentials in groundwater.

Elements sensitive to redox changes are: (1) those that exist in nature in more than one valence state, such as, Fe and Mn; and (2) those that make up most organic and some inorganic substances, C, N, S, H and O. In natural water the equilibrium species in the C—N—S—H$_2$O system at 25°C and 1-atm. total pressure are CH_4, CO_3^{2-}, HCO_3^-, H_2CO_3, NH_4OH, NH_4^+, N_2, NO_3^-, H_2S, HS^- S^{2-}, HSO_4^- and SO_4^{2-} (Thorstenson, 1970). Most of these same constituents are found in water from many landfills. Fields of dominance of constituents of interest are shown on a standard Eh—pH diagram (Fig.2). The sequence of reactions from oxidizing to reducing which generate these species and others are given in Table I. This sequence, has been observed in marine systems by many, including Richards (1965), Claypool et al. (1973), and Martens and

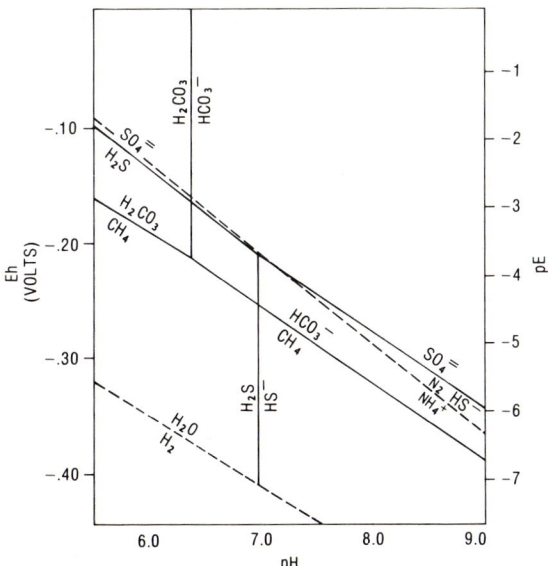

Fig.2. Eh—pH diagram showing fields of predominance of dissolved species in the system C—N—S—H—O at 25°C and 1-atm. total pressure; for C, N and S concentrations of 10^{-3}, 10^{-3} and 10^{-2} m, respectively (from Thorstenson, 1970).

Berner (1977). The sequence of coexisting phases during oxidation with increasing Eh and pH < 7 should be CH_4–H_2S–NH_4^+; HCO_3^-–H_2S–NH_4^+; HCO_3^-–SO_4^{2-}–NO_3^-. The NH_4^+–NO_3^- boundary is not shown on Fig.2, but at pH 7 it would be at an Eh of about +0.35 V. Because the pH range of marine sediments is generally 7–8 while many landfills are in the 6–7 range this sequence of reducing reactions will occur at an Eh higher in landfills than in anaerobic marine sediments (Fig.2).

In addition to thermodynamic stabilities of species mentioned above, Fe and Mn are important in the sequence of redox reactions. Reactions for reduction and mobilization of Fe and Mn in natural groundwater systems are given in Table I. Concentration and distribution of Fe and Mn in groundwater near a landfill is similar to that in both marine and lacustrine environments. Although both elements are more soluble in their reduced forms (Mn^{2+} and Fe^{2+}), the manganous ion is less sensitive than the ferrous ion to redox changes and remains more mobile. The pronounced difference in behavior of Fe and Mn is clearly indicated (Fig.3) where the field of dominance of Mn^{2+} covers a far wider range of Eh and pH values than does Fe^{2+}. In anaerobic water with a pH range of 6–8 the predominant species are: Fe^{2+} and Mn^{2+}; in the presence of sulfide, FeS_2 is dominant; and with high concentration of metals and CO_2 species, both $MnCO_3$ and $FeCO_3$ can form. At higher Eh values oxides and hydroxides become dominant species. The enhanced mobility of Mn has been demonstrated on the basis of fluxes of dis-

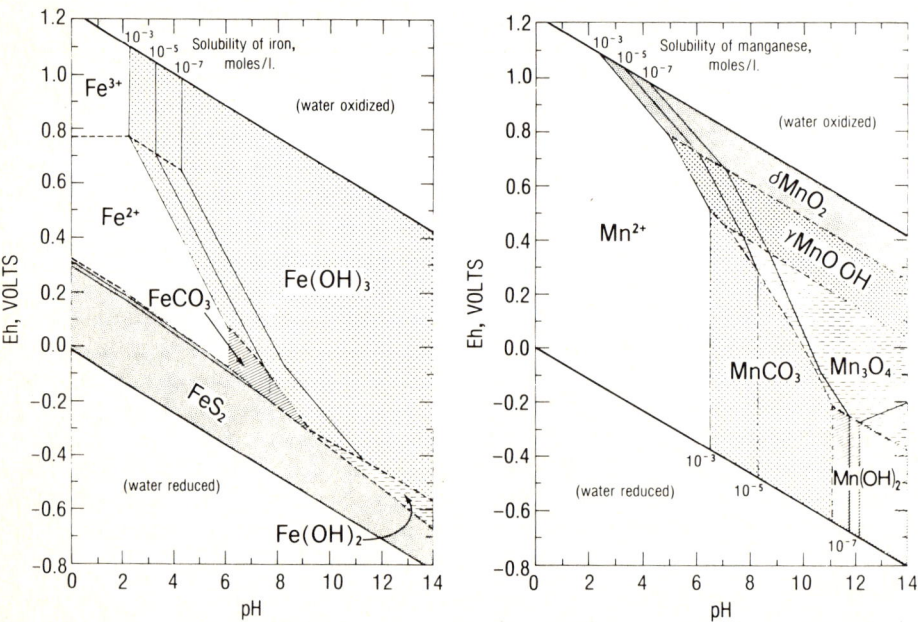

Fig.3. Eh–pH diagrams for Fe and Mn phases; for 25°C, 1-atm. total pressure, and total C and S concentrations of 10^{-3} and 10^{-4} m, respectively (from Jones and Bowser, 1979).

solved components from ocean sediments (Manheim, 1976) and is also an important consideration in formation of ferromagnesian nodules.

The ability of Fe—Mn-oxides to sorb and selectively coprecipitate other transition metals means these oxides are important controls in exchange of trace metals and nutrients between sediments and water. It is this ability to sorb and coprecipitate other metals that makes the Fe—Mn reactions of such great importance at landfills. These oxides, particularly Mn, can either release or attenuate trace metals. In both contaminated and natural soils taken from beneath a landfill Mn-rich oxides have about a tenfold higher heavy-metal percentage than Fe-rich oxides (Suarez and Langmuir, 1976). A lower pH (6—7) such as encountered in most landfills apparently enhances the sorptive capacity of Mn-oxides relative to Fe-oxides. Because of this greater coprecipitation potential and the higher solubility of Mn- over Fe-oxides, heavy metals can be transported farther in Mn-rich soils and aquifer materials which are in contact with leachate. During dissolution of Mn^{2+} by anaerobic leachate, associated heavy metals, especially Co, Ni, Cu and Zn are also mobilized. Another process of partitioning between Fe and Mn which may occur at some landfills is removal of Fe by flocculation of Fe—organic acid colloids as observed in estuaries by Sholkovitz (1978).

If these reactions involving C, N, S, O and H species and Fe and Mn proceed as predicted by thermodynamics, the stepwise reaction sequence discussed would be observed as Eh and pH varies. But, because organic matter is involved in most of these reduction—oxidation reactions and both kinetic and thermodynamic factors control processes, several reactions probably occur simultaneously. Thus, there may be considerable overlap of reaction zones in both marine sediments and stressed groundwater systems.

CONCEPT OF ZONES RELATED TO GROUNDWATER FLOW

The concept of redox zonation in regional aquifer systems was demonstrated in the Atlantic Coastal Plain by Back and Barnes (1965) where they observed a progressive lowering of Eh along direction of groundwater flow. The decrease in Eh was attributed to groundwater contact with lignite which caused Fe reduction. This concept was further developed by Edmunds (1973) in the Lincolnshire Limestone where a sharp decrease in Eh was associated with loss of oxygen and appearance of HS^- and Fe^{2+}. Champs et al. (1979) compared such redox zones of four separate aquifer systems, including the two previously mentioned and named biochemical zones on the basis of redox processes: oxygen—nitrate, iron—manganese and sulfide.

The similarity of reactions that occur in waste-disposal areas and organic-rich marine sediments can be further explored by considering the effect of movement of constituents away from the area of generation. Although bioturbation is an important process in the upper few centimeters of marine sediments (Goldhaber et al., 1977), at lower depths diffusion of solutes in the pore water is the main transport process that results in vertical concentration gradients. This is in contrast to waste-disposal sites where dilution and dispersion by lateral movement of groundwater are the most important transport processes.

A great deal of mineralogic and geochemical work has been done to understand processes that generate biogeochemical zones in marine sediments and to identify reactions that occur within them that control the concentrations and distribution of redox-sensitive constituents. These zones in a vertical section of a reducing sedimentary marine environment are primarily a consequence of ecological succession as shown in a simplified sketch (Fig.4). The major zones are: two anaerobic zones, carbonate- and sulfate-reducing; an aerobic zone which occurs in the upper sediments and water up to the zone where photosynthesis occurs. The boundaries of these zones are not rigidly fixed and in organic-rich sediments the anaerobic zone may include the lower portion of the water column. The carbonate-reducing zone has been depleted in S by sulfate reduction at an earlier stage before the zone became so highly reducing, and is now characterized by CH_4 and NH_4^+ which are stable in more reducing environments. In the upper part of the anaerobic section organic particulate matter obtains oxygen for its decomposition from SO_4^{2-}. This reaction (Fig.4) releases Fe, Mn and associated metals and generates H_2S and HCO_3^-. The H_2S can combine with Fe to form an iron sulfide layer while releasing H^+ to form H_2 in the reducing zone. Part of the released Fe and Mn migrate upward to the oxygenated zone where Fe—Mn-oxide layers and nodules can form. The generated HCO_3^- can diffuse upward to provide a sufficient concentration for the interstitial water or ocean water to become supersaturated with respect to calcite and cause the precipition of calcareous crusts.

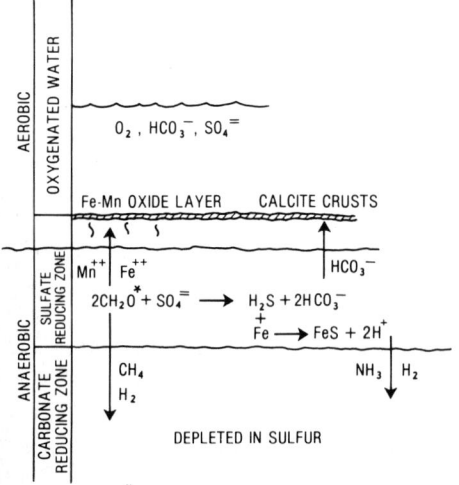

Fig.4. Schematic zonation with associated reactions in organic-rich marine sediments.

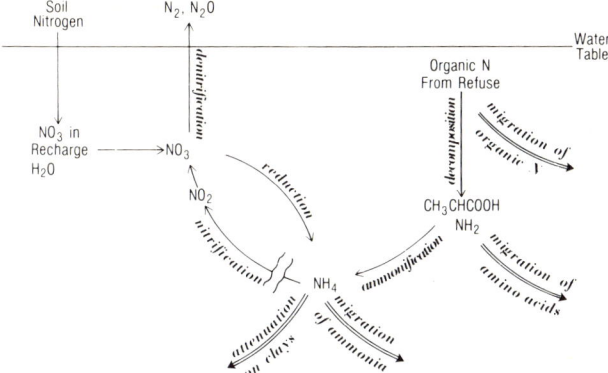

Fig.5. Schematic flow-chart showing reactions of N-containing compounds below the water table in an anaerobic environment (from Baedecker and Back, 1979).

This vertical biogeochemical zonation is similar in many respects to the lateral zonation downgradient from waste-disposal sites. For example, a study of disposal pits (Leggat et al., 1972) showed that groundwater pollution, resulting from waste, mobilized minerals originally stable in the floodplain deposits. After the pits had been in operation a few years, chemical oxygen demand and organic C concentrations showed the same general increase as specific conductance and chloride in downgradient wells. The data indicated that the movement of Mn and Fe "fronts" provided strong evidence of actual movement of leachate. As conditions became more reducing, H_2S produced by reduction of sulfate appeared in wells. Mapping concentrations of Mn^{2+}, Fe^{2+} and H_2S showed the position of each front and provided a basis for predicting arrival of the next reducing front.

Three biochemical zones — an anaerobic zone, a transition zone and an aerobic zone — were identified around landfills in Germany (Matthess, 1972; Golwer et al., 1975). The boundaries of these zones in space and time are controlled by dilution and self-purification processes which include: (1) microbiological decomposition of organic material where micro-organisms obtain their oxygen from nitrates and sulfates in the reducing zone; (2) precipitation of iron sulfide and coprecipitation of trace elements with Fe- and Mn-hydroxides in the transition zone; and (3) escape of gases and oxidation reactions resulting from availability of oxygen from soil air or from groundwater in the aerobic zone.

These zones were also observed in and downgradient from a landfill in Delaware (Baedecker and Back, 1979). Boundaries for the zones were determined on the basis of dissolved oxygen and using Kjd N/$NO_3^{(*)}$ as a redox couple. The reaction path of N species in anaerobic water (Fig.5) shows that

---

*Kjd N or Kjeldahl nitrogen refers to the analytical procedure that determines reduced nitrogen, i.e., organic N and NH_4^+.

in absence of nitrification the main sinks for N are NH_4^+ and organic N. As leachate mixes with oxygenated groundwater downgradient, the ratio $KjdN/NO_3$ decreases (Fig.6). High Fe and Mn concentrations were also observed in the anoxic zone and were related to redox zonation.

Slight oxidation of Fe^{2+} ions causes precipitation of ferric compounds while Mn^{2+} remains in solution. One consequence of this relative solubility is that as leachate moves downgradient Mn can move farther into oxygenated groundwater. This is shown by the decrease of Fe/Mn ratios (Fig.7) downgradient from the landfill. This is similar to the effects of slow oxidation of Fe and Mn transported to lakes as organic and inorganic complexes. As the metals are released, relative solubility is demonstrated by a decrease of Fe/Mn ratios in lake waters away from rivers discharging into them (Jones and Bowser, 1979).

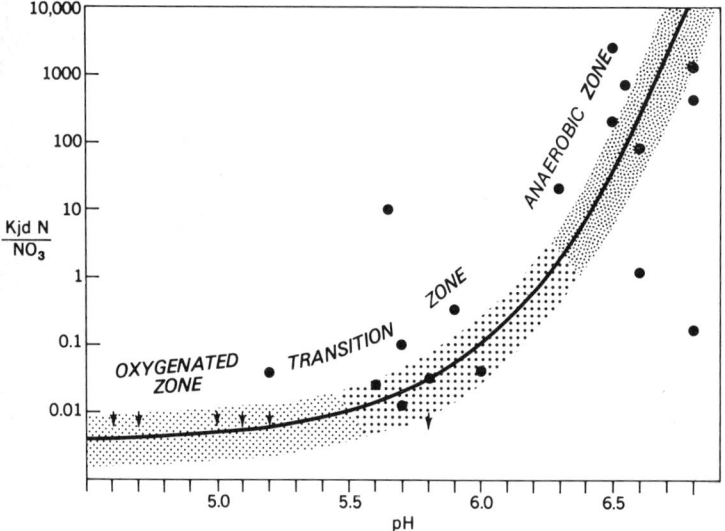

Fig.6. Relationship of nitrogen index ($KjdN/NO_3$) and pH for water in and downgradient from a landfill (from Baedecker and Back, 1979).

Consideration of field investigations such as those above, along with the sequence of reactions (Table I) and fields of dominance of ionic species controlled by Eh and pH (Figs. 2 and 3) it is possible to construct a hypothetical lateral zonation downgradient from a landfill which is analogous in many respects to the marine sequence.

In the first stage (Fig8A) as the leachate seeps from the landfill into the highly oxygenated environment of either the unsaturated or saturated zones the organic compounds decompose to form the most oxidized species which are stable at the highest redox values. That is, by observing the relative positions of the boundaries between the reduced and oxidized species on Eh—pH

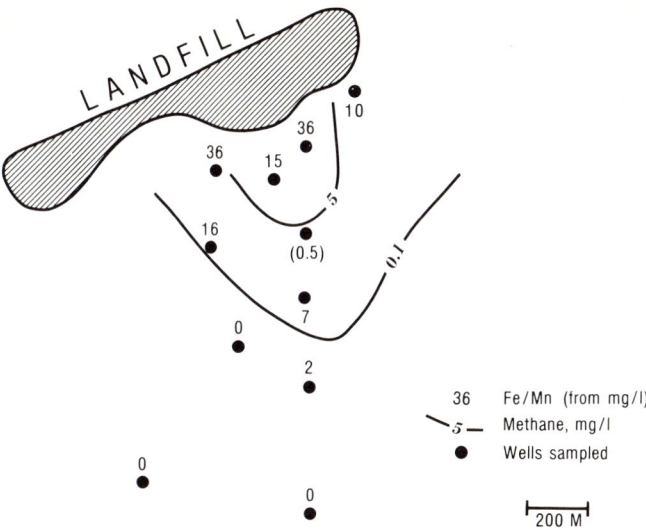

Fig.7. Ratio of Fe to Mn in a reducing plume downgradient from a landfill. Zones of CH_4 concentrations >5; $0.1-5$ and <0.1 mg/l define the reducing plume. Value in parentheses was calculated using lower limit of detection for Fe.

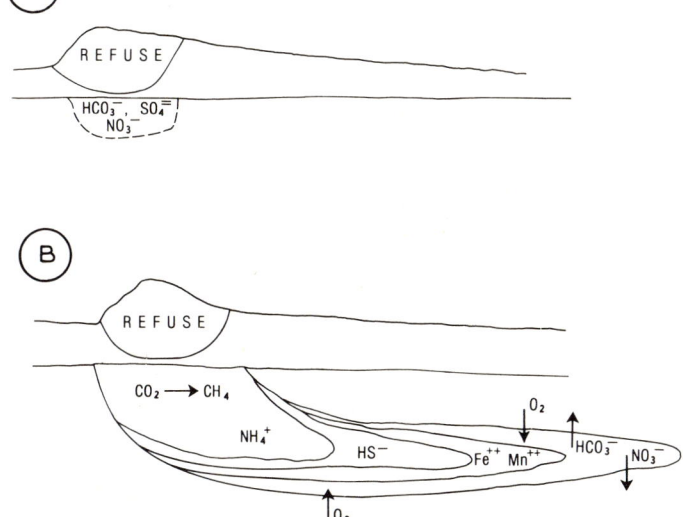

Fig.8. Chemical evolution of hypothetical redox zones near a landfill: (A) early stage; and (B) late stage (physical configuration of plume after Pickens et al., 1978).

diagrams (Figs. 2 and 3) the order of zonation can be predicted. Therefore, the first leachate slug to migrate will undergo aerobic respiration to degrade C compounds to CO_2 which during solution of minerals will convert to HCO_3^- at the pH (6.5) of many landfills. As indicated by the high redox position of the NH_4^+—NO_3^- boundary, N compounds will be oxidized to NO_3^- (Table I). S species will be oxidized to SO_4^{2-}. Because of the large Eh-pH range in which Mn^{2+} is the dominant manganese ion, its mobilization will be an early reaction. As reactions consume available free oxygen and the plume becomes more reducing, decomposition of additional organic compounds reduce ferric hydroxides and Fe^{2+} becomes mobile.

During later stages continued degradation of organic compounds causes greater reduction of redox levels to where NH_4^+ is the dominant N species; H_2S is the dominant S species at pH of less than about 7 and HS^- at slightly higher pH values. The reduction of SO_4^{2-} uses the last source of oxygen, other than organic material itself, and organic compounds then degrade by processes of fermentation to form CO_2, NH_4^+ and CH_4. Combined effects of the above reactions and groundwater flow results in the idealized zonation (Fig.8B).

As the plume migrates through the aquifer certain mineralogic alterations similar to those in marine sediments occur. For example, Fe-oxides (or hydroxides) are deposited as cements or coatings on aquifer material downgradient from landfills. Depending on ionic concentrations, pH and Eh, it is also likely that siderite ($FeCO_3$) or pyrite (FeS) is deposited downgradient from some landfills. These reactions are similar to those discussed previously for marine sediments (Fig.4). In a landfill environment the sources of SO_4^{2-} are from solution of gypsum or oxidation of pyrite and sources of Fe^{2+} are from reduction of oxide coatings, pyrite or from the trash itself. That is, in the flow regime upgradient from a landfill formation of SO_4^{2-} from dissolution of pyrite causes the water to become somewhat reducing and mobilizes the ferrous ion. The Fe^{2+} and SO_4^{2-} can migrate to mix with the leachate where the sulfate-reducing bacteria generate sulfide which permits the reprecipitation of pyrite downgradient. Depending on the pH, concentrations of iron, carbonate and sulfide, siderite could form in a similar manner in zones highly charged with CO_2. In more reduced zones Fe-sulfides may precipitate, and in more oxygenated zones Fe-oxides may form. Considering the dramatic redox change in aquifers as the plume migrates it is possible that metals, particularly Fe, recycle through several mineral phases as oxides, sulfides and carbonates.

Although this zonation is idealized and hypothetical, certain aspects of it have been observed downgradient from landfills. For example, zonation is indicated by various fronts (Fig.9) in which the highest concentrations of NH_4, CH_4, HCO_3^-, Fe and Mn are closest to the landfill. Farther downgradient, is the front showing the last occurrence of NH_4 and CH_4, and beyond that, the front showing the last of the Fe and Mn. The HCO_3^- line indicates where its concentration has decreased to that of normal background values.

Because of oxidation of NH_4^+ and mixing with oxygenated water the NO_3^- values increase downgradient and the NO_3^- front shows where the values become equal to background levels. No sulfide zone existed at this site because of the absence of S minerals in the natural sediemnts or S-containing compounds in the trash. In other areas, SO_4 sorption in the unsaturated zone (Wood, 1978) may limit the amount of SO_4 that reaches the water table and, therefore, limits the formation of sulfides.

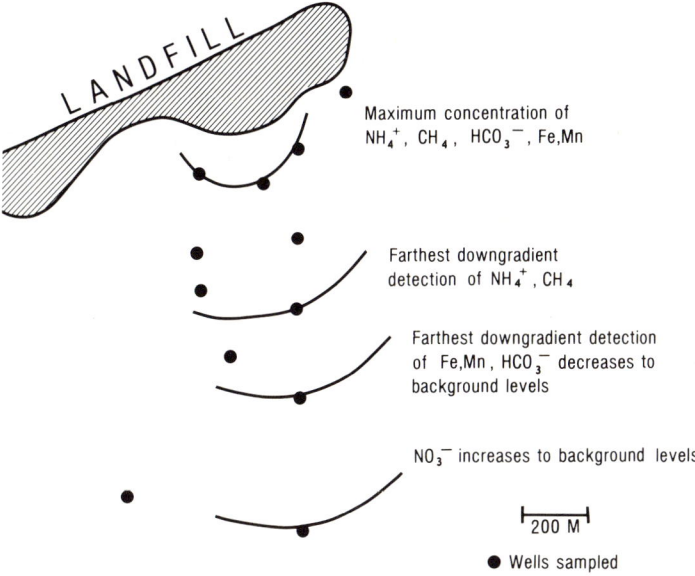

Fig.9. Fronts for chemical species in a reducing plume downgradient from a landfill in Delaware.

CONCLUSIONS

The analogy between landfills and marine sediments is based on the similarity of chemical reactions and processes involving organic matter in environments of restricted circulation with limited exposure to oxygen. Redox zones develop in response to decomposition of organic material and its effects on behavior of inorganic constituents. The boundaries and extent of these zones are controlled by competing rates of reactions and transport. The sequence of reactions which dominates within a zone is, in part, determined by the abundance of constituents that serve as sources of oxygen — O_2, NO_3^-, SO_4^{2-}, O-containing organic compounds and metal oxides. The far greater concentration of organic material in landfills means respiration (aerobic) and fermentation (anaerobic) are the dominant reactions, whereas in marine sediments, dissolution of calcareous minerals and sulfide generation may be dominant reactions.

In summary, the fronts of the biogeochemical zones are more transient in groundwater systems than in marine sediments because the position of marine fronts are fixed by rates of sedimentation, rates of biochemical reactions, and rates of diffusion. However, in groundwater systems the fronts will not be identically fixed in space around all landfills because of the wide range in groundwater velocities and dispersive characteristics of aquifer materials. That is, the positions of the fronts, width of each zone and the degree of zone overlap are controlled by the distribution of head and head change within the flow regime. Because of the direct relationship between head and flow velocity within any single landfill environment the homogeneity of the zonation will be a consequence of the relative rate of groundwater movement and relative rates of reactions within each zone. At a landfill where groundwater flow is rapid in comparison to reaction kinetics the fronts will migrate farther and the zones will be wider with more clearly identifiable fronts. At a landfill where the flow is extremely slow, relative to reaction rates, the zonation may be essentially non-existent or at least nonidentifiable.

ACKNOWLEDGEMENTS

Many of the ideas expressed in this paper have developed over a period of years by one of us (M.J.B.) through discussions with Ian Kaplan, and others in his group at UCLA. We also are most appreciative of beneficial discussions with our colleagues including Don Thorstenson, Donald W. Fisher, Tyler Coplen, and Elliott Spiker of U.S. Geological Survey; Ron Stouffer and Michael Apgar, Delaware State Division of Environmental Control; Keros Cartwright, Illinois State Geological Survey; and John A. Cherry, University of Waterloo. We also thank Eric Sundquist, USGS, Don Thorstenson, and Keros Cartwright for their thoughtful and critical review of the manuscript, and Blair Jones for making Fig.3 available.

REFERENCES

Apgar, M.A. and Langmuir, D., 1971. Groundwater pollution potential of a landfill above the water table. Ground Water, 6: 76—96.
Back, W. and Barnes, I., 1965. Relation of electrochemical potentials and iron content to groundwater flow patterns. U.S. Geol. Surv., Prof. Pap. 498-C, 16 pp.
Baedecker, M.J. and Back, W., 1979. Hydrogeological processes and chemical reactions at a landfill. Ground Water, 17(5) (in press).
Berner, R.A., 1971. Principles of Chemical Sedimentology. McGraw-Hill, New York, N.Y., 244 pp.
Carothers, W.W. and Kharaka, Y.K., 1978. Aliphatic acid anions in oil-field waters — Implications for origin of natural gas. Am. Assoc. Pet. Geol. Bull., 62: 2441—2453.
Champs, D.R., Gulens, J. and Jackson, R.E., 1979. Oxidation—reduction sequences in ground-water flow systems. Can. J. Earth Sci., 16: 12—23.
Cherry, J.A., Shaikh, A.U., Tallman, D.E. and Nicholson, R.V., 1979. Arsenic species as an indicator of redox conditions in groundwater. In: W. Back and D.A. Stephenson (Guest-Editors), Contemporary Hydrogeology — The George Burke Maxey Memorial Volume, J. Hydrol., 43: 373—392 (this volume).

Chian, E.S.K., 1977. Stability of organic matter in landfill leachates. Water Resour. Res., 11: 225—232.

Chian, E.S.K. and DeWalle, F.B., 1977. Characterization of soluble organic matter in leachate. Environ. Sci. Technol., 11: 158—163.

Claypool, G.E. and Kaplan, I.R., 1974. The origin and distribution of methane in marine sediments. In: I.R. Kaplan (Editor), Natural Gases in Marine Sediments, Plenum, New York, N.Y., pp. 99—139.

Claypool, G.E., Presley, B.J. and Kaplan, I.R., 1973. Gas analysis in sediment samples from Legs 10, 11, 13, 14, 15, 18, and 19. In: Initial Reports Deep Sea Drilling Project, Vol. 19, U.S. Government Printing Office, Washington, D.C., pp. 879—884.

Degens, E.T. and Mopper K., 1976. Factors controlling the distribution and early diagenesis of organic material in marine sediments. In: J.P. Riley and R. Chester (Editors), Chemical Oceanography, Vol. 6, Academic Press, New York, N.Y., pp. 59—113.

Dunlap, W.F., Shew, P.C., Scalf, M.R., Cosby, R.L. and Robertson, J.V., 1976. Isolation and identification of organic contaminants in ground water. In: L.H. Keith (Editor), Identification and Analysis of Organic Pollutants in Water, Ann Arbor Science, Ann Arbor, Mich., pp. 453—477.

Edmunds, W.M., 1973. Trace element variations across an oxidation—reduction barrier in a limestone aquifer. In: E. Ingerson (Editor), Proc. Symp. on Hydrogeochemistry and Biogeochemistry, Tokyo, 1970, Clarke Co., Washington, D.C., pp. 500—527.

Fritz, P., Matthess, G. and Brown, R.M., 1976. Deuterium and oxygen-18 as indicators of leachwater movement from a sanitary landfill. In: Interpretation of Environmental Isotope and Hydrochemical Data in Groundwater Hydrology, I.A.E.A., Vienna, pp. 131—142.

Games, L.M. and Hayes, J.M., 1976. On the mechanisms of CO_2 and CH_4 production in natural anaerobic environments. In: J.O. Nriagu (Editor), Environmental Biogeochemistry, Ann Arbor Science, Ann Arbor, Mich., pp. 51—73.

Giger, W. and Roberts, P.V., 1978. Characterization of refractory organic carbon. In: R. Mitchell (Editor), Water Pollution Microbiology, Vol. 22, Wiley—Interscience, New York, N.Y., pp. 135—175.

Goldhaber, M.B., Aller, R.C., Cochran, J.K., Rosenfeld, J.K., Martens, C.S. and Berner, R.A., 1977. Sulfate reduction, diffusion, and bioturbation in Long Island Sound sediments. In: Report of the FOAM Group, Am. J. Sci., 277: 193—237.

Golwer, A., Matthess, G. and Schneider, W., 1975. Effects of waste deposits on groundwater quality. Proc. Symp. on Groundwater Pollution, Moscow 1971. Int. Assoc. Hydrol. Sci. Publ., 103: 159—166.

Griffin, R.A., Frost, R.R., Au, A.K., Robinson, G.D. and Shimp, N.F., 1977. Attenuation of pollutants in municipal landfill leachate by clay minerals, Part 2. Heavy metal adsorption. Ill. Geol. Surv., Environ. Geol. Notes, No. 79, 47 pp.

Hanshaw, B.B. and Back, W., 1979. Major geochemical processes in the evolution of carbonate aquifer systems. In: W. Back and D.A. Stephenson (Guest-Editors), Contemporary Hydrogeology — The George Burke Maxey Memorial Volume, J. Hydrol., 43: 287—312.

Huang, W.H. and Keller, W.D., 1970. Dissolution of rock forming silicate minerals in organic acids: simulated first-stage weathering of fresh mineral surfaces. Am. Mineral., 55: 2076—2094.

Hughes, G.M., Schleicher, J.A. and Cartwright, K., 1976. Supplement to the final report on the hydrogeology of solid waste disposal sites in Northeastern Illinois. Ill. Geol. Surv., Environ. Geol. Notes, No. 80.

Jeris, J.S. and McCarty, P.L., 1965. Biochemistry of methane fermentation using C^{14} tracers. J. Water Pollut. Fed., 37: 178—192.

Jones, B.F. and Bowser, C.J., 1979. The mineralogy and related chemistry of lake sediments. In: A. Lerman (Editor), Lakes, Chemistry, Geology, Physics. Springer, New York, N.Y., pp. 179—236.

Kunkle, G.R. and Shade, J.W., 1976. Monitoring ground-water quality near a sanitary landfill. Ground Water, 14: 11—20.

Langmuir, D., 1972. Controls on the amounts of pollutants in subsurface waters. Earth Miner. Sci., 42: 9—13.

Leenheer, J.A. and Huffman, Jr., E.W.D., 1976. Classification of organic solutes in water using macroreticular resins. U.S. Geol. Surv., J. Res., 4: 737—751.

Leggat, E.R., Blakey, J.F. and Massey, B.C., 1972. Liquid-waste disposal at the Linfield disposal site, Dallas, Texas. U.S. Geol. Surv., Momo. Rep., 33 pp.

Manheim, F.T., 1976. Interstitial waters of marine sediments. In: J.P. Riley and R. Chester (Editors), Chemical Oceanography, Vol. 6, Academic Press, New York, N.Y., pp. 115—186.

Martens, C.S. and Berner, R.A., 1977. Interstitial water chemistry of anoxic Long Island sediments, 1. Dissolved gases. Limnol. Oceanogr., 22: 10—25.

Matthess, G., 1972. Hydrogeologic criteria for the self-purification of polluted groundwater. Int. Geol. Congr., 24th Sess., Sect. II Hydrogeol., pp. 296—304.

Nissenbaum, A., Presley, B.J. and Kaplan, I.R., 1972. Early diagenesis in a reducing fjord, Saanich Inlet, British Columbia, I. Chemical and isotopic changes in major components of interstitial water. Geochim. Cosmochim. Acta, 36: 1007—1027.

Pickens, J.F., Cherry, J.A., Grisak, G.E., Merritt, W.F. and Risto, B.A., 1978. A multilevel device for ground-water sampling and piezometric monitoring. Ground Water, 16(5): 322—327.

Pytkowicz, R.M., 1965. Rates of inorganic calcium carbonate nucleation. J. Geol., 73: 146—199.

Richards, F.A., 1965. Anoxic basins and fjords. In: J.P. Riley and G. Skirrow (Editors), Chemical Oceanography, Vol. 1, Academic Press, New York, N.Y., pp. 611—645.

Robez, J., 1973. Mass spectrometry in water analysis. In: L.L. Ciaccio (Editor), Water and Water Pollution Handbook, Vol. 4, Marcel Dekker, New York, N.Y., pp. 1557—1613.

Sholkovitz, E.R., 1978. The flocculation of dissolved Fe, Mn, Al, Cu, Ni, Co, and Cd during estuarine mixing. Earth Planet. Sci. Lett., 41: 77—86.

Smith, P.H. and Mah, R.A., 1966. Kinetics of acetate metabolism during sludge digestion. Appl. Microbiol., 14: 368—371.

Suarez, D.L. and Langmuir, D., 1976. Heavy metal relationships in a Pennsylvanian soil. Geochim. Cosmochim. Acta, 40: 589—598.

Takai, Y., 1970. The mechanisms of methane fermentation in flooded paddy soil. Soil Sci. Plant Nutr. (Tokyo), 16: 238—243.

Thorstenson, D.C., 1970. Equilibrium distribution of small organic molecules in natural waters. Geochim. Cosmochim. Acta, 34: 745—770.

Toerien, D.F. and Hattingh, W.H.J., 1969. Anerobic digestion, I. The microbiology of anaerobic digestion. In: Water Research, Vol. 3, Pergamon, New York, N.Y., pp. 385—416.

Wasserburg, G.S., Mazor, E. and Zartman, R.E., 1963. Isotopic and chemical composition of some terristrial natural gases. In: J. Geiss and E.D. Goldberg (Editors), Earth and Science Meteoritics, North-Holland Publishing Co., Amsterdam, pp. 219—240.

Weiss, A., 1969. Organic derivatives of clay minerals, zeolites, and related minerals. In: G. Eglinton and M.T.J. Murphy (Editors), Organic Geochemistry, Springer, New York, N.Y., pp. 737—775.

Willey, L.M., Kharaka, Y.K., Presser, T.S., Rapp, J.B. and Barnes, I., 1975. Short chain aliphatic acid anions in oil field waters and their contribution to the measured alkalinity. Geochim. Cosmochim. Acta, 39: 1707—1711.

Wolin, M.J., 1974. Metabolic interaction among intestinal microorganisms. Am. J. Clin. Nutr., 27: 1320—1328.

Wood, W.W., 1978. Use of laboratory data to predict sulfate sorption during artificial ground-water recharge. Ground Water, 16: 22—31.

[2]

TIME-DEPENDENT SORPTION ON GEOLOGICAL MATERIALS

PAUL R. FENSKE

Desert Research Institute, Water Resources Center, University of Nevada System, Reno, NV 89506 (U.S.A.)

(Accepted for publication February 7, 1979)

ABSTRACT

Fenske, P.R., 1979. Time-dependent sorption on geological materials. In: W. Back and D.A. Stephenson (Guest-Editors), Contemporary Hydrogeology — The George Burke Maxey Memorial Volume. J. Hydrol., 43: 415—425.

The transport of radionuclides through geologic materials has been predicted using the assumption that sorption on rocks along the transport path occurs instantaneously. This assumption is valid if equilibrium sorption is reached rapidly relative to groundwater velocity. Laboratory experiments with large pieces of rock and also varying particle sizes indicate that equilibrium sorption may be reached too slowly under some conditions for the assumption of instantaneous sorption to be valid.

An empirical equation has been devised from the laboratory experiments, to describe the time-dependent sorption process for ^{137}Cs on breccia and andesite. The hypothetical example is then worked out for flow through granular materials (pebble and cobble size) and flow through fractured rocks considering a time-dependent sorption. Dispersion is not considered in this analysis. The hypothetical prediction shows that for usual groundwater velocities, transport considering time-dependent sorption is not significantly greater than transport assuming equilibrium sorption. At velocities of groundwater above 30.5 cm/hr: (1 ft./1 hr.) the effect of the time-dependent sorption term can easily be seen in the analysis. The analysis also shows that sorption during transport through granular (pebble and cobble size) materials is less, with less retardation, than sorption during transport through fractured rocks. This is because the mass of rock/volume of water ratio is far greater in fractured materials than in granular materials. Consideration of time-dependent sorption will probably not increase predictive capability for radionuclide transport.

INTRODUCTION

Prediction of transport of radionuclides in groundwater is an important aspect of short- and long-term safety assessment of radioactive waste repositories for low-level waste, and high-level waste as well as inadvertant repositories created by underground nuclear testing. A radionuclide transport prediction requires knowledge of the hydrologic system, groundwater velocity, dispersivity and sorption.

Only sorption is considered in this paper. In radionuclide transport predictions sorption is assumed to be instantaneous. In fractured-rock aquifers and in intergranular systems containing relatively large particles such as gravel beds, rubble, and some valley fill alluvium, however, the time required to attain equilibrium sorption may be long compared to the groundwater residence time. Under these conditions the use of the equilibrium assumption for predicting reduction of cation concentrations resulting from sorption may lead to underestimates of actual cation concentrations. This paper investigates whether a better estimate of reduction in cation concentration in groundwater flowing through such aquifers may be made by considering time-dependent rather than equilibrium sorption so that the sorptive process can be related to groundwater velocity. Since it is impractical to obtain inter-fracture blocks from an aquifer and impractical to carry out equilibrium distribution coefficient measurements on these blocks if they could be obtained, a method is derived for converting laboratory measurements to parameters required to describe field conditions.

In this paper an empirical expression for the time-dependent distribution coefficient is presented and necessary constants are approximated from laboratory measurements. Time-dependent sorption is then used to estimate reduction in concentration of ^{137}Cs during groundwater transport through fractured and intergranular rock aquifers. Only sorption on geological materials, and, in particular, sorption on large masses of similar mineralogy such as inter-fracture blocks, boulders, cobbles and pebbles will be considered.

DISTRIBUTION COEFFICIENTS

Sorption of cations has been quantified by an equilibrium distribution coefficient (Higgins, 1959):

$$K_d = \frac{V}{M} \frac{A_s}{A_w} = \frac{V}{M} \left(\frac{C_0}{C} - 1 \right) \tag{1}$$

where

K_d	= distribution coefficient	(ml/g)
V	= volume of water	(ml)
M	= mass of solid	(g)
A_s	= amount of radioactivity sorbed on solid phase	
A_w	= amount of radioactivity remaining in solution	
C_0	= initial concentration of radionuclide in water	
C	= concentration of radionuclide in water after sorption	

The assumption made here is that the sorbed radionuclide is distributed evenly throughout the mass of the solid at equilibrium. The equilibrium distribution coefficient is proportional to the ratio between the amount of sorbed and the amount of non-sorbed cation. The proportionality constant is the ratio between the volume of water and the mass of the rock and this ratio makes the equilibrium coefficient a constant for all rocks of the same mineralogical type regardless of their physical states. Equilibrium distribution coefficient measurements show:

(1) The equilibrium distribution coefficient increases with decrease in particle size. This relationship is illustrated in Table I for cesium sorption on rocks from Amchitka Island, Alaska.

(2) The equilibrium distribution coefficient tends to remain constant for particles larger than some critical size.

(3) The time required for equilibrium sorption to be reached increases with particle size.

TABLE I

Distribution coefficient of ^{137}Cs on STS "A" drill core from UAe-1, Amchitka Island

Core No.	Depth* (ft.)	Rock type	K_d (ml/g)		
			500—4000 µm	62—500 µm	< 62 µm
22	5,125	breccia	133	305	598
24	5,236	breccia	207	287	527
27	5,340	basalt	337	519	662
28	5,413	breccia	173	288	539
31	5,530	breccia	118	187	262
34	5,636	breccia	247	684	983
36	5,645	sandstone	278	407	636
39	5,656	siltstone	168	458	558
43	5,693	breccia	130	357	528
47	5,751	breccia	136	274	504
52	5,832	breccia	742	1,020	2,670
54	5,940	breccia	83	122	141
56	6,043	andesite	218	823	1,210
60	6,141	breccia	192	342	513
61	6,197	breccia	110	160	197
63	basalt	basalt	199	564	1,110
64	6,379	breccia	137	313	690

Experimental conditions: 72 hr.; temperature 25°C; 5 g rock; 20 ml tracer solution.
*1 ft. = 0.3048 m.

The largest particles in Table I are of the order of 1 cm in diameter and several orders of magnitude smaller than an average inter-fracture block. The data from equilibrium sorption measurements suggest that for a given rock type the equilibrium distribution coefficient is proportional to the aggregate surface area of the individual grains composing the rock and not upon the mass or external surface area of the rock. For very small particles sorption appears to depend upon the surface area of the crystals plus new surfaces created by pulverizing.

EXPERIMENTAL DATA

Pieces of rock were obtained from three different cores from the Cannikin emplacement hole, Amchitka Island, Alaska. They consisted of an andesite weighing 1912 g, a breccia weighing 1626 g, and two small pieces of andesite from the same core weighing 409 g. These pieces of rock were placed in three containers with a solution of ^{137}Cs. At approximately exponentially-spaced time intervals the concentration of ^{137}Cs in the water was measured and the time-dependent distribution coefficient calculated, using:

$$K_t = \frac{V}{M}\left(\frac{C_0}{C_t} - 1\right) \qquad (2)$$

K_t = distribution coefficient at time t (ml/g)
V = volume of water (ml)
C_0 = initial concentration of radionuclide in water
C_t = concentration of radionuclide in water at time t

The total duration of the experiment was 1702 hr. Sorption data for the three core samples are tabulated on Table II, and presented graphically in Fig. 1.

TABLE II

Summary K_t and C/C_0 for large rocks (^{137}Cs)

Time (hr.)	Sample 1		Sample 2		Sample 3	
	K_t	C/C_0	K_t	C/C_0	K_t	C/C_0
1	0.14	0.965	0.24	0.943	0.40	0.910
2	0.34	0.921	0.33	0.923	0.76	0.841
4	0.65	0.860	0.69	0.852	1.42	0.738
7	0.86	0.824	1.14	0.778	2.24	0.641
22	2.26	0.639	2.92	0.578	6.80	0.370
47.5	5.33	0.429	7.36	0.352	15.8	0.202
118	15.2	0.209	27.4	0.127	35.7	0.101
286	39.9	0.0912	78.4	0.0485	64.5	0.0584
550	71.2	0.0532	133	0.0292	116	0.0332
1,702	146	0.0267	191	0.0205	200	0.0195

The general trend of all of the data is similar and nearly linear at a slope of 1 for times shorter than 50—100 hr. Note, also, that the largest sample, sample 1, displays the best linearity. Sample 3 is the smallest sample consisting of two pieces and departs from linearity the earliest. Since the equilibrium distribution coefficient is assumed to be approached asymptotically, the departure from linearity is expected.

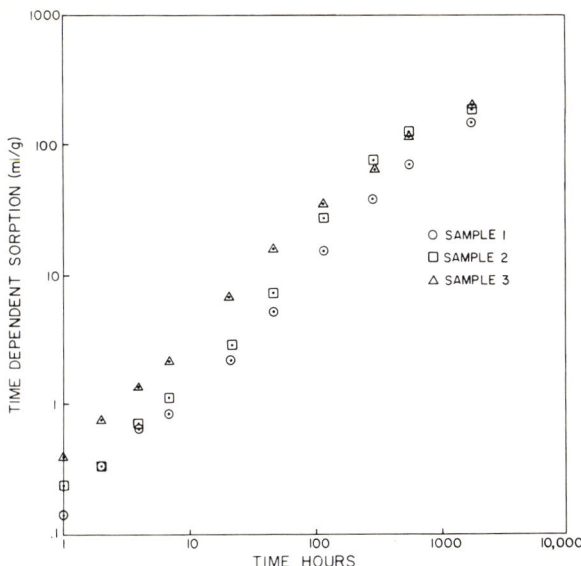

Fig. 1. Time-dependent sorption for large pieces of rock.

EMPIRICAL EXPRESSION FOR TIME-DEPENDENT SORPTION

On the basis of the trend of the data, the slope was assumed to be 1.0. The early-time data can then be described with a linear equation with a slope of 1.0, an intercept of 0.0, and K_t as a function of K_d and time such as:

$$K_t = K_d \beta t \tag{3}$$

K_t = time-dependent distribution coefficient (ml/g)
K_d = equilibrium distribution coefficient (ml/g)
β = reaction rate constant (hr.$^{-1}$)
t = time

The time-dependent distribution coefficient probably approaches the equilibrium distribution coefficient asymptotically. Therefore, the first equation must be an approximation, for small values of β and time, of a function describing this phenomenon. The asymptotic relationship may be described by eq. 4 as a first approximation to the relationship between the time-dependent distribution coefficient and the equilibrium distribution coefficient:

$$K_t/K_d = 1 - \exp(-\beta t) \tag{4}$$

Although other equations could be used, this equation is the simplest one that satisfies the boundary conditions. The exponential form is consistent with a first-order reaction. Hendricks (1972), however, describes the solid diffusion and sorption process with second-order reaction kinetics. Helfferich (1962) notes that as a rule no "order" can be attributed to ion-exchange processes and many rate equations are suggested in the literature that apply specifically to systems investigated but are not universal in application. Because cations must move approximately radially toward the centers of rock pieces, eq. 4 is probably too simple a description of the sorption process for small pieces of rock. Also, a more complex sorption history would be expected to be associated with rocks composed of two physical phases such as the breccia, sample 2. However, eq. 4 would probably be a good approximation for sorption into the walls of a fracture surface of large areal extent.

Eq. 4 is arranged in dimensionless form and a type curve of βt vs. K_t/K_d was constructed. The K_d and β were then determined for each sample by curve matching. This was done by obtaining the best fit for the late-time data rather than for the early-time data which might be more heavily influenced by sample irregularities. In addition, because of the logarithmic nature of the plot, graphical departures from the early-time data are less significant in an absolute sense than graphical departure from the late-time data. The sorption equilibrium equation can also take the form of eq. 5 by substituting eq. 4 into eq. 2:

$$K_d = \frac{V}{M_t}(C_0/C_t - 1) \tag{5}$$

V = volume of water
M_t = $M[1 - \exp(-\beta t)]$ = imaginary mass participating in sorption phenomenon at end of time, t
C_0 = initial cation concentration
C_t = cation concentration at time, t

The concept of imaginary time-dependent effective mass, M_t, which participates in the sorption phenomenon at the end of any time is introduced here. The derivative of M_t with respect to time as time approaches zero as a limit can be thought of as a specific mass accumulation rate times the external surface area of the rock, $m_0 A_0$. By differentiating the expression for M_t with respect to time under eq. 6, we have an expression for the sorption reaction rate constant in terms of the product of mass and β. Equating the two values of the derivative provides an expression for β:

$$(dM_t/dt)_{t \to 0} = M\beta = m_0 A \tag{6}$$

m_0 = specific mass accumulation rate
A = external surface area of rock
β = $m_0 A/M = m_0 a$
a = A/M = specific surface area of rock

The water containing the cation is initially in contact only with the external surface of the rock. This provides us with the data and the method for applying laboratory measurements to the field. The equilibrium distribution coefficient and the specific mass accumulation rate are determined in the laboratory.

The specific external surface area of the rocks composing the aquifer are estimated from field measurements.

The empirical equations presented have been used to calculate the sorption reaction rate constant, the specific mass accumulation rate, and equilibrium distribution coefficients for the three samples (Table III). It is surprising that the specific mass accumulation rate is similar for all samples. In particular, the andesites, samples 1 and 3, although substantially different physically, have nearly the same specific mass accumulation rates. The lower specific mass accumulation rate is associated with the higher K_d as might be expected. Since the external surface area of the samples was not measured directly but only estimated from the weight and dimensions of the samples part of this correspondence between specific mass, accumulation rates may be fortuitous.

TABLE III

The equilibrium distribution coefficient, the sorption reaction rate constant, the specific surface area of rock and the specific mass accumulation rate for the three samples

No.	Sample	K_d (ml/g^{-1})	β (10^{-3} hr.$^{-1}$)	a (cm^2 g^{-1})	m_0 (10^{-3} g cm^{-2} hr.$^{-1}$)
1	andesite	210	0.71	0.38	1.87
2	breccia	200	1.85	0.35	5.28
3	andesite	220	1.35	0.92	1.47

TIME-DEPENDENT SORPTION DURING TRANSPORT

In the past section, sorption was investigated as a function of time in what is known as a batch process. That is, the water containing a radionuclide, ^{137}Cs, in this case, could be in contact with the rocks for any desired length of time, and therefore, the equilibrium K_d could be reached. For the case of sorption during transport, however, the water containing the radionuclide is in contact with the rock only for some relatively short period of time. The mass of rock and volume of water involved in the process are fixed by the porosity of the aquifer through which transport of ^{137}Cs occurs. First, equilibrium sorption during transport will be examined and then consideration of the time-dependent sorption process, as discussed earlier in this paper, will be added.

A restatement of eq. 1 that is more appropriate for consideration of transport in an aquifer is as follows:

$$S/W = [\rho(1-\theta)/\theta]K_d \tag{7}$$

S = fractional activity sorbed on solid
W = fractional activity dissolved in water
ρ = density of solid
θ = porosity of aquifer
$S + W$ = 1

First of all, it is evident from eq. 7 that, since the porosity of the rock appears essentially in the denominator, the lower the porosity, the higher the sorptive capacity during transport through that rock under conditions of an equilibrium K_d. Since fractured rocks normally have at least several orders of magnitude less porosity than granular rocks, the sorptive capacity of fractured rocks during transport will be much higher than the sorptive capacity of granular rocks under equilibrium conditions.

Secondly, it is evident from eq. 7 that, since $S + W = 1$, the equation expresses, or can be thought to express, the probability of an ion of ^{137}Cs being either on the solid or in the water. If the transport path then is broken into equal-sized compartments, normal to the transport direction, the distribution of the ^{137}Cs in these compartments on the solid and in the water can be expressed by the binomial distribution (Higgins, 1959; Levy, 1972). The expression is $(S + W)^n$, where n is one less than the total number of compartments that are used along the transport path. For example, if water containing a slug of ^{137}Cs the length of one compartment has moved through five compartments, the distribution of ^{137}Cs through each of those five compartments is calculated using eq. 8, the binomial distribution, which shows the fractional amount of the radionuclide in each of the five compartments:

$$S^4 + 4S^3W + 6S^2W^2 + 4SW^3 + W^4 = 1 \tag{8}$$

If the fractional amount sorbed on the solids is desired, this total distribution would be multiplied by S. If the fractional amount dissolved in the water is desired, this total distribution would be multiplied by W.

Although the size selected for these compartments can be somewhat arbitrary, it should be noted that the compartment must be large enough to include a representative amount of rock. Therefore, in granular materials, the compartments could be much shorter than they would be in fractured materials, where the length of a compartment would have to be at least of the order of a fracture spacing.

Eq. 8 is applicable to equilibrium sorption described in eq. 7. It assumes that the ratio, S/W is a constant during transport. To consider the time-dependent sorption, eq. 5 may be restated similar to eq. 7.

$$S/W = [\rho(1-\theta)/\theta]K_d[1 - \exp(-\beta t)] \tag{9}$$

In eq. 9, the ratio of sorbed radionuclide to the dissolved radionuclide will increase as a function of time. As a first approximation, the process that would take place in a three-compartment model (for simplicity) similar to the five-compartment model discussed under equilibrium sorption is as follows:

$$S_1S_2 + S_1(W_1 + W_2) + W_1^2 = 1 \tag{10}$$

Subscripts refer to residence time increments. Water containing a slug of dissolved ^{137}Cs of exactly the same length as the compartment length is introduced into compartment 1. The length of the compartment divided by the water velocity gives residence time of this water within that compartment. For this residence time, according to eq. 9, a certain fraction of ^{137}Cs will remain in solution. During transport into compartment 2, the fractional amount left in compartment 1 is S_1, and the fractional amount transported to compartment 2 is W_1. Now, as the ^{137}Cs in solution moves into compartment 2, uncontaminated water moves into compartment 1. In compartment 2 the distribution coefficient is determined by the residence time of the ^{137}Cs slug in that compartment. In compartment 1 the sorbed ^{137}Cs has been in contact with the rock twice as long and therefore, its distribution coefficient is higher than in compartment 2. In compartment 1, then, the fractional amount sorbed is $S_2 S_1$ and the fractional amount in solution available for transport to compartment 2 is $W_2 S_1$. In compartment 2 the amount sorbed is $S_1 W_1$ and the amount in solution available for transport is W_1^2. This process can be continued in the same manner for any number of compartments desired. In eq. 10 the process is described for a three-compartment model.

A simple algorithm that reproduces the characteristics of eq. 10, using a computational procedure similar to that classically used in computing Pascal's triangle, was derived and programmed on a computer. This program was then used to calculate the movement of ^{137}Cs through 500 compartments. The results of these calculations for a granular aquifer and a fractured aquifer are illustrated in Figs. 2 and 3. The granular aquifer is assumed to be composed of rubble with dimensions similar to the rock materials in the original laboratory measurements. The porosity assigned was 20% and β was $1 \cdot 10^{-3}$/hr. Smaller particles were not considered because the large surface area coupled with the small particle volume would represent conditions adequately described by equilibrium sorption. The fractured aquifer was assumed to consist of cubical inter-fracture blocks that measured approximately 1 m on the side. For this type of system using the data developed from the laboratory measurements, the β would be $3.91 \cdot 10^{-5}$ hr.$^{-1}$ and the porosity would probably be in the neighborhood of 10^{-3}. Calculations were made for equilibrium sorption, for water velocity of 30.5 cm/day (1 ft./day), and a water velocity of 30.5 cm/hr. (1 ft./hr.). Comparison of Fig. 2 and Fig. 3 shows that at equilibrium sorption, reduction in concentration of ^{137}Cs is about an order of magnitude greater in the fractured rocks than in the intergranular rocks. For a velocity of 30.5 cm/day (1 ft./day), there is not a great change. When the velocity is increased to 30.5 cm/hr. (1 ft./hr.), which is a relatively high groundwater velocity transport of ^{137}Cs through the granular aquifer significantly exceeds transport of ^{137}Cs through the fractured aquifer. At a groundwater velocity of 30.5 cm/hr. (1 ft./hr.) for example, a peak relative concentration of ^{137}Cs of $2.5 \cdot 10^{-4}$ would be transported about 18.3 m /yr. (60 ft./yr.) in an aquifer composed of pebble and cobble size material. In a fractured rock aquifer, however, a peak concentration of $6.7 \cdot 10^{-5}$ would not be transported out of compartment 1 indicating less than 5.2 m/yr. (17 ft./yr.) of transport. In both cases, the groundwater would have moved 2.67 km (8760 ft.). This difference in transport is attributable to much larger porosity of the granular rocks which more than offsets the greater specific surface area for these rocks as incorporated in the parameter, β. Aquifers composed of pebble and cobble size

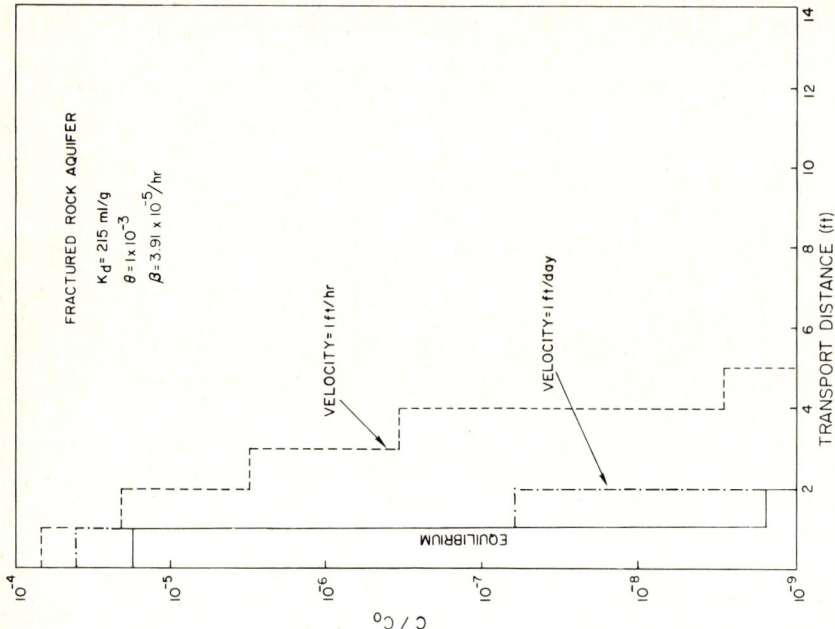

Fig. 3. Transport of ^{137}Cs in fractured aquifer.

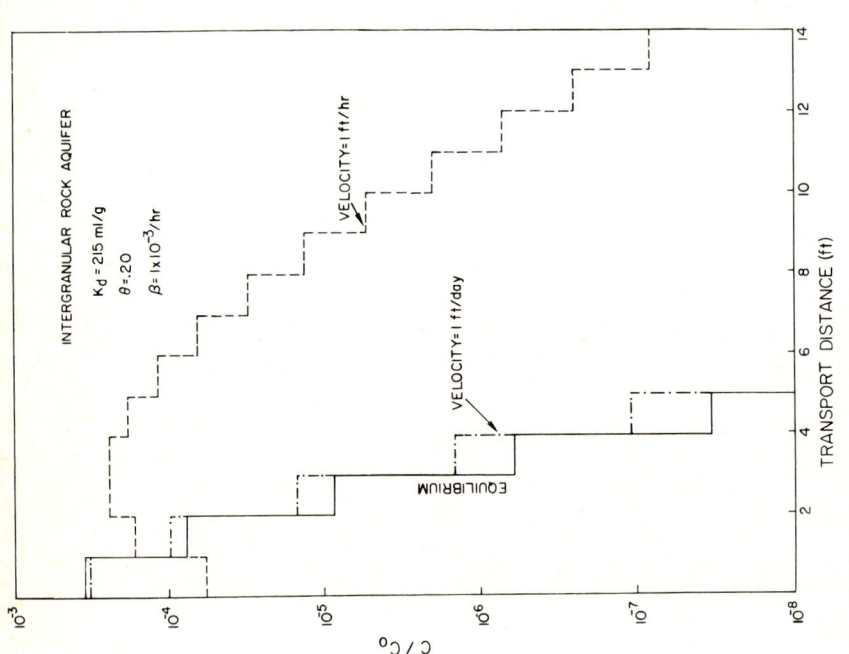

Fig. 2. Transport of ^{137}Cs in intergranular aquifer.

materials probably also have a higher hydraulic conductivity than either aquifers composed of sand size materials or fractured rock, and groundwater velocities are potentially higher and transport of radionuclides probably greater in these materials than in either sandy intergranular aquifers or fractured rock aquifers.

SUMMARY AND CONCLUSIONS

Sorption has generally been considered a process in which equilibrium between ions sorbed on the solid and ions dissolved in water is reached instantaneously. Data presented in this paper show that the sorption process, particularly for larger pieces of rock, is significantly time-dependent. On the basis of laboratory data empirical equations have been developed to describe the time-dependent process. Numerical analysis of transport of ^{137}Cs through both granular (pebble and cobble size) rock and fractured rock aquifers shows that transport of ^{137}Cs is more highly retarded relative to the movement of groundwater in the fracture rock aquifer than the granular rock aquifer. However, a relatively high groundwater velocity had to be used to induce a significant movement of the ^{137}Cs. Probably the greatest transport of ^{137}Cs will occur in aquifers composed of pebble and cobble size particles such as gravel bars or rubble. In general, only small portions of transport paths would rarely include this kind of material. For the normal range of groundwater velocities approximately 1.5 m/day (5 ft./day) to 1.5 m/yr. (5 ft./yr.) it would appear that the error caused by not considering time-dependent sorption is less than other errors involved in predicting transport of radionuclides in groundwater systems.

ACKNOWLEDGEMENTS

The laboratory work in support of this study was done during 1970 and 1971 by Edward H. Essington, now at Los Alamos Scientific Laboratories, during the time he worked at Teledyne Isotopes, Palo Alto Laboratories.

The research, both at Desert Research Institute and the Palo Alto Laboratories of Teledyne Isotopes was supported by funds from the Department of Energy. Work at Desert Research Institute was done under Department of Energy Contract EY-76-6-08-1253.

REFERENCES

Helfferich, F., 1962. Ion Exchange. McGraw-Hill, New York, N.Y., pp. 251—319.
Hendricks, D.W., 1972. Sorption in flow through porous media. In: I.A.H.R., Fundamentals of Transport Phenomena in Porous Media. Elsevier, Amsterdam, pp. 384—392.
Higgins, G.H., 1959. Evaluation of the groundwater contamination hazard from underground nuclear explosions. Lawrence Radiation Lab., Lawrence Livermore Lab., Livermore, Calif., Rep. UCRL-5538 (also, J. Geophys. Res., 64: 1509).
Levy, H.B., 1972. On evaluating the hazards of groundwater contamination by radioactivity from an underground nuclear explosion. Lawrence Livermore Lab., Livermore, Calif., Rep. UCRL-51278.

[2]

THE VOLUME-AVERAGED MASS-TRANSPORT EQUATION FOR CHEMICAL DIAGENETIC MODELS

P.A. DOMENICO and V.V. PALCIAUSKAS

Department of Geology, University of Illinois at Urbana-Champaign, Urbana, IL 61801 (U.S.A.)

(Accepted for publication May 10, 1979)

ABSTRACT

Domenico, P.A. and Palciauskas, V.V., 1979. The volume-averaged mass-transport equation for chemical diagenetic models. In: W. Back and D.A. Stephenson (Guest-Editors), Contemporary Hydrogeology — The George Burke Maxey Memorial Volume. J. Hydrol., 43: 427—438.

The mass-transfer equation as appropriate for chemical diagenesis problems is developed from the viewpoint of a local volume-averaging theorem. The averaging theorem employed is presented from the physical viewpoint of a divergence theorem. Specific features of the averaged equations include some well known results, such as the structure of a diffusion coefficient in a porous sediment and the emergence of dispersion as a dominant mixing phenomenon. Other results reflect the inclusion for provisions for an upper moving boundary for accumulating sediments and a suggested structure for heterogeneous reactions at phase interfaces. Specific forms for the fluxes at interfaces are determined on the basis of conceptual models of the reaction process.

INTRODUCTION

Numerous diagenetic models have been proposed over the years to assess the effects of diffusion, advection, sediment accumulation, reaction rates, and compaction on the chemical composition of pore water in marine sediments undergoing isothermal diagenesis. The most common approach to such problems is to view the porous solid and contained fluids as a continuum and then to simply employ the usual differential equations of mass transfer, appropriately modified by the porous structure. This is not entirely satisfactory for several reasons. First, for any single-phase medium, diffusion takes place in the liquid phase enclosed in a porous solid, with heterogeneous reactions occurring at the liquid—solid interface. Hence, we are immediately faced with the prospect of providing additional relationships or equations to account for reactions at surfaces, the first of several necessary modifications. In addition, a bulk diffusion coefficient must be defined so as to account for the hindering of diffusion by collision with sediment particles. This is no major problem if all are in agreement on what constitutes a diffusion coefficient in a porous sediment and this agreement is substantiated in experimental studies. This, apparently, is not always the case. Berner (1975, 1976a), for example, for purposes of clarity, states that the diffusion coefficient in his diagenetic models includes the effects of tortuosity, but does not include adsorption or ion exchange; we might thus assume that some investigators have included these concentration attenuation processes in the diffusion coefficient. Such chemical processes are indeed concentration attenuation, or thinning, mechanisms in mass transfer, but their inclusion in a "diffusion coefficient" precludes the definition of coefficients that can be related to measurable physical quantities of the porous medium.

Another requirement for the direct utilization of equations designed to describe transport in fluids is the insertion of the medium's porosity in the appropriate terms, often without the benefit of derivation. This operation is necessary to transform equations that describe fluids into corresponding equations that describe fluids in porous solids. A more serious objection, however, is that velocities and concentrations associated with any single-phase medium are not defined in terms of the velocities and concentrations in the fluid phase of a porous medium. With a single-phase medium (fluid), the quantities are obviously averaged within the fluid volume whereas in a water-saturated sediment, the quantities often require averaging in the porous part only.

Most of the problems alluded to above can be eliminated by utilizing a volume-averaging approach to describe mass transport in sediments. The first part of this paper will thus deal with an averaging theorem that can be applied to obtain the spatial averages of the various gradients and divergence of vector fields common to transport equations. Although this is essentially the averaging theorem of Whitaker (1967), some physical interpretations are provided that demonstrate that the averaging theorem is merely the average of the divergence of some property. The second part of the paper deals with the application of this theorem in the derivation of a mass-transport equation suitable for diagenesis studies in a compacting medium. The third part of the paper examines some specific forms for heterogeneous reaction as suggested by the area-integral terms that emerge from the averaging procedure.

A PHYSICAL INTERPRETATION OF THE AVERAGING THEOREM

The mass-transport equation for the ith constituent that holds at any point in a fluid is:

$$-\nabla \cdot J_i - \nabla \cdot (v c_i) + r_i = \partial c_i / \partial t \qquad (1)$$

where J_i is the diffusive mass flux of the ith dissolved constituent with respect to the solvent per unit time per unit area; c is the concentration of the dissolved species in mass per unit volume of water; v is the velocity of the fluid which in this approximation is equal to the center of mass velocity; t is time; and r_i is a source term representing the production of the ith species by homogeneous reaction. Our initial objective is to associate a local volume average of the various scalar and vector fields with every point in a porous medium. Consider the point z (which can be in the fluid or solid phase) and associate it with a volume V and an enclosing surface S (Fig. 1). For simplicity, it is assumed this is a spherical volume concentric to z and of sufficient size such that the radius of the sphere is much larger than the characteristic dimensions of the pores. If V_f denotes the volume occupied by the pores in the

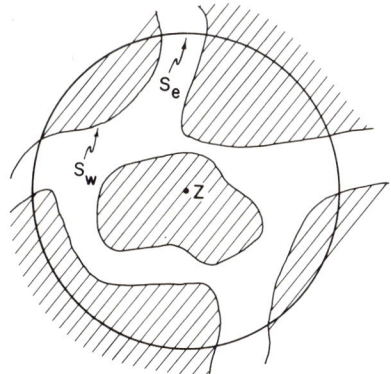

Fig. 1. The averaging volume, where S_w is the total pore surface and S_e is the intersection of the pore volume with the spherical surface bounding the averaging volume.

averaging volume V, and V_s denotes the volume occupied by the solids, then:

$$V = V_f + V_s \tag{2}$$

where V is the total averaging volume. The averaging volume is assumed constant, but V_f, V_s and the porosity can depend on position.

Let B represent any field. The average of this property B over the volume V is defined as:

$$(1/V) \int_V B dV = (V_f/V)(1/V_f) \int_{V_f} B dV_f + (V_s/V)(1/V_s) \int_{V_s} B dV_s \tag{3}$$

or

$$<B> = \psi <B>_f + (1-\psi)<B>_s \tag{4}$$

where the brackets $<>$ indicate an averaged quantity and ψ is the porosity. Eq. 3 states that the average of some property over the volume V is equal to the average over the pore volume within V plus the average over the solid volume within V. Examples of such properties include temperature T, density ρ, velocity v, or some diffusive mass flux vector J. By analogy to eqs. 3 and 4, the average of the divergence of some property B becomes:

$$(1/V) \int_V (\nabla \cdot B) dV = (1/V) \int_{V_f} (\nabla \cdot B_f) dV_f + (1/V) \int_{V_s} (\nabla \cdot B_s) dV_s \tag{5}$$

or

$$<\nabla \cdot B> = \psi <\nabla \cdot B>_f + (1-\psi)<\nabla \cdot B>_s \tag{6}$$

Since B is a vector, the divergence expressed in eq. 5 has the physical meaning of a net outflow per unit volume, with the first term on the right-hand side representing total outflow from the pore volume within the averaging volume, and the second term representing total outflow from the solid volume. The total outflow from the pore volume must equal the total outflow from the surface enclosing the pore volume within the averaging volume (Fig. 1). This surface is simply $S_w + S_e$, where S_w is the total pore wall surface within the averaging volume and S_e represents the intersection of the pore volume with the spherical surface S of the total volume. The net outflow of B through the pore walls is simply:

$$\text{(net outflow through } S_w) = \int_{S_w} (B \cdot n) dS_w \tag{7}$$

where n is the unit outward normal to the pore volume. The outflow through the pores themselves, that is, through the surface S_e is simply:

$$\text{(net outflow through the surface } S_e) = \nabla \cdot \int_{V_f} B dV_f = \nabla \cdot V_f <B>_f \tag{8}$$

Summing eqs. 7 and 8:

$$\int_{V_f} (\nabla \cdot B) dV_f = \nabla \cdot \int_{V_f} B dV_f + \int_{S_w} (B \cdot n) dS_w \tag{9}$$

or, dividing both sides by the total volume V:

$$\psi <\nabla \cdot B>_f = \nabla \cdot (\psi <B>_f) + (1/V) \int_{S_w} (B \cdot n) dS_w \tag{10}$$

This result states that the total net outflow from the pore volume equals the sum of the net outflows through the pore channels and through the pore walls.

Eq. 10 is perfectly equivalent to Whitaker's (1967) averaging theorem when we choose to average over the fluid volume only, as might be the case to obtain a continuity equation of fluid flow, or the mass transport of some species through the pore space. For mass transfer, the last term represents the net rate at which some chemical species is produced at the liquid—solid (pore wall) interface by heterogeneous chemical reaction, where $(B \cdot n)$ is the flux of the species across the interface. For heat flow, the surface integral represents the net rate of heat exchange between the solid and the fluid. For chemical or thermal equilibrium, this term is exactly zero.

The analog result of eq. 10 for the solid phase is:

$$(1 - \psi) <\nabla \cdot B>_s = \nabla \cdot (1 - \psi) <B>_s + (1/V) \int_{S_w} (B \cdot n) dS_w \tag{11}$$

where $(1 - \psi) <\nabla \cdot B>_s$ is the total net outflow from the solid volume. Eq. 11 is equivalent to Whitaker's (1967) averaging theorem when we choose to average over the solid phase only. Summing eqs. 10 and 11 and combining with eq. 6:

$$\langle \nabla \cdot B \rangle = \nabla \cdot [\psi \langle B \rangle_f + (1-\psi) \langle B \rangle_s] + (1/V) \int_{S_w} [B] \cdot n \, dS_w \tag{12}$$

where $\langle \nabla \cdot B \rangle$ now represents net outflow from the total volume. Here, the surface integrals have been combined where $[B]$ represents the value of B in the fluid region minus the value of B in the solid. This point was made by Whitaker (1967), who averaged over the total volume only. Eq. 12 is the averaging theorem for the divergence of the total volume.

By replacing the vector field B by a scalar ϕ times a constant vector A, $B = \phi A$ and inserting this expression into eqs. 10–12 yields the following results for the local volume average of a gradient:

$$\psi \langle \nabla \phi \rangle_f = \nabla (\psi \langle \phi \rangle_f) + (1/V) \int_{S_w} \phi n \, dS_w$$

$$(1-\psi) \langle \nabla \Phi \rangle_s = \nabla (1-\psi) \langle \phi \rangle_s + (1/V) \int_{S_w} \phi n \, dS_w \tag{13}$$

$$(1-\psi) \langle \nabla \Phi \rangle_s = \nabla (1-\psi) \langle \phi \rangle_s + (1/V) \int_{S_w} [\Phi] n \, dS_w$$

where $[\Phi]$ is the value of the scalar in the fluid region minus the value of the scalar in the solid.

THE VOLUME-AVERAGED MASS-TRANSPORT EQUATION FOR DIAGENETIC STUDIES

Averaging over the fluid volume only, eq. 1 is expressed by eq. 5:

$$(1/V) \int_{V_f} [-\nabla \cdot J - \nabla \cdot (vc) + r - (\partial c/\partial t)] \, dV_f = 0 \tag{14}$$

where the subscript i has been dropped and J, v and c refer to the fluid phase only. The first term on the left-hand side of eq. 14 becomes, by the averaging theorem of eqs. 9 and 10:

$$\psi \langle \nabla \cdot J \rangle_f = \nabla \cdot (\psi \langle J \rangle_f) + (1/V) \int_{S_w} (J \cdot n) \, dS_w \tag{15}$$

We now have two terms to deal with, one treating an average flux in the fluid phase, and the other an area-integral term. As mentioned previously, the area-integral term is interpreted as the rate at which some species is produced at the liquid–solid interface by heterogeneous chemical reaction, and is designated r_s. This will be a point for further discussion. As we are averaging over the fluid volume only, the flux J in eq. 15 is readily expressed through Fick's law in the fluid continuum:

$$J_f = -D \nabla c_f \tag{16}$$

where D is the diffusion coefficient in the fluid. By substitution, the first term on the right-hand side of eq. 15 now becomes:

$$\nabla \cdot (\psi <J>_f) = -\nabla \cdot (\psi <D\nabla c>_f) = -\nabla \cdot (\psi D<\nabla c>_f) \qquad (17)$$

In this development, the expression $D\psi <\nabla c>$ is recognized as Fick's law for the diffusion of a constituent in the fluid phase of a porous medium, where D represents the diffusion coefficient in a fluid. From the averaging theorem of eq. 13:

$$\psi <\nabla c>_f = \nabla(\psi <c>_f) + (1/V) \int_{S_w} cn\, dS_w \qquad (18)$$

so that, combining eqs. 17 and 18:

$$-\nabla \cdot (\psi D<\nabla c>_f) = \nabla \cdot D[\nabla(\psi <c>_f) + (1/V) \int_{S_w} cn\, dS_w] \qquad (19)$$

The area-integral term reflects the influence of the local geometry of the porous medium upon the diffusion of the constituent. This geometric influence is related to the tortuosity of the medium. A tortuosity vector is thus defined as:

$$\boldsymbol{\tau} = \int_{S_w} cn\, dS_w \qquad (20)$$

From eqs. 15, 19 and 20 the volume average of the first term of eq. 14 now becomes:

$$\nabla \cdot D[\nabla(\psi <c>_f) + \boldsymbol{\tau}(1/V)] + r_s = 0 \qquad (21)$$

where r_s is the area-integral term of eq. 15. This would be the starting place to describe steady-state diffusion plus heterogeneous reaction in the fluid phase of a porous medium.

The volume average of the second term on the left-hand side of eq. 14 is straightforward. Applying the averaging theorem of eqs. 9 and 10:

$$\psi <\nabla \cdot vc>_f = \nabla \cdot (\psi <vc>_f) + (1/V) \int_{S_w} cn \cdot v_f dS_w \qquad (22)$$

where, in the area-integral term, v_f is the velocity with respect to a fixed coordinate system evaluated at the interface. The third term on the left-hand side of eq. 14 merely results in a volume-averaged homogeneous reaction $\psi <r>$.

The volume average of the last term on the left-hand side of eq. 14 would likewise be straightforward were it not for the fact that the diagenesis problem is usually studied in compacting media. Recognizing that compaction results in the movement of solid grains, the volume average of the time derivative becomes:

$$\psi <\partial c/\partial t>_f = \partial(\psi <c>_f)/\partial t - (1/V) \int_{S_w} cn \cdot v_s dS_w \qquad (23)$$

where the area integral term accounts for compaction in a deforming media, with v_s as the velocity of the solids at the fluid—solid interface. As the velocity of the fluids at the interface equals approximately the velocity of the solids at the interface, the area integral terms of eqs. 22 and 23 cancel each other. Collecting all appropriate terms, the volume-averaged mass-transport equation becomes:

$$\nabla \cdot D[\nabla(\psi <c>_f) + (1/V)\tau] - \nabla \cdot (\psi <vc>_f) + r_s + \psi <r> = \partial(\psi <c>_f)/\partial t \quad (24)$$

The procedures for incorporating the convective effects known as mechanical dispersion are straightforward and well known. Of the various model studies of diagenesis, however, only one to our knowledge makes an attempt to evaluate the effects of dispersion in mass transport. Stiller et al. (1975) in a discussion of transport through Lake Kinneret sediments correctly recognize that mechanical dispersion is a mixing phenomenon caused by variations in velocity through a complex pore structure. Incorrectly, however, they dismiss these effects on the argument that such effects have been demonstrated to be insignificant on the laboratory scale where the dispersivity is taken as equal to the mean grain size in the flow apparatus from which the measurements are taken. As has been reasonably well established, the dispersivity or mixing length on the field scale is more on the order of tens of meters.

The incorporation of the convective effect known as mechanical dispersion requires some approximations between the average of a product and the product of two averages. In that fluid velocity and concentration cannot be considered independently in the convective term of eq. 24, the average of the product, as expressed, is taken as equal to the product of the averages plus some deviation. That is:

$$<vc>_f = <v>_f <c>_f + <\delta v \delta c>_f \quad (25)$$

The last term of eq. 25 represents the effect of local velocity fluctuations on the transport of the constituent, and is interpreted as the mechanical dispersion vector Ω. Substitution in eq. 24 gives:

$$\nabla \cdot D[\nabla(\psi <c>_f) + (1/V)\tau - \psi <\Omega>/D] - \nabla \cdot \psi <v>_f <c>_f + r_s + \psi <r> =$$
$$\partial(\psi <c>_f/\partial t) \quad (26)$$

where

$$<\Omega> = <\delta v \delta c>_f \quad (27)$$

Further manipulations with eq. 26 are possible by expanding the time derivative and convective terms and incorporating the conservation of fluid mass equation for a constant-density fluid:

$$\partial \psi/\partial t = -\nabla \cdot (\psi <v>_f) \quad (28)$$

The convective term of eq. 26 is expressed:

$$\nabla \cdot \psi <v>_f <c>_f = <c>_f \nabla \cdot \psi <v>_f + \psi <v>_f \cdot \nabla <c>_f \quad (29)$$

Substituting eq. 28 into 29 and incorporating this result in eq. 26 gives:

$$\nabla \cdot D[\nabla(\psi <c>_f) + (1/V)\tau - \psi <\Omega>/D] - \psi <v>_f \cdot \nabla <c>_f + r_s + \psi <r> = \psi \partial <c>_f/\partial t \tag{30}$$

It is understood here that $<v>_f$ is the velocity of the fluid relative to a fixed coordinate system. In diagenesis problems, a change of variables is often incorporated to account for the upper moving boundary in depositional basins. This point has been discussed in several papers by Imboden (1975) and Berner (1976a, b). The relationship required is:

$$<v>_f = <v>_T + w \tag{31}$$

where $<v>_T$ is the velocity of the fluid with respect to the top of the sedimentary pile and w is the rate of deposition minus the rate of compaction, in essence the rate of burial of sediment grains below the sediment—water interface.

From the form of eq. 30 we recognize that the transport coefficients are incorporated in the following terms:

$$D[\nabla(\psi <c>_f) + (1/V)\tau - \psi <\Omega>/D] \tag{32}$$

where D is the diffusion coefficient in a fluid environment unhindered by pores, and the tortuosity vector is given in eq. 20. In the absence of fluid flow, and assuming porosity is constant:

$$D\psi[\nabla <c>_f + (1/V\psi)\tau] \tag{33}$$

which is a rigorous statement of Fick's first law for diffusion in the fluid phase of a porous sediment. The form of eq. 33 has prompted the definition of a bulk diffusion coefficient D' which includes the effects of tortuosity on free diffusion, giving the qualitatively correct statement most commonly referred to as Fick's law for diffusion in sediments:

$$J = -D'\psi \nabla <c>_f \tag{34}$$

By qualitatively correct, we mean that eq. 34 is empirically designed in a form identical to Fick's original law (eq. 16), but with assurances that the proportionality constant $\psi D'$ is less than the corresponding proportionality constant D for a fluid environment. In the absence of an evaluation for D', which is generally the case, an effective diffusion coefficient D_e is generally approximated so that:

$$J = -D_e \nabla <c>_f \tag{35}$$

Approximations for D_e are of the form:

$$D_e = DD^* \tag{36}$$

where D^* is some characteristic constant which is less than one, and serves the purpose of a retardation factor in order that $D_e < D$; that is, in order that the effective diffusion coefficient in a porous sediment is less than the diffusion coefficient in a totally fluid environment. In most cases, D^* is a function of both porosity and tortuosity (see, e.g., Evans et al., 1961; Li and Gregory, 1974).

When fluid flow takes place, additional approximations are required for the dispersion vector of eq. 32. In that dispersion is a spreading phenomenon, the effects of which are identical to diffusion, it is reasonable to treat a dispersive flux analogously to a diffusive flux; that is, proportional to a concentration gradient. This has prompted the use of some approximate gradient dependent mixing length analysis such that:

$$<\Omega> = <\delta v \delta c>_f \approx -D_m \nabla <c>_f \tag{37}$$

where D_m is the coefficient of mechanical dispersion. This means that eq. 32 becomes:

$$D_e \nabla <c>_f + D_m \nabla <c>_f = D_h \nabla <c>_f \tag{38}$$

where D_h is the coefficient of hydrodynamic dispersion (Bear, 1972), and incorporates the effects of mechanical dispersion and molecular diffusion. In the absence of fluid flow, D_h reduces to D_e. In its most tractable form, eq. 30 can be expressed:

$$\nabla \cdot [D_h \nabla <c>_f] - \psi <v>_f \cdot \nabla <c>_f + r_s + \psi <r> = \psi \partial <c>_f / \partial t \tag{39}$$

THE AREA-INTEGRAL TERM

The area-integral term of eq. 15 can be expressed as:

$$r_s = (\xi/S_w) \int_{S_w} J \cdot n \, dS_w = \xi <J \cdot n> \tag{40}$$

where J has already been defined as mass per unit area per unit time, $<J \cdot n>$ is the average flux across the boundary; and ξ is the specific surface, defined as the surface area of the pores per unit bulk volume of porous material. Eq. 40 involves no assumptions about the mechanism of reaction. If there is a finite rate of solute transfer from the solid material to the solution, or vice versa, it can be assumed that the rate depends on the pertinent parameters of the problem. These include the concentration of all components in the mobile phase and in the solid, the velocity, temperature, and density of the moving fluid, and other parameters which may determine whether the reaction is rate- or diffusion-controlled. In short, a conceptual model of reaction is a first requirement.

One such conceptual model may consist of as many as five steps in series: (a) diffusional transport of solute molecules to the interface; (b) adsorption at the interface; (c) reaction at the surface; (d) desorption of products at the interface; (e) diffusional transport of products from the interface. In this conceptual model, steps (b), (c) and (d) constitute an interaction between the solid material and the liquid and are processes that occur at the surface of the solid. If these steps are considerably more rapid than those controlling transport [(a) and (e)], the overall reaction is considered to be transport-controlled. On the other hand, if the adsorption—desorption process at the interface is slower than the transport, the overall reaction is considered to be surface controlled. In either case, an empirical relation is required for the average flux across the boundary.

For transport (diffusion) control two approaches are possible. In one case we might expect that the mass flux is some function of a concentration gradient where ∇c must somehow include the concentration at the fluid—solid interface. This suggests that we incorporate the ideas of A. Noyes and W. Nernst (Kingery, 1959, pp. 3—5) for a transport-controlled process. Namely, that when dissolving solids in a liquid: (1) the reaction rate at the surface is sufficiently rapid that an equilibrium concentration is maintained at the interface; (2) there is a stationary layer attached to the surface; and (3) the rate of diffusion of reactants through the stationary layer controls the rate of reaction. A reasonable estimate of ∇c on the interface is thus $(c_{eq} - \langle c \rangle)/\sigma$ where c_{eq} denotes the saturation concentration and σ is taken as the thickness of the stationary layer in a single pore. By definition, then:

$$\langle J \cdot n \rangle \xi = D_i (c_{eq} - \langle c \rangle) \xi / \sigma \qquad (41)$$

where D_i is the diffusion coefficient of the ith constituent. In reality, the thickness of the surface layer is determined by hydrodynamic conditions, the layer presumably getting thinner with increased velocity. All other things equal, this speeds up the diffusion-controlled reaction. Eq. 41 has been employed in transport equations to examine the physical and kinetic factors that tend to promote an approach to equilibrium in carbonate systems (Palciauskas and Domenico, 1976).

Yet another approach to diffusion-controlled reactions is the surface renewal (or penetration) model. The postulate here is that a fluid element comes into contact with a surface for a certain time period, during which unsteady-state diffusion occurs between the fluid element and the surface. At the end of this time period, the fluid element moves away from the surface and is replaced by a new element from the bulk fluid. The mean value for the mass flux over the time interval t_e is (Higbie, 1935, p. 370; Szekely and Themelis, 1971, p. 429):

$$\langle J \cdot n \rangle = 2(D_i/\pi t_e)^{1/2}(c^* - \langle c \rangle) \qquad (42)$$

where c^* designates the concentration at the surface. By definition, then:

$$\langle J \cdot n \rangle \xi = 2\xi (D_i/\pi t_e)^{1/2}(c^* - \langle c \rangle) \qquad (43)$$

The important parameter here is the contact time t_e, which in most cases will be the ratio of some characteristic length (such as a mean grain diameter) and the fluid velocity. Clearly, as the contact time gets small due to a large velocity, the diffusion-controlled reaction increases. The effect is then the same as the velocity related stationary layer of eq. 41. In fact, the coefficients

D_i/σ and $(D_i/\pi t_e)^{1/2}$ of eqs. 41 and 43 can be replaced by a mass-transfer coefficient k, where k has the dimensions of a mass flux/mass concentration difference, or length/time. Further, if c^* of eq. 43 is approximated by c_{eq}, the only remaining difference between eqs. 41 and 43 is due to different conceptual models of the mass-transfer process.

Surface-controlled reactions have been defined as processes of adsorption whereby a component is extracted from the liquid stream and becomes part of the solid matrix, and a reverse process, desorption, where a soluble component of the solid becomes part of the solute. Adsorption and desorption are thus processes that take place at the interface and require that we somehow describe the process as a mass flux between the bulk fluid and the solid. The mass flux in the most practical case is described as the product of a mass-transfer coefficient and a concentration difference. In the general case, then:

$$<J \cdot n>\xi = \xi(k_1<c> - k_2 N) \qquad (44)$$

where N is the surface concentration of the solute in the solid phase; and k_1 and k_2 are forward and backward rate constants, respectively. Eq. 44 states that adsorption and desorption occur simultaneously and at different rates. The utilization of this reversible adsorption process is widespread in transport problems (Amundson, 1950; Lapidus and Amundson, 1952; Deju, 1971). Other reaction forms have been reviewed by Domenico (1977).

With regard to all these empirical relations, some authors employ the surface area of the material per unit volume of water in place of ξ, the specific surface. This may be appropriate when describing individually submersed crystals or spheres, but the averaging procedure indicates that the surface area per unit bulk volume is the appropriate parameter when attempting to treat reactions in sedimentary bodies. This fact is well established in commercial leaching operations where increasing the specific surface through grinding never fails to increase the dissolution rate (Peters and Skrivanek, 1962).

CONCLUDING STATEMENT

It is suggested that the volume-averaged mass transport equation provides one of the more complete and informative descriptions of the diagenetic process. Several reasons may be cited for this, most of which relate to the opening remarks in the introduction to this paper. Included here is the emergence of factors like porosity and tortuosity in the structure of the averaged equations, as well as information on the diffusion coefficient and reactions at phase interfaces. It is noted further that such reactions conveniently emerge as source terms in the differential equation. In addition, the phenomena of dispersion is not overlooked as is common in most mass-transport models applied to diagenesis. Indeed, if a main purpose of diagenetic studies is to disentangle the intruding physical effects of transport in order to obtain chemical kinetic information, errors can occur when dispersion is ignored. For example, when employing a curve-fitting procedure which matches mathematical solutions with actual concentration vs. depth profiles in marine sediments, the concentration attenuation or thinning caused by dispersion can be erroneously attributed to reaction. This follows from the fact that diffusion, dispersion and chemical reaction all act in the direction of concentration attenuation.

REFERENCES

Amundson, N.R., 1950. Mathematics of adsorption in beds, II. J. Phys. Colloid Chem., 54: 812—820.
Bear, J., 1972. Dynamics of Fluids in Porous Media. American Elsevier, New York, N.Y., 764 pp.
Berner, R.A., 1975. Diagenetic models of dissolved species in the interstitial waters of compacting sediments. Am. J. Sci., 275: 88—96.
Berner, R.A., 1976a. Inclusion of adsorption in the modelling of early diagenesis. Earth Planet. Sci. Lett., 29: 333—340.
Berner, R.A., 1976b. Distinguishing reference frames, comments on a paper by D.L. Graf and D.E. Anderson. J. Geol., 84: 115—118.
Deju, R.A., 1971. A model of chemical weathering of silicate minerals. Geol. Soc. Am. Bull., 82: 1055—1062.
Domenico, P.A., 1977. Transport phenomena in chemical rate processes in sediments. Annu. Rev. Earth Planet. Sci., 5: 287—317.
Evans, R.B., Watson, G.M. and Mason, E.A., 1961. Gaseous diffusion in porous media at uniform pressure. J. Chem. Phys., 35: 2076—2083.
Higbie, R., 1935. The rate of adsorption of a pure gas into a still liquid during short periods of exposure. Trans. Am. Inst. Chem. Eng., 31: 365—389.
Imboden, D.M., 1975. Interstitial transport of solutes in nonsteady state accumulating and compacting sediments. Earth Planet. Sci. Lett., 27: 221—228.
Kingery, W.D., 1959. Introduction, in W.D. Kingery (Editor), Kinetics of High Temperature Processes. M.I.T. Press, Cambridge, Mass., pp. 1—7.
Lapidus, L. and Amundson, N.R., 1952. Mathematics of adsorption in beds, VI. J. Phys. Chem., 56: 984—988.
Li, Y.H. and Gregory, S., 1974. Diffusion of ions in sea water and in deep sea sediments. Geochim. Cosmochim. Acta, 38: 703—714.
Palciauskas, V.V. and Domenico, P.A., 1976. Solution chemistry, mass transfer, and the approach to chemical equilibrium in porous carbonate rocks and sediments. Geol. Soc. Am. Bull., 87: 207—214.
Peters, E. and Skrivanek, J., 1962. On the mass transport mechanism of leaching processes. Can. J. Chem. Eng., 40: 1—5.
Stiller, M., Carmi, I. and Munnich, K.O., 1975. Water transport through Lake Kinneret sediments traced by tritium. Earth Planet. Sci. Lett., 25: 297—304.
Szekely, J. and Themelis, N., 1971. Rate Phenomena in Process Metallurgy. Wiley, New York, N.Y., 784 pp.
Whitaker, S., 1967. Diffusion and dispersion in porous media. Am. Inst. Chem. Eng. J., 13: 420—427.

[2]

PROBLEMS OF LARGE-SCALE GROUNDWATER DEVELOPMENT

S. MANDEL

Center for Groundwater Research, Hebrew University, Jerusalem (Israel)

(Accepted for publication February 7, 1979)

ABSTRACT

Mandel, S., 1979. Problems of large-scale groundwater development. In: W. Back and D.D. Stephenson (Guest-Editors), Contemporary Hydrogeology — The George Burke Maxey Memorial Volume. J. Hydrol., 43: 439—443.

The difficulties encountered in the planned development of groundwater resources are analysed by comparing typical "scripts" for surface and groundwater resources, respectively. The suggested remedies are: (a) campaigns of systematic field investigations in any area where the intensive exploitation of groundwater is envisaged; (b) standardization and codification of modern research techniques so that they can more readily be used by any investigator; (c) the construction of water supply networks serving more than one user in order to stimulate cooperation and mutual control among water users and to facilitate the centralized regulation of groundwater abstraction.

INTRODUCTION

In 1975 the late G.B. Maxey and the author were engaged in studies of groundwater mining. A collection of case histories revealed that the overexploitation of groundwater is a very common practice, especially, though not exclusively, in dry areas. Generally, with very few exceptions, overexploitation develops unintentionally and is only belatedly recognized. Available data refer mainly to areas where attempts are being made to rationalize groundwater mining and to plan rescue schemes. Data from areas where groundwater mining "just happens", or where it has run its full destructive course, generally, remain inaccessible. Thus the available data are probably indicative of a problem that will become acute, on a global scale, within the next two or three decades, unless the present trends in groundwater development are reversed.

The respective merits of sustained yield exploitation versus mining may be arguable in each particular area. The wide-spread uncontrolled development of irreplaceable water resources is certainly an undesirable state of affairs. The present paper attempts to analyse the reasons for this situation and to suggest remedies against its further spread.

SCRIPTS FOR SURFACE WATER AND GROUNDWATER DEVELOPMENT

The rational development of a natural resource requires information as input to planning, and planning as guide to action. However, the actual course of development tends to follow a "script" generated by the interaction of human attitudes, technical, and societal factors. A comparison of typical scripts for surface water and groundwater development shows why it is relatively easy to follow rational procedures in the first case, while it is difficult to enforce them in the second one.

A typical script for surface water development

(1) The drainage basin of the river is mapped and some quantitative data are collected long before large-scale diversions are thought of.

(2) The design criteria of dams, spillways and intake structures are dictated by the stochastic hydrological characteristics of the river, not only by the discharges that are actually going to be diverted. As a consequence, larger works are able to supply water at a smaller unit cost, and the resultant "economy of scale" (Hall and Dracup, 1970) forces competitors for water supplies to cooperate already during the planning phase.

(3) Legal doctrines recognize community rights on river water and refer to the watershed as basic natural unit.

(4) Feasibility studies, based on a thorough investigation of the resource are mandatory because of the large investment that is involved.

(5) The scope of investigations is widened in response to public concern with the environmental inpact of the project.

(6) Quantitative investigations of the resource are based on flow gaugings and on meteorological observations, i.e. data that can be collected prior to any construction work. The long lead time between the study and the implementation of the project makes it possible to collect adequate data.

(7) The completed scheme is operated by a statutory authority exercising physical control over the installations.

(8) Water users accept the necessity of regulation, since water abstraction by any one user visible affects water availability to all other users.

(9) Even gross mismanagement is unlikely to damage the resource.

A typical script for groundwater development

(1) Well drilling and groundwater abstraction precede systematic data collection.

(2) Well drilling is dictated solely by the exigencies of demand. The "economy of scale" that may, perhaps, be achieved by the simultaneous drilling of many wells is negligible as compared to the advantage of development according to immediate requirements. Well owners compete for water supplies from the same common pool without even realizing it.

(3) Legal doctrines regarding groundwater rights are vague and based on antiquated concepts.

(4) The people who invest their money in well drilling are concerned only with the availability of groundwater at particular sites. Investigations concerning the resource as a whole are carried out, on a routine basis, and on a meagre budget, by state or national agencies, if at all. Intensive investigations are deferred until the water-supply situation visibly deteriorates.

(5) Public concern with environmental impacts is activated only in rare cases, for example, when intensive groundwater abstraction diminishes the flow of a nearby stream or, very belatedly, when soil subsidence occurs.

(6) Quantitative information on the resource is derived mainly from observations in exploited well fields, i.e. as a kind of "by-product" of exploitation. Due to the short lead time between the discovery of the groundwater resource and its intensive development, information trails more and more behind the actual situation in the field.

(7) Attempts are made to control groundwater abstraction by stringent legal and administrative measures.

(8) Water users resent regulation as unwarranted interference in their private business and stultify effective control.

(9) Continued overexploitation damages the resource irreversibly.

TECHNICAL FACTORS AND SOCIETAL ASPECTS

A change of the script for groundwater development requires the modification of many more factors than those that are usually considered in the "planning space" of a water supply project (Wiener, 1972). We shall try to distinguish between technical factors and between societal aspects although both are intertwined and cannot be completely separated.

Technical factors

The information needed in order to plan the crucial, early phase of groundwater development comprises: mapping of the aquifer, elucidation of the mechanisms of natural replenishment and of natural drainage, reliable (though not necessarily precise) data on average annual replenishment, aquifer constants and exploitable reserves, evaluation of constraints on exploitation, and of the critical water levels corresponding to each constraint.

It should be noted that the list does not include items that can be obtained only from extensive records, such as stochastic elements, or the time scale of hydrological processes. Therefore, the requisite information can, in principle, be assembled by a systematic campaign of field investigations, including well drilling, within a relatively short time, say 2—3 years.

In areas that are being developed by international finances, such as the U.N. special fund, such campaigns have become common practice. The budgets that have to be allocated for this purpose are seen to be only a small fraction of the aggregate value of the project. Affluent nations tend to cling to the traditional views that the investigations for engineering projects are the investors' concern, and confine themselves to the role of defending the public interest by administrative controls. In the last analysis it is this traditional attitude which so often prevents the rational development of groundwater, and which has to be modified.

The present state of the art of field investigations leaves much to be desired in spite of the fact that many new techniques are available, such as improved geochemical methods, environmental isotopes, artificial tracers, remote sensing, etc. However, there is a wide gap between a viable scientific method and its routine application by non-specialists. The field investigator needs precisely specified working procedures, standardized instruments and, above all, he needs reliable information on the applicability and limitations of each technique. In short what he needs is an armory of sophisticated but standardized diagnostic tools akin to the one that is at the disposal of any modern medical practitioner. Efforts in this direction may accomplish a major breakthrough.

Societal aspects

Legal doctrines should enable the regulation of groundwater abstraction and they should recognize groundwater basins and aquifers as natural units. Whether this necessity calls for the nationalisation of groundwater resources is a different matter. However, legislation is of little value unless backed up by effective administrative measures which, in the case of groundwater resources, are notoriously difficult to enforce.

The most obvious control measure, licensing of well drilling and of pumping aggregates, regulates actual groundwater abstraction only within limits. Criteria for well licensing have to be based on peak demand that is supposed to occur only during a small part of the year. Subsequently, the peak demand season tends to become more and more prolonged due to the increase of irrigated areas, to changes in cropping patterns, and to the construction of storage reservoirs. Even when these developments are detected, for example, by regular overflights of the area, it is almost impossible to reverse them. Control by water meters is still less practicable. Infuriated well owners will find ways to tamper with them, or to threaten or bribe the unlucky official who is sent to read them.

The problem of control is, essentially, a problem of societal structure. A farming area, where each farm derives water from its own well strictly for its own purposes is practically uncontrollable. The farmers-well owners cherish their independence above everything else, they do not wish to realize that they are all in the same boat and may even organize for the purpose of defeating any "enemy" who threatens to shatter this illusion. Under these conditions legal doctrines and sweeping administrative powers are quickly reduced to the status of fictions.

In a developing area steps should be taken to prevent this situation. Already at a very early phase of development water users should be encouraged to construct small water-supply networks, serving more than one user. In this case, accounting procedures make the installation of water meters mandatory and some measure of mutual control is enforced. It is far easier to regulate groundwater abstraction by a small number of water supply utilities than by a multitude of privately owned wells. Incidentally such an arrangement also facilitates the more efficient exploitation of the aquifer since well sites can be selected according to hydro-geological criteria rather than on the basis of land ownership. Each network may be jointly owned and operated by the users it supplies, the point being only to prevent the one to one correspondence of individual water users and individual well owners.

The case history of Israel (Wiener, 1972, pp. 401—405; Mandel, 1977, pp. 44—51) illustrates this procedure. Many areal water-supply networks, each one supplying several farms, were constructed long before the infrastructure for the enforcement of the stringent Israeli water law had been built up. Thus many, though not all, water users were accustomed to the idea that the regulation of water supplies is a necessary evil. On this basis the transition to a strictly controlled water economy — in a country that derives 2/3 of its supply from groundwater — was accomplished with little friction.

REFERENCES

Hall, W.A. and Dracup, J.A., 1970. Water Resources Systems Engineering. McGraw-Hill, New York, N.Y., 384 pp.
Mandel, S., 1977. The overexploitation of groundwater in dry regions. In: Arid Zone Development. Bullinger, Cambridge, Mass., pp. 31—51.
Wiener, A., 1972. The Role of Water in Development. McGraw-Hill, New York, N.Y., 496 pp.

[2]

THE IMPACTS OF COAL STRIP MINING ON THE HYDROGEOLOGIC SYSTEM OF THE NORTHERN GREAT PLAINS: CASE STUDY OF POTENTIAL IMPACTS ON THE NORTHERN CHEYENNE RESERVATION

WILLIAM W. WOESSNER[1], CHARLES B. ANDREWS[2] and THOMAS J. OSBORNE[2]

[1] *Water Resources Center, Desert Research Institute, Las Vegas, NV 89109 (U.S.A.)*
[2] *Northern Cheyenne Research Project, Lame Deer, MT 59043 (U.S.A.)*

(Accepted for publication April 25, 1979)

ABSTRACT

Woessner, W.W., Andrews, C.B. and Osborne, T.J., 1979. The impacts of coal strip mining on the hydrogeologic system of the Northern Great Plains: case study of potential impacts on the Northern Cheyenne Reservation. In: W. Back and D.A. Stephenson (Guest-Editors), Contemporary Hydrogeology — The George Burke Maxey Memorial Volume. J. Hydrol., 43: 445—467.

50% of the coal production of the U.S.A. will be obtained from the western coal fields by 1990. The majority of this coal will be produced from large-scale strip mining of the Tertiary Fort Union and Wasatch Formations of Wyoming, Montana and North Dakota. The rapid escalation of coal strip-mining activities in the Northern Great Plains, where groundwater is the principal source of domestic and agricultural supply, threatens to alter significantly the local and regional hydrologic regime.

At the request of the Northern Cheyenne Tribe of southeastern Montana, a study of the potential impacts of strip mining on their water resources was designed and implemented. After a basic hydrogeologic study was conducted, a hypothetical mine study site was selected for evaluation. The pre-mining hydrologic system was defined from a monitoring well network, aquifer testing, water-quality sampling and stream gaging. Saturation extract and leachate analyses were conducted on cuttings of overburden from wells in the mine area and results were used to predict the concentration of total dissolved solids (TDS) in spoil-water discharge. A material-balance model was developed which described the quantity and quality of groundwater and recharge for zones of an unmined coal seam, spoil area, clinker and alluvium. A complementary material-balance model was also developed for a stream receiving post-mining discharge.

The TDS of the spoil groundwater outflow was determined to be 4100 mg/l by averaging saturation extract data. Analyses of modeling results indicated that groundwater downgradient from the mined area could be increased in TDS by 300—2070 mg/l depending on the spoil recharge rate. The quality of the receiving stream with a mean annual flow of 14.2 m³/s could be increased in TDS by 1-26 mg/l. The site-specific mine-impact TDS changes were projected over an area of similar hydrogeology along 30 km of stream length in the Tongue River Valley. Strip mining of the entire minable area would have a major impact on the regional groundwater quality and a measurable impact on the quality of the receiving stream. Analysis of projected hydrologic properties of the post-mining system indicated that water-quality impacts will last for hundreds of years.

INTRODUCTION

Coal production within the last decade has been shifting from the eastern U.S.A. to six western states: Montana, Wyoming, Utah, North Dakota, Colorado and New Mexico. The increased emphasis on western coal is a result of many factors including the passage of air-pollution laws, the rapid population growth and development of the west, increased emphasis on coal as an energy resource, and the economics of strip mining. Prior to 1972, coal production in the western region had never been more than 7% of the national coal production, or $13 \cdot 10^6$ t (metric tons). In 1977, the six major coal-producing western states produced $107 \cdot 10^6$ t, 17% of the national production (USDI, 1978b). The U.S. Department of Energy has projected that by 1990 western coal production will be 50% of the national coal production (USDOE, 1978). The Northern Great Plains will become the largest single producing section of the country if the U.S. Department of Energy projections are realized. Much of this coal would be produced from the Ter-

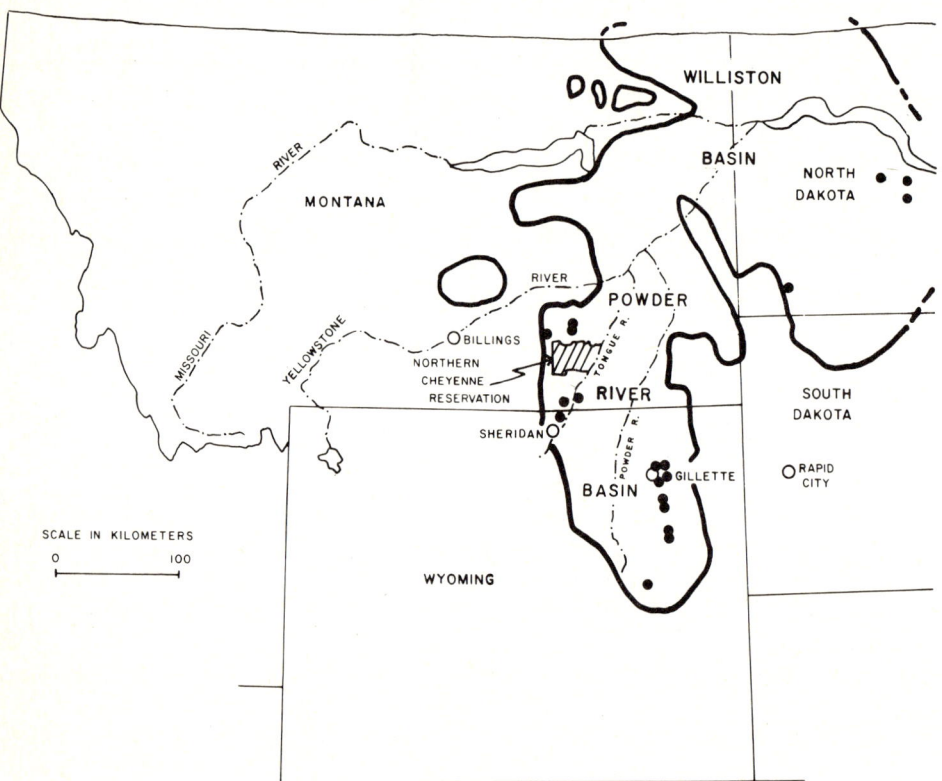

Fig.1. The Northern Great Plains Region showing the location of the Fort Union Formation in the Powder River Basin and the Williston Basin, and the location of the Northern Cheyenne Reservation. The location of operating coal strip mines producing over 10^6 t annually are shown by the *dots*.

tiary Fort Union and Wasatch Formations (Fig.1) of the Powder River Basin which is projected to supply 26% of the national coal production by 1990. An estimated 270 km² area (USDI, 1978b, pp. 5—18) will have been disturbed by coal strip mining in the Powder River Basin by 1985.

The extensive strip mining of the Northern Great Plains will cause considerable changes in the amount, distribution and quality of water in mined areas. The potential alterations in the hydrologic regime are a major concern in the water-short Northern Plains where groundwater is the principal source of domestic and livestock supply, and where crop yields from sub-irrigated alluvial valleys are the mainstay of the agricultural economy. Most of the groundwater is produced from near-surface alluvial, sandstone and coal aquifers, aquifers that are very susceptible to damage from strip mining.

Many of the potential impacts of strip mining on the hydrologic system can be alleviated by proper mine placement and modern methods of strip mining. Strip mining, though, will almost always entail a lowering of the water table and a degradation in groundwater quality caused by the water movement through spoils. The dewatering impacts, the drying up of wells and springs in the vicinity of the mine and the lowering of the water table in adjacent and downgradient alluvial valleys, can be predicted with a fair degree of certainty before mining using conventional techniques. The Surface Mining Control and Regulation Act (91 STAT 445) requires that these types of impacts be predicted beforehand, and that these impacts be minimized. Water-quality impacts have been little studied, and cannot be predicted beforehand with a fair degree of certainty.

The big unknowns regarding strip mining in the west, in addition to long-term changes in water quality, are the long-term conjunctive and regional consequences of mining. Site-specific studies of the impacts of mining in the west have been conducted since the early 1970's, but little work has been done on the long-term and regional issues.

The Northern Cheyenne Tribe owns $4.5 \cdot 10^9$ t of strippable low-sulfur sub-bituminous coal and a total coal reserve estimated at $21 \cdot 10^9$ t (Woessner et al., 1978). The research on which this paper was based was conducted to inform the Northern Cheyenne Tribe what the long-term hydrologic impacts of strip mining would be on the water resources of the Reservation, and what impacts strip mining in the vicinity of the Reservation would have on the Reservation's water resources.

PREVIOUS WORKS AND GENERAL MINING IMPACTS

The impacts of coal strip mining on the hydrologic resources of the Northern Great Plains have only been studied in detail since the early 1970's when state and federal laws required the preparation of an environmental impact statement before approving new mining operations. The majority of mining-impact research efforts have taken place in Montana, North Dakota and Wyoming, where seven of the largest strip-mining operations in the U.S.A.

are located, including AMAX Coal Company's Belle Ayr Mine in Wyoming which produced over $16 \cdot 10^6$ t in 1978. Early studies attempted to determine pre-mining hydrology and predict the effect of mine dewatering on the groundwater potentials in the mine area. Work in the past four years has been focused on mine dewatering analyses, post-mining water quality, mine spoil hydrology and methods of predicting long-term impacts to the local and regional hydrologic systems. But as in the eastern part of the U.S.A. (Rogowski et al., 1977), there presently are no standard techniques available to assess beforehand the degree of impact and the long-term effects of strip mining on a given area.

W.A. Van Voast, of the Montana Bureau of Mines and Geology, has performed the most extensive site-specific hydrogeologic mining-impact research in the region. Research by Van Voast and others has included studies of pre-mining hydrologic conditions, dewatering calculations, monitoring of hydrologic changes during mining and post-mining and spoil hydrology research (Van Voast, 1974; Van Voast and Hedges, 1975; Van Voast et al., 1975, 1976, 1977, 1978). Other hydrogeologic mining-impact research has been conducted in Montana by Arnold and Dollhopf (1977), Dollhopf et al. (1977), Woessner et al. (1978); in Wyoming by Rahn (1976), Davis and Rechard (1977); in North Dakota by Moran et al. (1976, 1978), Croft et al. (1978), Groenewold et al. (1979); and in the three-state region by Anonymous (1978). The modeling and predicting of post-ming impacts has been attempted by Van Voast and Hedges (1975), McWhorter et al. (1976), Van Voast et al. (1976, 1977, 1978), Moran et al. (1978), and Groenewold et al. (1979). Additional efforts to assess site-specific impacts in the form of environmental impact statements have been prepared by federal and state agencies (USDI, 1974a, 1975, 1976a, b, c, d, e, f, 1977a, b, 1978c, d, e, f, 1979).

Evaluation of the regional impacts of existing and proposed mining operations on the hydrologic system of the Northern Great Plains has been attempted by regional study groups (NGPRP, 1974; USDI, 1974a, 1978d; Northern Powder River Basin Environmental Assessment, pers. commun., 1979). However, these programs have been qualitative in nature. More quantitative work has been funded by the U.S. Bureau of Mines (Hittman and Associates, Inc., 1977, 1978) to develop a digital model of the near-surface groundwater system in the Powder River Basin. The U.S. Environmental Protection Agency has also supported work to assess and identify water-quality problems associated with coal strip mining in the west and the development of groundwater monitoring programs (Slawson, 1978).

Based on a review of the literature and current research efforts, a discussion of the general hydrologic impacts of surface coal mining on the hydrologic system of a region is presented in three parts: (1) pre-mining; (2) mining; and (3) post-mining.

Pre-mining impacts

Unconfined, semi-confined and confined groundwater systems within 60—120 m of the surface are common in the western coal region. The impacts of the accepted practice of the pre-mining drilling of four to five holes per section over a large area during the exploration phase and mine plan drilling on 15-m centers have not been addressed in the existing environmental impact statements. When pre-mining drilling perforates both shallow and deep groundwater systems in an area, avenues for increased rates of groundwater migration and mixing are created. In the western region, large areas may be explored which will not be mined for various reasons, but, as a result of exploration drilling, measureable groundwater impacts may occur. Wells used to determine pre-mining hydrogeologic conditions are usually emplaced in conjunction with or after mine plan drilling. If mixing of groundwaters and head equalization begins before observation wells are emplaced to establish impacts of mining, these wells will yield data which represent an altered hydrologic system. Without extensive pre-mine drilling information the extent or severity of this impact cannot be fully assessed.

Impacts during mining

Once mine plan drilling is completed, ammonium nitrate explosives are used to fracture the overburden and the initial mine cut is made. The natural groundwater flow is intercepted, and groundwater is released from storage. As mining proceeds, additional groundwater can enter the mine from upward leakage through newly fractured underlying beds below the coal if the proper hydraulic gradient exists (USDI, 1974a). Reversals of the hydraulic gradient which result from dewatering operations can induce additional groundwater movement to the mine from nearby reservoirs and streams if proper geologic conditions permit (Riordan et al., 1978). Eventually a new equilibrium will be established which will control inflow to the mine. Water levels in wells and spring flow surrounding the mine may decline if they come within the influence of the cone of depression associated with the operation. Van Voast et al. (1976) stated that, generally, the hydrostatic pressure decline is not preceptable more than 100 m from a typical western strip mine and that changes are very gradual. However, as discovered in his work, the rate of decline and area of influence is dependent on site-specific hydraulic properties of the materials. At the West Decker mine in southeastern Montana, a water-level decline of three meters was noted 1500 m from the mine within the first year (Van Voast and Hedges, 1975). Two years later, the decline had doubled at the same distance. Van Voast et al. (1976) noted that after four years of dewatering, rates of water-level decline are diminishing.

The quality of the mine effluent has been found to be similar to the pre-mining groundwater with a few exceptions. Van Voast et al. (1976) noted that besides the natural groundwater constituents, the mine effluent at the

West Decker mine contained nitrate concentrations ranging from less than 1 to 100 mg/l. These unnatural concentrations of nitrate were attributed to the ammonium nitrate explosive compounds used in blasting and/or nitrate compounds being dissolved from the overburden.

Post-mining impacts

At the completion of a mining operation, the once-saturated coal seam and permeable overlying rocks of the mine area are replaced by mine spoil. Van Voast et al. (1977) describe the character of the spoil as wasted coal at the mine base, with coarse spoil material which has rolled down spoil piles during emplacement, thin portions of consolidated material between mine cuts, abandoned gravel haul roads, and blasted overburden material. Groundwater will begin to resaturate the spoil materials as pumping of intercepted groundwater ceases. Direct recharge by precipitation will also provide water for resaturation. Once sufficient saturation has occurred and groundwater is no longer moving into the spoils from all directions, water will begin to move through the spoils and continue downgradient.

Mining operations physically alter the properties of the pre-mining earth materials which affected the occurrence, movement and quality of groundwater in the mine area. If the mining operation has only affected a small portion of a large groundwater flow system, the direction of the groundwater moving through the spoils and out of the mine site will be in the same general direction as the larger system. However, if the mining operation affects the major area of a flow system, post-mining paths may be altered depending on the availability of recharge, and hydrologic properties of the spoils. Analyses of the post-mining properties of mine spoils has been attempted by Van Voast (1974), Verma and Thames (1975), Farmer and Richardson (1976), McWhorter et al. (1976), Rahn (1976), Arnold and Dollhopf (1977), and Van Voast et al. (1977).

Identification of the hydraulic properties of spoil-filled mine areas has been attempted by conducting pump tests on older resaturated spoils, infiltration tests on the spoils, and by laboratory permeameter tests on spoil samples. Pump-test hydraulic conductivities of spoils ranged from 0.021 to 18.3 m/day. Field infiltrometer tests and laboratory permeability analyses at eight mine sites indicated an average spoil hydraulic conductivity of 0.61 m/day with a range of 0.4–2.1 m/day. Storativity values calculated from pump tests at different sites have been in the range of 0.10–0.23. Van Voast et al. (1976) noted that wells, finished in saturated spoils at the old Rosebud Mine at Colstrip, were reacting to barometric pressure which indicated a storativity of 10^{-5}. They postulated that the rubble zone at the base of the spoil may be acting as a confined system similar to the pre-mining coal system. It is postulated by most researchers that the spoils probably have a slightly higher hydraulic conductivity than the undisturbed coal, sandstone and shale.

Recharge rates to the spoils from precipitation have not been documented. Work by Arnold and Dollhopf (1977) and Dollhopf et al. (1977) has indicated that with only two years of data, new spoils at Colstrip were receiving recharge from a basin in which surface runoff collects in the study area, but spoils at two other sites receiving only direct precipitation had not received measurable recharge. Soil moisture studies revealed that recharge would move to depths of 7—10 m below the surface of the spoils only to be lost by evapotranspiration which reversed the downward movement.

The magnitude of post-mining impacts on groundwater quality are highly variable and difficult to predict. The variability in quality is a function of the quality of groundwater resaturating the spoils, the amount of recharge from precipitation which has reached the water table and the type, distribution and leachability of spoil materials. The processes governing water-quality changes are poorly understood and the rate controlling processes are unknown. Van Voast et al. (1977) generalized their observations about spoil water quality as follows:

(1) Young mine spoils (a few years old) contained water of similar chemical character as the surrounding undisturbed material.

(2) Older mine spoils (30 years and more) contained water more mineralized than the undisturbed material.

(3) Water quality varied with depth.

(4) The common ions of calcium, magnesium and sulfate, and trace metals, such as lead and nickel, were found in higher concentration than in the undisturbed material.

(5) Water quality varied spatially in the spoils and also in the undisturbed material.

Post-mining flow system modeling has been attemped, using the material-balance approach by Van Voast and Hedges (1975), and Van Voast et al. (1976, 1977). Deterministic digital modeling of post-mining hydrology has been attempted by Anonymous (1978) with little success because of insufficient data on post-mining aquifer parameters and chemical transport definition (R.W. Davis, pers. commun., 1979), and with mixed success by McWhorter et al. (1976), because of over parameterization. Croft et al. (1978) and Moran et al. (1978) have attempted to provide a geochemical framework for modeling post-mining hydrologic impacts at North Dakota sites.

Post-mining impacts are not restricted to the mine site. Mine spoil water will move away from the site in the direction of decreasing hydraulic head. Poor-quality groundwater resulting from spoil leaching can degrade down-gradient groundwater supplies and surface-water bodies at points of groundwater discharge. The spoil water may also move into underlying groundwater systems if hydraulic conditions permit. Results of studies attempting to determine hydrologic parameters of spoil material and the associated groundwater system indicates water-quality impacts may last for long periods of time after mining operations cease.

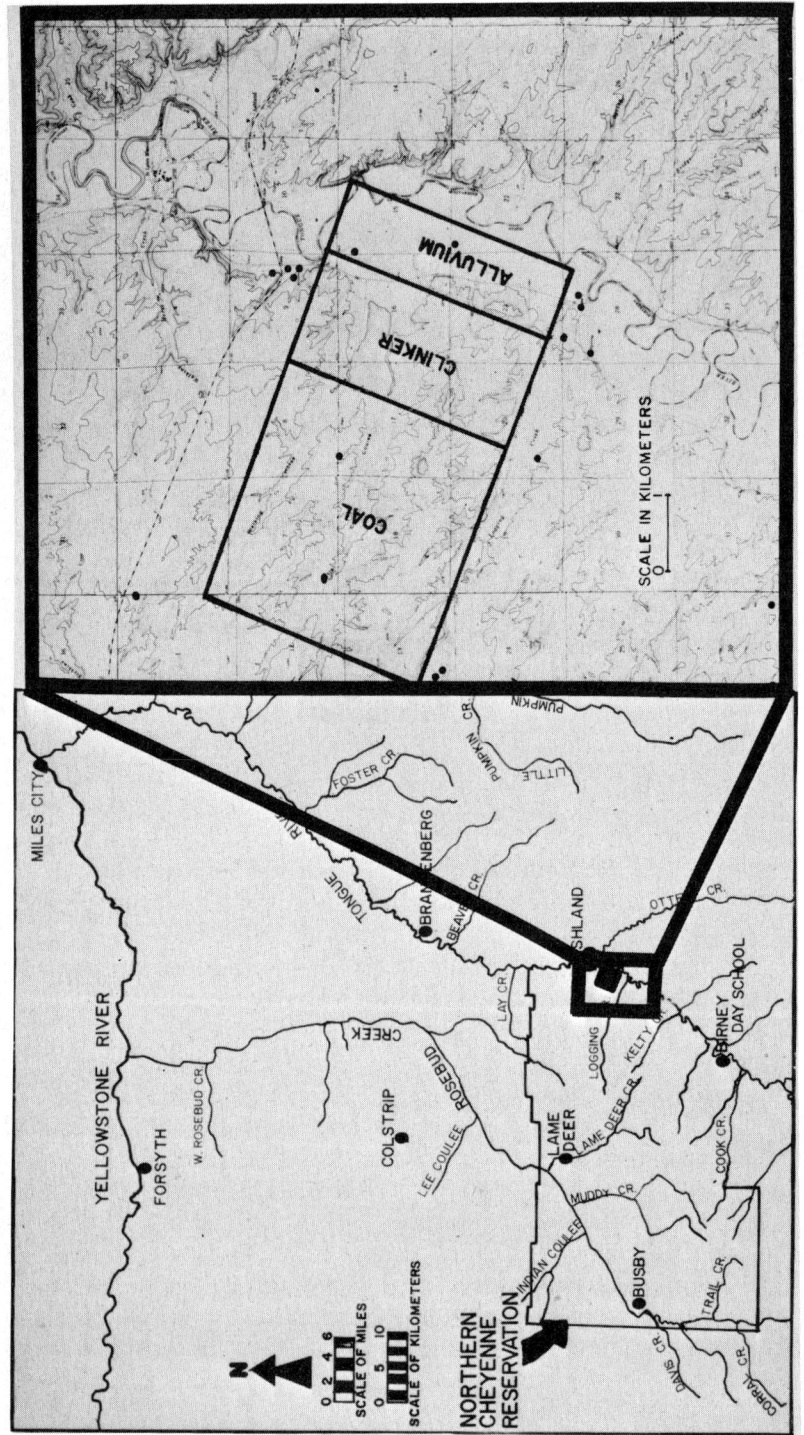

Fig. 2. The hypothetical Logging Creek Mine Site adjacent to the Tongue River on the Northern Cheyenne Reservation. Monitoring wells are depicted by the *dots* on the larger scale map.

SITE DESCRIPTION

The 1754.08-km² Northern Cheyenne Indian Reservation is located in the Powder River Basin in southeastern Montana. The Reservation area is an upland which has been deeply dissected by the Tongue River and Rosebud Creek and their tributaries. All drainages on the Reservation are in the Yellowstone River Basin (Fig.2).

The climate of the Northern Cheyenne Indian Reservation is continental and semi-arid. Average annual precipitation varies from 30 cm in the stream valleys to 48 cm in the higher central portions of the Reservation.

The region is underlain by sandstone, shale, clinker and eleven subbituminous coal seams of the Paleocene Tongue River Member of the Fort Union Formation. Younger unconsolidated alluvial deposits overlie the Fort Union Formation in the major stream valleys. Groundwater is present in the unconsolidated sands and gravels of the valleys and adjacent and underlying sandstones, coal and clinker. Porous fractured clinker, which covers over 30% of the land surface on the Reservation, consists of the baked shale, sandstone, and coal ash which forms during the burning of coal seams. A stacked groundwater flow system is present in which extensive clinker deposits, lenticular sandstones, thick continuous coal seams, and alluvial materials transmit the major portion of flow.

Currently, there is no large-scale mining on the Northern Cheyenne Reservation. Mining was eminent on the Northern Cheyenne Reservation in the early 1970's, at which time coal companies had permits to explore and options to lease 56% of the Reservation. The Tribe in 1973 petitioned the Secretary of Interior to void the leases and permits, for, among other reasons, not educating the Tribe as to the impacts mining would have on the Reservation. The Secretary in 1974 declared the leases and permits to be held in abeyance. The status of the leases and permits is still uncertain. The Tribe hopes that the permits and leases will soon be voided. To ascertain potential hydrogeologic impacts of mining, a hypothetical mine site, the Logging Creek Mine Site, located in the extensive Knobloch coal deposits of the Tongue River Valley was selected for mining-impact analyses (Fig.2). The mine site which would sustain an annual yield of $9 \cdot 10^6$ t for thirty years was selected assuming an average coal thickness of 18.3 m and a 3:1 stripping ratio (Fig.3).

METHODS

Quantification of impacts from potential mining on the hydrologic system of the Reservation was attempted only with respect to post-mining impacts on the total dissolved solids (TDS) concentration of groundwater and surface water on the Reservation. Prediction of more specific impacts during mining is not possible unless detailed mine operation plans are known. Examination of other post-mining hydrogeologic impacts was beyond the scope of this project

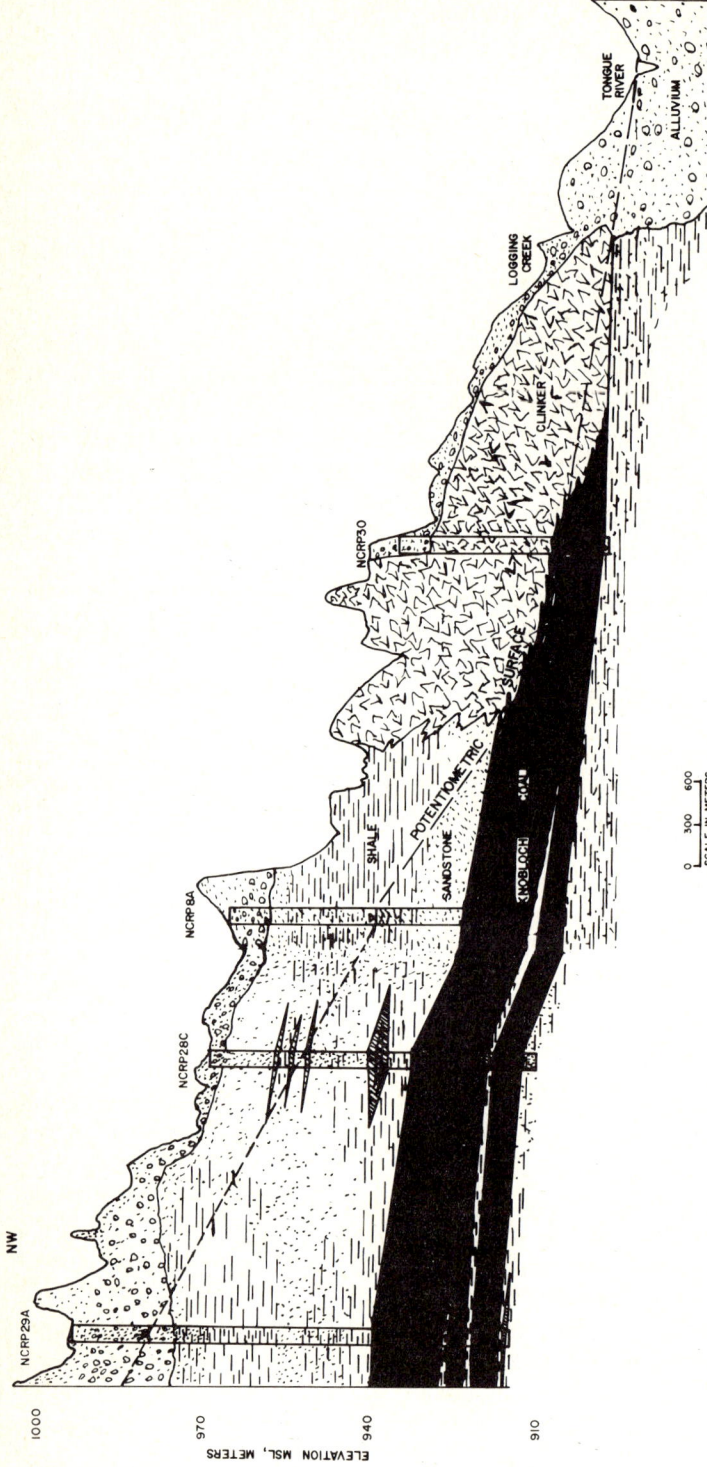

Fig. 3. Geologic cross-section of the Logging Creek Mine Site. The section is along the southern boundary of the hypothetical site shown in Fig. 2. (Adopted from E. Heffern, pers. commun., 1978.)

Material-balance model

A material-balance model was used to determine the changes in the TDS of the groundwater discharging to the streams in the vicinity of the hypothetical mine and the changes in Tongue River water quality.

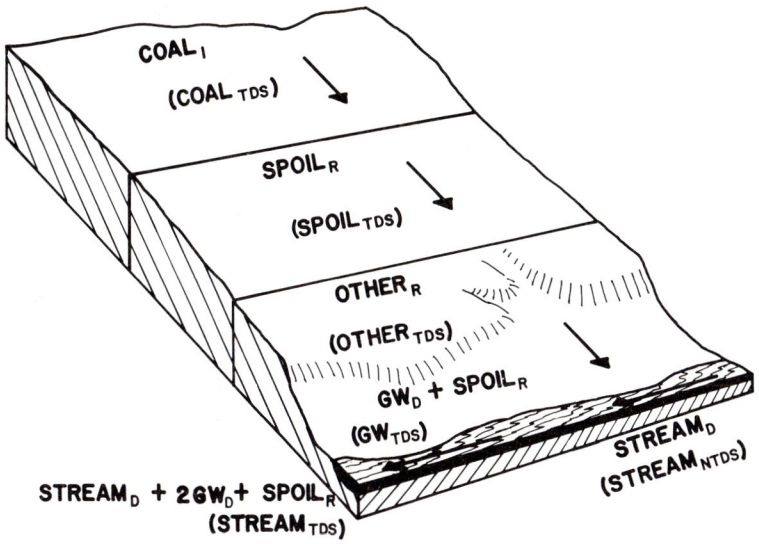

Fig.4. Schematic of the modeled groundwater system.

The following equations were used to determine water-quality changes in the conceptual system (Fig.4) which assume that the system is at steady state and that no chemical reactions occur downgradient of the spoils:

$$GW_{TDS} = \frac{(COAL_I + SPOIL_R) \times SPOIL_{TDS} + OTHER_R \times OTHER_{TDS}}{(GW_D + SPOIL_R)} \quad (1)$$

where

GW_{TDS}	= TDS concentration of groundwater outflow from the groundwater system after equilibrium is established (mg/l)
$COAL_I$	= natual groundwater inflow to the coal seams prior to mining (m³/day)
$SPOIL_R$	= recharge that occurs to the spoils (m³/day)
$SPOIL_{TDS}$	= TDS of the water leaving the spoils and entering the undisturbed groundwater system (mg/l)

OTHER$_R$ = groundwater that enters the system downgradient from the mine site, which includes recharge to the alluvium, clinker, and inflow from deep systems

OTHER$_{TDS}$ = average TDS of the groundwater that enters the system downgradient from the mine

GW$_D$ = pre-mining outflow from the groundwater system (mg/l)

All parameters in eq.1 were determined from field data except OTHER$_{TDS}$:

The following equation was used to determine OTHER$_{TDS}$:

$$\text{OTHER}_{TDS} = \frac{\text{GW}_D \times \text{GW}_{NTDS} - (\text{COAL}_I \times \text{COAL}_{TDS})}{\text{OTHER}_R} \qquad (2)$$

where

COAL$_{TDS}$ = TDS concentration of groundwater flowing in the coal seam (mg/l)

GW$_{NTDS}$ = pre-mining groundwater quality of the water discharging to the stream in the vicinity of the mine (mg/l)

Water quality changes in the receiving stream were then determined with the following equation:

$$\text{STREAM}_{TDS} = \frac{\text{STREAM}_D \times \text{STREAM}_{NTDS} + (\text{GW}_D + \text{SPOIL}_R) \times \text{GW}_{TDS} + \text{GW}_D \times \text{GW}_{NTDS}}{\text{STREAM}_F + \text{SPOIL}_R + 2\text{GW}_D} \qquad (3)$$

where

STREAM$_{TDS}$ = TDS concentration in the stream, into which the groundwater affected by the mine discharges, downstream from the affected area (mg/l)

STREAM$_D$ = streamflow upstream from the mine area (m³/day)

STREAM$_{NTDS}$ = TDS concentration in the stream upstream from where the post-mining groundwater enters the stream (mg/l)

Required parameters

The physical properties required as input parameters for the material-balance model are: flow of stream upstream from the mined area, TDS concentrations of the stream, natural groundwater inflow into the stream in the section influenced by mining, natural TDS of groundwater discharging into the stream, groundwater inflow into the mine site, TDS of the groundwater inflow, other groundwater inflow to the system, water quality of groundwater leaving the spoils, and the recharge rate to the spoils.

A detailed hydrologic investigation was conducted from 1975 to 1977 by the staff of the Northern Cheyenne Research Project (Woessner et al., 1978) in the vicinity of the Logging Creek Mine Site area to define the parameters needed to model the impacts of mining. The basic monitoring network consisted of stream-gaging stations on the Tongue River at the northern and southern Reservation boundaries and on Logging Creek near the mouth, and 24 wells finished in the various water-bearing strata in the area. Water levels were monitored monthly, and water quality samples were taken monthly from 20 of the wells and analyzed for 20 standard constituents and a total of 16 trace metals. The specific techniques used for defining the parameters were:

(1) $STREAM_D$. Stream flows in the vicinity of the modeled area were interpolated from the gaging stations on the Tongue River located at the southern and northern Reservation boundaries.

(2) $STREAM_{NTDS}$. The natural total dissolved solids concentration of the streams in the vicinity of the modeled area were measured monthly during 1976 and 1977 and during low-flow periods. The fifty-percentile TDS concentration and mean TDS were calculated from the monthly values.

(3) GW_D. Groundwater inflow to the Tongue River was determined from seepage runs.

(4) GW_{NTDS}. The natural TDS of groundwater discharging to the Tongue River in the vicinity of the modeled area was estimated by averaging the TDS concentrations of wells located in the alluvium adjacent to the river.

(5) $COAL_I$. The natural groundwater flow through the coal seams was estimated using Darcy's law. The potentiometric gradient was determined from wells finished in the Knobloch coal seam, the thickness of the seams were determined from borings, and the hydraulic conductivity was determined by eleven short-term one-well pump tests. It is assumed that after mining is completed the groundwater inflow to the spoils will be equal to the groundwater flow through the coal seams prior to mining.

(6) $COAL_{TDS}$. The TDS concentrations of the groundwaters flowing into the mine area were calculated by averaging the TDS concentrations which were determined from water samples taken from wells finished in coal seams located in the hypothetical mine site or upgradient from the mine site.

(7) $OTHER_R$. The quantity of groundwater entering the system by recharge downgradient from the proposed mine area and by discharge from deep flow systems was determined by subtraction. $OTHER_R$ includes groundwater recharge to the clinker and recharge to the alluvium and inflow to the alluvium from deep flow systems. The recharge rates were calculated from well hydrographs, and the discharge from the deeper systems was determined by subtraction. This breakdown was not necessary for purposes of calculating changes in stream TDS, as it does not improve the predictive capabilities of the model.

(8) $SPOIL_{TDS}$. Saturation extract analyses of 27 samples of drill cuttings from the hypothetical mine site areas were run by D.B. McWhorter at Colorado State University (pers. commun., 1978). The saturation extract TDS values were averaged to obtain a TDS value which is assumed to be representative of the quality of the spoil recharge as it enters the spoil groundwater system. Composite leachate columns of drill cuttings were leached with groundwater from the mine site and with distilled water to simulate recharge. It was determined that the spoils are readily leached by both the pre-mining groundwater and distilled water.

Saturation paste extracts have previously been used by McWhorter et al. (1976) to determine spoil water quality from a strip-mine operation in Colorado, and by Van Voast and Hedges (1975) to forecast spoil water quality at the Decker mines, Montana. D.B. McWhorter (pers. commun., 1978) suggested that the average saturation-paste TDS values for a series of composite samples of drill cuttings of the overburden may not allow prediction of the spoil water quality at an individual well finished in the spoil, but that it would be representative of the general quality of spoil water.

(9) $SPOIL_R$. A range of spoil recharge rates from 0.0031 to 0.031 m/yr. was used. The lower rate was based on the rate of recharge calculated using well hydrographs for the period 1975—1977 for unconfined sandstone, shale and coal. The higher rate was employed as a worse-case figure.

IMPACTS IN THE LOGGING CREEK AREA

Pre-mining system

Groundwater movement through the site is from the northwest to the southeast, discharging into the clinker and alluvial materials and eventually into the Tongue River (Fig.5). 359 m³/day of groundwater discharges from the southeastern mine boundary to the clinker zone. This water moves through the clinker zone where 426 m³/day of recharge enters the system. A total of 785 m³/day of water either seeps out as springs at the clinker boundary and runs downgradient and re-enters the alluvial material or migrates downward through fractured basal clinker to sandstones and shales in direct

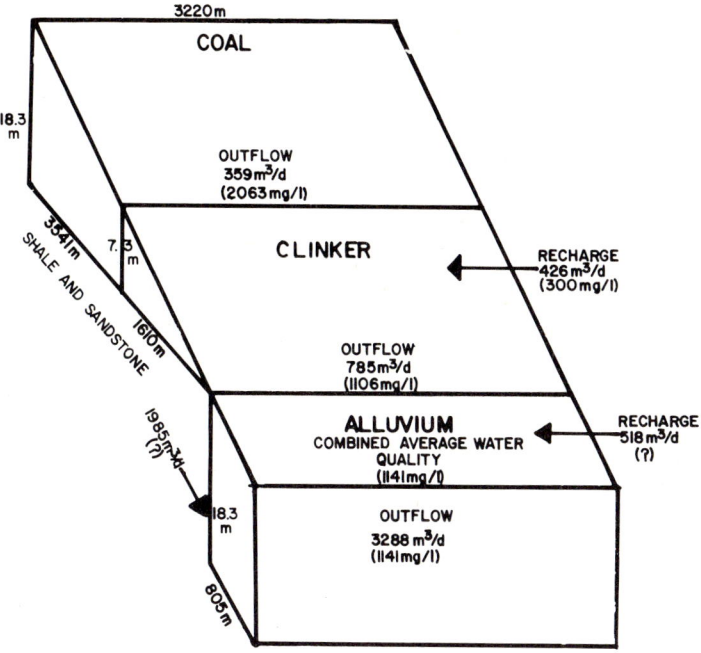

Fig.5. Groundwater flows and water quality in the Logging Creek area prior to mining.

contact with alluvium. Water enters the alluvium from the clinker zone, from deeper flow systems and by recharge. A total of 3288 m³/day of groundwater discharges from the alluvium to the Tongue River. The water discharging to the Tongue River has an average TDS of 1141 mg/l.

Post-mining system

At the completion of mining, spoil material to a depth of 61 m will fill the $11.4 \cdot 10^6$-m³ mine site. Groundwater in the coal and clinker surrounding the site will flow to the site under the existing gradients and resaturation of the spoils will begin.

The post-mining material-balance model of the Logging Creek mine area is presented in Fig.6. The parameters used in the model are listed in Table I. The post-mining outflow from the mine site increases from the pre-mining value 359 m³/day to values of 456—1327 m³/day depending upon the recharge rate. The net outflow to the Tongue River from the groundwater system increases by 30.1—301 m³/day^{-1} km^{-1} depending on the recharge rate used. The TDS of the groundwater discharging to the Tongue River increases to 1442—1986 mg/l. The change in TDS of the Tongue River is dependent

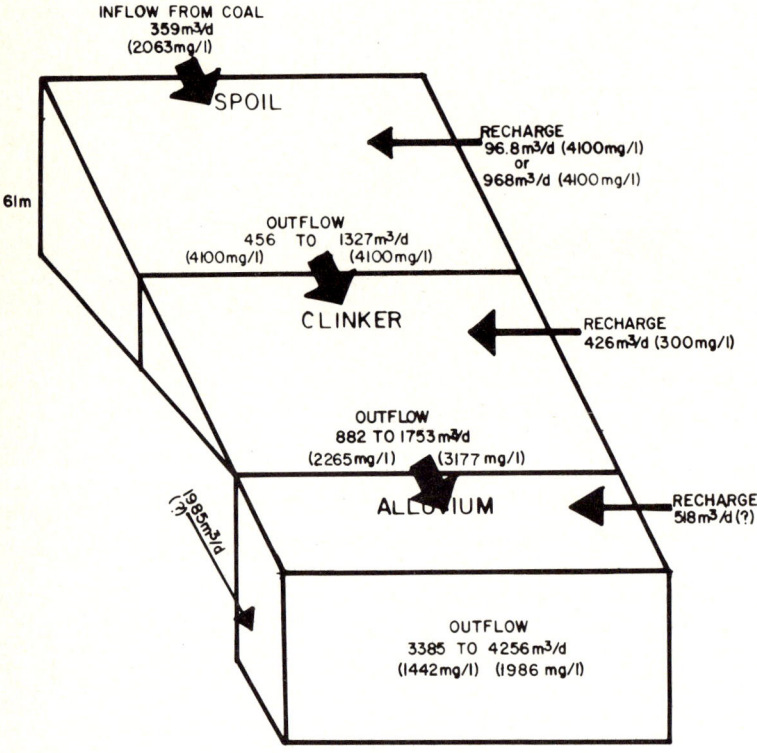

Fig.6. Groundwater flows and water quality in the Logging Creek area after mining has occurred and the flow system has been re-established.

on the flow rate. Water-quality changes in the Tongue River and the groundwater discharging to the river are listed in Table II.

Analyses of the water-quality data indicate that a single mine in the Logging Creek area will have a significant impact on the groundwater quality and a minimal impact on the quality of the Tongue River. In both cases, however, water-quality impacts will not be short-term. Analyses of leaching experiments showed that the majority of available leachable salts will be removed from the spoils during the passage of the first pore volume of water through the material. Based on recharge rates of 0.0031—0.031 m/yr. and an assumed effective spoils porosity of 0.1, it would take 140—1400 yr. to pass one pore volume of recharge through unsaturated spoils (it is assumed that the lower 18.2 m of the spoils is resaturated by groundwater). The leaching process will continue long after the 30-yr. life of the mine. Impacts to the water quality of the clinker and alluvial units downgradient from the mine and Tongue River will not be apparent immediately after the completion

TABLE I

Parameters used in the material-balance model of the Logging Creek mine site

$STREAM_F$ (m³/day)		$STREAM_{NTDS}$ (mg/l)	
average flow	$1.23 \cdot 10^6$	at average flow	270
50% flow	$0.49 \cdot 10^6$	at 50% flow	623
low flow	$0.14 \cdot 10^6$	at low flow	836
GW_D (m³/day)	3,288	GW_{NTDS} (mg/l)	1,141
$OTHER_R$ (m³/day)	2,929	$OTHER_{TDS}$*¹ (mg/l)	1,028
$COAL_I$ (m³ day)	359	$COAL_{TDS}$ (mg/l)	2,063
$SPOIL_R$ (m³/day)	96.8—968	$SPOIL_{TDS}$*² (mg/l)	4,100

*¹ Calculated using eq.2.
*² Based on drill cuttings from NCRP wells 8A, 27 and 29.

of mining. The time required to resaturate the mine spoils and establish a flow system is unpredictable at this time. Based on Darcy's law with convective transport only, groundwater leachate travel times can be predicted once the system has been re-established (Fig.7). Changes in water quality would be discernable in the clinker and the alluvium within a very short time after a groundwater flow system had been re-established. However, the low hydraulic conductivity of the spoils, the small rates of recharge, and the large volume of spoils would result in the production and outflow of high-TDS water from the mine site for hundreds of years.

TABLE II

Impacts of the hypothetical Logging Creek mine on water quality of Tongue River and on quality of groundwater discharging to the Tongue River

	Natural TDS	Post-mining TDS	
		0.0031-m/yr. recharge	0.031-m/yr. recharge
Discharging groundwater TDS (mg/l)	1,141	1,442	1,986
Streamflow TDS (mg/l) at			
mean annual flow	275	276	278
50-percentile flow	630	632	638
measured low flow	850	857	876

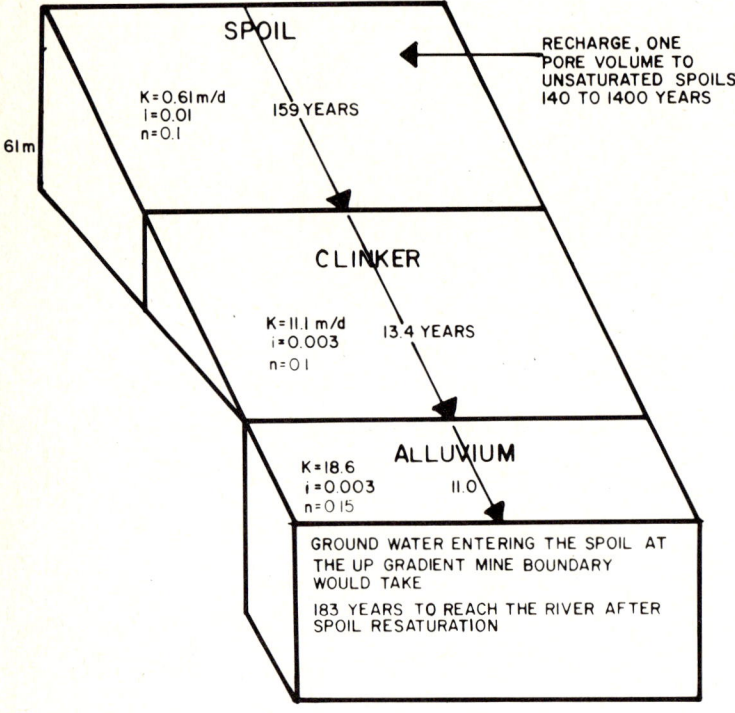

Fig.7. Groundwater travel times in the Logging Creek area after mining has occurred and the flow systems have been re-established.

REGIONAL IMPACTS

The extensive coal deposits found in the Tongue River Valley both on and off the Reservation could support large-scale mining in the near future. Mining operations similar to the Logging Creek Mine Site could exist for 20—40 km on either side of the river in this region. It is estimated that $4.3 \cdot 10^9$ t of low-sulfur strippable coal lie in the Tongue River Valley on and adjacent to the Reservation (Fig.2). The Northern Cheyenne Tribe has no intentions of mining the coal on the west side of the river at this time. However, MONCO mining company is planning to commence mining some of the coal on the east side of the river in 1982 if the necessary permits are obtained.

Mining of a large portion of the valley would create spoil water of low quality as predicted at the Logging Creek Mine Site. Most of the shallow

TABLE III

Predicted TDS concentrations in the Tongue River at the Northern Reservation boundary if the 30-km section of the Tongue River Valley on and adjacent to the Northern Cheyenne Reservation is mined

	Natural TDS (mg/l)	Post-mining TDS (mg/l)	
		0.0031-m/yr. recharge	0.031-m/yr. recharge
Tongue River flow			
mean annual flow	284	301	349
50-percentile flow	644	683	788
measured low flow	875	995	1,278

groundwater in the mined areas, plus part of the down-stream regions, would be of marginal use for stock and unattractive for other uses. Sub-irrigated crops in the alluvial valley may suffer if TDS rises above predicted levels. The quality of the Tongue River would also be affected if a large portion of the Tongue River Valley were mined.

Stream impact analysis for a 30-km section of the valley was initiated at a point 20 km upstream from the Logging Creek Mine Site. It was assumed that both sides of the river valley were mined. Although the thickness and extent of the coal varies widely in the valley, post-mining groundwater discharge rates were based on pre-mining Logging Creek area coal seam and alluvial flow rates, recharge data, and a recharge rate to the spoils at 0.0031 and 0.031 m/yr. Based on the two recharge rates to the spoil, groundwater discharge to the Tongue River was calculated to be $2.1 \cdot 10^4$ and $2.6 \cdot 10^4$ m^3 day^{-1} km^{-1} with TDS values of 1442 and 1986 mg/l, respectively. Calculated quality impacts to the Tongue River at the downstream boundary of the 30-km section are listed in Table III.

Extensive mining adjacent to the Tongue River will cause a significant change in the water quality of the river. However, calculated long-term changes in water quality are not great enough to render the water unfit for livestock watering or irrigation. If the changes calculated for mining the area on and adjacent to the Reservation are coupled to changes that will be occurring upstream on the Tongue River due to strip mining, the conclusion is that water quality in the river could be severely degraded in the long term. Presently, 60 km upstream from the southern Reservation boundary, near Decker, Montana, over $11 \cdot 10^6$ t are being mined annually from pits located adjacent to the river and this rate could double within the next decade. Adding the impacts likely to be caused by these mining activities, the calculated TDS changes for the Tongue River as shown in Table III double. Exacerbating these calculated changes in water quality is the increasing use of Tongue River water upstream in Wyoming for irrigation which will further degrade the quality of water in the Tongue River.

Strip mining in the Tongue River Valley will necessarily imply a long-term degradation of water quality in the Tongue River. The changes will not occur rapidly, but will happen gradually over a period of many years.

CONCLUSIONS

Strip mining in the Northern Great Plains Region will have a long-term impact on groundwater and surface-water quality. Evaluation of a hypothetical $9 \cdot 10^6$-t/yr. mine in the Tongue River Valley leads to the conclusion that the impacts of a single mine on groundwater quality are significant, but that the impacts on surface-water quality are small. The cumulative impact from the extensive mining which is likely to occur in the Tongue River Valley of the Powder River Basin will be, though, a significant deterioration in surface-water quality on the long term. This deterioration will last for centuries.

The material-balance approach based on an assumed relation between saturation paste extracts and spoil water quality used in this paper is simplistic. However, it is the only viable approach available at this time for quantifying site-specific and regional long-term impacts from coal strip mining. Process oriented deterministic models of the chemical processes occurring in the spoils and downgradient from the spoils, of recharge to the spoils, and of water movement through spoils are needed to refine further the impacts of mining. Future studies of coal strip mining should concentrate on these areas of research.

ACKNOWLEDGEMENTS

The authors thank the staff of the Northern Cheyenne Research Project for their assistance, especially Ed Heffern, staff geologist, for mapping the geology of the Tongue River Member in the study area. Thanks are also extended to Allen Rowland, President of the Northern Cheyenne Tribe, and the Northern Cheyenne Tribal Council for their encouragement of this research effort, and to Dr. David Stephenson for helping conceptualize this research project and for his assistance. This research was funded by the Northern Cheyenne Tribe through a grant from the U.S. Environmental Protection Agency (R803566).

REFERENCES

Anonymous, 1978. A cooperative program to evaluate surface and groundwater problems associated with potential strip mine sites. Mont. State Univ.—Mont. Coll. Miner. Sci. Technol.—N.D. State Univ.—Univ. N.D.—Univ. Wyo., Consort. Proj. Mont. State Univ., Bozeman, Mont., U.S. Environ. Prot. Agency Grant No. R803727 (unpublished).

Arnold, S.B. and Dollhopf, D.J., 1977. Soil water and solute in Montana strip mine spoils. Mont. Agric. Exp. Stn., Mont. State Univ., Bozeman, Mont., Res. Rep. 102, 129 pp.

Croft, M.G., Fisher, D.W., Thorstenson, D. and Crawley, M.E., 1978. Hydrology and geochemistry of the Gascoyne lignite mine, North Dakota. EOS (Trans. Am. Geophys. Union), 59(12): 1067 (abstract).

Davis, R.W. and Rechard, P.A., 1977. Effects of surface mining upon shallow aquifers in the Eastern Power River Basin, Wyoming. Water Resour. Res. Inst., Univ. Wyo., Laramie Wyo., Water Res. Ser. 67, 47 pp.

Dollhopf, D.J., Jenson, I.B. and Hodder, R.L., 1977. Effects of surface configuration in water pollution control on semiarid mined lands. Mont. Agric. Exp. Stn., Bozeman, Mont., Res. Rep. 114, 179 pp.

Farmer, E.E. and Richardson, B.Z., 1976. Hydrologic and soil properties of coal mine overburden piles in southeastern Montana. 4th Symp. on Surface Mining and Reclamation, Natl. Coal Assoc.—Bitum. Coal Res., Inc., Louisville, Ky., pp. 120—130.

Groenewold, G.H., Cherry, J.A., Hemish, L., Rehm, B, Meyer, G. and Winczewski, L., 1979. Geology and geohydrology of the Knife River Basin and adjacent areas in west central North Dakota. N.D. Geol. Surv., Rep. Invest. 64.

Hittman Associates, Inc., 1977. Monitoring and modeling of shallow groundwater in the Powder River Basin. Phase I Rep. (prepared under U.S. Bur. Mines, Contract No. J0265050).

Hittman Associates, Inc., 1978. Monitoring and modeling of shallow groundwater in the Powder River Basin. Annu. Tech. Rep. (prepared under U.S. Bureau of Mines Contract No. J0265050).

McWhorter, D.B., Rowe, J.W., Van Liew, M.W., Chandler, R.L., Skogerboe, R.K., Sunada, D.K. and Skogerboe, G.V., 1976. Surface and subsurface water quality hydrology in surface mined watersheds. Ind. Environ. Res. Lab., U.S. Environ. Prot. Agency, Cincinnati, Ohio, 357 pp.

Moran, S.R., Cherry, J.A., Ulmer, J.H., Peterson, W.M., Somerville, M.H., Schafer, J.K., Lechner, D.O., Triplett, C.L., Loken, G.R. and Fritz, P., 1976. An environmental assessment of a 250MMSCFD dry ash Lurgi coal gassification facility in Dunn County, North Dakota. Univ. N.D. Eng. Exp. Stn., Rep. 76-12-EES-01.

Moran, S.R., Groenewold, G.H. and Cherry, J.A., 1978. Geologic, hydrologic, and geochemical concepts and techniques in overburden characterization for mined-land reclamation. N.D. Geol. Surv., Rep. Invest. 63, 152 pp.

NGPRP (Northern Great Plains Resource Program) 1974. Shallow groundwater in selected areas in the Fort Union Coal Region. N. Great Plains Resour. Progr., Groundwater Subgroup, U.S. Geol. Surv., Open File Rep. No. 74-371.

Rahn, P.H., 1976. Potential of coal strip-mine spoils as aquifers in the Powder River Basin. Old West Reg. Comm. Billings, Mont. 108 pp.

Riordan, P.J., Wilson, J.L., Schreiber, R.P. and Venyke, C.P., 1978. Application of groundwater model to study Malakoff—Cyuga lignite field, Texas. EOS (Trans. Am. Geophys. Union), 59(12): 1067 (abstract).

Rogowski, A.S., Pionke, H.B. and Broyan, J.G., 1977. Modeling the impact of strip mine and reclamation processes on quality and quantity of water in mined areas: a review. J. Environ. Qual., 6(3): 237—244.

Slawson, Jr., G.C., 1978. Development of groundwater quality monitoring program for coal strip mining. EOS (Trans. Am. Geophys. Union), 59(12): 1067 (abstract).

USDI (United States Department of Interior), 1974a. Final Environmental Statement, proposed plan of mining and reclamation Big Sky Mine, Peabody Coal Company coal lease M15965, Colstrip, Montana. U.S. Geol. Surv., Final Environ. Statement, 74-12, 1-2: 4 8 pp.

USDI (United States Department of Interior), 1974b. Final Environmental Statement, proposed development of coal resources in Eastern Powder River Basin of Wyoming. U.S. Bur. Land Manage.

USDI (United States Department of Interior), 1975. Final Environmental Statement, proposed plan of mining and reclamation, Belle Ayr Mine, Campbell County, Wyoming. U.S. Geol. Surv.

USDI (United States Department of Interior), 1976a. Environmental Analysis Record, Velva Mine coal leases, ERA-MT020-6-91, Consolidation Coal company, North Dakota. U.S. Bur. Land Manage.

USDI (United States Department of Interior), 1976b. Environmental Analysis Record, Glenharold mine coal leases, ERA-MT020-7-2, Consolidation Coal Company, North Dakota. U.S. Bur. Land Manage.

USDI (United States Department of Interior), 1976c. Final Environmental Statement, proposed 20 year plan of mining and reclamation, Westmorland Resources Tract III, Crow Indian Ceded Area, Montana. U.S. Geol. Surv., 396 pp.

USDI (United States Department of Interior), 1976d. Environmental Analysis Record, Falkrik mine coal lease, M-31-05-3, North Dakota. U.S. Bur. Land Manage.

USDI (United States Department of Interior), 1976e. Final Environmental Statement, proposed plan of mining and reclamation, Cordero Mine, Campbell County, Wyoming. U.S. Geol. Surv.

USDI (United States Department of Interior), 1976f. Final Environmental Statement, proposed plan of mining and reclamation, Eagle Butte Mine, Campbell County, Wyoming. U.S. Geol. Surv.

USDI (United States Department of Interior), 1977a. Final Environmental Statement, proposed plan of mining and reclamation, East Decker and North Decker extension mines, Decker Coal Company, Big Horn County, Montana. U.S. Geol. Surv.—Mont. Dep. State Lands, 758 pp.

USDI (United States Department of Interior), 1977b. Draft Environmental Statement, proposed plan of mining and reclamation, East Gillette Mine, Campbell County, Wyoming. U.S. Geol. Surv.

USDI (United States Department of Interior), 1978a. Draft Environmental Statement, proposed development of coal resources in the Eastern Powder River Basin of Wyoming. U.S. Bur. Land Manage.

USDI (United States Department of Interior), 1978b. Draft Environmental Statement, federal coal management program. U.S. Bur. Land Manage., 654 pp.

USDI (United States Department of Interior), 1978c. Draft Environmental Statement, proposed plan of mining and reclamation, Caballo Mine, Campbell County, Wyoming. U.S. Geol. Surv.

USDI (United States Department of Interior), 1978d. Draft Environmental Statement, proposed plan of mining and reclamation, Pronghorn Mine, Campbell County, Wyoming. U.S. Geol. Surv.

USDI (United States Department of Interior), 1978e. Draft Environmental Statement, proposed mining and reclamation plan, Spring Creek Mine, Big Horn County, Montana. U.S. Geol. Surv.—Mont. Dep. State Lands.

USDI (United States Department of Interior), 1978f. Draft Environmental Statement, proposed expansion of mining and reclamation plan, Big Sky Mine, Rosebud County, Montana. U.S. Geol. Surv.—Mont. Dep. State Lands.

USDI (United States Department of Interior), 1979. Draft Environmental Statement, proposed plan of mining and reclamation, Coal Creek Mine, Campbell County, Wyoming. U.S. Goel. Surv., FES 79-1.

USDOE (United States Department of Energy), 1978. Federal coal leasing and 1985 and 1990 coal production forecasts. 26 pp.

Van Voast, W.A., 1974. Hydrologic effects of strip mining in southeastern Montana — Emphasis: one year of mining near Decker. Mont. Bur. Mines Geol., Bull. 93, 24 pp.

Van Voast, W.A. and Hedges, R.B., 1975. Hydrogeologic aspects of existing and proposed strip coal mines near Decker, southeastern Montana. Mont. Bur. Mines Geol., Bull. 97, 31 pp.

Van Voast, W.A., Hedges, R.B. and Pagenkopf, G.K., 1975. Hydrologic impacts of coal mine effluents and spoil leachates. In: Fort Union Coal Field Symposium, Vol. 3, Mont. Acad. Sci., Billings, Mont., pp. 289—303.

Van Voast, W.A., Hedges, R.B. and McDermott, J.J., 1976. Hydrologic aspects of strip mining the subbituminous coal fields of Montana. 4th Symp. on Surface Mining and Reclamation, Natl. Coal Assoc. Bitum. Coal Res., Inc., Louisville, Ky., pp. 160—172.

Van Voast, W.A., Hedges, R.B. and McDermott, J.J., 1977. Hydrogeologic conditions and projections related to mining near Colstrip, southeastern Montana. Mont. Bur. Mines Geol., Bull. 102, 43 pp.

Van Voast, W.A., Hedges, R.B. and McDermott, J.J., 1978. Hydrologic characteristics of coal mine spoils, southeastern Montana. Mont. Univ. Joint Water Resour. Res. Cent., Rep. 94, 34 pp.

Verma, T.R. and Thames, J.L., 1975. The rehabilation of land disturbed by surface mining coal in Arizona. J. Soil Water Conserv., 30: 129—131.

Woessner, W.W., Osborne, T.J., Heffern, E.L., Whiteman, J., Spotted Elk, W. and Morales-Brink, D., 1978. Hydrologic impacts from potential coal strip mining Northern Cheyenne Reservation. Draft U.S. Environ. Prot. Agency Cincinnati, Ohio, 448 pp.

[2]

CONNECTOR WELLS, A MECHANISM FOR WATER MANAGEMENT IN THE CENTRAL FLORIDA PHOSPHATE DISTRICT

PHILIP E. LAMOREAUX

P.E. LaMoreaux & Associates, Inc., Tuscaloosa, AL 35401 (U.S.A.)

(Accepted for publication May 28, 1979)

ABSTRACT

LaMoreaux, P.E., 1979. Connector wells, a mechanism for water management in the Central Florida Phosphate District. In: W. Back and D.A. Stephenson (Guest-Editors), Contemporary Hydrogeology — The George Burke Maxey Memorial Volume. J. Hydrol., 43: 469—490.

Connector wells, a mechanism for water management in the Central Florida Phosphate District, have proven to be an effective means of moving good-quality groundwater from one formation downward to another under gravity flow. There results beneficial recharge to the underlying Floridan Aquifer and a solution to a dewatering problem that enhances the mining of the phosphate ore.

Three unique recharge systems have been developed in the phosphate district of Florida and include: recharge through connector wells by gravity flow of water from the overburden sand aquifer to the Floridan Aquifer; a system of siphon wells recharging water from the overburden to the underlying Floridan Aquifer system, and the recharge of surplus surface water from streams to the Floridan Aquifer by connector wells.

The siphon well recharging system has also been used effectively to relieve head or hydrostatic pressures built up in dams and dikes.

INTRODUCTION

Groundwater in southwest Georgia and central Florida is a major resource and, at times, a major problem. The provision of water for expanding municipal, agricultural, and industrial needs in central Florida has resulted in extensive development of water from wells. The increasing demand for water during the past few years has been met largely through the production of more and more water from wells developed principally in the Floridan Aquifer.

The Floridan Aquifer is a sequence of calcareous Cenozoic limestones and dolomites ranging in thickness from about 600 to 1800 m (1970 to 5900 ft.). These rocks underlie Florida and extend northward into southern Georgia and Alabama. Parts of the Floridan Aquifer are riddled by solution cavities and interconnected cavernous systems. The Floridan is one of the most productive aquifers in the world and is recharged by rainfall, streams and lakes.

The demand for water in central and southwest Florida is not uniform over the State. Some of the greatest demands for water are for municipal use and large withdrawals of water from wells occur in the major municipal and industrial well fields; for example, in the upper Tampa Bay area. Also, there is extensive pumpage during parts of the year for irrigation of truck crops such as strawberries or vegetables, and for irrigation of the citrus fruit trees. Industry demands many more millions of cubic meters of water a day. It is estimated that by about 1985 the water demand in the Southwest Florida Water Management District will nearly equal the average annual replenishment from rainfall. Thereafter, unless some type of augmentation of water into the district is accomplished, large-scale mining of water will be required.

Some types of augmentation could include: (1) the recycling of industrial water, (2) the use of sewage effluents, (3) the desalination of brackish or salt water, (4) the importation of water from other water basins, (5) extensive development of recharge programs to the Floridan Aquifer from available sources of good-quality surface water and groundwater, or (6) a combination of all of these and a large-scale water conservation program.

Much of Florida is a water-rich area because of the very high rainfall. Statewide the annual rainfall averages about 140 cm (55 in.). Evaporation and transpiration annually require approximately 102 cm (40 in.) or more than 70% of the rain that falls. There remains, however, a very substantial amount of water for recharge to the gigantic Floridan Aquifer System. It is estimated that in central Florida about $934 \cdot 10^4$ m^3 day^{-1} ($64 \cdot 10^4$ gal.day^{-1} mi.$^{-2}$) are recharged to this large artesian system. However, when we consider the projected use of water in this area, it must be recognized that problems will develop of a greater and greater magnitude in the future.

A specific illustration of present withdrawals is the citrus industry in central Florida which uses $\sim 75.6 \cdot 10^4$ m^3 day^{-1} ($2 \cdot 10^8$ gal. day^{-1}); the phosphate industry, $\sim 94.5 \cdot 10^4$ m^3 day^{-1} ($2.5 \cdot 10^8$ gal. day^{-1}); and water for industrial and municipal use. Large-capacity wells, some capable of pumping 315—630 l s^{-1} (5,000—10,000 gal. min.$^{-1}$) each supply these demands. There has resulted, therefore, a substantial impact on the potentiometric surface of the Floridan Aquifer. Fig. 1 is a generalised geohydrologic cross-section showing the Floridan Aquifer.

Recognizing the problem of declining water levels in the region, the Southwest Florida Water Management District and some of the industries of the area began to consider ways in which the problem could be alleviated or minimized. An attempt has been made by many industries to more efficiently use the water pumped from the ground.

Research has also resulted in the development of an innovative solution of recharge wells. These wells drain water by gravity from the overburden and/or streams downward into the Floridan Aquifer. The water recharged is carefully studied and monitored, and is as good or of a better quality than the water in the Floridan Aquifer. For example, water in the overburden overlying the phosphate rock occurs in relatively unconsolidated deposits of sand, gravel and clay. The water is soft and generally low in total dissolved solids.

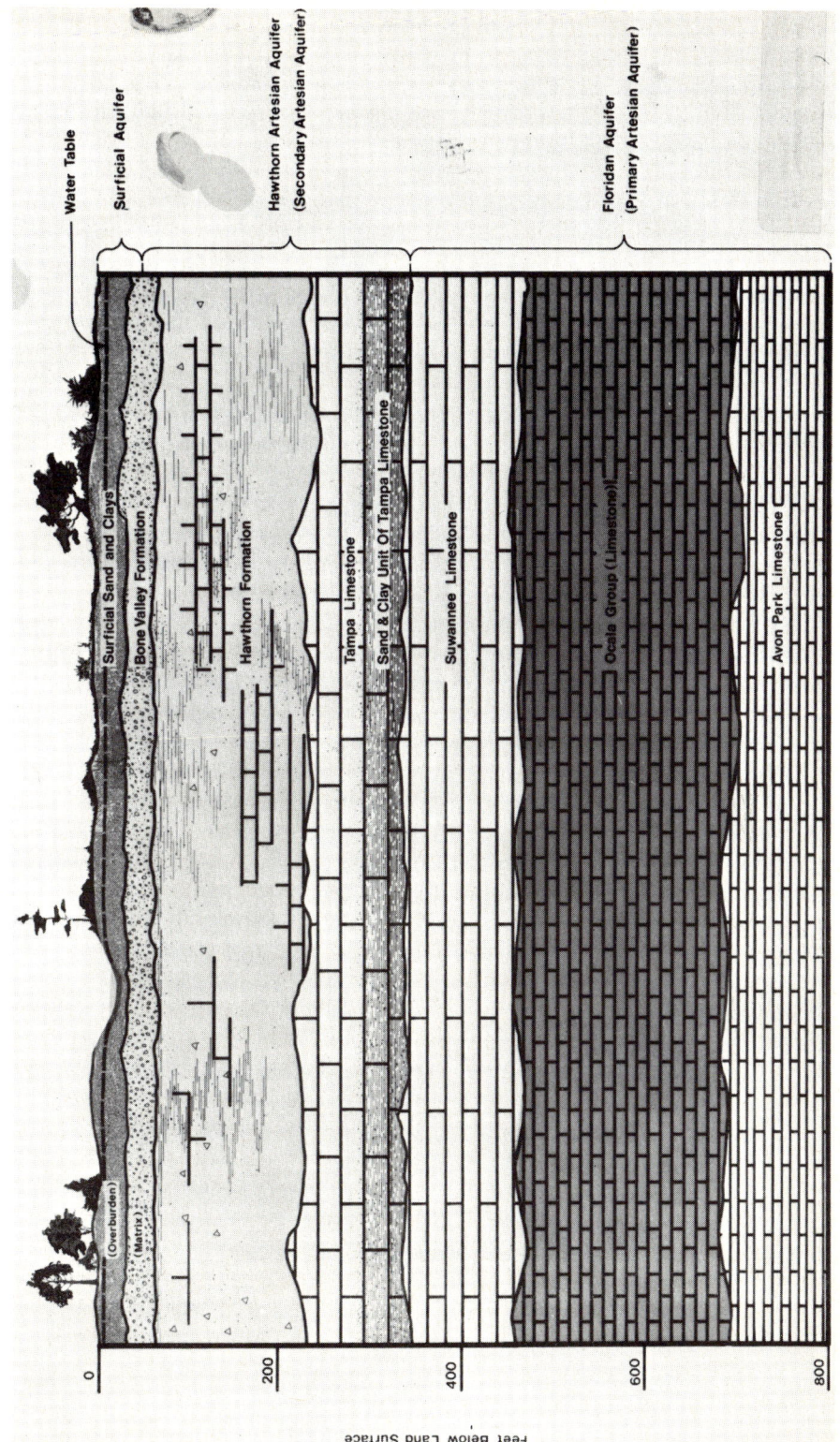

Fig. 1. Geohydrologic cross-section.

In much of the Central Phosphate District, the water table in the shallow aquifer is near land surface. In fact, the water can be seen in many places to stand in ponds and, as a result, evaporation rates are high and transpiration from plants, trees and grass is great. In these areas, rainfall runs off rapidly to streams and the recharge by infiltration of rainfall is low.

A UNIQUE HYDROGEOLOGIC SITUATION FOR RECHARGE

Surface geologic studies supported by test drilling in the area show that there are three distinct hydrogeologic units: an upper unit of unconsolidated material consisting of clay, sand and gravel; a middle unit consisting of the ore matrix, or phosphate ore zone; and a lower bedrock unit made up of a sequence of limestone formations.

In parts of the area the overburden ranges from 9 m (30 ft.) to as much as 18 m (60 ft.) thick. Much of these overburden sediments is saturated with fresh groundwater.

The feasibility of recharge connector wells varies with the hydrogeology of any area. In the Central Phosphate District, there are areas in which the potentiometric surface of the Floridan Aquifer is from a few to many tens of meters below the contact between the ore matrix and the underlying limestone. The Floridan is extensively cut by solution cavities that have a great capacity to accept water from a recharge connector well system. Therefore, water can be drained under gravity from properly constructed wells developed in the overburden into the underlying Floridan Aquifer. This mechanism allows a way to manage surface water and water stored in the surficial alluvial aquifer system during mining. At the same time, it allows the recharge of large quantities of water from either the shallow water-table aquifer or from streams, lakes, and ponds to the Floridan Aquifer to help counteract the impact of the heavy withdrawals for municipal, agricultural and industrial purposes in this heavy-water-use area.

It is imperative that a thorough knowledge of the hydrogeology of the area should be available as a basis for recharge projects and that a carefully established monitoring program be carried out to protect the quality and measure the quantity of recharge.

There follow the principal considerations in establishing these unique systems in the Southwest Florida Water Management District. The systems involve:

(1) Recharge through connector wells by gravity flow of water from the overburden to the Floridan Aquifer.

(2) A system of siphon wells recharging water from the overburden to the underlying Floridan Aquifer system.

(3) The recharge of surplus surface water from streams of an area to the Floridan Aquifer.

A MULTIDISCIPLINE APPROACH TO THE DEVELOPMENT OF RECHARGE SYSTEMS

The potential application of artificial recharge techniques requires a thorough knowledge of the hydrology and geology of the surface-water regime, the water-table aquifer, and the Floridan Aquifer. Careful consideration must be given to soils, vegetation, and cultural activities in an area, as well as to climatological, environmental and regional influences to determine the most desirable long-term program.

In the U.S.A., the application of artificial recharge methods for dewatering and water conservation purposes dates back many years. However, it was not until the 1930's that specific attention was directed by the U.S. Geological Survey to artificial recharge techniques in the U.S.A. for the purpose of replenishing the groundwater resources.

In the Central Florida Phosphate District, the potential for artificial recharge was recognized by P.E. LaMoreaux & Associates as early as 1970 as a method of recharging the artesian Floridan Aquifer with water from the overlying water table aquifer. Since that time many recharge connector wells have been installed by the phosphate industry, and these wells have increased the efficiency of open-pit mining techniques by dewatering the ore matrix, which resulted in recharging the Floridan Aquifer by many thousands of cubic meters of water each day (LaMoreaux et al., 1970, 1975; LaMoreaux & Associates, Inc., 1975, 1977; LaMoreaux, 1977a,b). Good-quality water that would ordinarily be lost to runoff and evapotranspiration is transferred deeper into the groundwater system where it can be withdrawn later for domestic, agricultural and industrial uses (Hutchinson and Wilson, 1974; Cawley, 1975; Knochenmus, 1975; United Nations, 1975; Tolson and Doyle, 1977).

Soil parameters

Several types of soils occur in the phosphate district. Soils maps are used to delineate the most favorable areas for recharge wells. For example, the soils associated with areas of nearly level topography, commonly referred to as the pine and palmetto flatwoods, are strongly acid soils with a thin surface covering of organic matter and a hardpan layer composed of poorly to well-indurated quartz sand, fine organics and iron oxides. Because of the presence of the hardpan, which varies considerably in depth and thickness, these soils promote perched water-table conditions and overland runoff. Many of these areas are the least favorable for application of artificial recharge techniques.

There are, however, interspersed throughout the area deep, sandy soils of the Pamlico, Scranton and St. Lucie soil types. These soils are characterized by low-profile ridges, crests, or knolls. They have a very thin surface covering of organic matter and are acidic, generally do not possess a hardpan layer, but have rapid internal drainage. Vegetation is commonly sparse to moderate; surface drainage is almost entirely absent; and the infiltration of water is rapid. These soils are favorable for the installation of recharge-connector wells.

WATER-TABLE AQUIFER

The water-table aquifer is composed predominantly of an upper sand unit comprising the sandy part of the overburden and a lower phosphorite unit that contains the ore matrix zone (Fig. 1). Permeability varies greatly from place to place in both of these units. However, generally, the overburden is more permeable than the matrix because of its lower clay content and the greater amount of permeable sands.

Because of variability in thickness, concentration of fine-grained sediments, intermittent hardpan layers, clay lenses, and permeable sand stringers, the overburden and matrix are generally hydraulically connected and respond as a unit. On the basis of extensive core-logging data, the upper sand unit generally ranges from less than 1.5 m (5 ft.) to more than 18 m (60 ft.) in thickness in the phosphate district. The lower phosphorite unit can generally be recognized by gray to grayish-green phosphatic clay with clayey sand lenses and sand stringers. The variability in thickness and composition indicate that the hydraulic characteristics of the water-table aquifer are also highly variable. A sand/clay ratio map (Fig. 2), as well as a map showing the configuration of top of matrix (Fig. 3), shows this lithology. The hydrologic variables are shown in Fig. 4.

Water-table aquifer test

A test well and four observation wells were constructed in the NW 1/4 NE 1/4 SW 1/4 NW 1/4 Section 27, Township 34 South, Range 23 East, in Hardee County. During March 10—13, 1976, a detailed pumping test was conducted to determine the water-bearing characteristics of the water-table aquifer at that location (Fig. 5).

The test well, MCRW-1, was pumped with a submersible pump at a rate of 79.5 l min.$^{-1}$ (21 gal. min.$^{-1}$) for 28 hr. Drawdowns in the test well and observation wells were recorded at frequent intervals.

The pumping test data was analyzed by the Boulton (1954, 1963, 1964) delayed yield formula. Table I summarizes the results of the computations.

In addition to the detailed pumping test conducted on well MCRW-1, four other short-term tests were conducted during July 1976 at other locations in the project area (Fig. 5). The results of the data evaluation from these tests are given in Table II.

The transmissivity values obtained at each test location were related to the saturated thickness of the aquifer. Table III summarizes the results of these tests.

Data from Table III show that the hydraulic properties of the aquifer vary greatly over the property. The variability is more a function of stratigraphy and sedimentary characteristics than saturated thickness of the aquifer. The data also show that the water-table aquifer underlying the property exhibits hydraulic conditions more conducive to the application of recharge-connector well techniques in the western-most part of the property than compared to that underlying the remaining part of the property where the overburden is relatively thin and permeabilities are low.

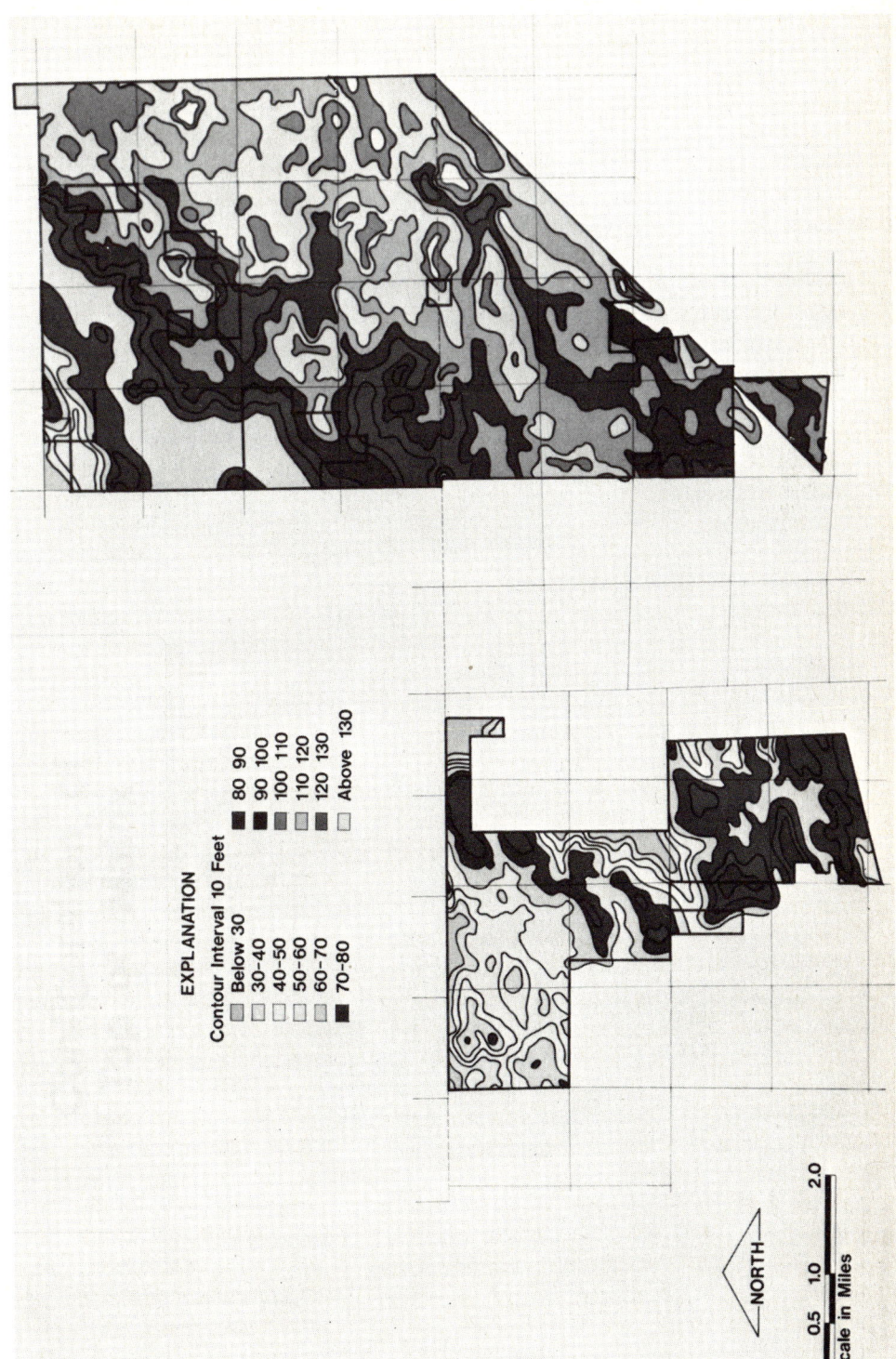

Fig. 2. Configuration of top of matrix in an area in Hardee County, Florida. (See Fig. 5 for location.) A map used with the thickness of overburden map to determine volume of sediments to be dewatered.

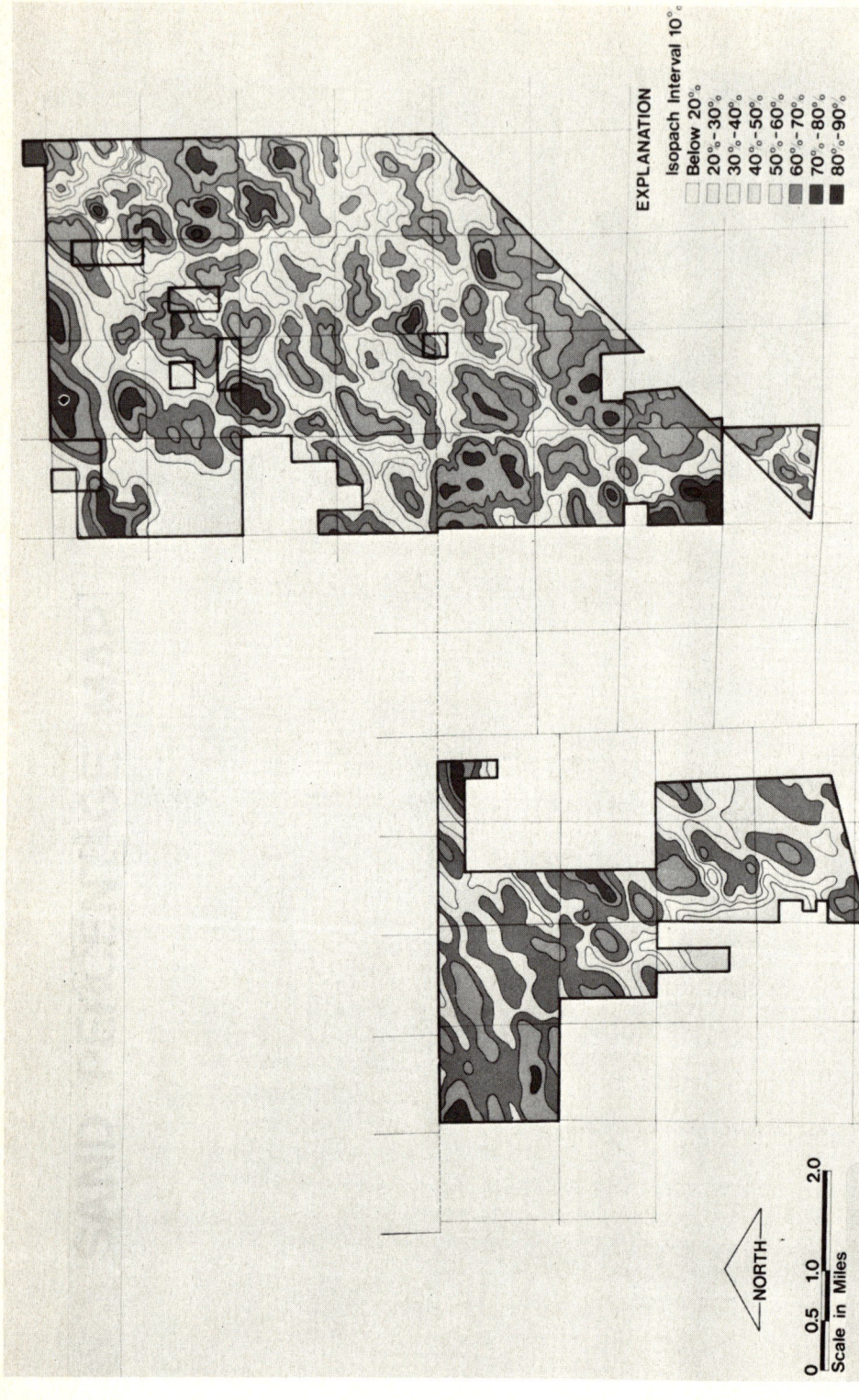

Fig. 3. Isolith of the upper sand unit of the overburden (sand percentage map) in an area in Hardee County, Florida. (See Fig. 5 for location.)

477

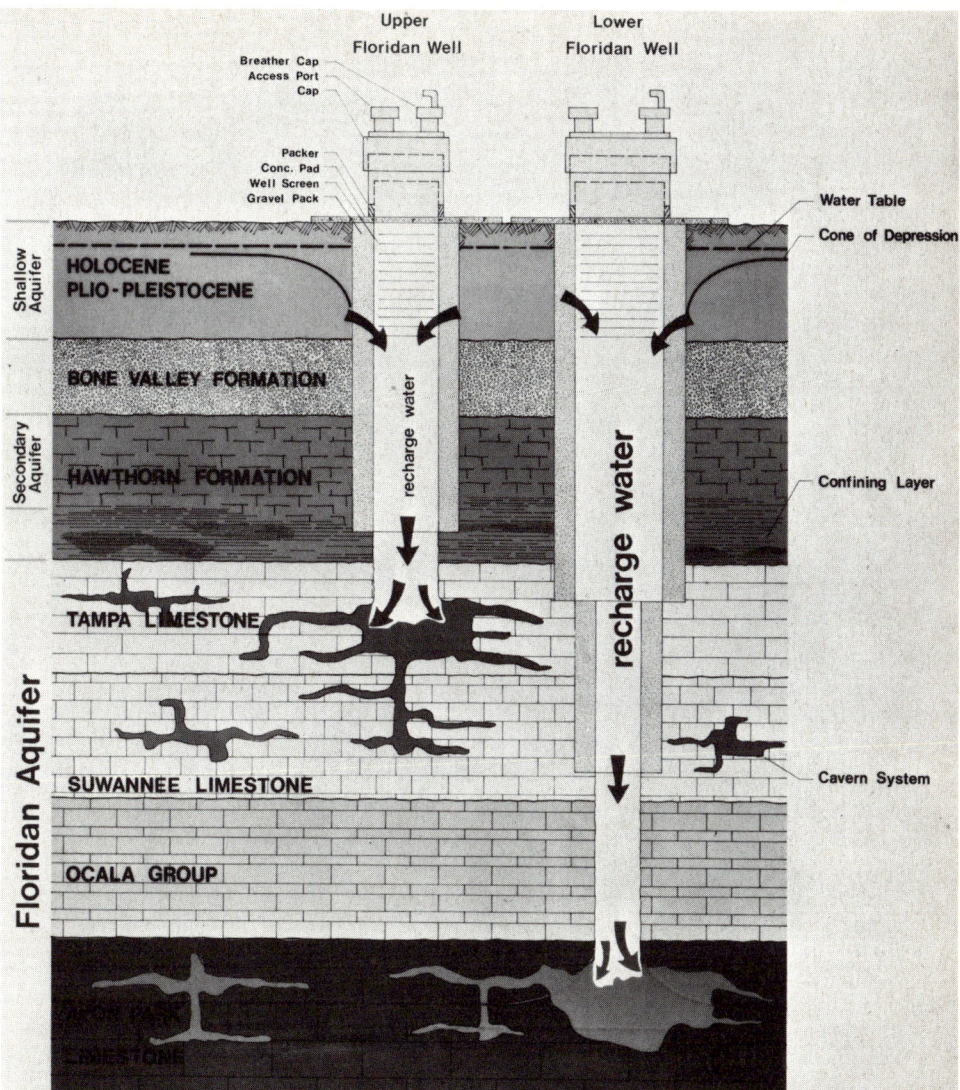

Fig. 4. Pilot recharge-connector well system of the Floridan Aquifer (not to scale).

Seasonal variation of the water-table aquifer

The water table varies seasonally in response to many factors including climatological, geohydrological and biological influences. The lowest water levels commonly occur during the dry season of March, April and May, and the highest levels commonly occur during the wet season of August,

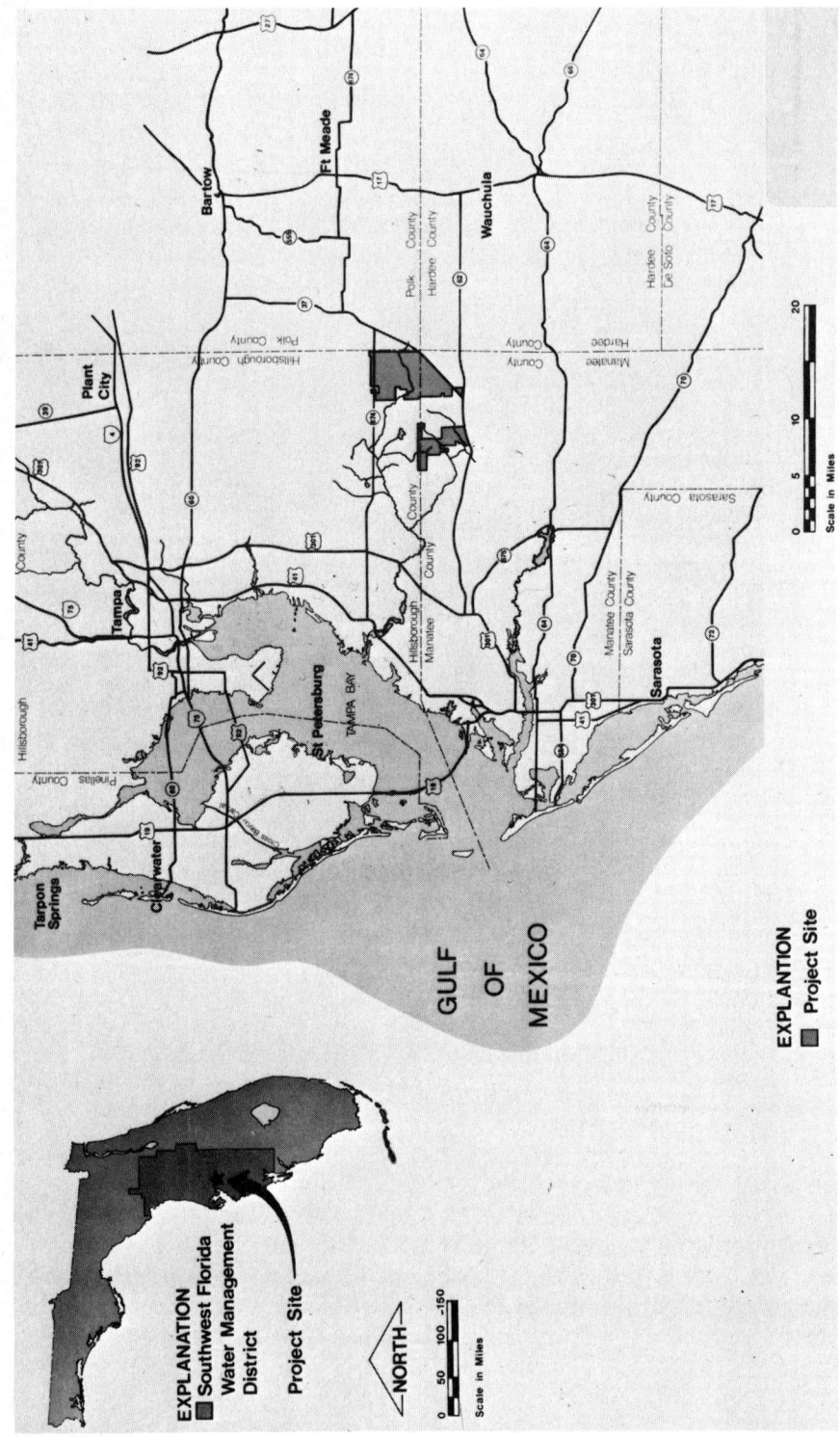

Fig. 5. Location of pumping test sites.

TABLE I

Results of a pumping test on well MCRW-1

Well pumped	Well observed	Part of hydrograph analyzed	Transmissivity (gal. day^{-1} ft.$^{-1}$)
MCRW-1	MCRO-1	drawdown	11,500
MCRW-1	MCRO-2	drawdown	13,100
MCRW-1	MCRO-3	drawdown	10,200
Average	—	—	12,000

1 gal.day^{-1} ft.$^{-1}$ = 1.2418·10^{-2} m^2 day^{-1}.

TABLE II

Results of short-term pumping tests

Well pumped	Average transmissivity (gal. day^{-1} ft.$^{-1}$)
MCSA-7	11,000
MCSA-9	1,600
MCSA-15	3,300
MCSA-16	3,100

1 gal.day^{-1} ft.$^{-1}$ = 1.2418·10^{-2} m^2 day^{-1}.

TABLE III

Summary of pumping test results

Test site	Average transmissivity (gal. day^{-1} ft.$^{-1}$)	Saturated thickness (ft.)	Unit permeability (gal. day^{-1} ft.$^{-1}$)
MCRW-1	12,000	27	444
MCSA-7	11,000	17	647
MCSA-9	1,600	17	94
MCSA-15	3,300	17	194
MCSA-16	3,100	17	182

1 gal.day^{-1} ft.$^{-1}$ = 1.2418·10^{-2} m^2 day^{-1}.

September and October. The time of occurrence and duration of the wet and dry seasons can vary from year to year. Contour maps of water-table elevations during dry and wet months of 1976 are shown in Fig. 6. These maps indicate an average fluctuation of about 1.5 m (5 ft.) for the period of observation.

480

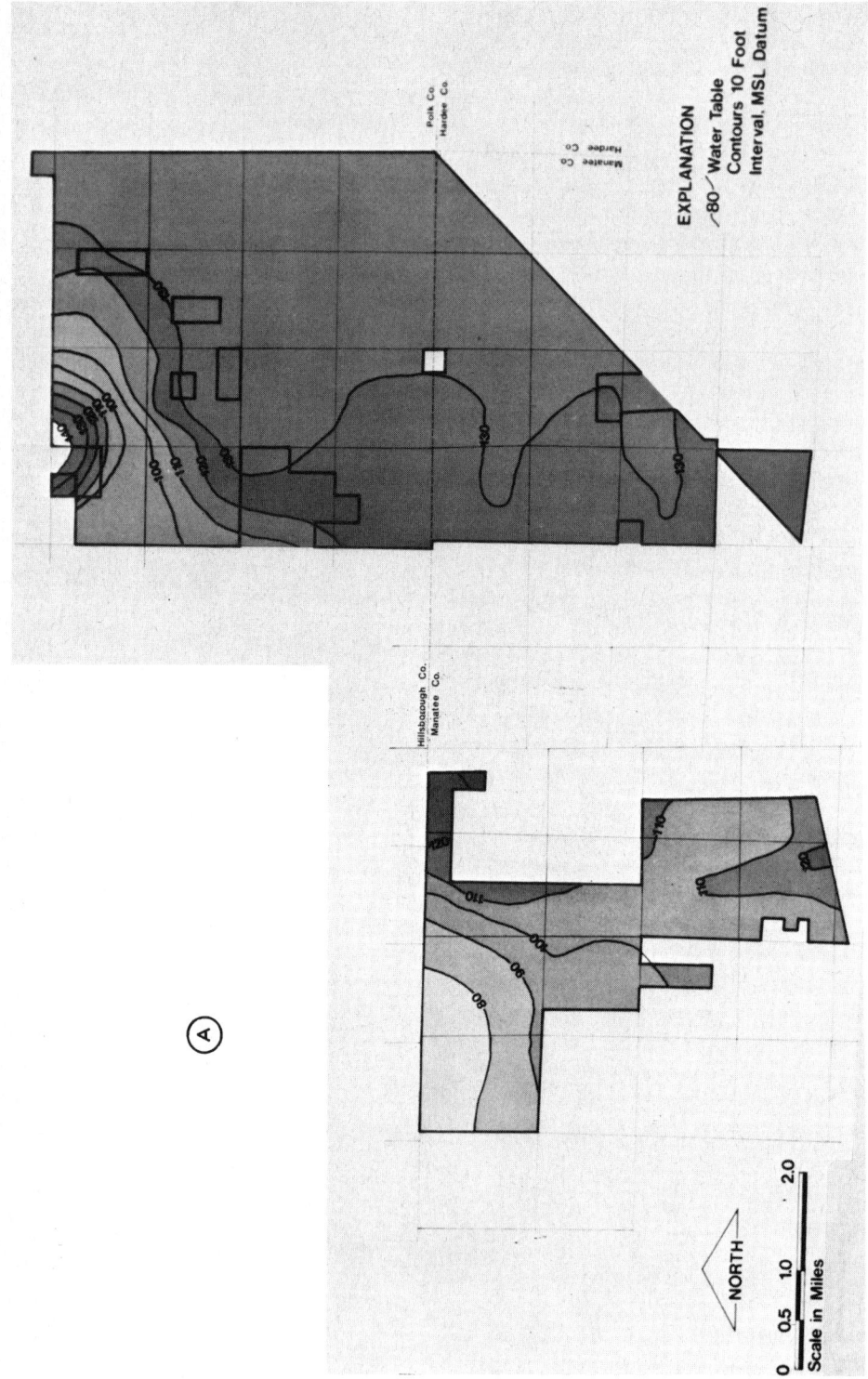

481

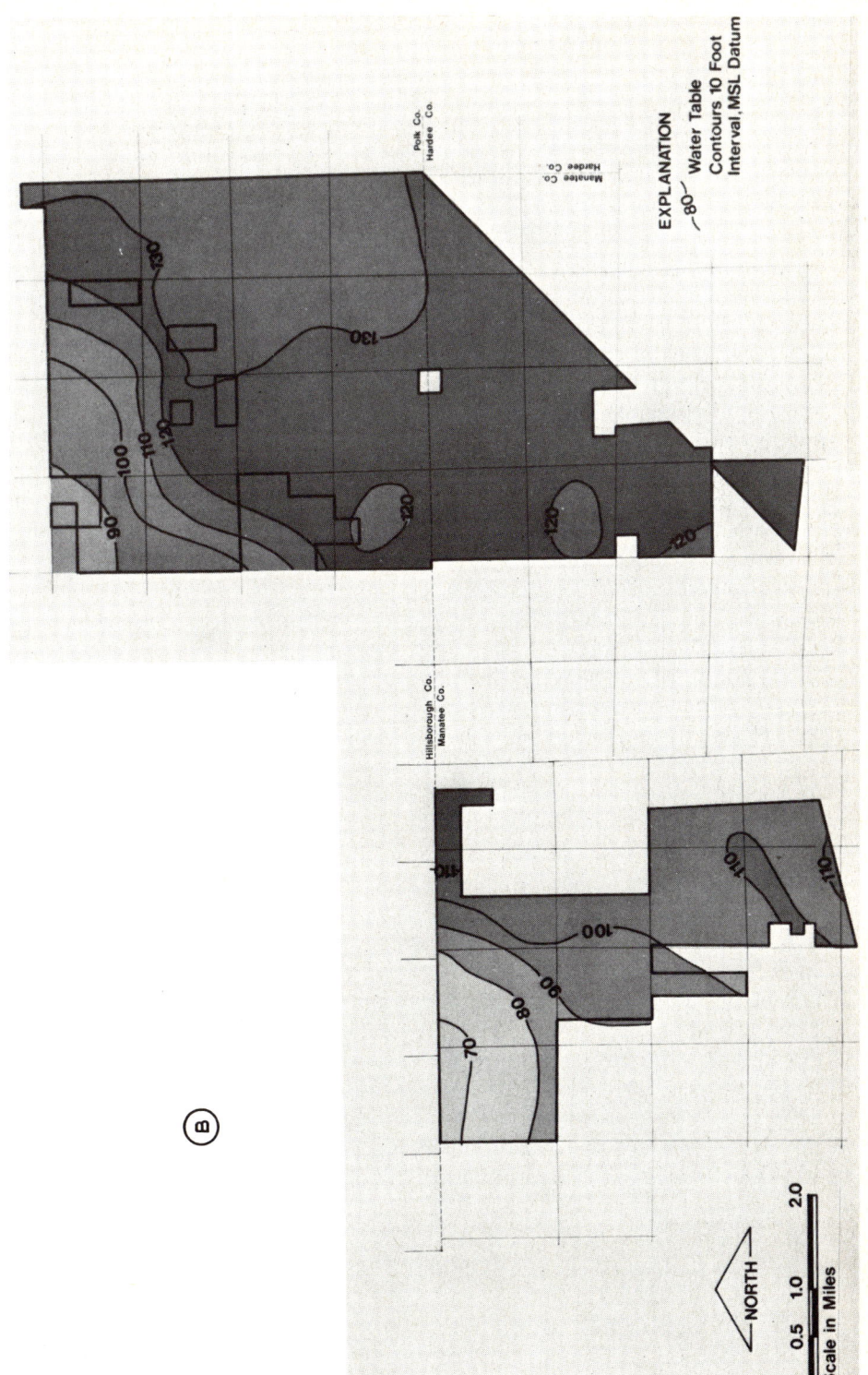

Fig. 6. Water-table elevations: (A) May 12–14, 1976 (wet season); and (B) July 26, 1976 (dry season). (See also Fig. 5.).

FLORIDAN AQUIFER

The artesian aquifers at the property can be subdivided into four major geohydrologic units. In increasing depth, these units are the first confining bed, the Upper Unit of the Floridan Aquifer, the second confining bed, and the Lower Unit of the Floridan Aquifer (Fig. 1).

The first confining bed underlies the water-table aquifer. This confining layer consists of clay, sandy clay, marl and dense limestone comprising parts of the matrix, and the upper, dense part of the Hawthorn Formation, and has very low permeability. It is as much as 80 m (262 ft.) in thickness, but generally averages about 55 m (180 ft.) in thickness.

The Upper Unit of the Floridan Aquifer underlies the first confining bed. It is composed of permeable limestones of the Hawthorn Formation and Tampa Limestone. Although this aquifer is reported to be about 45 m (148 ft.) thick (Wilson, 1975) in Hardee County, only a zone having a thickness of about 12 m (40 ft.) readily yields water at the proposed plant site.

Pumping tests indicated that the Upper Unit of the Floridan Aquifer has an average transmissivity of about 806 m^2 day^{-1} ($6.5 \cdot 10^4$ gal.day^{-1} ft.$^{-1}$) and a coefficient of storage of $1.66 \cdot 10^{-2}$. No leakage through the first confining bed was observed during the pumping test (Ferris et al., 1962).

Dense clay and fine-grained sand at the base of the Tampa Limestone comprise the second confining bed. As a unit, these heterogeneous materials, which are about 40 m (130 ft.) thick, comprise an effective confining bed having a very low permeability. Wilson (1975) reports that the leakance (P'/m') through this confining bed, as determined from aquifer tests, was $1.5 \cdot 10^{-4}$ gal. day^{-1} ft.$^{-3}$ (Jacob and Lohman, 1952).

The Lower Unit of the Floridan Aquifer consists of limestone and dolomitic limestone of the Suwannee Limestone, the Ocala Limestone and the Avon Park Limestone, and is about 230--275 m (755—900 ft.) thick. Although it consists of three separate geologic formations, it reacts during pumping as one geohydrologic unit. Wilson (1975) reports that a transmissivity of 25,048 m^2 day^{-1} ($2.02 \cdot 10^6$ gal.day^{-1} ft.$^{-1}$) and a coefficient of storage of $3.0 \cdot 10^{-5}$ was obtained from a pumping test in northeastern DeSoto County.

Analysis of data from pumping tests carried out on the Mississippi Chemical Company property (LaMoreaux, 1977a,b) indicated that the Lower Unit of the Floridan Aquifer had an average transmissivity of ~14,880 m^2 day^{-1} (~$1.2 \cdot 10^6$ gal.day^{-1} ft.$^{-1}$) and that the coefficient of storage ranged from $4.2 \cdot 10^{-4}$ to $4.2 \cdot 10^{-2}$. Analysis of the pumping test indicated that there was no leakage through the second confining bed during the pumping period (Theis, 1935; Jacob and Lohman, 1952; Hantush, 1962; Lohman, 1972).

Shape and fluctuation of the potentiometric surface for the Floridan Aquifer

Prior to the development of the Floridan Aquifer, the normal range between high-water levels at the end of the rainy season and low-water levels at the end of the dry season was ~3 m (~10 ft.) (Stringfield, 1966). As a result of development for agricultural, industrial and municipal water supplies, the seasonal variation is greater. The present seasonal variation in the potentiometric surface is shown by comparison of the wet season potentiometric surface for September 1976 (Fig. 7A) with the dry-season potentiometric surface for May 1977 (Fig. 7B). These maps indicate that the maximum seasonal fluctuation in the potentiometric surface reaches as much as 10 m (30 ft.) in the heavily pumped areas, but generally ranged from about 3—6 m (10—20 ft.) in much of the Southwest Florida Water Management District (SWFWMD.)

Because of the difference in the hydrostatic head of about 18 m (60 ft.) between the water-table aquifer and the Floridan Aquifer, hydrologic conditions are suitable for the use of recharge-connector wells to drain water by gravity flow from the water-table aquifer to the underlying Floridan Aquifer.

Sources of water for recharge to aquifer systems

During an average year approximately 137 cm (54 in.) of rain falls in the vicinity of the Mississippi Chemical Corporation property. Approximately 60%, or 81 cm (32 in.), of this rainfall is received during the months of June through September. The remaining 56 cm (22 in.) of rain is distributed from October through May. On an annual basis, rainfall (P), evapotranspiration (ET) and runoff (R) are estimated by the SWFWMD to be distributed approximately as follows:

P (137 cm or 54 in.) = ET (101 cm or 40 in.) + R (26 cm or 14 in.)

Evapotranspiration returns about 75% of the rainfall to the atmosphere. Much of the evapotranspiration is probably evaporation of rejected recharge when the water table is at or near the land surface. Research has shown that direct evaporation from the water table occurs to a depth of 1.5—2.5 m (5—10 ft.) below land surface. Since the depth to the water table on the property is generally 1.5 m (5 ft.) or less, much of the evapotranspiration on the property may be direct evaporation from the water table as rejected recharge. Runoff, which accounts for only about 25% of the total rainfall, represents both overland runoff to surface streams and groundwater recharge.

Aquifer transmissivity — Recharge well capabilities and spacing

The results of the pumping test on the water-table aquifer (Table I), show that the transmissivity averages 148.8 m^2 day^{-1} (12,000 gal.day^{-1} $ft.^{-1}$). It is recognized that this value of transmissivity does not represent the weighted average, owing to changes in thickness of the sand (Fig. 2), in the western part of the mining area; nor does the value of transmissivity represent the average transmissivity at the average water-table stage in January because the pumping test was carried out in April 1976 during a relatively low stage of the water table. The average transmissivity of 148.8 m^2 day^{-1} (12,000 gal.day^{-1} $ft.^{-1}$) as determined from the pumping test of April 1976 was used in the following analyses of the effectiveness of a recharge-connector well system.

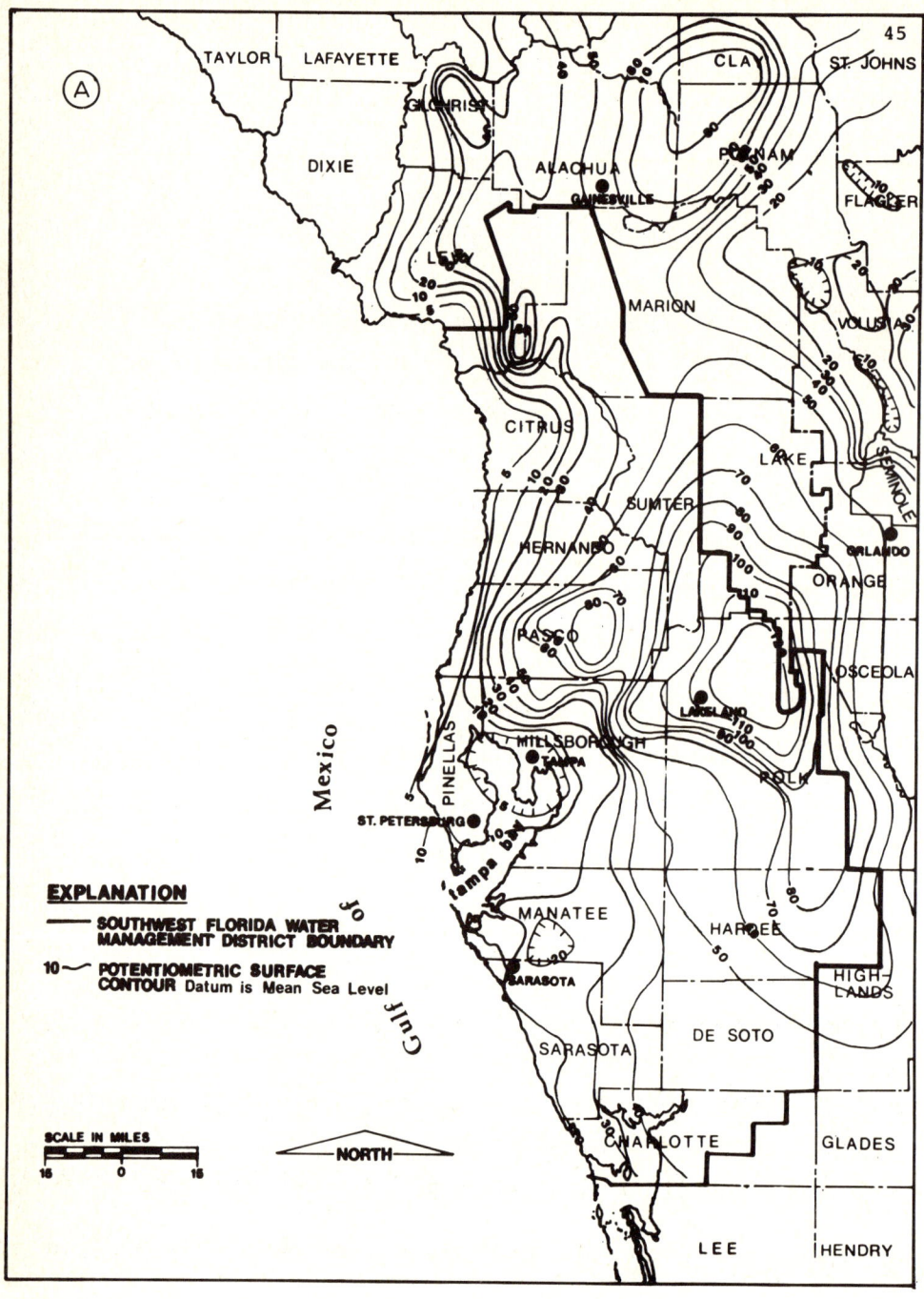

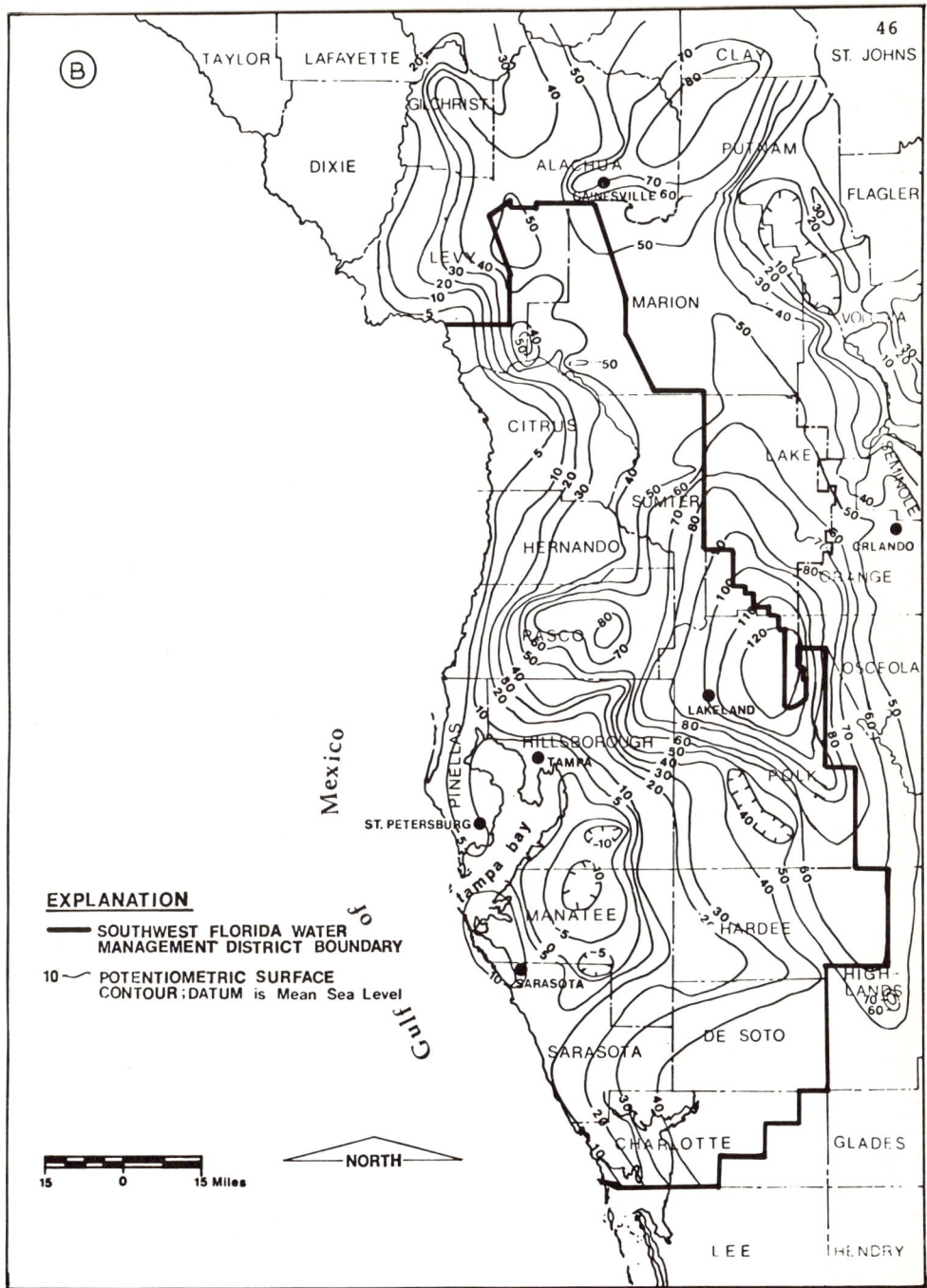

Fig. 7. Potentiometric surface of Floridan Aquifer for: (A) September, 1976 [modified from Ryder et al. (1976)]; and (B) May, 1977 [modified from Ryder et al. (1977)].

Table IV gives the well spacing, the discharge per well, number of wells per square mile for different values of capture rate and effective drawdown, and for various well radii and saturated thicknesses. The construction of a typical recharge connector well is shown in Fig. 8. Well spacing increases as the capture rate and the effective drawdown decreases, and as the well radius and the saturated thickness of the aquifer increase. The well radius is relatively insignificant in determining well spacing.

If the capture rate is 30.48 cm yr.$^{-1}$ (1.0 ft. yr.$^{-1}$), about 7.9 l s^{-1} (125 gal. min^{-1}) would be available for recharging the Floridan Aquifer; if the capture rate is 45.72 cm yr.$^{-1}$ (1.5 ft. yr.$^{-1}$) about 12.6 l s^{-1} (200 gal. min.$^{-1}$) would be available for recharging the Floridan.

TABLE IV

Variation in well field schematics and resulting capture rates [transmissivity (T) = 12,000 gal. day^{-1} ft.$^{-1}$; storage (S) = 0.15]

C_o (ft. yr.$^{-1}$)	s_e (ft.)	L (ft.)	Q (10^3 ft.^{3} day^{-1})	Q (gal. min.$^{-1}$)	N_w (mi.$^{-2}$)
s_w = 20 ft.; r_w = 0.5 ft.:					
1.0	5	1,680	7.73	40.2	10
1.0	10	1,145	3.59	18.6	21
1.5	5	1,385	7.88	40.9	15
1.5	10	950	3.71	19.3	31
s_w = 20 ft.; r_w = 1.0 ft.:					
1.0	5	1,760	8.49	44.0	9
1.0	10	1,230	4.15	21.5	18
1.5	5	1,480	9.00	46.8	13
1.5	10	1,050	4.53	23.5	25
s_w = 40 ft.; r_w = 0.5 ft.:					
1.0	5	2,690	19.8	103.0	4
1.0	10	2,325	14.8	76.9	5
1.5	5	2,210	20.1	104.3	6
1.5	10	1,920	15.2	78.7	8
s_w = 40 ft.; r_w = 1.0 ft.:					
1.0	5	2,800	21.5	111.6	4
1.0	10	2,400	15.8	82.0	5
1.5	5	2,300	21.7	112.9	5
1.5	10	2,000	16.4	85.4	7
1.0	1	3,100	26.3	136.8	3
1.0	20	1,680	7.73	40.2	10

L = well spacing; Q = well discharge; N_w = number of wells per square mile; C_o = capture rate; s_e = effective drawdown; s_w = available drawdown (equal to aquifer thickness); r_w = well radius.

1 ft. = 0.3048 m; 1 ft.^{3} day^{-1} = 2.8317 · 10^{-2} m^3 day^{-1}; 1 gal. min.$^{-1}$ = 0.0631 l s^{-1}; 1 mi^{-2} = 0.3861 km^{-2}.

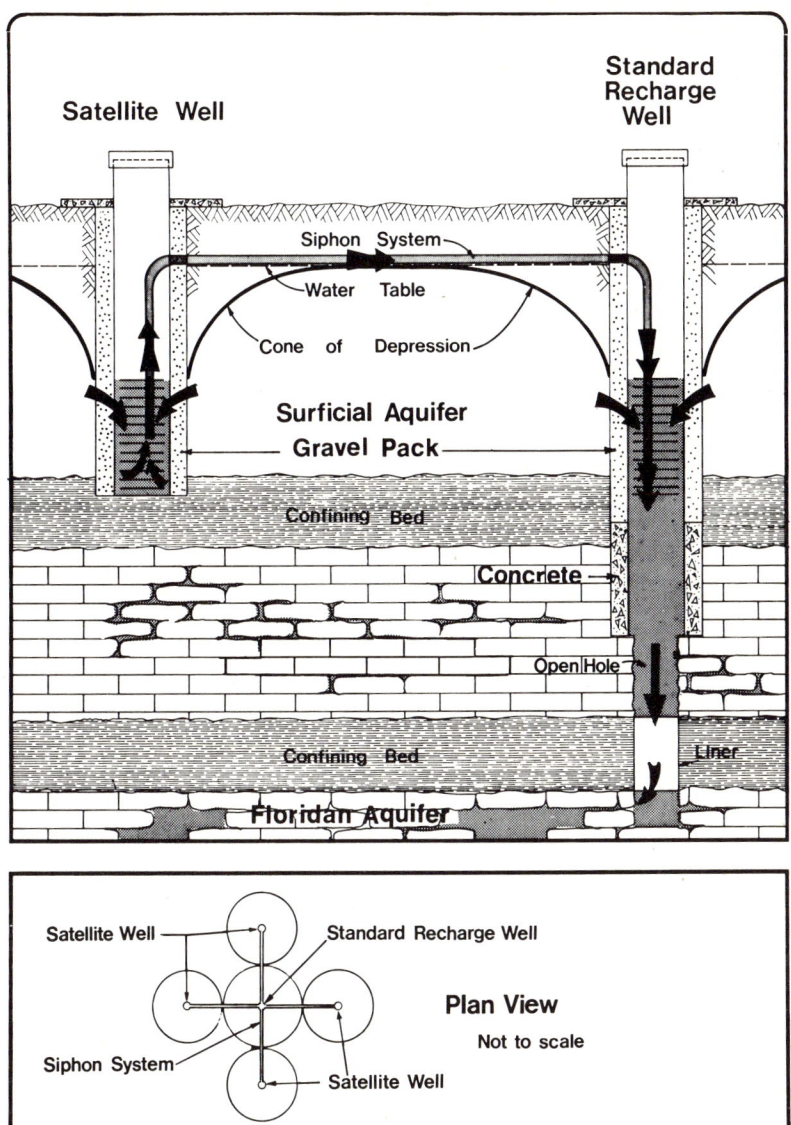

Fig. 8. Connector siphon well system.

Based on the results in Table IV, only two to four recharge-connector wells would be required to capture the available water. An increase in the number of recharge wells would be more effective in lowering the water table for dewatering purposes, but the discharge per well will be reduced and recharge wells might not be economically or hydrologically feasible (Edwards, in press).

Siphon well system for dewatering overburden on limestone

The collector siphon well system is recommended as a supplement to recharge-connector wells where the hydrogeologic conditions are favorable. It is dependable and more economical than single recharge-connector wells alone.

Study of existing soil maps, water-table elevations and core drilling logs in the installation area indicated a typical transmissivity of 24.8 m^2 day^{-1} (2000 gal.day^{-1} ft.$^{-1}$) or less and saturated thicknesses of 2—8.5 m (7—28 ft.). Artificial recharge to the surficial aquifer by seepage from the retention pond was occurring throughout the installation area.

The hydrogeology of the area indicates that gravity drainage from the surficial water-table aquifer to the underlying karstic artesian Floridan Aquifer is feasible. The surficial aquifer consist of 3—15 m (10—50 ft.) of quartzose and phosphorite sand and sandy clay with transmissivities ranging from 24.8 to 74.4 m^2 day^{-1} (2000 to 6000 gal.day^{-1} ft.$^{-1}$). The Floridan Aquifer consists of approximately 244 m (800 ft.) of limestone and dolostone with highly developed solution channelling and has an average transmissivity of 9610 m^2 day^{-1} (775,000 gal.day^{-1} ft.$^{-1}$).

The system has been operating successfully for more than one year. Vacuum gages and analog flowmeters are installed to further evaluate performance.

The most efficient design was determined to be locating the recharge-connector well at the center of the pipeline with flow from both directions; however, radial designs would be more efficient. Outlet valve design is important in operating the system. The design must allow clearance for flow-logging tools to bypass the valve in the recharge-connector well to obtain maximum drawdown in the satellite wells.

Localized water quality caused bacteria growth in some pipelines, increasing pipe friction losses, but installation of "oversized pipelines" more than compensated for the increased friction losses. Chlorination and the surge of restarting the siphon system clean the pipeline to almost new condition, and should constitute the only regularly scheduled maintenance for the system.

The pilot system layout is shown in Fig. 8. The location of satellite wells was caused by site limitations. The satellite wells were installed at 45.7-m (150-ft.) intervals. This spacing was determined by observation of the drawdown effects in the recharge-connector well. Spacing of satellite wells was difficult to determine because of the undetermined amount of seepage from the retention pond. Test pumping of each well yielded an expected initial pipeline flow from which the steady-state flow was estimated. Starting and steady-state flow capacities were calculated for each well, and a pipe size was selected which could handle the total flow with a minimum of loss.

The wells were installed with sufficient blank casing below the screens to enable maximum drawdown in each well. PVC casing was selected for wells and the pipeline because of its resistance to corrosion, ease of handling and installation and low friction characteristics. Initially, the siphon outlet elevation was installed at the highest inlet, eliminating any possibility of air entering the system. The system was designed so water would flow from the most distant well with a gradual slope to the recharge-connector well outlet.

The outlet was designed with a valve which is closed during starting of the siphon. A vacuum was created in the pipeline to begin the gravity flow. A section of clear PVC was installed in the pipeline to observe the flow.

The velocity and quantity of flow can be calculated from the Bernoulli equation. The assumption of full flow must be checked because of the performance limitation of 1 atm. (10.4 m of water). As the summit (minimum) pressure decreases, dissolved gases in the water come out of solution and help form intermittent discontinuities as the pressure approaches a vacuum. A break in the siphoning action occurs at a point less than the theoretical limit as the summit (minimum) pressure continues to decrease.

RECHARGE BY CONNECTOR WELLS FROM A STORAGE BASIN CONTAINING SURPLUS RUNOFF

The concept of using surplus runoff from streams to recharge the Floridan Aquifer has been a subject of substantial study over a period of several years. It requires a detailed analysis of surface water flow characteristics in an area as well as a very carefully conducted water-quality study to determine the type of surface water available, compatibility with water in the receiving aquifer, carefully engineered storage basin and filtration system, and a connecting pipeline system to a recharge-connector well or series of recharge-connector wells.

The ultimate objective of such a system would be to take advantage of surplus storm surface water, or water that would be normally lost to evaporation and/or transpiration in an area. This recharge technique is being considered for managing water resources in the future. These wells would recharge water to the Floridan Aquifer that is normally lost to the oceans as runoff during periods of high flow. Runoff is stored in a recharge basin where it is filtered to make it equal to or better in quality compared to water in the Floridan Aquifer. The filtered water is then collected in an underdrain system and directed to a large-diameter recharge well where the water will enter the Floridan Aquifer.

Although not in use at present, research necessary to implement this type of recharge well is underway.

REFERENCES

Boulton, N.S., 1954. Unsteady radial flow to a pumped well allowing for delayed yield from storage. Int. Assoc. Sci. Hydrol., Assem. Gen. Rome, Vol. II, pp. 472—477.
Boulton, N.S., 1963. Analysis of data from nonequilibrium pumping tests allowing for delayed yield from storage. Proc. Inst. Civ. Eng., 26(6693: 469—482.
Boulton, N.S., 1964. Discussions on the analysis of data from nonequilibrium pumping test allowing for delayed yield from storage. Proc. Inst. Civ. Eng., 28 : 603—610.

Cawley, C.L., 1975. International Minerals & Chemical Corporation's recharge connector well system. In: Hydrogeology of West Central Florida, Southeast. Geol. Soc. Publ. No. 17, 19th Field Conf., Tampa, Fla., pp. 66—71.

Edwards, J.R., in press. Collector well system — New technology for artificial recharge. Int. Assoc. Hydrogeol. Monogr. II.

Ferris, J.G., Knowles, D.B., Brown, R.H. and Stallman, R.W., 1962. Theory of aquifer tests. U.S. Geol. Surv., Water-Supply Pap. 1536-E, pp. 69—174.

Hantush, M.S., 1962. Validity of the Depuit—Forchheimer well-discharge formula. J. Geophys. Res., 67(6): 2417—2420.

Hutchinson, C.B. and Wilson, W.E., 1974. Evaluation of a proposed connector well, northeastern DeSoto County, Florida. U.S. Geol. Surv., Water-Resour. Invest. 5-74, 41 pp.

Jacob, C.E. and Lohman, S.W., 1952. Nonsteady flow to a well of constant drawdown, in an extensive aquifer. Trans. Am. Geophys. Union, 33(4) : 559—569.

Knochenmus, D.D., 1975. Hydrologic concepts of artificially recharging the Floridan Aquifer in Eastern Orange County, Florida — A feasibility study. Fla. Dep. Nat. Resour., Bur. Geol., Rep. Invest. 72, 36 pp.

LaMoreaux, 1977a. Water resources evaluation for Mississippi Chemical Corporation, Hardee County, Florida. P.E. LaMoreaux and Assoc., Inc., Tuscaloosa, Ala.

LaMoreaux, 1977b. Supporting report for consumptive-use permit application for Mississippi Chemical Corporation, Hardee County, Florida. P.E. LaMoreaux & Assoc., Inc., Tuscaloosa, Ala.

LaMoreaux, P.E. & Associates, Inc., 1975. Water resources evaluation for W.R. Grace & Company Hookers Prairie Tract, Polk County, Florida, Vol. 2, Appendix I. P.E. LaMoreaux and Assoc., Inc., Tuscaloosa, Ala., 86 pp.

LaMoreaux, P.E. & Associates, Inc., 1977. Application for development approval for Mississippi Chemical Corporation. P.E. LaMoreaux and Assoc., Inc., Tuscaloosa, Ala.

LaMoreaux, P.E., Raymond, D. and Joiner, J., 1970. Hydrology of limestone terranes, annotated bibliography of carbonate rocks. Ala. Geol. Surv. Bull. 94, Part A, 242 pp.

LaMoreaux, P.E., Legrand, H.E., Stringfield, V.T. and Tolson, J.S., 1975. Progress of knowledge about hydrology of carbonate terranes. Ala. Geol. Surv. Bull. 94, Part E, 168 pp.

Lohman, S.W., 1972. Ground-water hydraulics. U.S. Geol. Surv., Prof. Pap. 708, 70 pp.

Ryder, P.D., Laughlin, C.P. and Mills, L.R., 1976. Potentiometric surface of the Floridan aquifer, southwest Florida Water Management District, September 1976. U.S. Geol. Surv., Open-file Rep. 77—353, one map.

Ryder, P.D., Laughlin, C.P. and Mills, L.R., 1977. Potentiometric surface of the Floridan aquifer, southwest Florida Water Management District and adjacent areas, May 1977. U.S. Geol. Surv., Open-file Rep. 77—552, one map.

Stringfield, V.T., 1966. Artesian water in Tertiary limestone in the southeastern United States. U.S. Geol. Surv. Prof. Pap. 517, 226 pp.

Theis, C.V., 1935. The relation between the lowering of the piezometric surface and the rate and duration of discharge of a well using ground-water storage. Am. Geophys. Union, Trans., 16th Annu. Meet., Part 2, pp. 519—524.

Tolson, J.S. and Doyle, F.L. (Editors), 1977. Karst hydrogeology: Int. Assoc. Sci. Hydrol. 12th Congr., Huntsville, Ala., Sept. 20—22, 1976, 578 pp.

United Nations, 1975. Ground water storage and artificial recharge. Nat. Resour., Water Ser. No. 2, 270 pp.

Wilson, W.E., 1975. Ground-water resources of DeSoto and Hardee Counties, Florida. U.S. Geol. Surv., Open-File Rep. No. 75-428, 189 pp.

[2]

SIMULATED CHANGES IN POTENTIOMETRIC LEVELS RESULTING FROM GROUNDWATER DEVELOPMENT FOR PHOSPHATE MINES, WEST-CENTRAL FLORIDA

WILLIAM E. WILSON and JAMES M. GERHART

U.S. Geological Survey, Tampa, FL 33614 (U.S.A.)

(Accepted for publication May 10, 1979)

ABSTRACT

Wilson, W.E. and Gerhart, J.M., 1979. Simulated changes in potentiometric levels resulting from groundwater development for phosphate mines, west-central Florida. In: W. Back and D.A. Stephenson (Guest-Editors), Contemporary Hydrogeology — The George Burke Maxey Memorial Volume. J. Hydrol., 43: 491—515.

A digital model of two-dimensional groundwater flow was used to predict changes in the potentiometric surface of the Floridan aquifer resulting from groundwater development for proposed and existing phosphate mines during 1976—2000. The modeled area covers 15,379 km² in west-central Florida.

In 1975, groundwater withdrawn from the Floridan aquifer for irrigation, phosphate mines, other industries and municipal supplies averaged about 28,500 l/s. Withdrawals for phosphate mines are expected to shift from Polk County to adjacent counties to the south and west, and to decline from about 7,620 l/s in 1975 to about 7,060 l/s in 2000.

The model was calibrated under steady-state and transient conditions. Input parameters included aquifer transmissivity and storage coefficient; thickness, vertical hydraulic conductivity, and storage coefficient of the upper confining bed; altitudes of the water table and potentiometric surface; and groundwater withdrawals.

Simulation of November 1976 to October 2000, using projected combined pumping rates for existing and proposed phosphate mines, resulted in a rise in the potentiometric surface of about 6 m in Polk County, and a decline of about 4 m in parts of Manatee and Hardee counties.

INTRODUCTION

Long-range projections of water use in west-central Florida suggest that substantial increases in groundwater withdrawals will occur for municipal supplies, irrigation and phosphate mines. In the mid-1970's, principal interest was focused on the phosphate industry, whose mines and associated chemical plants used hundreds of millions of liters of groundwater per day for processing. In 1975, phosphate mining was confined to Polk County (Fig. 1), but as the ore became depleted, mining companies, through permit applications to regulatory agencies, were seeking to expand operations into Hardee, Hillsborough, DeSoto and Manatee counties over the next several decades.

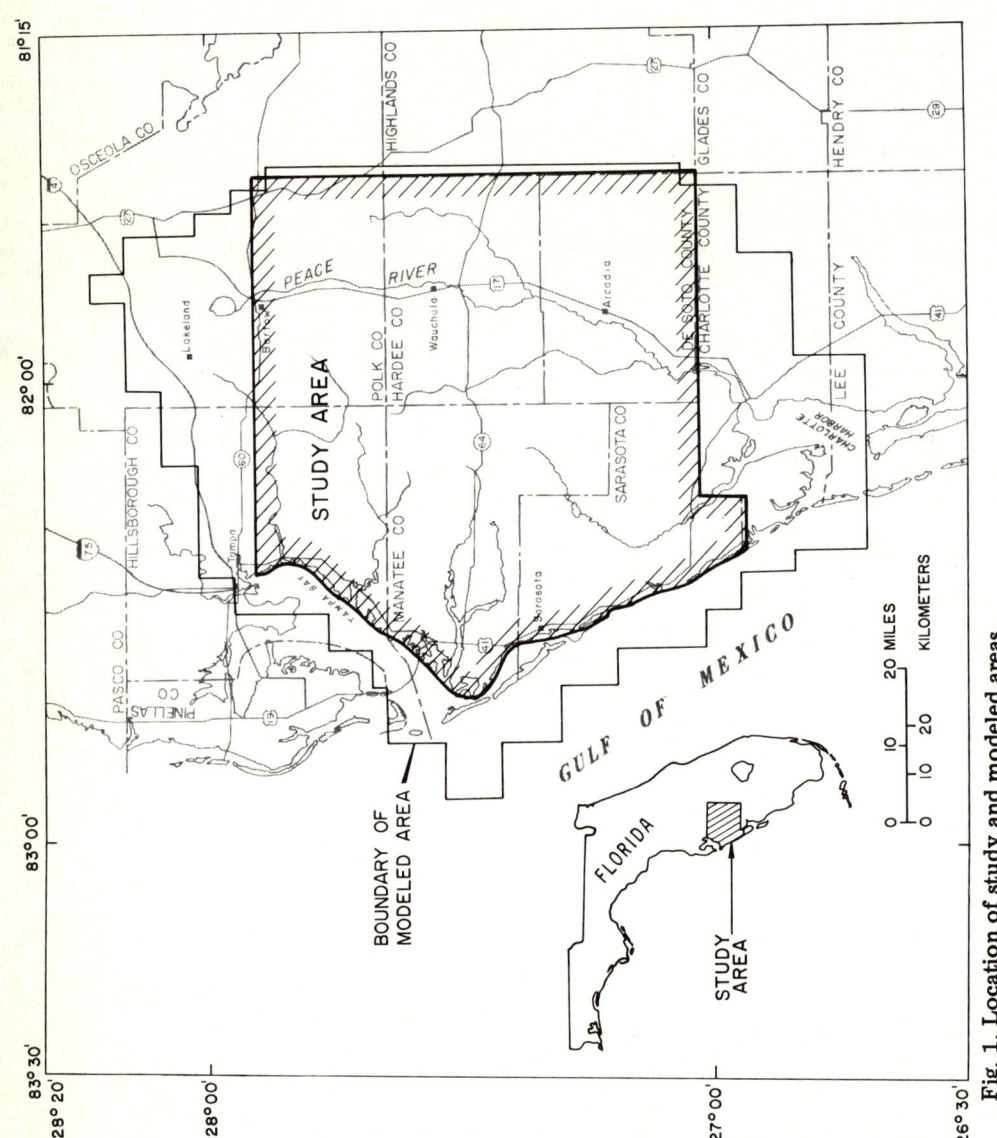

Fig. 1. Location of study and modeled areas.

Most demands for water will be met by groundwater from the Floridan aquifer. The combined withdrawals could have major effects on the hydrology of the area. In 1975, the U.S. Geological Survey (USGS) entered into a cooperative agreement with the Southwest Florida Water Management District to determine the regional hydrologic effects of anticipated groundwater withdrawals by major users, including municipalities, irrigators and the phosphate industry.

In 1976, the President's Council on Environmental Quality directed the U.S. Environmental Protection Agency (EPA) to develop an Areawide Environmental Impact Statement to analyze cumulative interrelated impacts of present and proposed phosphate development in central Florida. The Impact Statement was to be based primarily on existing data and would utilize results of a USGS investigation for groundwater aspects. In 1977, the USGS published preliminary findings on the effects of withdrawals by the phosphate industry (Wilson, 1977) in order to provide timely results in support of the EPA investigation.

This paper revises and updates the preliminary findings of the 1977 report. The objective is to evaluate the amount of change in the potentiometric surface of the Floridan aquifer to be expected as a result of proposed or anticipated groundwater withdrawals for phosphate mines in west-central Florida. The objective is accomplished principally through calibration and application of a regional digital model of groundwater flow. This paper presents one aspect of a larger report that is in preparation, in which the effects of withdrawals for municipal supplies, irrigation, and other industries are considered as well as withdrawals for phosphate mines.

The study area covers 9,150 km^2 in west-central Florida south and east of Tampa (Fig. 1). To determine effects of groundwater development in the area, hydrogeologic data were evaluated and a groundwater flow model was calibrated for a larger region. The modeled area covers 15,379 km^2 (Fig. 1).

The authors are grateful for information obtained from many sources during this investigation. The EPA and their contractors, Texas Instruments, Inc., Geraghty and Miller, Inc., and Thomasino and Associates, Inc., provided valuable information. Many consulting firms, including P.E. LaMoreaux and Associates, William F. Guyton and Associates, Dames and Moore, Inc., and Richard C. Fountain and Associates, provided, through their clients, results of detailed site investigations in the study area. The Florida Phosphate Council provided detailed groundwater pumpage records for the phosphate industry.

HYDROGEOLOGY

Groundwater in the study area occurs in two principal aquifers, the surficial aquifer and the Floridan aquifer. Groundwater in the surficial aquifer is unconfined, and groundwater in the Floridan aquifer is confined. Recharge to and discharge from the Floridan aquifer is principally by leakage through a confining bed that separates the aquifer from the overlying surficial aquifer.

The surficial aquifer underlies most of the study area and consists predominantly of fine to very fine sand and clayey sand. The aquifer includes deposits commonly referred to as surficial sand, terrace sand, phosphorite, Caloosahatchee Marl (Pleistocene and Pliocene), and parts of the Bone Valley Formation (Pliocene).

The principal sources of groundwater are highly transmissive zones in the Floridan aquifer. The Floridan aquifer was originally defined by Parker (Parker et al., 1955, p. 189) to include all or parts of the Lake City Limestone (Eocene), Avon Park Limestone (Eocene), Ocala Limestone (Eocene), Suwannee Limestone (Oligocene), Tampa Limestone (Miocene), and "permeable parts of the Hawthorn Formation that are in hydrologic contact with the rest of the aquifer". This definition is closely followed in this report, except for differences in the identity of the top and base of the aquifer.

Confining beds bound the Floridan aquifer above and below. As used in this report, the upper confining bed of the Floridan aquifer is the full clastic-carbonate sequence between the surficial aquifer and the Floridan aquifer. This sequence includes, in different parts of the study area, all or part of the Bone Valley Formation, Tamiami Formation (Pliocene), Hawthorn Formation (Miocene) and Tampa Limestone. Although these formations commonly contain permeable zones, especially in the southern and eastern parts of the study area, well yields are generally substantially less than those from the underlying carbonate section of the Floridan aquifer. For modeling purposes, the carbonate-clastic section is considered to be a single confining unit overlying the Floridan aquifer. The base of the Floridan aquifer is considered to be at the first occurring intergranular anhydrite and gypsum in the carbonate rocks, generally in the lower part of the Avon Park Limestone or upper part of the Lake City Limestone. The permeability and porosity of the section of carbonates containing intergranular evaporites is significantly lower than where evaporites are absent.

The Bone Valley Formation is one of the world's most important sources of phosphate, and hundreds of millions of liters of groundwater are used each day in the extraction and processing of phosphate ore. The ore deposit underlies about 5,180 km^2 in central Florida and is a shallow-water, marine and estuarine phosphorite of Pliocene age (Altschuler et al., 1964). Phosphate occurs in the form of grains of fluorapatite in a deposit of pebbly and clayey sands. The ore generally occurs 3—20 m below the land surface and is 1—15 m thick (Fountain et al., 1971). The ore is mined from open pits. Water is used to transport a matrix slurry to beneficiation plants, to separate the phosphate from the matrix, and to convert the phosphate into useful products.

GROUNDWATER WITHDRAWALS

Groundwater from the Floridan aquifer is the major source of supply in the modeled area. Withdrawals from the aquifer in the modeled area averaged about 28,500 l/s in 1975 (Table I), principally for public-supply, agricultural and industrial purposes.

TABLE I

Groundwater withdrawal rates (l/s) from the Floridan aquifer

County[*1]	1975						2000		
	municipal supplies[*2]	irrigation		self-supplied industry		total 1975	phosphate mines		
		irrigation season average[*3]	average annual[*4]	phosphate mines[*5]	other[*6]		existing[*7]	proposed[*8]	total
Charlotte	0	964	508	0	0	508	0	0	0
DeSoto	33.3	7,140	3,780	0	21.0	3,830	0	414	414
Hardee	39.9	4,860	2,570	0	57.4	2,670	0	2,680	2,680
Hillsborough	521	4,560	2,410	35.9	2,030	5,000	28.5	1,110	1,140
Manatee	0	2,710	1,430	0	28.5	1,460	0	1,830	1,830
Polk	1,300	1,850	977	7,580	4,600	14,500	464	539	1,000
Sarasota	311	381	202	0	0	513	0	0	0
Total (rounded)	2,200	22,500	11,900	7,620	6,740	28,500	492	6,570	7,060

[*1] Includes only those parts of the counties within the modeled area.
[*2] From Healy (1977).
[*3] November 1, 1975—May 12, 1976.
[*4] November 1, 1975—October 31, 1976; assumes no irrigation withdrawal during May 13 through October 31.
[*5] Based on data from the Florida Phosphate Council (B. Barnes, pers. commun., 1977).
[*6] Data from Florida Phosphate Council (B. Barnes, pers. commun., 1977) and from information obtained during USGS water-use inventory (Leach, 1977).
[*7] Rates based on 1975—1977 pumping data from Florida Phosphate Council, and on projected life spans of existing mines (EPA, 1978).
[*8] Based on projected life spans and pumping rates of proposed mines (EPA, 1978; J. Heuer, pers. commun., 1978).

Locations and amounts of withdrawal for municipal and non-phosphate industrial supplies were obtained from the 1975 water-use inventory of the USGS (Leach, 1977). Amounts and locations of withdrawals for phosphate mines and chemical plants were obtained mostly from data supplied by the Florida Phosphate Council. These data included locations, average pumping rates, and hours pumped during 1975 for wells at each phosphate mine and chemical plant. The pumping rates included all Floridan-aquifer withdrawals reported to the Florida Phosphate Council by mining and chemical companies. Withdrawal rates for irrigation are based on an inventory of irrigation rates, times and locations conducted during this investigation.

The largest single use of groundwater is for irrigation of crops, principally citrus, vegetables and pasture. In 1975—1976, irrigation withdrawals averaged about 11,900 l/s (Table I). Unlike other major users, withdrawals for irrigation are highly seasonal. During the rainy summer months, little or no withdrawals are made. During the 1975—1976 irrigation season (November 1, 1975— May 12, 1976), withdrawals for irrigation averaged about 22,500 l/s (Table I). Largest withdrawal rates were in DeSoto and Hardee counties.

Most municipalities depend on groundwater from the Floridan aquifer for public supplies. Table I lists by county the 1975 daily withdrawal rates based on average annual withdrawals for 13 municipalities. In 1975 about 2,200 l/s was withdrawn for municipal supplies in the modeled area.

Groundwater is withdrawn for a variety of industrial uses, principally phosphate mining, phosphate chemical plants and citrus processing. Table I shows that in 1975 more than one-half (7,620 l/s) of the industrial withdrawals were for phosphate mining. Almost all withdrawals for mining were in Polk County.

Table I shows projections of groundwater withdrawal rates in the year 2000 for both existing and proposed phosphate mines. Existing mines are those that were permitted as of August 1, 1976; proposed mines are those that have or plan to begin operations after that date. Proposed mines include seven listed by the EPA (1978, fig. 2.2) as "DRI mines" (those for which Development of Regional Impact applications were pending on August 1, 1976) and 13 planned for later development. Withdrawal rates for existing mines in Table I reflect the expected phasing out of mines in Polk County. The potential shift of mining activity to the south and west of Polk County is reflected by the higher rates for proposed mines for DeSoto, Hardee and Manatee counties in 2000. Total withdrawal rates for phosphate mining, including proposed mines, are projected to increase until about 1985 (EPA, 1978), and then decrease to 7,060 l/s in 2000 (Table I).

DESCRIPTION OF HYDROLOGIC MODEL

A digital simulation model was used to compute hydraulic head changes in time and space in the Floridan aquifer in response to applied hydraulic stresses. The model utilizes a finite-difference method in which differential equations describing groundwater flow are solved numerically. The equations require that hydraulic properties, boundaries, and stresses be defined for the area modeled. The digital model of two-dimensional flow used in this study was a modified version of one described by Trescott et al. (1976).

A 35,245-km^2 area of west-central Florida and the adjoining Gulf of Mexico was subdivided into a rectangular finite-difference grid within which the modeled area was defined (Fig. 2). Block sizes in the grid range from 3.2 km × 3.2 km to 16 km × 16 km. The node at the center of each grid block is designated by row and column number; for example, the node at row *20*, column *5* is expressed as *20-5*.

Within the gridded area, model boundaries were selected to coincide as closely as possible with hydrologic boundaries. In order to simulate both steady-state and transient boundary conditions without shifting the positions of model boundaries, the two-dimensional flow model described by Trescott et al. (1976) was modified to include a head-controlled flux boundary condition. This condition was utilized during all simulations for all lateral model boundaries.

In designating nodes adjacent to the boundaries as head-controlled flux boundary nodes, it was assumed that beyond each boundary node there exists a point where the head in the Floridan aquifer does not change. These constant-head points were assumed to be 32—64 km beyond the model boundaries, or far enough from the modeled area so that they would not significantly alter simulation results. Between boundary nodes and constant-head points, hydrologic properties were assumed to be uniform and the same as properties in the boundary nodes.

A change in potentiometric head in a head-controlled flux boundary node causes lateral flow across that boundary in an amount determined by the magnitude of head change. Boundary flow is calculated as the product of the head change and a coefficient. The coefficient is obtained for each boundary node using an analytical solution of the partial differential equation describing one-dimensional steady-state flow in the region between the model boundary and the constant-head point. This boundary condition is based on a steady-state solution and therefore can only be used in simulations of sufficient duration to ensure that changes in aquifer storage are negligible by the ends of the simulations.

Major assumptions made in the model analysis are as follows:

(1) Groundwater movement in the Floridan aquifer occurs horizontally in a single-layer isotropic medium.

(2) Water moves vertically into or out of the Floridan aquifer through the upper confining bed. No leakage occurs through the lower confining bed.

(3) Head in the surficial aquifer does not change in response to imposed stress.

(4) Movement of the saltwater—freshwater interface is assumed to have little or no effect on calculated heads.

The groundwater flow system is shown schematically in Fig. 3. Regionally the system approximates the assumed conditions, although locally, deviations

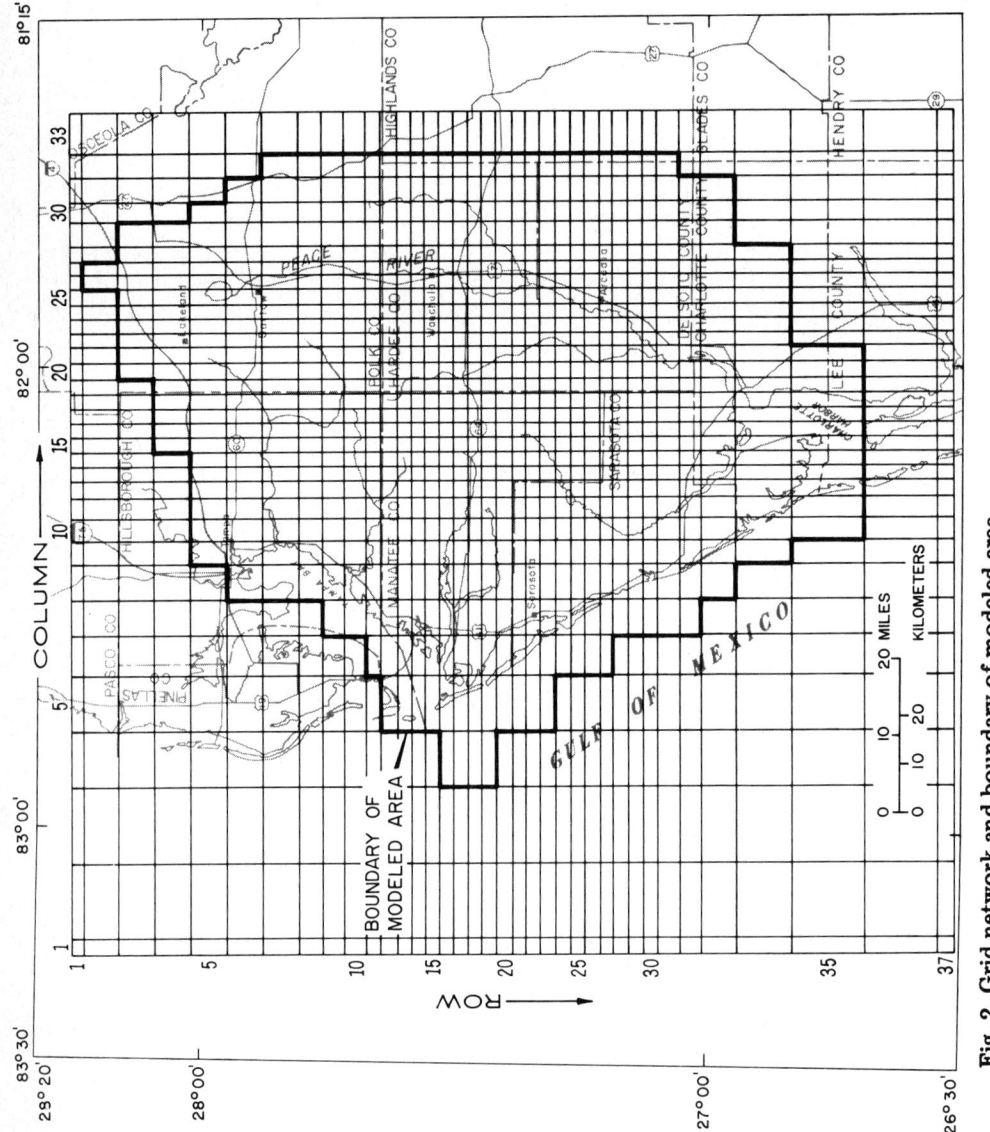

Fig. 2. Grid network and boundary of modeled area.

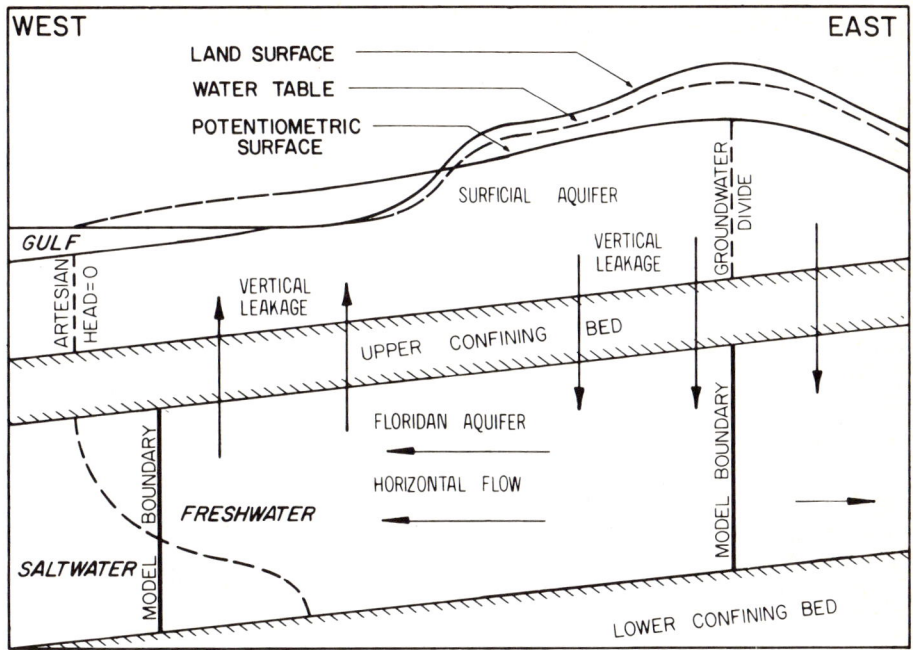

Fig. 3. Conceptual model of steady-state flow.

occur. Most wells are finished as open holes and tap most of the thickness of the Floridan aquifer. A confining bed overlies the Floridan aquifer throughout the area, and natural aquifer discharge and recharge occur principally as vertical leakage through this bed. Aquifer tests have shown the lower confining bed (Lake City Limestone) to be relatively impermeable. The seasonal range of fluctuation of the water table in the surficial aquifer is generally less than 1 m, and in most of the study area the water table is little affected by withdrawals from the Floridan aquifer.

On the other hand, the model greatly oversimplifies a complex system. The model is inadequate to simulate vertical flow components in recharge and discharge areas, multiple zonation of the Floridan aquifer, and movement of the saltwater—freshwater interface. Some errors remaining in the calibrated model are due to these inadequacies. Nonetheless, the model used was the most appropriate one available at the time the investigation began, considering the size of study area, objectives of the investigation, and state of knowledge of the hydrogeology.

MODEL CALIBRATION

The model was calibrated before simulating effects of projected changes in groundwater withdrawals. In this report, calibration refers to the process of adjusting input hydrologic parameters to the model until differences between model simulations and field observations are within acceptable limits. Calibration was checked by comparing model computations with different sets of field observations, namely by comparing simulated and observed potentiometric surfaces. The model was calibrated under steady-state and transient conditions.

Steady-state model

In calibration of the steady-state model, a simulated potentiometric surface was compared to the observed September 1975 potentiometric surface (Fig. 4), which was assumed to reflect steady-state conditions. A steady-state condition exists when there are no changes in aquifer storage with time. Such a condition was approximated in September 1975. Hydrographs indicated that in September 1975, the potentiometric surface was near the end of the summer recovery period and was changing little with time. Principal stresses on the aquifer system in September were withdrawals for municipal supplies, phosphate mining, and other industrial supplies. Pumping rates for these uses vary during the year, but the variations are generally too small to have much impact on the regional fluctuations of the potentiometric surface. Withdrawals for irrigation were assumed to be insignificant in September 1975. Field checks indicated that little irrigation occurred during and immediately following the summer rainy season. Hydrographs indicated that in most areas the potentiometric surface in 1975 did not start declining as a result of fall irrigation until October or early November.

Input parameters to the steady-state model included pumping, water-table altitude, aquifer transmissivity, and confining-bed thickness and vertical hydraulic conductivity.

Withdrawals in September 1975 were assumed to be the same as average 1975 groundwater withdrawal rates (Table I). No irrigation pumpage was included. Average altitudes of the September 1975 water table were estimated for each node from USGS topographic quadrangle maps (scale 1:24,000; contour interval 5 ft.). The water table was assumed to be about 1 m below land surface in flat swampy areas, river flood plains, and near lakes; depths of 1—6 m below land surface were assumed for sand-ridge areas.

Modeled transmissivity of the Floridan aquifer is shown in Fig. 5. The map is a revision of a preliminary map by Wilson (1977, fig. 5). Boundaries of the map units were modified slightly to conform to block boundaries, and transmissivity in the southeastern part of the area was changed from 83,600 to 46,500 m^2/day (Fig. 5), based on model recalibration. The value and distribution of transmissivity (Fig. 5) were based on results of aquifer tests, variations in the gradient of the potentiometric surface for September 1975 (Fig. 4), and on adjustments resulting from model calibration.

A preliminary map of the thickness of the upper confining bed was prepared for the model. The map was later revised for separate publication

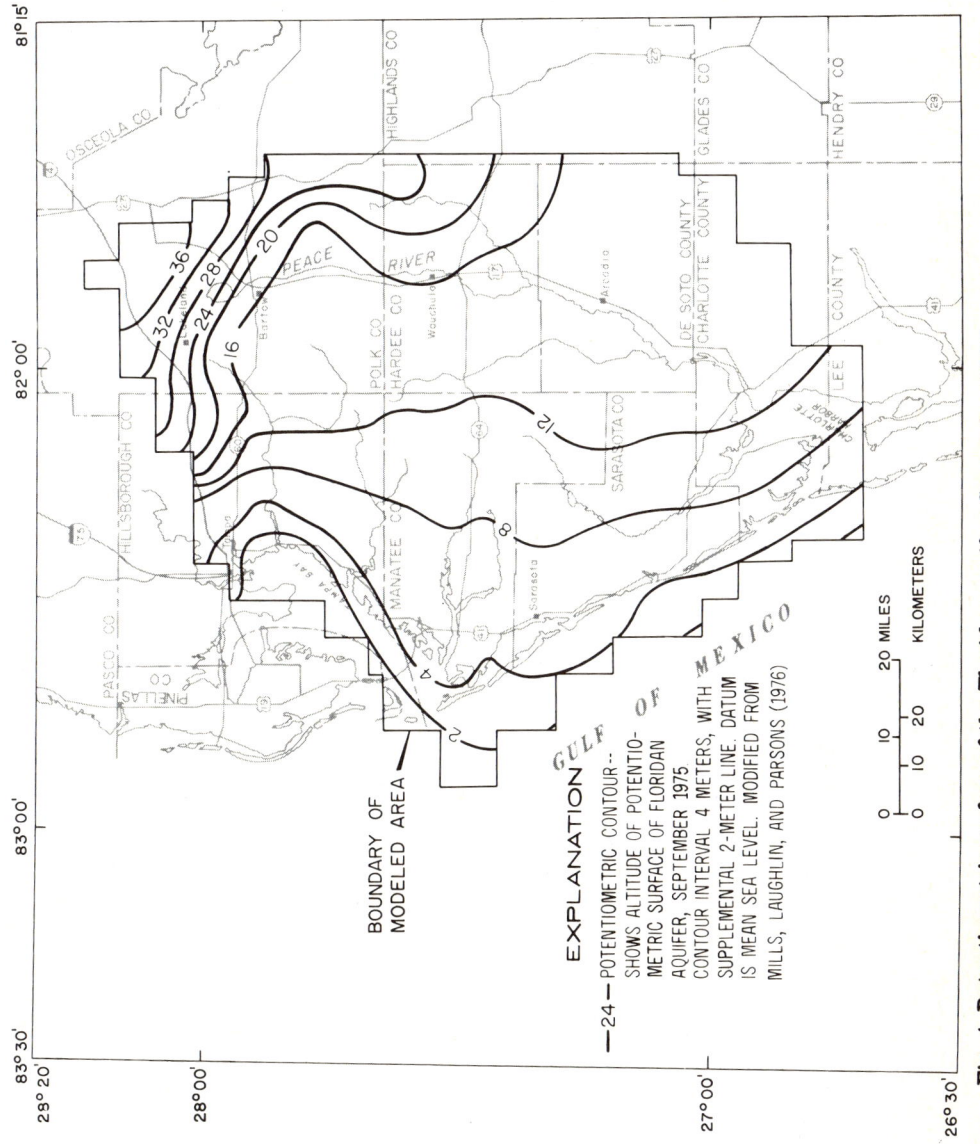

Fig. 4. Potentiometric surface of the Floridan aquifer, September 1975.

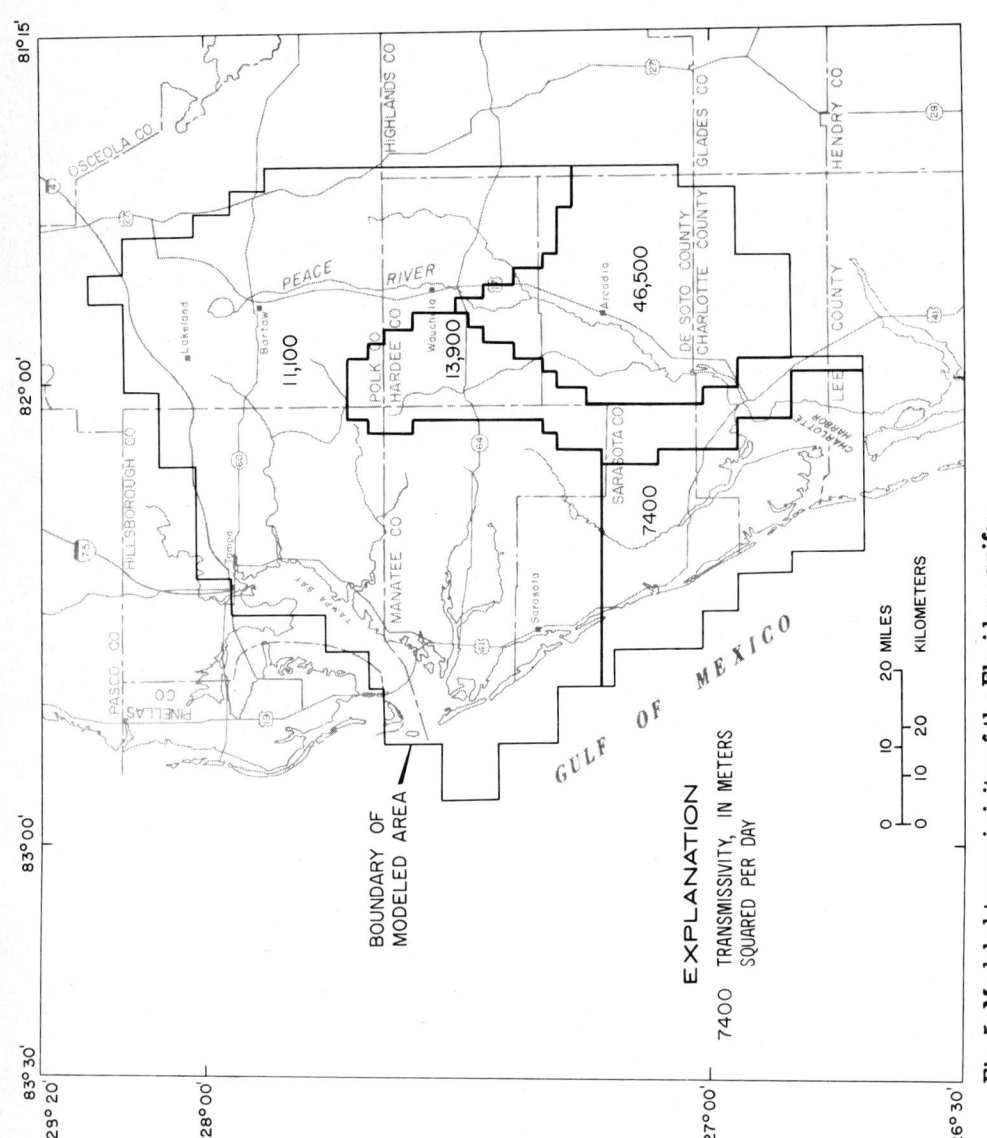

Fig. 5. Modeled transmissivity of the Floridan aquifer.

(R.M. Wolansky, pers. commun., 1978). Thickness ranged from about 6 m in the northern part of the modeled area to about 240 m in the southern part.

Modeled vertical hydraulic conductivity of the upper confining bed is shown in Fig. 6. Vertical hydraulic conductivity (K') was determined initially by multiplying leakance coefficients (K'/b') estimated from aquifer tests by confining-bed thickness (b') at each aquifer test site. The initial map, based on these results, was modified during model calibration. Many of the parameter adjustments during calibration were made with vertical hydraulic conductivity, because it was considered to be the least reliably known input parameter. The resulting map of vertical hydraulic conductivity is primarily a calibration map, but values are within the probable range of error of original aquifer-test estimates. The relatively low value of $2.6 \cdot 10^{-4}$ m/day in the northwestern part of the area (Fig. 6) is probably due to the lesser amount of permeable carbonate rock in the confining bed in that area compared to elsewhere.

Modeled leakance coefficient of the upper confining bed is shown in Fig. 7. This parameter is the ratio of the confining-bed vertical hydraulic conductivity (Fig. 6) and confining-bed thickness. Because vertical hydraulic conductivity varies areally, trends in leakance coefficient do not necessarily parallel those of confining-bed thickness.

The steady-state potentiometric surface simulating September 1975 conditions is shown in Fig. 8. This surface may be compared with the observed September 1975 potentiometric surface as mapped in Fig. 4. Differences between computed and observed heads at nodes of the model grid ranged from 0 to 10.3 m. The difference was less than 3 m at 89% of the nodes, and more than 6 m at 2% of the nodes. All differences greater than 4 m were in the northern part of the modeled area, mostly in Polk County, and occurred in nodes where withdrawals were occurring, where the potentiometric gradient was relatively steep, and adjacent to boundaries.

Transient model

Model calibration was extended to transient flow. In the transient model, computed hydraulic head is a function of starting conditions and time, and therefore storage coefficients were incorporated into the model. Following a procedure similar to the steady-state calibration, a simulated May 1976 potentiometric surface was compared to the observed May 1976 surface (Fig. 9). The simulated surface was obtained by computing drawdowns from the simulated steady-state potentiometric heads, after simulating a 194-day pumping period representing the irrigation season (November 1, 1975—May 12, 1976). Computed drawdowns were then subtracted from the observed September 1975 potentiometric map to obtain the simulated May map. In this analysis, it was assumed that the potentiometric surface on November 1, 1975, was approximately the same as that of September 1975. These assumptions are generally borne out by hydrographs of wells in the study area.

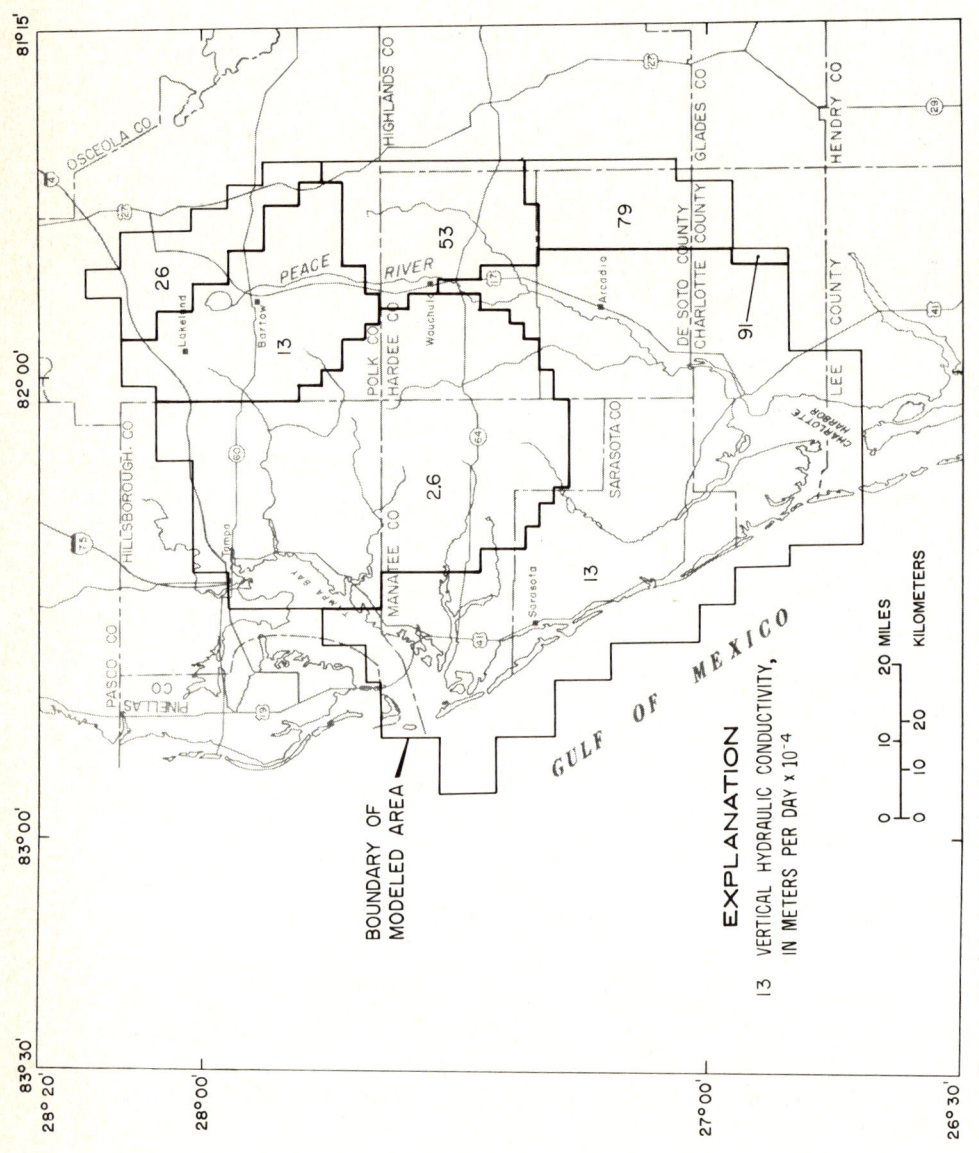

Fig. 6. Modeled vertical hydraulic conductivity of the upper confining bed of the Floridan aquifer.

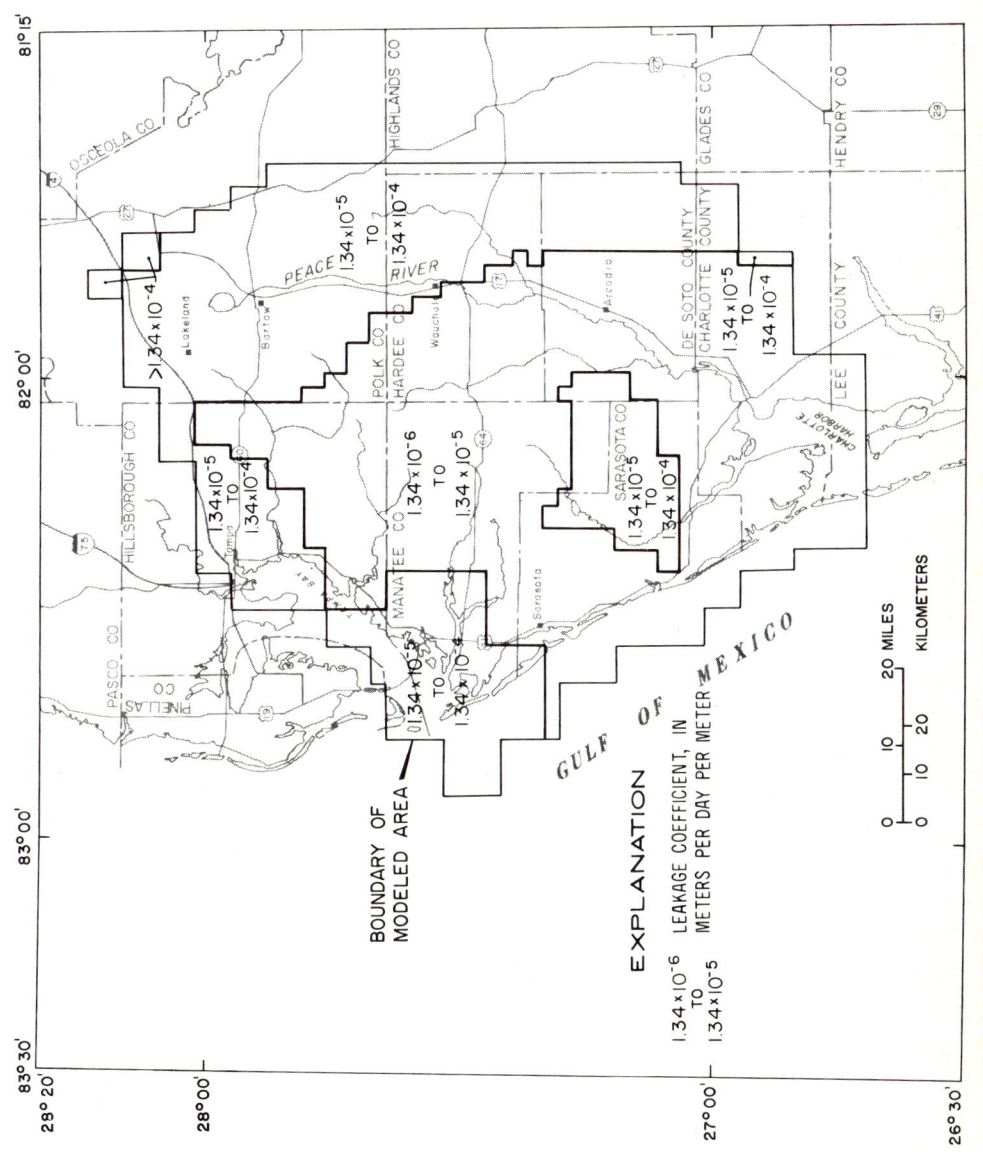

Fig. 7. Modeled leakance coefficient of the upper confining bed of the Floridan aquifer.

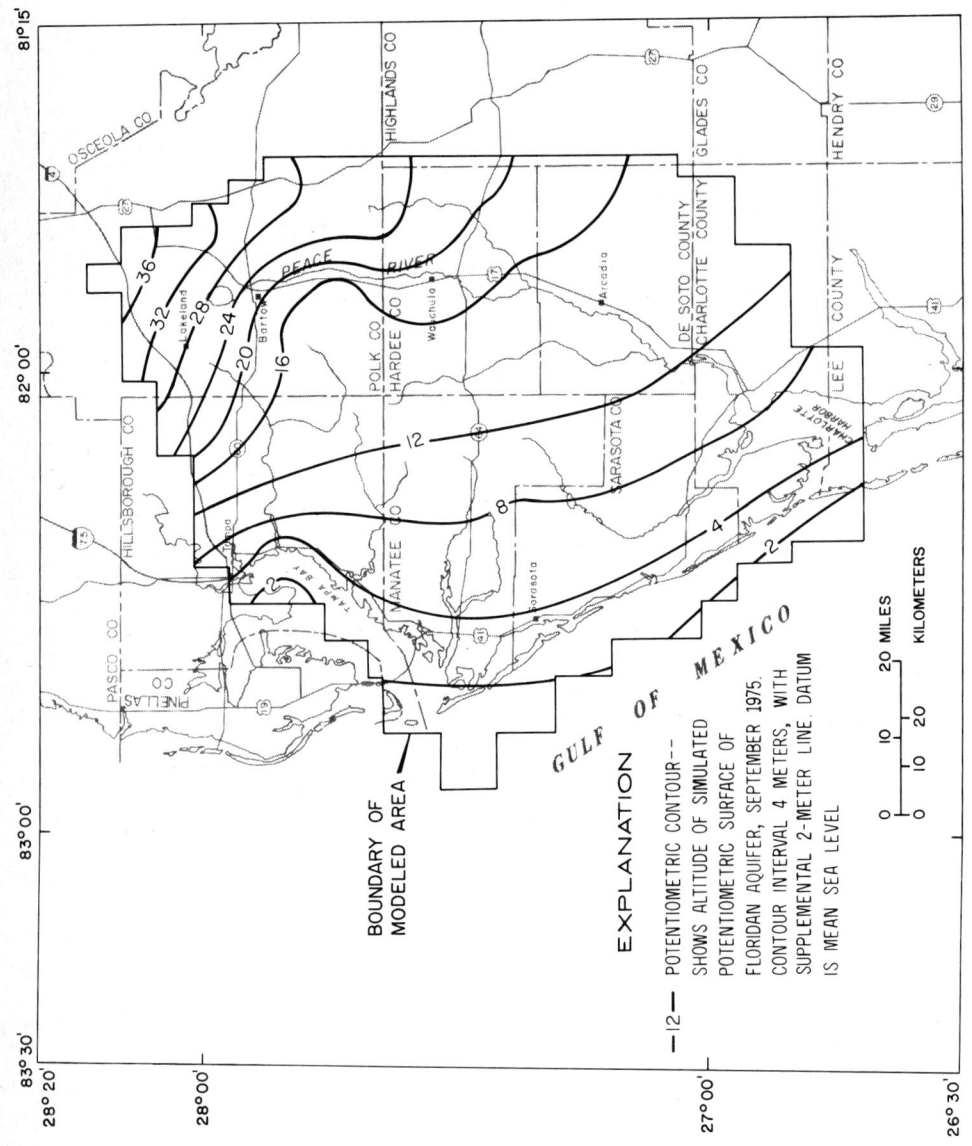

Fig. 8. Simulated steady-state potentiometric surface of the Floridan aquifer, September 1975.

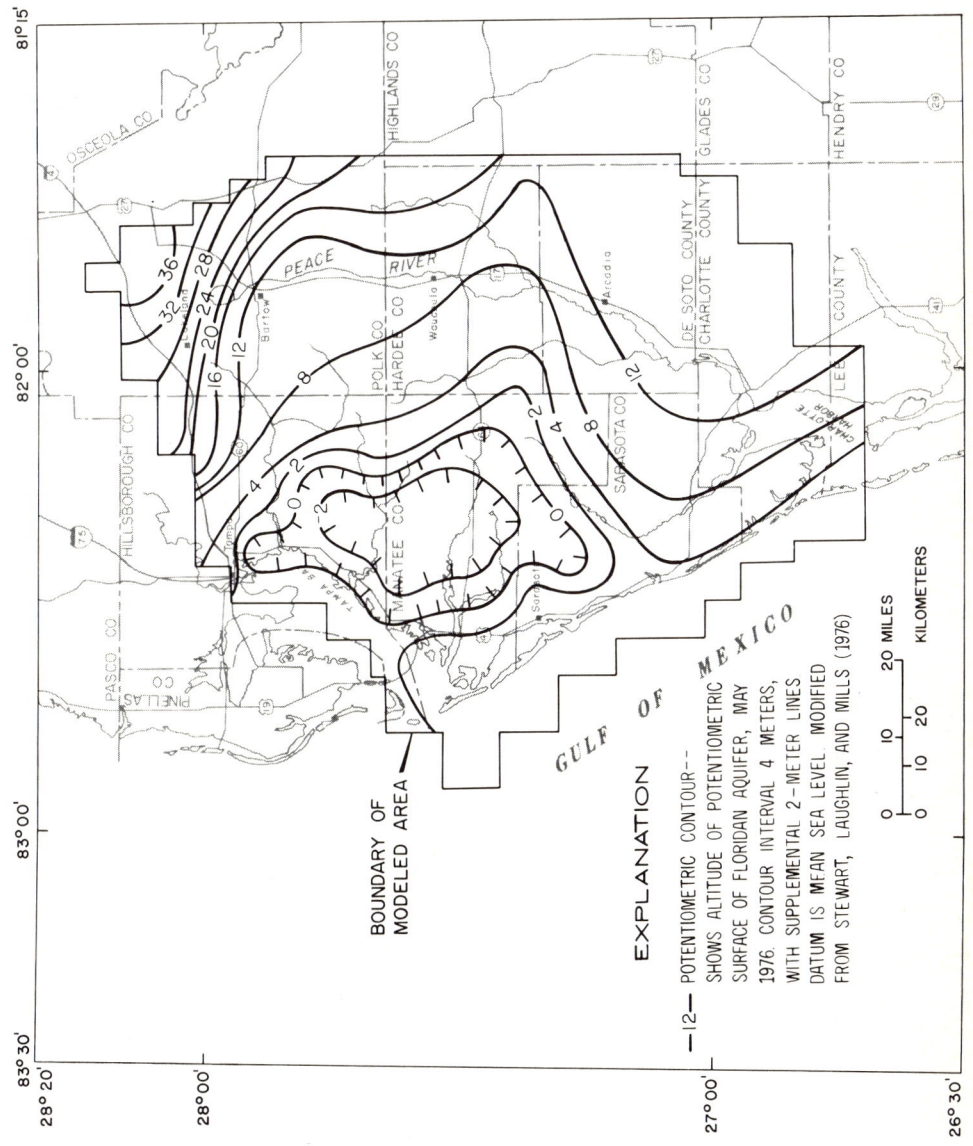

Fig. 9. Potentiometric surface of the Floridan aquifer, May 1976.

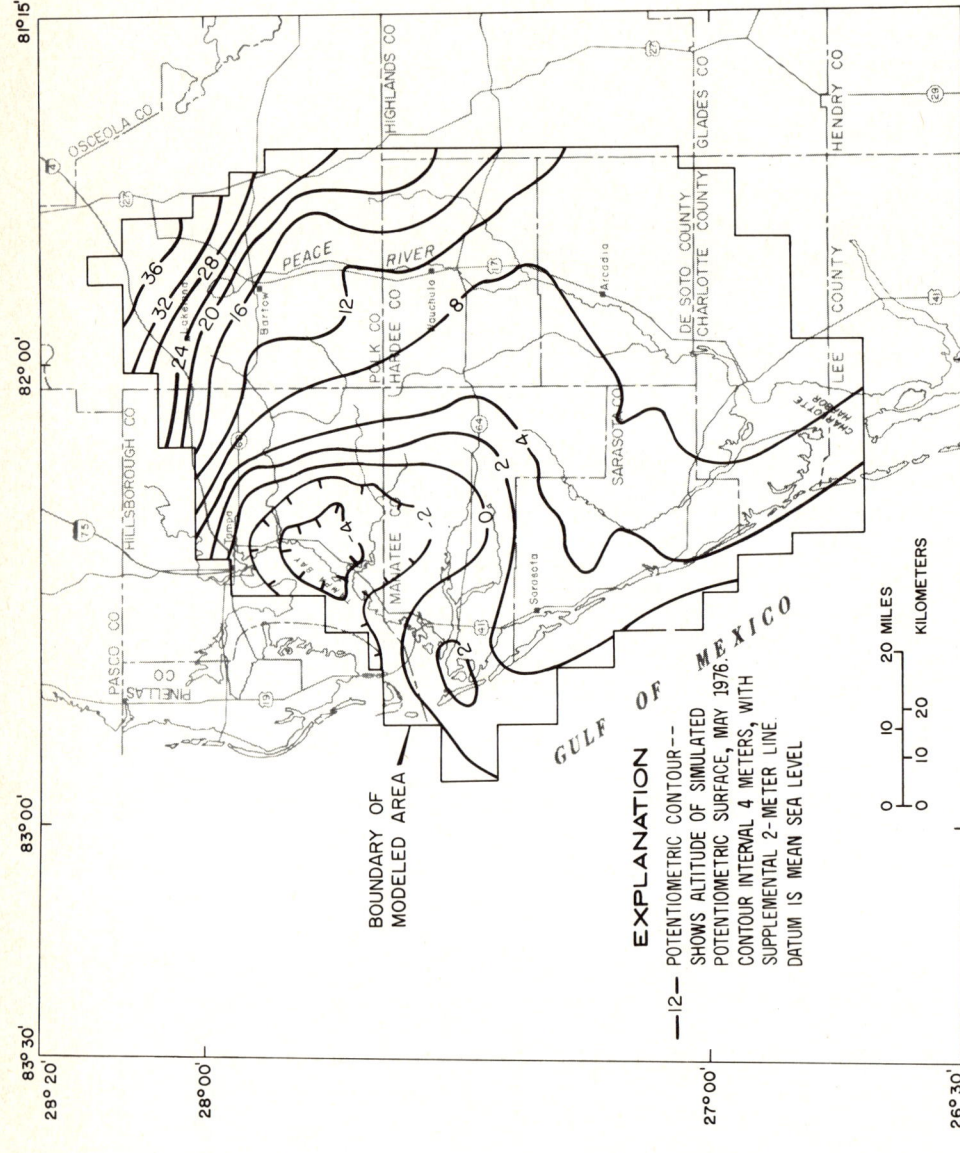

Fig. 10. Simulated potentiometric surface of the Floridan aquifer, May 1976.

Input parameters for the transient calibration were the same as for the steady-state model, with the addition of aquifer and confining-bed storage coefficients. Initially, values of aquifer transmissivity and vertical hydraulic conductivity were not changed; in succeeding runs, these values were modified to improve calibration. Appropriate values of groundwater withdrawal rates and water-table altitudes were used to fit changing conditions from November 1975 to May 1976.

Irrigation pumpage was simulated by using values obtained during an irrigation inventory. Withdrawal rates for municipal supplies and for self-supplied industries were the same as for the September 1975 steady-state calibration. Average 1975 and 1976 withdrawal rates were used for phosphate mines. Some new mines that began operation during the calibration period were included.

Throughout the study area, the water table of the surficial aquifer during the irrigation season was assumed to average about 0.3—1.0 m lower than in September 1975. Values were selected based on 1975—1976 hydrographs of observation wells in the surficial aquifer.

Storage coefficient of the Floridan aquifer was determined from the product of an assumed average specific storage of $3.3 \cdot 10^{-6}$ m^{-1} times thickness of the Floridan aquifer. Storage coefficients ranged from $8.8 \cdot 10^{-4}$ in the north to $1.9 \cdot 10^{-3}$ in the south.

Storage coefficient of the upper confining bed of the Floridan aquifer was determined for each node from the product of an estimated average specific storage of $3.3 \cdot 10^{-5}$ m^{-1} and confining-bed thickness. Values ranged from $2.0 \cdot 10^{-4}$ in the northern part of the area to $7.8 \cdot 10^{-3}$ in the southeastern part.

The transient-model potentiometric surface simulating May 1976 conditions is shown in Fig. 10. This surface may be compared to the observed May 1976 potentiometric surface (Fig. 9). Differences between computed and observed heads at nodes in the model grid ranged from 0 to 4.2 m. The difference was less than 3 m at 78% of the nodes. The most significant difference between the simulated surface and observed May 1976 potentiometric surface is in the position of the depression in the western part of the modeled area. The depression in the simulated surface is centered in Hillsborough County rather than Manatee County.

SIMULATED EFFECTS OF GROUNDWATER WITHDRAWALS

Transient model analyses were used to simulate changes in the potentiometric surface during 1976—2000 resulting from projected groundwater withdrawals for phosphate mines. The effects of withdrawals for existing and proposed mines were evaluated separately and in combination; withdrawal rates for municipal supplies, irrigation and other self-supplied industries were not changed. Results are presented as contour maps showing simulated changes in potentiometric levels. Positions of lines of equal change on these maps are based on linear interpolations between data points, plotted at the centers of nodal blocks.

Each existing phosphate mine is expected to continue groundwater withdrawals until the ore underlying the mine property is depleted. Projected withdrawal rates of existing mines are based on 1975 and 1976 inventories of phosphate pumpage provided by the Florida Phosphate Council. In the simulation, withdrawal rates at each mine were held constant during the life of the mine. As existing mines phase out, withdrawal rates are expected to decline to about 492 l/s by 2000 (Table I). Also included was about 964 l/s of aquifer recharge through connector wells in 1975; recharge amounts for each mine were held constant during the life of the mine. Data for projected life spans of existing mines were provided by Texas Instruments, Inc. (W. Underwood, pers. commun., 1977).

Simulated changes in the potentiometric surface resulting from projected changes in withdrawal rates for existing mines are shown in Fig. 11. The map shows a rise throughout most of the area; maximum rise is more than 8 m in southwestern Polk County. The rise would be expected because of projected declines in withdrawal rates of existing mines.

At least 20 new mines are proposed to begin mining operations before 2000, mostly in Hardee, DeSoto and Manatee counties. Withdrawal rates and life spans for proposed mines are based on data provided by Texas Instruments, Inc. (William Underwood, pers. commun., 1977) and by the Southwest Florida Water Management District (J. Heuer, pers. commun., 1978). Most rates are based on an assumed requirement of about 21.5 m^3 of groundwater per ton of phosphate mined (EPA, 1978, p. 2.16). By the end of 2000, withdrawal rates for proposed mines are expected to be 6,570 l/s (Table I).

Assignment of nodes to proposed mine withdrawal sites was determined by overlaying the model grid on a map showing areas of proposed mines. Where a mine was in more than one node, a single node was selected to represent the mine withdrawals. Actual well locations may differ from those selected, but this difference should not significantly affect the regional distribution or amount of head change.

Simulated changes in potentiometric head resulting from withdrawals for proposed mines are shown in Fig. 12. The map shows a decline throughout most of the area. Maximum decline is about 6 m in eastern Manatee County and western Hardee County.

Fig. 13 shows the combined effects of existing and proposed mines. In this simulation, withdrawal rates and durations of proposed mines were superimposed on those of existing mines. The map shows a rise of the potentiometric surface in Polk County (maximum of about 6 m), and a decline elsewhere (maximum of about 5 m). The areal extent and magnitude of both the rise and decline are smaller than when the effects of pumping for existing and proposed mines are considered separately.

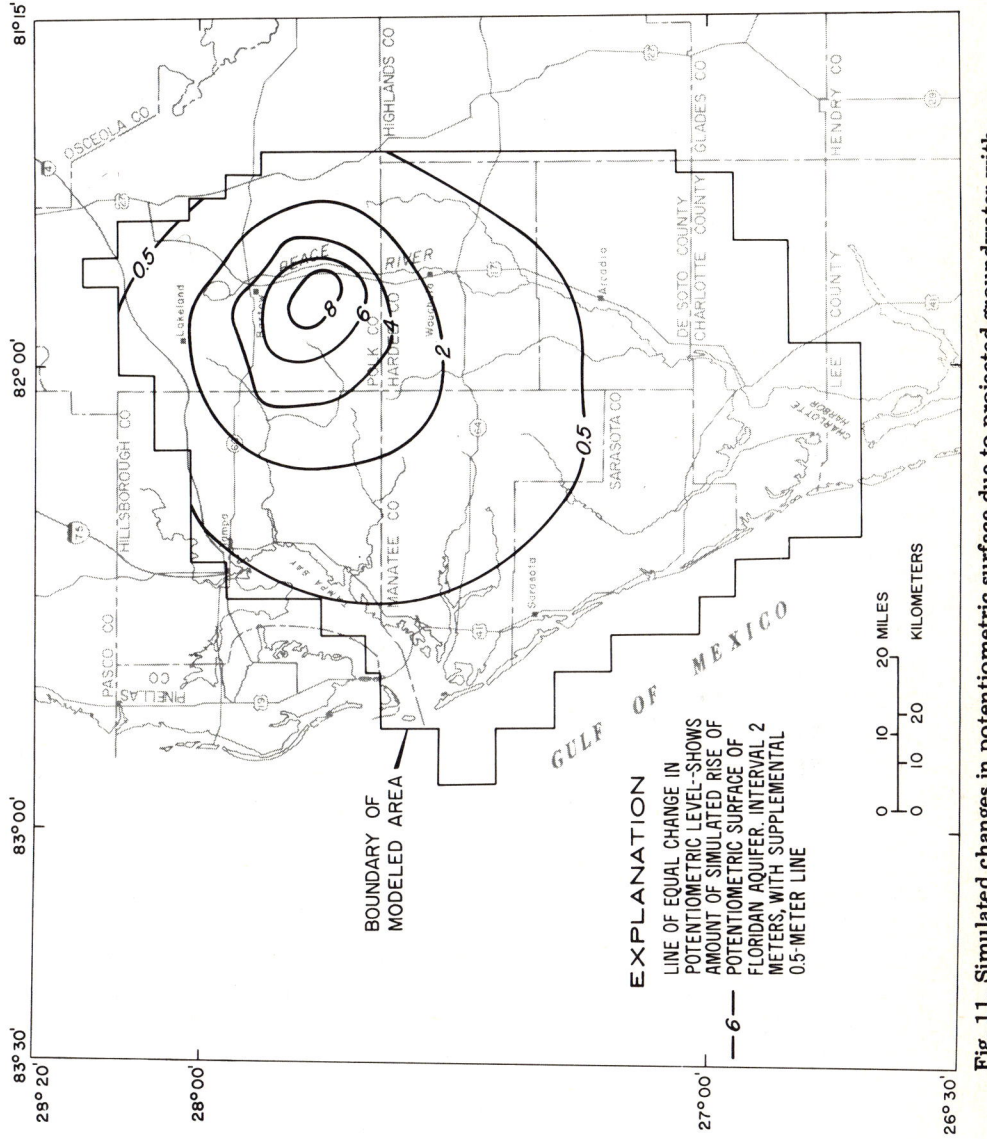

Fig. 11. Simulated changes in potentiometric surface due to projected groundwater withdrawals for existing phosphate mines, 1976—2000.

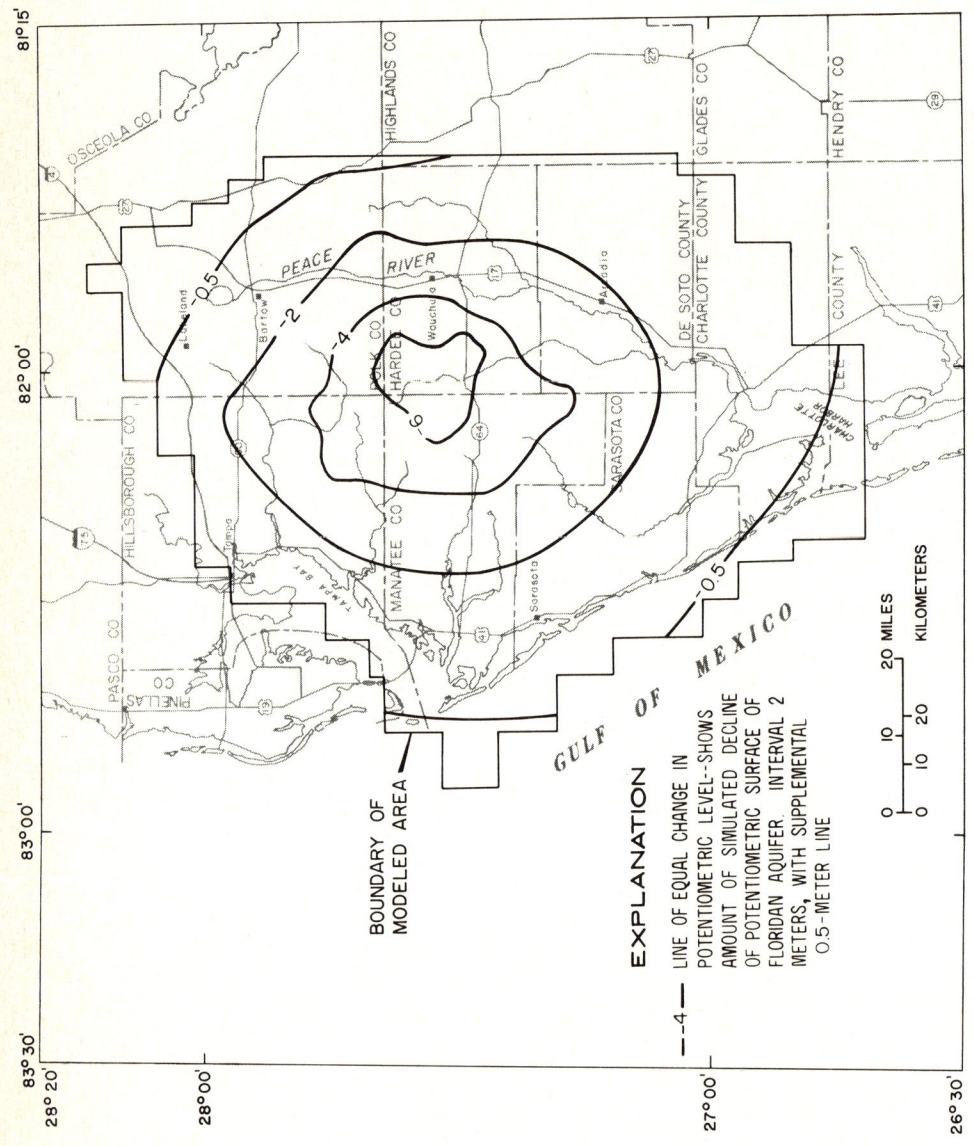

Fig. 12. Simulated changes in potentiometric surface due to projected groundwater withdrawals for proposed phosphate mines, 1976–2000.

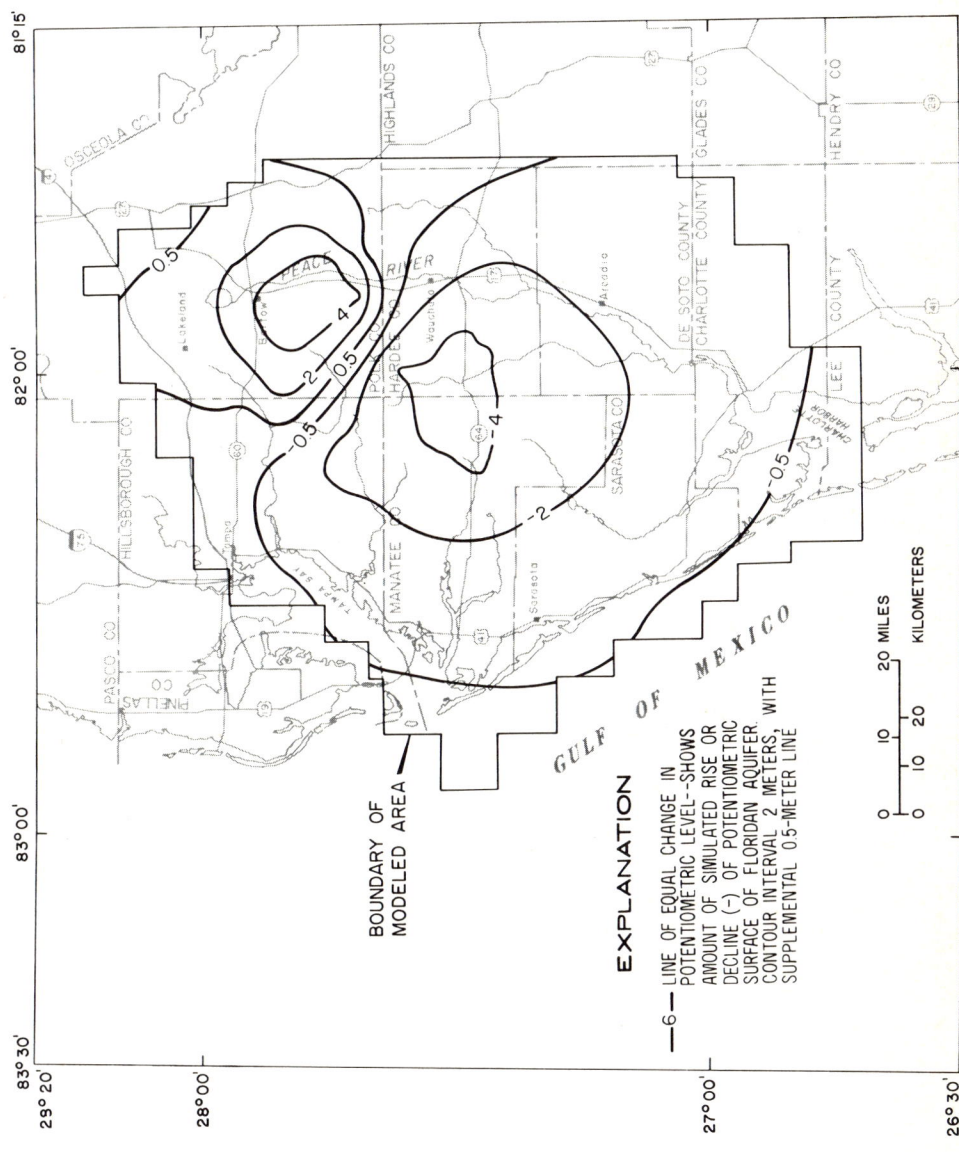

Fig. 13. Simulated changes in potentiometric surface due to projected groundwater withdrawals for existing and proposed phosphate mines, 1976–2000.

APPRAISAL OF RESULTS

The objective of this investigation was to predict potentiometric head changes to be expected as a result of groundwater development for phosphate mines in west-central Florida. The effects of development for other uses, such as irrigation and municipal supplies, were not assessed.

A two-dimensional digital model of groundwater flow was used to predict head changes. The modeling activity represents an initial effort to integrate hydrologic parameters that affect potentiometric head changes, and to determine effects of withdrawals on a regional scale. The model used was the most advanced and appropriate one available at the time the investigation began. Nonetheless, certain assumptions underlying use of the model were not fully met by field conditions. For example, in some areas vertical components of flow exist within the Floridan aquifer, the aquifer is anisotropic, some leakage probably occurs through the lower confining bed, and the water table fluctuates seasonally and in response to pumping stresses in the Floridan aquifer. Boundary conditions can only be approximated by the model, and the effect of a moving saltwater—freshwater interface on the distribution of heads in coastal areas cannot be assessed by the model. All these limitations may serve to introduce errors in calibration and in predicted head changes.

The model was calibrated by simulating heads from potentiometric maps, under steady-state and transient conditions. In the calibrations, uncertainties were introduced by using pumpage data, especially for irrigation, that may not be very reliable. Inadequate irrigation withdrawal data probably account for the difference in locus of the cone of depression in the May 1976 observed and simulated maps (Figs. 9 and 10). The best combination of parameters evaluated, that is, the one producing the least error in the calibration process, still simulated heads that in places substantially differed from observed heads.

In the prediction analysis, the water table was held constant from year to year on the assumption that the surficial aquifer could be fully recharged each year. If, however, pumping from the Floridan aquifer were to result in a long-term or seasonal decline of the water table, leakage would be reduced. Additional drawdown of the potentiometric surface would then be required to sustain leakage at a rate sufficient to supply the water being discharged by pumping.

Model results can be used to obtain a sense of the magnitude of changes in potentiometric levels that could be expected on a regional basis if the projected pumping schemes are carried out. These results suggest that the effects of projected pumping rates are on the order of a few meters and not tens of meters. The impact of these effects on the environmental system cannot be addressed by this model. However, predicted potentiometric changes can provide the basis for future impact analysis.

REFERENCES

Altschuler, Z.S., Cathcart, J.B. and Young, E.J., 1964. Geology and geochemistry of the Bone Valley Formation and its phosphate deposits. Annu. Meet., Geol. Soc. Am., November 19—20, 1964, Guideb. Field Trip No. 6, 68 pp.
EPA (U.S. Environmental Protection Agency), 1978. Draft areawide environmental impact statement, central Florida phosphate industry, EPA 904/9-78-006, Region IV, Atlanta.
Fountain, R.C., Bernardi, J.P., Gardner, C.H., Gurr, T.M. and Zellars, M.E., 1971. The central Florida phosphate district. 7th Forum on the Geology of Industrial Minerals, 41 pp., Fla. Bur. Geol. Field Trip Guideb.
Healy, H.G., 1977. Public water supplies of selected municipalities in Florida, 1975. U.S. Geol. Surv. Water-Resour. Invest., 77-53, 309 pp.
Leach, S.D., 1977. Water-use inventory in Florida, 1975. U.S. Geol. Surv. Open-File Rep. 77-577, 57 pp.
Mills, L.R., Laughlin, C.P. and Parsons, D.C., 1976. Potentiometric surface of Floridan aquifer, Southwest Florida Water Management District, September 1975. U.S. Geol. Surv., Open-File Rep.
Parker, G.G., Ferguson, G.E., Love, S.K., Hay, N.D., Schroeder, M.C., Norren, M.A., Bogart, D.B., Jonker, C.C., Langbein, W.B., Brown, R.H. and Spicer, H.C., 1955. Water resources of southeastern Florida, with special reference to the geology and groundwater of the Miami area. U.S. Geol. Surv., Water-Supply Pap. 1255, 965 pp.
Stewart, J.W., Laughlin, C.P. and Mills, L.R., 1976. Potentiometric surface of Floridan aquifer, Southwest Florida Water Management District, May 1976. U.S. Geol. Surv., Open-File Rep.
Trescott, P.C., Pinder, G.F. and Larson, S.P., 1976. Finite-difference model for aquifer simulation in two dimensions with results of numerical experiments. U.S. Geol. Surv., Tech. Water-Resour. Invest., Book 7, Ch. 1, 116 pp.
Wilson, W.E., 1977. Simulated changes in ground-water levels resulting from proposed phosphate mining, west-central Florida — preliminary results. U.S. Geol. Surv., Open-File Rep. 77-882, 46 pp.

[2]

DEPRESSURIZATION OF A MULTI-LAYERED ARTESIAN SYSTEM FOR WATER AND GROUT CONTROL DURING DEEP MINE-SHAFT DEVELOPMENT

WILLIAM M. GREENSLADE and GEORGE W. CONDRAT

Dames & Moore, Phoenix, AZ 85009 (U.S.A.)
Dames & Moore, Salt Lake City, UT 84111 (U.S.A.)

(Accepted for publication March 6, 1979)

ABSTRACT

Greenslade, W.M. and Condrat, G.W., 1979. Depressurization of a multi-layered artesian system for water and grout control during deep mine-shaft development. In: W. Back and D.A. Stephenson (Guest-Editors), Contemporary Hydrogeology — The George Burke Maxey Memorial Volume. J. Hydrol., 43: 517—536.

Uranium in the Grants Mineral Belt of northwestern New Mexico is explored from progressively greater depths. As mining depths increase, more aquifers and higher hydrostatic heads are encountered. Water control during mine shaft sinking is critical to successful shaft development.

Field testing procedures were designed to produce data on aquifer coefficients, groutability and rock strength appropriate to the short-term construction period involved in shaft sinking. Design coefficients were selected on the basis of field and laboratory test results and regional geohydrology.

The uranium ore is located between 915 and 1220 m (3000 and 4000 ft.) below the surface. The geology of the area studied comprises alternating marine and non-marine sandstone, siltstone and shale. The ore is located in the lowermost of six major aquifers identified.

A depressurizing system was designed to reduce the hydrostatic pressure and shaft water inflow from each of the aquifers. The groundwater velocity across the shaft was maintained below 0.61 m/day (2 ft./day) and minimized movement of grout away from the shaft.

INTRODUCTION

Uranium mining in northwestern New Mexico began in the 1950's. As the search for new reserves continues it is necessary to mine at increasingly greater depths. Early mines were generally less than 244 m (800 ft.) in depth and water was removed from shafts and workings with sump pumps. New ore discoveries are at depths of 610—1220 m (2000—4000 ft.) and aquifers are under 336 m (1100 ft.) or more of hydrostatic head. Water control during shaft sinking is necessary for efficient and safe operation.

An investigation was undertaken to assess the geohydrologic conditions at a proposed 1068 m (3500 ft.) deep mine. Two shafts, approximately 92 m (300 ft.) apart were to be sunk into the ore-bearing zone. The ore zone was known to be an aquifer and several overlying formations were also thought to be water-bearing.

The investigation was designed to evaluate the depth and water yielding capabilities of the ore horizon and overlying formations. If required, a water-control system was to be designed.

GEOLOGY

The site is located in the San Juan Basin, a structural depression that occupies a 64,750-km^2 (25,000-mi.^{2}) area in northwestern New Mexico and adjacent parts of Colorado, Arizona and Utah. Approximately 4575 m (15,000 ft.) of sedimentary rock are present in the deepest part of the basin. The San Juan structural and topographic basins are approximately coincident in areal extent.

The geology of the southern and western portions of the San Juan structural basin, in which the site is located, is characterized by a thick sequence of sandstones and shales, generally dipping to the northeast. The basin was formed during late Cretaceous through Eocene time. The average negative structural relief from rim to central low-point is nearly 1525 m (5000 ft.). The mine site is located on Chaco Slope, a structural subdivision of the basin in which sedimentary rocks dip northeastward from the Zuni uplift toward the Central Basin, as shown in Fig. 1.

In the site area subsurface strata dip approximately 5° in a N20°E direction. No major faulting or folding has been detected. Minor faulting and folding may exist but has not been confirmed by drilling.

The oldest (and deepest) strata that will be encountered in mine construction and development is the Westwater Canyon Member of the Morrison Formation (late Jurassic). The Westwater Canyon Member is a continental deposit consisting of coalesced alluvial channel systems. This sheet of coalesced channel deposits is more than 101 km (63 mi.) wide in a NW—SE direction, approximately perpendicular to the paleocurrents (streams) from which most of the sediments were deposited. Paleocurrents flowed generally northeastward from a highland area located southwest of the area. The complexity of the depositional pattern is illustrated in Fig. 2.

The uppermost unit of the Morrison Formation is the Brushy Basin Shale Member. It is a hard, non-marine shale that conformably overlies the Westwater Canyon. The Brushy Basin generally forms an effective confining layer for water in the Westwater Canyon Member.

In Cretaceous time, deposition commenced with southwest transgression of the Dakota Sandstone over the Morrison Formation. During most of the Cretaceous four separate facies were being deposited at any one time: (1) marine shale in the seaward phase (to the northeast); (2) offshore sand deposits near the shoreline; (3) coals, silts, sands and shales in the lagoonal

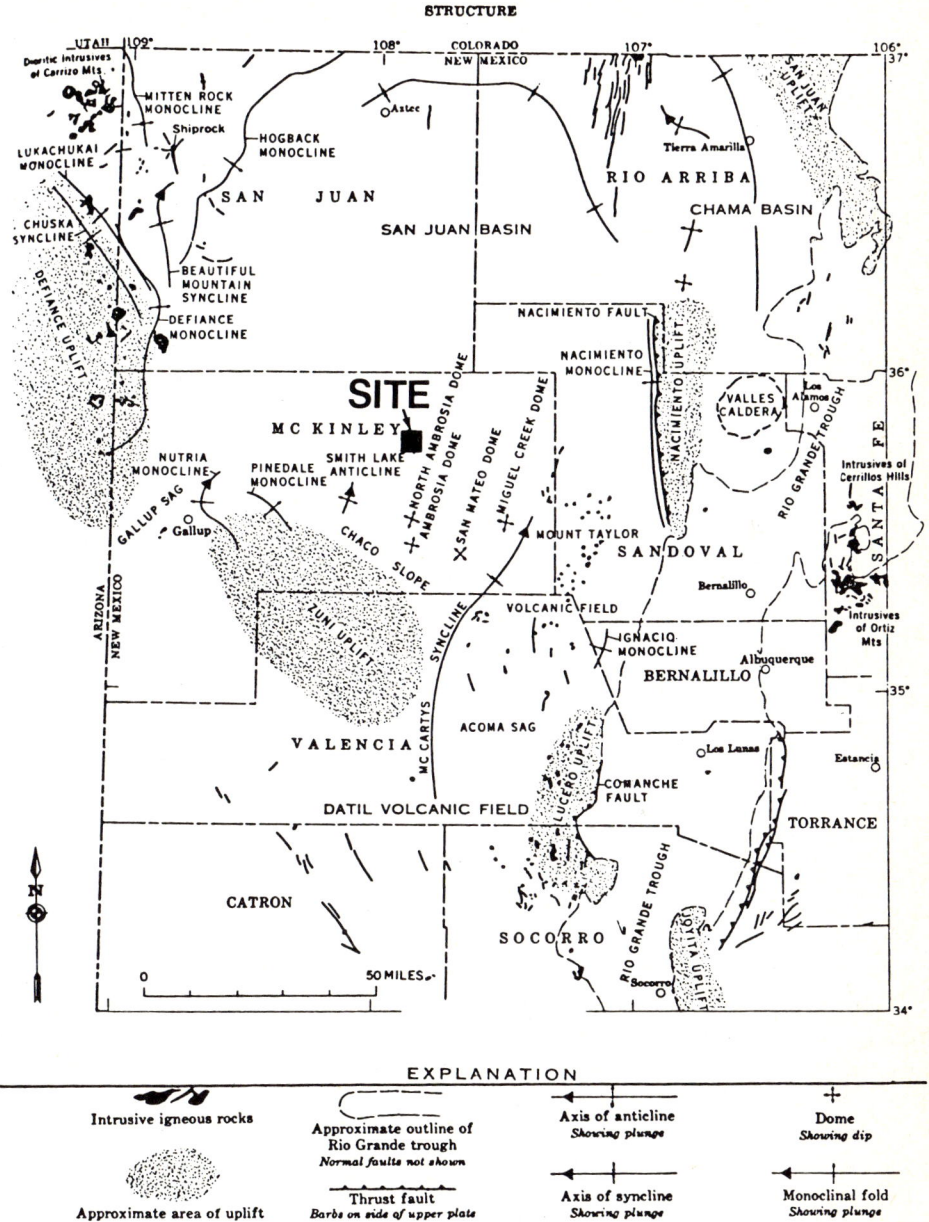

Fig. 1. Tectonic map of San Juan Basin and adjacent areas, northwestern New Mexico.

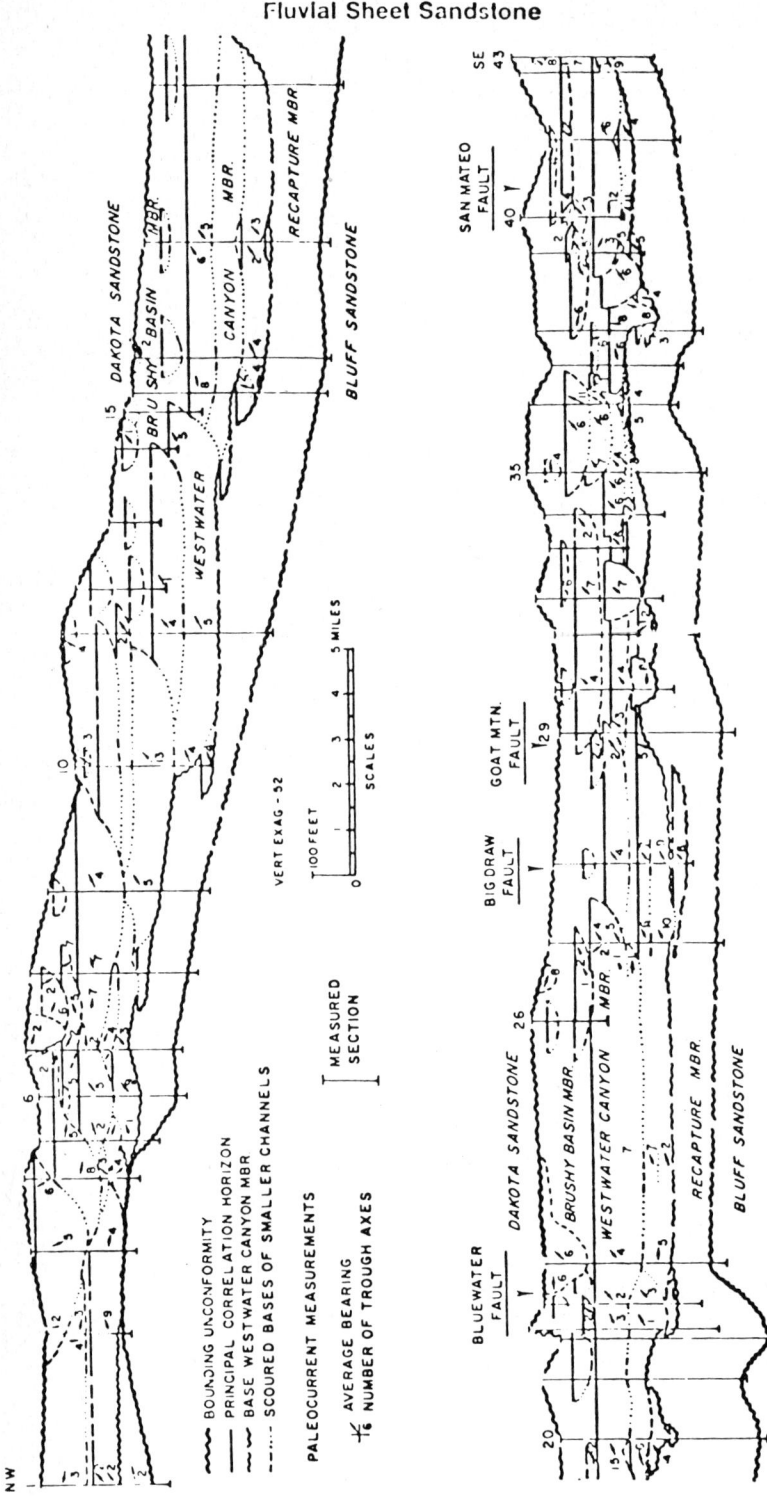

Fig. 2. Stratigraphic cross-section showing relations of Westwater Canyon Member to enclosing rocks.

to brackish-water environment; and (4) sands, silts and shales largely devoid of coal in the continental floodplain environment. As the shoreline retreated and advanced, this depositional suite of four facies also retreated or advanced. A complicated series of interfingering shales, mudstones and coals resulted. In ascending order these include the Dakota Sandstone, Mancos Shale, Gallup Sandstone, Crevasse Canyon Formation, Hosta Tongue of the Point Lookout, Satan Tongue of the Mancos Shale, Point Lookout Sandstone and Menefee Formation.

The stratigraphy of the area is shown in Fig. 3.

FIELD INVESTIGATIONS

Core hole

The study purpose, in addition to an evaluation of site hydrologic characteristics, included the investigation of rock strength and engineering properties. Evaluation of these latter properties required drilling of a continuous core hole through the complete section to be penetrated by shafts. A hydrologic study plan was designed to make maximum use of the data obtainable from this core hole. In addition to the usual field and laboratory strength tests, the core was subjected to grain-size analyses, and permeability tests. Visual logging of the core for hydrologic properties included lithology, fracture intensity and cementation.

A set of geophysical logs were obtained. These included caliper, density, temperature, self-potential, resistivity, porosity and three-dimensional velocity.

Formation stratigraphy was based on detailed core examination and comparison with regional stratigraphy. Potential water-bearing zones were identified from hydrologic properties logged from the core and from geophysical logs. Criteria used were lithology, grain-size distribution and cementation.

Six aquifers were identified as described below.

Menefee Sandstone. The Menefee Formation consists of interbedded dark- to light-gray medium- to coarse-grained sandstone, shale, siltstone and coal layers. Generally, the sandstone is well-cemented. The Menefee Formation is exposed at the land surface and extends to a depth of 146 m (478 ft.) near the shafts. The interval between 31 and 43 m (100 and 140 ft.) contained the most permeable sandstone within the Menefee.

Point Lookout Sandstone. The Point Lookout Sandstone of Cretaceous age consists of gray massive uniform sandstone beds with some thin interbedded shale layers. It is moderately well-cemented.

The Point Lookout is 52 m (171 ft.) thick and occurs between 146 and 198 m (479 and 650 ft.) below the surface.

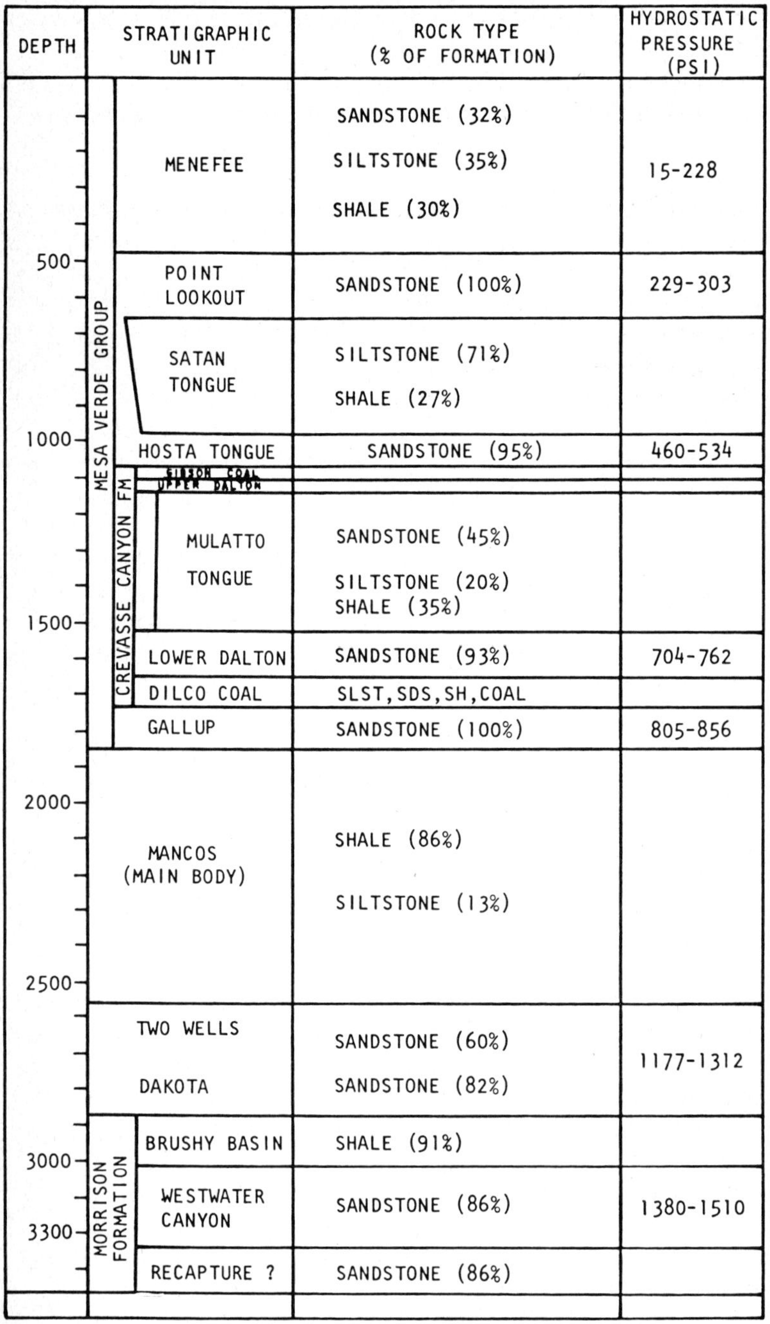

DEPTH	STRATIGRAPHIC UNIT			ROCK TYPE (% OF FORMATION)	HYDROSTATIC PRESSURE (PSI)
			MENEFEE	SANDSTONE (32%) SILTSTONE (35%) SHALE (30%)	15-228
500—		MESA VERDE GROUP	POINT LOOKOUT	SANDSTONE (100%)	229-303
			SATAN TONGUE	SILTSTONE (71%) SHALE (27%)	
1000—			HOSTA TONGUE	SANDSTONE (95%)	460-534
			GIBSON COAL / UPPER DALTON		
		CREVASSE CANYON FM	MULATTO TONGUE	SANDSTONE (45%) SILTSTONE (20%) SHALE (35%)	
1500—			LOWER DALTON	SANDSTONE (93%)	704-762
			DILCO COAL	SLST, SDS, SH, COAL	
			GALLUP	SANDSTONE (100%)	805-856
2000—	MANCOS (MAIN BODY)			SHALE (86%) SILTSTONE (13%)	
2500—					
	TWO WELLS			SANDSTONE (60%)	1177-1312
	DAKOTA			SANDSTONE (82%)	
3000—		MORRISON FORMATION	BRUSHY BASIN	SHALE (91%)	
3300—			WESTWATER CANYON	SANDSTONE (86%)	1380-1510
			RECAPTURE ?	SANDSTONE (86%)	

Fig. 3. Generalized stratigraphic section (1 psi = 1 lb. in.$^{-2}$ = 6.895·10^{-3} Pa).

Hosta Sandstone Tongue. The Hosta Sandstone Tongue of the Point Lookout is lithologically similar to the main body of the Point Lookout Sandstone. It is separated from the Point Lookout by 102 m (333 ft.) of shale and siltstone. The Hosta Sandstone is 26 m (86 ft.) thick and is encountered from 300 to 326 m (983 to 1069 ft.) below the surface.

Dalton Sandstone. The Dalton Sandstone of the Crevasse Canyon Formation consists of light-gray fine-grained medium- to thick-bedded sandstone with interbedded shale and siltstone layers. The Dalton is split into an upper and lower unit by the Mulatto Tongue of the Mancos. The upper Dalton occurs between depths of 336 and 345 m (1100 and 1131 ft.). It is a very fine-grained hard well-cemented sandstone and was not considered to be a significant water-producing zone. The lower Dalton Sandstone is encountered from 465 to 503 m (1525 to 1649 ft.) below the surface and has a more uniform grain-size distribution than the upper Dalton. The lower Dalton is 38 m (124 ft.) thick.

Gallup Sandstone. The Gallup Sandstone consists of dark buff to gray fine- to coarse-grained sandstone with interbedded shale and coal layers. Generally it is moderately- to well-cemented. The Gallup was penetrated from 528 to 567 m (1730 to 1860 ft.) below the surface in the shaft area. The Gallup can be divided lithologically into an upper and lower unit. The upper 12 m (40 ft.) is coarser grained and better sorted than the lower 24 m (80 ft.).

Dakota Sandstone. The Dakota Sandstone of early (?) Cretaceous age consists of light- to dark-gray fine- to coarse-grained sandstone, with numerous interbedded shale, siltstone and coal layers. At the site the sandstone is moderately-cemented and in some areas very friable. The Dakota (including the Two Wells Member) is encountered between 802 and 873 m (2630 and 2864 ft.) below the land surface. Total thickness is 71 m (234 ft.).

Westwater Canyon Sandstone. At the site, the Westwater Canyon Sandstone Member of the Morrison Formation consists of light- to medium-gray fine- to coarse-grained sandstone, with a few interbedded shale and siltstone layers. The sandstone is poorly- to well-cemented and is highly variable, both horizontally and vertically. Grain size and cementation vary abruptly and generally the coarser zones are more poorly-cemented. The Westwater Canyon Sandstone was penetrated between 918 and 1010 m (3010 and 3310 ft.) below the surface.

The Westwater Canyon Sandstone is of greatest interest, as it is the most prolific aquifer in the section and also the ore-bearing zone. All mining will take place in the Westwater Canyon Sandstone.

Discontinuity mapping

Since the presence of fractures, joints, or faults could significantly affect permeability, a program to map discontinuities was undertaken. The program consisted of logging fracture intensity and orientation in the core, mapping of joints and fractures on surface outcrops, and development of cross-sections through the shaft site area based on electric logs of surrounding exploration drill holes.

Results of the mapping program indicated that jointing is generally perpendicular to bedding planes and to each other. One set of joints is roughly parallel to regional strike, the other perpendicular to it. Both joint sets are essentially vertical. Fracture intensity determined from the core was low, generally less than three fractures per foot. Essentially all fractures were parallel to bedding planes. No faulting or folding was noted in the cross-sections. It was concluded that discontinuities were not a significant contributor to overall permeability.

Hydrologic test wells

The test program consisted of installation and testing of three observation wells and a test well. Cost considerations prevented the location of observation wells in all potential aquifers. The three potentially most prolific aquifers were selected: the Point Lookout, Dakota and Westwater Canyon sandstones. The test well was drilled through the same aquifers as the observational wells and, in addition, was perforated opposite the three other potential water-producing zones (the Gallup, Dalton and Menefee sandstones).

All observation wells were drilled to the top of the aquifer; casing was installed and the annulus cemented from the bottom to the surface. The hole was then advanced through the aquifer. The Point Lookout and Dakota observation wells were left open-hole below the casing. The Westwater Canyon observation well would not stay open and slotted casing was set opposite the aquifer.

All wells were under artesian pressure and flowed at the surface. The observation wells were equipped with a well head and a gate or butterfly valve. Two valves were attached to the casing top in order to permit installation of pressure recording devices and allow casing access when water levels were drawn down below the valves.

Observation wells were equipped with orifice-discharge measuring devices. Aquifer properties were determined from constant head, variable discharge tests performed on each well by allowing them to flow.

The hydrologic test well was drilled to the top of the Dakota Formation. Casing was installed and the annulus cemented from the bottom to the surface. An open-hole was then drilled through the Dakota Sandstone. After pump testing of the Dakota, the well was deepened to below the mining horizon in the Westwater Canyon Sandstone and slotted casing installed. The

TABLE I

Construction features of hydrologic test well and observation wells

Well name	Well No.	Aquifer(s) tapped	Depth of casing (ft.)	Slotted casing length (ft.)	Total depth (ft.)	Open internal (ft.)	Casing size (O.D.) (in.)	Hole size (in.)	Static water level above land surface (ft.)
Hydrologic test well	150	Dakota	2,630	—	2,864	234	7	$5\frac{7}{8}$	166
Hydrologic test well	150	Dakota and Westwater Canyon	2,630	350	3,310	680	7	$5\frac{7}{8}$	n.r.
Point Lookout observation well	179	Point Lookout	480	—	650	170	7	$5\frac{7}{8}$	50
Dakota observation well	178	Dakota	2,630	—	2,864	234	7	$5\frac{7}{8}$	166
Westwater Canyon observation well	180	Westwater Canyon	3,010	350	3,310	300	7	$5\frac{7}{8}$	176

1 ft. = 0.3048 m; 1 in. = 2.54 cm; n.r. = not recorded.

Dakota and Westwater Canyon aquifers were then pumped together.

After pumping of the Dakota and Westwater Canyon aquifers a wire-line bridge plug was set below the next overlying aquifer (Gallup). The casing was perforated with four holes per foot opposite the entire aquifer thickness and the zone tested. This sequence was repeated in the test well for each overlying aquifer. The construction features of the observation and test wells are given in Tables I and II.

A system was needed to pump each of the six aquifer zones in the test well. The large variation in head and expected pumping rates for each of the aquifers necessitated a flexible system. The use of a turbine pump would require several different pumps and the concomitant cost of purchasing, setting and pulling several pumps. An air-lift system was selected as it allowed flexibility at a reasonable cost. Each zone was pumped separately, except the Westwater Canyon which was pumped simultaneously with Dakota as both aquifers were open to the test well following pumping of the Dakota alone.

TABLE II

Data on perforated zones in test well

Sort of well	Well No.	Aquifer tapped	Depth of packer (ft.)	Interval tested (ft.)	Casing size (in.)	Static water level above land surface (ft.)
Hydrologic test well	150	Menefee	140	100—40	7	48
Hydrologic test well	150	Point Lookout	650	479—650	7	50
Hydrologic test well	150	Hosta	1,069	983—1,069	7	50
Hydrologic test well	150	Dalton	1,649	1,525—1,649	7	100
Hydrologic test well	150	Gallup	1,860	1,738—1,860	7	120

1 ft. = 0.3048 m; 1 in. = 2.54 cm.

Test-well discharge was measured with a specially constructed discharge tank. The water—air mixture was discharged into the tank and allowed to flow through expanded metal baffles. The discharge rate was measured by recording the water level above the bottom of a V-notch weir. A water-level recorder was installed in a stilling well in the tank.

Water levels in the observation wells were measured with 0.5% pressure gages and recorders attached to the well head. When water levels fell below the top of the casing, they were measured with electric tape. Water levels in the test well were measured with a nitrogen—air line system.

ANALYSIS OF AQUIFER TEST DATA

Observation wells flowed at the surface and were tested individually by opening the gate valve and measuring the decline in discharge with time. Shut-in pressure recovery tests were also performed.

TABLE III

Summary of hydrologic parameters from field testing program

Aquifer (descending order)	Well	Interval tested (ft.)	Airline setting in pumping well (ft.)	Transmissivity (gal. day^{-1} ft.$^{-1}$)	Storage coefficient	Method of analysis
Menefee	perforated zone in test well	100–140	100	942	—	Aron–Scott
		106–140	100	229	—	residual drawdown
Point Lookout	observation well	480–650	—	1,971	—	Aron–Scott
				2,654	—	Aron–Scott
	perforated zone in test well	479–650	440	2,900	$6.0 \cdot 10^{-6}$	Aron–Scott
			440	1,056	—	Aron–Scott
	observation well	480–650	440	1,148	—	residual drawdown
			440	1,172	—	Aron–Scott
Hosta	perforated zone in test well	983–1,069	600	498	—	Aron–Scott
			600	162	—	residual drawdown
Dalton	perforated zone in test well	1,525–1,649	800	5	—	Aron–Scott
			800	1	—	residual drawdown
Gallup	perforated zone in test well	1,738–1,860	500	35	$2.2 \cdot 10^{-3}$	Aron–Scott
			500	80	—	residual drawdown
			1,000	290	—	Aron–Scott
			1,000	426	—	residual drawdown
Dakota	hydrologic test well	2,630–2,864	1,000	562	$1.8 \cdot 10^{-8}$	Aron–Scott
			2,000	636	$4.0 \cdot 10^{-5}$	Aron–Scott
			—	207	$8.4 \cdot 10^{-5}$	distance drawdown
	Dakota observation well	2,630–2,864	1,000	381	—	residual drawdown
			2,000	508	—	residual drawdown
Westwater Canyon	hydrologic test well	2,630–3,310	2,000	825	$2.2 \cdot 10^{-3}$	Aron–Scott (combined test)
	Westwater Canyon observation well	3,010–3,310	2,000	660	$4.2 \cdot 10^{-5}$	Aron–Scott
			2,000	679	—	residual drawdown
			—	777	—	Aron–Scott
			—	1,015	$2.1 \cdot 10^{-9}$	Aron–Scott

1 ft. = 0.3048 m; 1 gal. day^{-1} ft.$^{-1}$ = 0.0124 m^2 day^{-1}.

The air-line method of test-well pumping resulted in maintaining pumping level at a depth determined by the amount of air available and air line submergence. Water levels measured in both pumped and observation wells declined during test-well pumping as did the discharge.

A method developed by Aron and Scott (1965) was used to calculate aquifer transmissivity. The specific drawdown (s/Q) is plotted against time on semilogarithmic paper, and:

$$T \cong \frac{2.3}{4\pi \Delta(s/Q)}$$

where s/Q is the specific drawdown per log cycle.

The aquifer coefficients were calculated from the recovery portions of the pumping test, using the Theis (1935) recovery method.

Field data from the perforated zones in the test well were affected by well loss. Well loss was estimated by constructing a theoretical distance drawdown plot for the aquifer characteristics calculated from an aquifer that had both an observation well and was perforated in the test well (Point Lookout Sandstone). The well loss estimated from this zone was extrapolated to other perforated zones not penetrated by an observation well, based on the assumption that well loss is proportional to the square of the discharge.

SELECTION OF DESIGN PARAMETERS

Field test results, laboratory permeability and grain-size determinations, and visual examination of rock core were used to select design parameters. Hydrologic parameters determined from field and laboratory testing programs are given in Tables III and IV, respectively. Transmissivity and permeability values varied considerably, reflecting the complex depositional pattern of the deposits. A range from the best estimate to a conservatively high value of transmissivity was estimated for the preliminary evaluation of potential inflow to the shafts and well field design. Field pumping test analyses were given the most weight in parameter selection, for these tests indicate any secondary, as well as primary, permeability effects and a much greater volume of aquifer is tested. Increased transmissivity values significantly higher than those obtained from Aron—Scott and residual drawdown plots of the field test data were selected for the Hosta, Dalton and Gallup aquifers. Use of these higher values was based upon the specific capacity of the test well, laboratory tests and lithologic similarities with other formations.

An estimate of $5 \cdot 10^{-3}$ for the storage coefficient was used for all aquifers except Point Lookout for which a value of 10^{-4} was used. Values were selected based upon field test results, and aquifer depth and thickness.

The potentiometric levels in the aquifers were obtained from observation wells and shut-in pressures at the test well.

Effective porosity, used to evaluate groundwater velocities, was estimated by assuming it to be 80% of the total porosity determined in the laboratory for the sandstone aquifers.

TABLE IV

Results of laboratory permeability tests on selected samples

Aquifer	Sample depth (ft.)	Permeability (gal. day^{-1} ft.$^{-2}$)	
		horizontal	vertical
Point Lookout	539.5	13	9.1
	618.4	2.7	1.97
Hosta	990.0	0.015	0.095
	1,063.0	1.95	1.6
Dalton	1,570.0	$4.9 \cdot 10^{-3}$	$3.8 \cdot 10^{-3}$
	1,642.0	0.92	1.9
Gallup	1,771.4	59	22
	1,830.0	$6.4 \cdot 10^{-3}$	$3.23 \cdot 10^{-3}$
Dakota (Two Wells Member)	2,581.3	$2.06 \cdot 10^{-3}$	$1.02 \cdot 10^{-3}$
	2,644.0	$4.9 \cdot 10^{-3}$	$4.45 \cdot 10^{-3}$
Dakota	2,729.3	7.5	2.9
	2,868.5	$5.3 \cdot 10^{-3}$	$8.05 \cdot 10^{-3}$
Westwater Canyon	3,027.3	11.2	8.9
	3,132.1	2.05	2.1
	3,264.2	6.9	7.4

1 ft. = 0.3048 m; 1 gal. day^{-1} ft.$^{-2}$ = 0.0124 m day^{-1}.

ESTIMATES OF INFLOWS TO SHAFTS

Estimates of maximum inflows to each shaft in the event that no water-control measures are taken during shaft sinking were made during initial studies in order to decide whether water-control measures were necessary and to aid in sizing emergency shaft pumping equipment. The estimates were made by treating each shaft as a large-diameter well and applying non-equilibrium well formula with adjustments for partial penetration. Since a maximum of 9.2 m (30 ft.) of shaft are to be unlined at any given time, exposed shaft intervals of 3.1, 6.1 and 9.2 m (10, 20 and 30 ft.) were evaluated.

The following modified form of the so-called Jacob equation (Cooper and Jacob, 1946) to estimate the specific capacity of the shaft was utilized since the factor u is small:

$$Q/s = \frac{T}{264 \log (Tt/1.87 r_w^2 S) - 65.5}$$

where r_w is the shaft radius in feet; S is the storage coefficient; T is the transmissivity in gal. day^{-1} ft.$^{-1}$; and t is the time after pumping started, in days. This theoretical full-penetration specific capacity was adjusted for partial penetration by the following equation (Turcan, 1963):

$$Q'/s' = Q/s \left[\frac{L}{M} \left\{ 1 + 7 \left(\frac{r_w}{2L} \right)^{\frac{1}{2}} \cos \frac{\pi L}{2M} \right\} \right]$$

where Q'/s' is the specific capacity of the partially penetrating shaft; L is the length of open hole; and M is the aquifer thickness. The adjusted specific capacity is then multiplied by the total available head to estimate water inflow. For thin aquifers, the equation is not valid and the full penetration specific capacity is used. No adjustments were made for well loss since these effects are expected to be minor and because maximum estimates were desired. The effects of dewatering near the aquifer face are minor under the initially great hydrostatic pressures and consequent steep hydraulic gradients. The results of the calculations are shown in Table V. On the basis of these results all aquifers except the Menefee Formation were thought to warrant water-control procedures.

TABLE V

Estimated range of inflows to proposed 22-ft. diameter shaft

Aquifer	Transmissivity (gal. day^{-1} ft.$^{-1}$)		Penetration (ft.)	Average flow rate first day (gal. min.$^{-1}$)
	minimum	maximum		
Menefee*	200	400	40	30—50
Point Lookout	2,000	3,000	10	300—400
			20	450—600
			30	550—800
Hosta*	500	1,000	86	450—850
Dalton	100	200	10	75—150
			20	100—200
			30	150—250
Upper Gallup*	500	700	40	750—1,000
Lower Gallup	200	300	80	350—500
Dakota	600	800	10	450—600
			20	650—850
			30	850—1,100
Westwater Canyon	1,000	1,500	10	550—750
			20	800—1,150
			30	1,000—1,450

1 gal. day^{-1} ft.$^{-1}$ = 0.0124 m^2 day^{-1}; 1 ft. = 0.3048 m; 1 gal. min^{-1} = 6.309·10^{-5} m^{3}s^{-1}.
*Thin aquifers not analyzed for partial penetration.

DEPRESSURIZING WELL FIELD DESIGN

Several water-control techniques were evaluated. These included grouting, freezing and pumping from deep wells. A system of deep wells and grouting was selected.

The sandstone aquifers cannot be completely dewatered with wells. The sandstones are artesian, deep, relatively thin, and have low transmissivities. If the water level is drawn below the top of the aquifer in a pumping well, very little additional drawdown at the shaft (compared to the total available drawdown) is gained and it is readily offset by the decrease in transmissivity at the pumping well due to the reduction of the saturated thickness of the aquifer. The principal benefit to be obtained by pumping from wells is a major reduction in artesian pressure; and while flow into a shaft is not eliminated it is reduced. Since the wells were not designed to dewater the aquifers they are referred to as depressurizing rather than dewatering wells.

Theis (1935) non-equilibrium constant-discharge well formulas were utilized in the depressurizing system analyses. Conservative (high-range) transmissivity values were used. These values and other aquifer parameters are shown in Table VI. Drawdowns produced from each well in the system were determined separately and were superimposed to determine total drawdown. Based upon the type of well construction, an estimated well loss was included in the calculations by assuming that a 30% reduction in specific capacity would occur as the result of well loss when a single well was pumping at maximum drawdown (to the top of the aquifer). In accordance with the approximate relationship that well loss varies with the square of the pumping rate (Jacob, 1947), well loss was reduced for pumping rates less than this maximum discharge rate. This approach facilitated comparison of various alternative depressurization systems. No recharge boundaries or barriers were identified near the shaft site in geologic information or in pump test data; therefore, no boundary conditions were included in the calculations. A check of the Reynolds number, utilizing flow velocity and mean grain diameter of the formation verified that the flow to the wells would be laminar.

To prevent excessive migration of grout injected for water control it was desired to maintain groundwater velocities in the shaft area below 0.61 m/day (2 ft./day). The groundwater velocity was calculated from the following form of Darcy's law:

$$V = KI/n_e$$

where V is the groundwater velocity; K is the permeability; I is the hydraulic gradient; and n_e is the effective porosity. For the designs proposed, the hydraulic gradient was calculated with drawdowns determined at numerous locations of interest near the shaft.

Design alternatives for the depressurizing system involved comparison of well-construction procedures, number of wells, field geometry and pumping duration. To reduce costs, careful consideration was made of the feasibility of completing depressurization wells in more than one aquifer and of

TABLE VI

Aquifer parameters used in design analyses

Aquifer	Depth interval (ft.)	Thickness (ft.)	Static water level above land surface (ft.)	Available drawdown	Transmissivity (gal. day^{-1} ft.$^{-1}$)	Storage coefficient ($\times 10^{-5}$)	Effective porosity (%)
Point Lookout	470–650	180	50	520	3,000	10	22.0
Hosta	983–1,069	86	50	1,033	1,000	5	20.0
Dalton	1,525–1,660	135	100	1,625	200	5	8.8
Gallup	1,738–1,850	112	120	1,858	1,000	5	18.0
Dakota	2,630–2,864	234	166	2,796	800	5	13.0
Westwater Canyon	3,013–3,400	387	176	3,189	1,500	5	17.0

1 ft. = 0.3048 m; 1 gal.day^{-1} ft.$^{-1}$ = 0.0124 m^2 day^{-1}.

deepening wells to lower aquifers when depressurization was no longer required. Completion of a well in a single aquifer is the most straightforward design and consists of a cased hole to the formation and a screened or open hole in the aquifer. Pump placement at the aquifer top was recommended since increment of additional yield obtained by drawing the water level below the top of the aquifer is quite small and problems of sanding and subsequent mandatory screening are avoided.

Multiple completions (in more than one aquifer) involve pumping larger quantities of water and are more complicated to construct. If the pumping level is to be drawn below the upper aquifer, cascading water will occur and larger diameter casing is needed for a pump shroud around the pump (for cooling). A screen and possibly gravel-packing of the upper aquifer may be required to eliminate potential sand inflow and caving which could result in loss of the pump and well. Also, if entrained air in the cascading water is significant, a gas separator would be required for the pump to prevent corrosion and cavitation. Although multiple completions with cascading water are technically feasible, increased difficulty, larger diameter, and potential problems do not favor these designs when reasonable alternatives are available. Multiple-aquifer completions where the pumping water level is not drawn below the top of the upper formation are favored as these avoid the problems of partially dewatered aquifers and cascading water.

A dual completion in the Dalton and Gallup aquifers was considered feasible without pumping below the top of the Dalton since these aquifers are located relatively close together and pumping lifts and volumes are not excessive. Dual completion of the Dakota and Westwater was not possible due to the combined high volume and lift requirements.

Deepening of wells is feasible if sufficient time is available for deepening between the end of the pumping period required for the upper aquifer and the required start of pumping in the lower aquifer. While there was insufficient time to deepen wells from any one aquifer to the next deepest one, sufficient time was available to deepen wells in the Point Lookout to a dual completion in the Dalton and Gallup formations, and to deepen Hosta wells to the Dakota Sandstone. Depending on the diameter requirements, deepening involves pulling screen, if any, and casing before deepening. Sufficient diameter to allow deepening without pulling casing was allowed in the Hosta wells since swelling of the overlying shales would likely prevent pulling the casing.

Selection of pumping duration prior to entering the aquifer with the shafts must allow for sufficient time to work out any problems in pumping-system mechanics and provide a reasonable reduction in head at the shaft. Neither the amount of depressurization nor the groundwater velocity are highly affected by moderate variations in duration of pumping; therefore, quite a bit of latitude in pumping duration was available. A 100-day pumping period prior to shaft sinking into each aquifer of interest was selected.

A number of symmetrical well arrangements with several pumping rate variations, including 1- through 8-well systems, were initially evaluated. A minimum distance of 46 m (150 ft.) from the center-line of each shaft was allowed because of congestion of drilling and grouting equipment, head frame, and other construction equipment near the shaft collar. Based upon depressurization requirements, groundwater velocity control, economic considerations, pump availability and well construction considerations, 4- and 6-well systems for all aquifers above the Westwater Canyon, and an 8-well system for the Westwater Canyon, were found to provide the most viable solutions. Various arrangements of wells were analyzed using digital computer calculations to determine critical groundwater velocities within 7.6 m (25 ft.) of shaft center-lines.

Recognizing the likely variations in aquifer permeability and porosity, and to provide a margin of safety, average groundwater velocity in the aquifer was designed to be at least one order of magnitude less than the 0.61 m/day (2 ft./day) required. The optimum arrangements of the systems are summarized in Fig. 4. Table VII gives the construction details for the recommended system.

It was recognized that the designs presented were preliminary and that some adjustments of recommended pumping rates, and, therefore, recommended pumping equipment, would be required. A program of testing of the first well completed in each aquifer was recommended to more narrowly define aquifer hydraulic parameters, to identify boundary conditions, if any, and to evaluate well losses and sand problems.

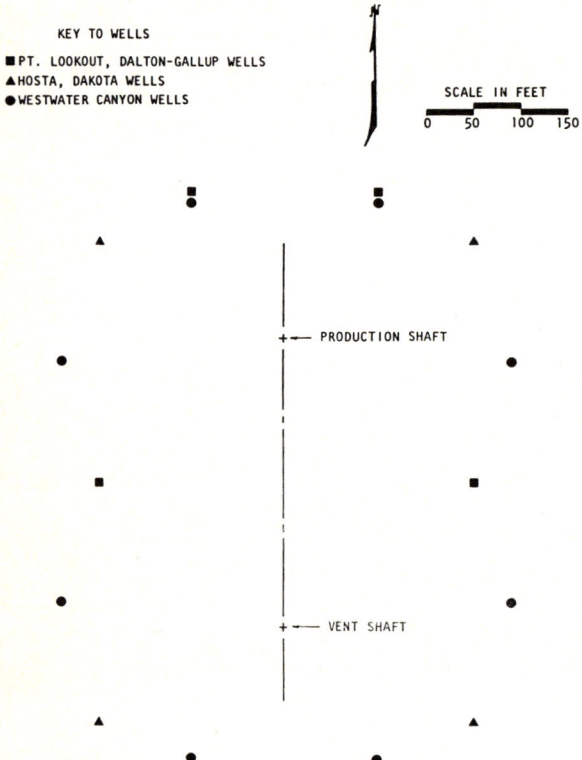

Fig. 4. Plot plan depressurization wells.

TABLE VII

Summary of recommended depressurizing system

Aquifer	Number of wells	Casing diameter (in.)	Well depth (ft.)	Open or screened interval (ft.)	Pumping rate per well (gal. min.$^{-1}$)	Total dynamic head* (ft.)	Head reduction (%)	Average velocity (ft. day^{-1})	Construction comments
Point Lookout	6	$8\frac{5}{8}$	650	470–650	200	570	83	0.08	single completion; no cementing
Hosta	4	$10\frac{3}{4}$	1,069	983–1,069	170	1,040	66	0.2	single completion; cement casing
Dalton–Gallup	6	$8\frac{5}{8}$	1,850	1,525–1,660 1,738–1,850	260	1,540	70	0.06 0.2	pull Point Lookout casing and screen; deepen to Dalton–Gallup for multiple completion
Dakota	4	$8\frac{5}{8}$	2,864	2,630–2,864	390	2,700	68	0.1	deepen Hosta wells to Dakota; single completion
Westwater Canyon	8	$8\frac{5}{8}$	3,400	3,013–3,400	440	2,900	78	0.06	single completion; cement casing

1 in. = 2.54 cm; 1 ft. = 0.3048 m; 1 gal. min.$^{-1}$ = 6.309·10^{-5} m^3 s^{-1}; 1 ft. day^{-1} = 0.3048 m day^{-1}.
*Includes discharge pressure of 13.8·10^{-2} Pa (20 lb. in.$^{-2}$) at well head.

PREDICTED vs. ACTUAL CONDITIONS

Partial results only are available, as shafts have progressed only through the upper two aquifers (Menefee and Point Lookout sandstones). Point Lookout Sandstone depressurizing wells have reduced hydrostatic pressures to, or below, predicted levels. Point Lookout wells began pumping on December 7, 1977. Average initial pumping rate was 11.5 l s^{-1} (182 gal. min.$^{-1}$) vs. a predicted 12.6 l s^{-1} (200 gal. min.$^{-1}$). Pumping levels were pulled down to, or near, pump intakes within the first few days of pumping and discharge valves had to be partially closed to prevent pumps from cavitating. Average well yield fell to 9.1 l s^{-1} (144 gal. min.$^{-1}$) after 140 days. The Point Lookout was grouted from the first shaft prior to sinking into the aquifer. On April 7, 1978, the shaft penetrated the aquifer and encountered a water inflow of 10.4 l s^{-1} (165 gal. min.$^{-1}$). Average yields from the wells dropped sharply to 8.2 l s^{-1} (130 gal. min.$^{-1}$) indicating that the grouting had not eliminated the hydraulic connection between the shaft and depressurizing wells. The second shaft encountered an aquifer on May 9, 1978, after 156 days of pumping. Inflow to this shaft was 1.26 l s^{-1} (20 gal. min.$^{-1}$). Average well pumping rate declined further, to 8.1 l s^{-1} (128 gal. min.$^{-1}$).

Analysis of the data from the Point Lookout aquifer indicated that depressurizing the system has effectively reduced hydrostatic heads in the shaft area to, or near, predicted levels. Well pumping rates are less than predicted, indicating that conservatism applied to transmissivity values was unnecessary. Measured inflows to the shafts from the Point Lookout aquifer were approximately one-half those predicted to occur without depressurization.

ACKNOWLEDGEMENTS

The authors wish to thank Phillips Uranium Corporation for permission to publish the data contained in this paper. Review of the manuscript by Pat Dominico and Gil Cochran is also appreciated.

REFERENCES

Aron, G. and Scott, V.H., 1965. Simplified solutions for decreasing flow in wells. Proc. Am. Soc. Civ. Eng., J. Hydraul. Div., 91(HY5): 1—12.
Cooper, Jr., H.H. and Jacob, C.E., 1946. A generalized graphical method for evaluating formation constants and summarizing well-field history. Am. Geophys. Union Trans., 27(4): 526—534.
Jacob, C.E., 1947. Drawdown test to determine effective radius of artesian well. Trans. Am. Soc. Civ. Eng., 112: 1047—1070.
Theis, C.V., 1935. The relation between the lowering of the piezometric surface and the rate and duration of discharge of a well using ground water storage. Am. Geophys. Union Trans., 16: 519—524.
Turcan, A.N., 1963. Estimating the specific capacity of a well. U.S. Geol. Surv., Prof. Pap. 450-E.

*1 gal.min^{-1} = 5.4432 m^3 day^{-1}.

[2]

GEOTHERMAL WELL TESTING

T.N. NARASIMHAN and P.A. WITHERSPOON

Earth Sciences Divison, Lawrence Berkeley Laboratory, University of California, Berkeley, CA 96720 (U.S.A.)
Department of Materials Science and Mineral Engineering, University of California, Berkeley, CA 96720 (U.S.A.)

(Accepted for publication April 27, 1979)

ABSTRACT

Narashimhan, T.N. and Witherspoon, P.A., 1979. Geothermal well testing. In: W. Back and D.A. Stephenson (Guest-Editors), Contemporary Hydrogeology — The George Burke Maxey Memorial Volume. J. Hydrol., 43: 537—553.

Just as in the case of hydrogeology and petroleum engineering, well testing is an invaluable tool in assessing the resource deliverability of geothermal reservoirs. While the techniques of production testing and interference testing already developed in hydrogeology and petroleum engineering provide a strong foundation for geothermal well testing, the latter is challenged by some special problems. These special problems stem primarily from the difficulties associated with the measurement of mass flow rate, pressure and temperature under the hostile environment prevalent within geothermal wells. This paper briefly looks into the state-of-the-art of geothermal well testing and provides a few illustrative field examples.

INTRODUCTION

Well test, or aquifer test, as it is commonly understood in hydrology, is mainly carried out to evaluate the hydraulic parameters of the groundwater reservoir. It essentially consists in producing water at controlled rates from one or more wells and simultaneously monitoring water level or fluid pressure changes in the producing well(s) and/or neighboring observation wells. The data so collected are then interpreted in terms of time, distance from the production well and other factors to arrive at quantitative estimates of the parameters of the reservoir: reservoir geometry; leakage of water from adjoining groundwater bodies; hydraulic efficiency of the well and so on.

Traditionally, hydrogeologists have been concerned with groundwater systems seldom hotter than 60°C, from which the required data can be collected relatively easily and interpreted. A large body of literature currently exists both in hydrogeology and in the allied discipline of petroleum engineering on designing, executing and interpreting such well tests.

In recent times, especially within the past decade, there has grown, in the U.S.A. and elsewhere, a tremendous interest in the exploration and exploitation of geothermal groundwater systems. These systems are generally characterized by temperatures of up to 360°C or more. They may either contain water entirely in the liquid phase (Raft River Valley, Idaho; Imperial Valley, California; Cerro Prieto, Mexico); or contain water entirely in the steam phase (The Geyser in northern California; Larderello in Italy); or may have water and steam coexisting within the reservoir (Wairakei, New Zealand during the exploitation phase). The elevated temperatures, the presence of more than one fluid phase and the high concentrations of dissolved gases and solids usually present in geothermal fluids render geothermal well testing an especially difficult field task. Although well-testing experience so far gained in hydrogeology and petroleum engineering has provided a strong foundation, considerable research and development remains to be carried out in order to meet the problems peculiar to geothermal reservoirs.

The purpose of this paper is to briefly evaluate the state-of-the-art of geothermal well testing. In scope, the paper will be restricted to hydraulic tests conducted after well completion and development.

NATURE OF REQUIRED FIELD DATA AND THEIR MEASUREMENT

From the point of view of reservoir dynamics, field data primarily sought after in geothermal well testing include: (a) mass flow rates from production or injection wells as a function of time; (b) variations in reservoir fluid pressure as a function of space and time; and (c) the temperature of the reservoir fluid as a function of space and time. Of the three categories of data listed, variation of reservoir temperature even after several months of well testing may generally be so small that for normal transient well-test analysis it is usually sufficient just to know the static distribution of temperature rather than its dependence on time. Recent studies by Lippmann et al. (1978) suggest that in geothermal reservoirs dominated by horizontal flows, the isothermal assumption is a reasonable basis for analysis. Data on mass flow rate and fluid pressure variations constitute, therefore, the most important well test data to be collected from geothermal systems. Additionally, it may be possible, in some cases, to collect such ancillary data as the content of dissolved solids or dissolved non-condensible gases in the geothermal fluids. While these ancillary data help in an improved understanding of the reservoir as a whole when considered in conjunction with the well test data, they are not essential for interpretation of reservoir dynamics.

Measurement of mass flow rates

Many geothermal wells, with water temperatures exceeding 150°C, are known to be self-flowing. The well-head pressures in these wells, while the well is discharging, may generally be less than the saturation pressure corresponding to the fluid temperature. As a result, the boiling effluent from the well is a two-phase mixture of steam and water, unless special efforts are taken to apply sufficient back pressure on the effluent to prevent boiling. The "steam quality" or the mass proportion of steam to that of water is a function of the water temperature and the exit pressure. The steam quality at the well head may vary from less than 1% in liquid-dominated systems to more than 99% in vapor-dominated systems. In these wells, the initiation of flashing may take place either in the discharge pipe, at the well head, or may take place at depths of several hundred meters below ground level, depending on the pressure—temperature regime of the flowing fluid.

In order to measure the mass flow rates of such two-phase effluents one could, when possible, maintain sufficient back pressure on the effluent through the use of orifice plates to assure single-phase flow. The flow rate can then be computed by measuring the pressure drop across the aperture. The measurement of flow rates of water by means of orifice plates in water wells (Anderson, 1977) as well as in gas wells (Frick and Taylor, 1962) are well known. Witherspoon et al. (1978) used the aforesaid technique for measuring hot-water flow rates (150°C water) in The Raft River Valley, Idaho. A similar technique is routinely used in The Geyser's geothermal field of northern California to measure steam flow rates.

When sufficient back pressure cannot be applied to prevent the formation of steam, one can measure flow rates by letting the hot-water flash at the discharge pipe, passing the two-phase mixture through a steam separator, running the separated steam and water phases through two distinct orifice meters and measuring the flow rates separately. A schematic diagram of this method, as used by the Republic Geothermal Inc. at East Mesa in the Imperial Valley of California is shown in Fig. 1. This technique was successfully used to measure total mass flow rates of up to 55 kg/s with steam quality of approximately 15%. In using orifice plates, it is customary to record the pressure differentials across the orifice on continuous recording charts providing a permanent document of flow variations.

A less accurate, but acceptable method of measuring flow rates may be to pass the water phase coming out of the separator through a V-notch weir and to evaluate the liquid-phase flow rate from the level of fluid at the V-notch. Simultaneously, the temperature and pressure are monitored at the separator, from which the steam quality is estimated. With steam quality and the flow water of the liquid fraction known, the total mass flow rate can be computed. Obviously, this method may not be very accurate, especially when steam quality is high.

An approximate but inexpensive and extremely useful method of measuring total mass flow rate and the heat content (enthalpy) of a two-phase geothermal effluent from a discharge pipe was suggested by James (1963—1964) and has been used in several geothermal fields. This method, developed from the concept of critical flow of fluids (that is, when a compressible fluid travels at the velocity of sound), consists in flowing the geo-

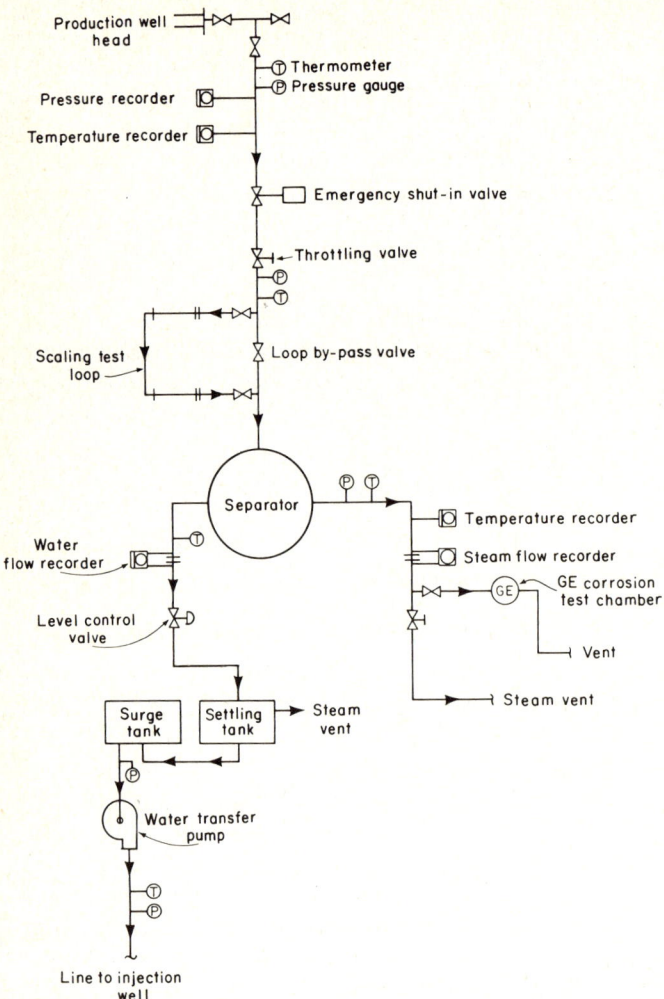

Fig. 1. Schematic of measuring mass flow rate of water and steam separately (courtesy of Republic Geothermal, Inc.).

thermal fluids to the atmosphere through a pipe of uniform cross-section and measuring the fluid pressure a fraction of a centimeter behind the lip of the discharge pipe. The lip pressure is then used to estimate the total mass flow rate, the mixture enthalpy and steam quality provided that the liquid rate can be measured (Ramey, 1978). A diagram of a setup for implementing the James technique is shown in Fig. 2.

In using the orifice meters of the James technique, a problem that is often encountered is that of scaling. Due to the sudden decrease of pressure downstream of the orifice, or due to shock phenomena accompanying critical flow, calcium carbonate and other materials may be deposited causing scale

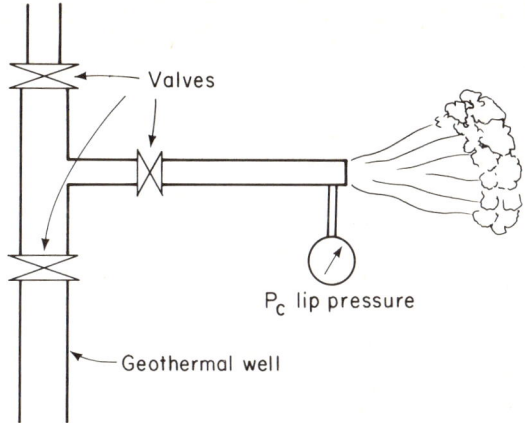

Fig. 2. Schematic setup for measuring mass flow rates by the James method.

formation. These scales will often reduce the aperture diameter and affect flow-rate computations. In order to minimize errors due to this problem during well tests, one could have two sets of orifice meters with a by-pass arrangement or use replaceable orifice plates.

The problem of variable flow rate

For ease of interpreting well-test it is often most desirable to conduct the well test at a constant rate or a step-wise rate, changing from one constant rate to another. However, it may not often be possible to maintain constant flow rates from geothermal wells. Some geothermal wells may no longer be self-flowing when they are shut-in and allowed to cool. In these wells, the increase in the weight of the cool-water column may be such that the natural flow is choked. Indeed, at Niland in the Imperial Valley of California where the geothermal brine has total dissolved solids in excess of $25 \cdot 10^4$ ppm, it is known (J. Morse, pers. commun., 1978) that on shutting-in and cooling the geothermal well, the fluid level can drop down to as much as 50 m within the well, creating a partial vacuum in the well casing between the fluid level and the well head. In order to make these cold wells flow once again, they may have to be stimulated by means of air lift or some other method. During the initial phases of the well production, therefore, there could be important departures from the desired constant flow rates and these departures have to be duly accounted for in interpreting the well-test data.

Measurement of fluid pressures

Other important data collected during a well test are the fluid pressures. In most well tests in hydrogeology, such data consist of water-level measurements either made manually with a steel tape or measured automatically with different types of water-level recorders.

It is quite well known that well-test analysis is based on the interpretation of pressure changes (drawdown, buildup) rather than the magnitude of absolute pressures. In a producing well where the temperature of the fluid column varies but little (as in most water wells) the pressure at the sand face is immediately reflected in water-level fluctuations in the well-bore storage effects) since the water density varies very little with depth in such wells. This, however, is not the case in a producing geothermal well. Indeed, due to the thermal gradients within the well bore and because of flashing within the well bore, the pressure changes at the sand face are considerably modified as they are transmitted to the well head. In fact, it is often noted that when a cold geothermal well begins to produce, well head pressure will actually rise for a time due to the lightening of the water column, before it begins to drop. Hence, to monitor reservoir pressure changes in the producing wells, well-head pressure measurements are of very little use and one has to rely almost exclusively on down-hole measurements. The most challenging problems of geothermal well-test instrumentation are directly related to this need for down-hole monitoring in producing geothermal wells.

Fortunately, such is not the case for non-producing observation wells. If such wells have positive well-head pressures and hence are completely filled with water, even small changes in the reservoir pressure are transmitted instantaneously to the surface due to the very low compressibility of water. Thus, in these wells, reservoir pressure changes can be accurately monitored by measuring well-head pressures. Witherspoon et al. (1978) were successful in using such well-head measurements on non-producing wells to obtain interference test data in The Raft River Valley, Idaho and at East Mesa California.

Down-hole pressure monitors

Currently, three different types of down-hole pressure monitors are used for down-hole pressure measurements in geothermal wells. The first of these is the "bomb-type" device (Amerada bomb; Kuster bomb) which is lowered by means of a wire line to any desired depth within the well. Essentially the bomb-type device consists of two parts; a pressure monitor and a timing device. The pressure monitor is a Bourdon® tube type instrument connected to a time-driven stylus. The stylus, driven by the clock along one axis and by the pressure sensor along the other, scratches the pressure—time curve on a small glass plate. On retrieval of the plate, the pressure—time data can be read off from the etched line with the help of a magnifier.

The bomb-type devices are compact (~ 2.5 cm in dia., and 1.8 m long) and are simple and robust enough to withstand temperatures as high as 300°C. They also measure the actual reservoir pressure variations at the point of observation. However, they have two disadvantages. Firstly, the resolution of the etched data is very much related to the total pressures (e.g., 2% of full scale). Thus, if the reservoir pressure is 17.2 MPa (2,500 lb. in.$^{-2}$). which can be expected in a 1800-m well, the accuracy is only about 0.03 MPa (5 lb. in.$^{-2}$). Secondly, once the bomb is lowered into the well, data will become available only on retrieval of the bomb and examination of the glass plate. Therefore, little is known about the success or progress of the test during the test itself. This is a significant handicap, since, ideally, one would like to modify the pattern of the test depending on the reservoir response observed during its progress.

A second type of down-mole pressure monitoring system is the Sperry-Sun® system which consists in monitoring the fluid pressure at the sand face but making the read-out at the land surface. This system is schematically shown in Fig. 3. A narrow tube (0.75—1.25 cm dia.) made of a corrosion-resistant alloy such as stainless steel or Inconel® is lowered to the desired depth in the well. At the bottom, the tube is attached to a chamber (~ 2.5 cm in dia.) through a Micropore® filter. The chamber, in turn, communicates with the well fluid through one or two openings of suitable size. At the well head, the tubing is connected to a pressure transducer.

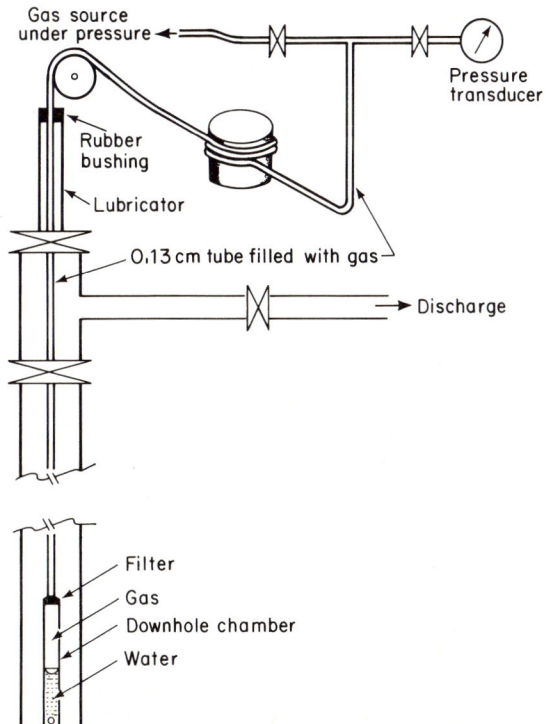

Fig. 3. Schematic of down-hole pressure monitoring setup using the gas-transmitter system.

In order to set up the system, the tubing is first lowered to the desired depth and a suitable gas (nitrogen or helium) is passed through it under pressures in excess of the expected reservoir pressure. When this is continued for several hours, the well fluids are expelled from the chamber and the gas will begin to bubble into the well. At this time, the supply of gas is cut off and the reservoir fluid, under existing pressure, compresses the gas and flows into the system. When carried out properly, the interface between the fluid and the gas will exist within the chamber. In this state, there exists a finite volume of gas in the tube—chamber system. As the reservoir pressure changes, the fluid—gas interface in the chamber will also change, causing the gas volume to increase or decrease. This volume change is reflected as a pressure change in the pressure transducer at the well head. Obviously, the pressure monitored at the well head will differ from the down-hole pressure by an amount equal to the weight of the column of gas and an appropriate correction has to be made in this regard. This pressure monitoring system has been used with success in geothermal wells in the Imperial Valley of California and elsewhere.

Because of its simplicity and the absence of any precision device downhole, this system has no temperature limitations, provided that the temperature gradient within the well is stable. However, during the early stages of production when temperatures are changing within the well, this system has some drawbacks.

Narasimhan et al. (1978) found during a geothermal well test at East Mesa, California, that the nitrogen pressure monitored by the Sperry-Sun® system indicated an increase in the fluid pressure as much as 0.86 MPa (125 lb. in.$^{-2}$) during the first few hours of production while in fact pressure would be expected to show a decline with production. This anomalous increase in pressure was attributed to the heating up of the nitrogen tubing concomitant with the well flow. To some extent this effect was later reduced in a subsequent test by increasing the tubing diameter from 0.8 to 0.13 cm (0.31 to 0.54 in.). Another way of minimizing the temperature effect on the gas is to replace the gas in the tubing with an inert liquid such as Silicone® oil, since oil has much lower thermal expansivity than gas. Nevertheless, even in this case the early pressure data is noticeably perturbed by thermal effects (R.C. Schroeder, pers. commun., 1979).

Miller and Haney (1978) have studied the transmission of a pressure-change signal in a fluid-filled capillary tube. They found that the signal is noticeably distorted if there is temperature change along the tubing or if the pressure transient is large. While even a small change in temperature (0.5°C) can accentuate the distortion, increasing the diameter of the tubing (from 0.07 to 0.25 cm) tends to decrease it. Of the two fluids, Silicone® oil and nitrogen, oil is preferable at high but steady temperatures while nitrogen is recommended when temperature varies significantly with time. In order that one could estimate sand-face pressure changes from well-head pressure data, not only must the exact time of flow-rate changes, but also the time-dependent variation of temperature within the well must be known.

The third category of pressure measuring devices is the precision quartz crystal device that has become practical in the past few years, thanks to the progress in electronic technology. This device makes use of the extremely predictable piezo-electric property of quartz crystal. Essentially, the pressure sensor consists of a carefully chosen natural untwinned quartz crystal. The crystal is cut and drilled to have a precise shape with reference to its crystallographic orientation. The resonance frequency of this crystal is very closely related to its shape. Hence, as the crystal is deformed under even very small pressures, its resonance frequency changes in a detectable and predictable fashion. Thus, the instrument consists of an electronic oscillator circuit whose frequency is controlled by the quartz sensor and an external frequency counter. The actual pressure measurement consists of counting the frequency of vibration of the crystal system at the existing pressure and reading the corresponding pressure from the calibration tables.

Inasmuch as the frequency of quartz is also dependent on temperature, an accurate knowledge of the temperature at the point of measurement is essential for an accurate pressure determination. Ideally, a simultaneous down-hole temperature measurement is very desirable in using the device, which has an accuracy of about 70 Pa (0.01 lb. in.$^{-2}$) over a pressure range of 70 MPa (10^4 lb. in.$^{-2}$).

The down-hole quartz crystal devices invariably have a certain number of electronic components associated with them. Although the quartz crystal itself possesses very regular frequency properties up to temperatures of 260°C or more, the electronic components can seldom withstand temperatures of more than 150°C. While reliable down-hole pressure measurements can be made at temperatures of up to 150°C (Witherspoon et al., 1978) over prolonged periods of time using these devices, such measurements at higher temperatures are not possible at present. Recent research at the Sandia Laboratories (Veneruso, 1977) indicates that the use of field effect transistors (FEP) instead of popular bipolar silicon transistors can extend the application of quartz crystal devices to temperatures of 325°C.

In the environment where they can be used, however, the quartz crystal devices can provide data of extreme precision at very frequent intervals which can enable some very sophisticated reservoir interpretations that have not been possible so far. The pressure data collected from an observation well located 1200 m from the producing well in the Raft River geothermal area of Idaho are presented in Fig. 4. The data are collected with a quartz crystal gauge placed at a depth of about 300 m below surface. In Fig. 4, one can clearly discern periodic fluctuations in pressure which are to be attribted to effects of earth tides (Witherspoon et al., 1978). In addition, one can also see the long-term pressure decline caused by the interference effect of the production well. The availability of such precise data as shown in Fig. 4 opens up the possibility of analyzing the response of aquifers to earth tides as a means of determining reservoir parameters.

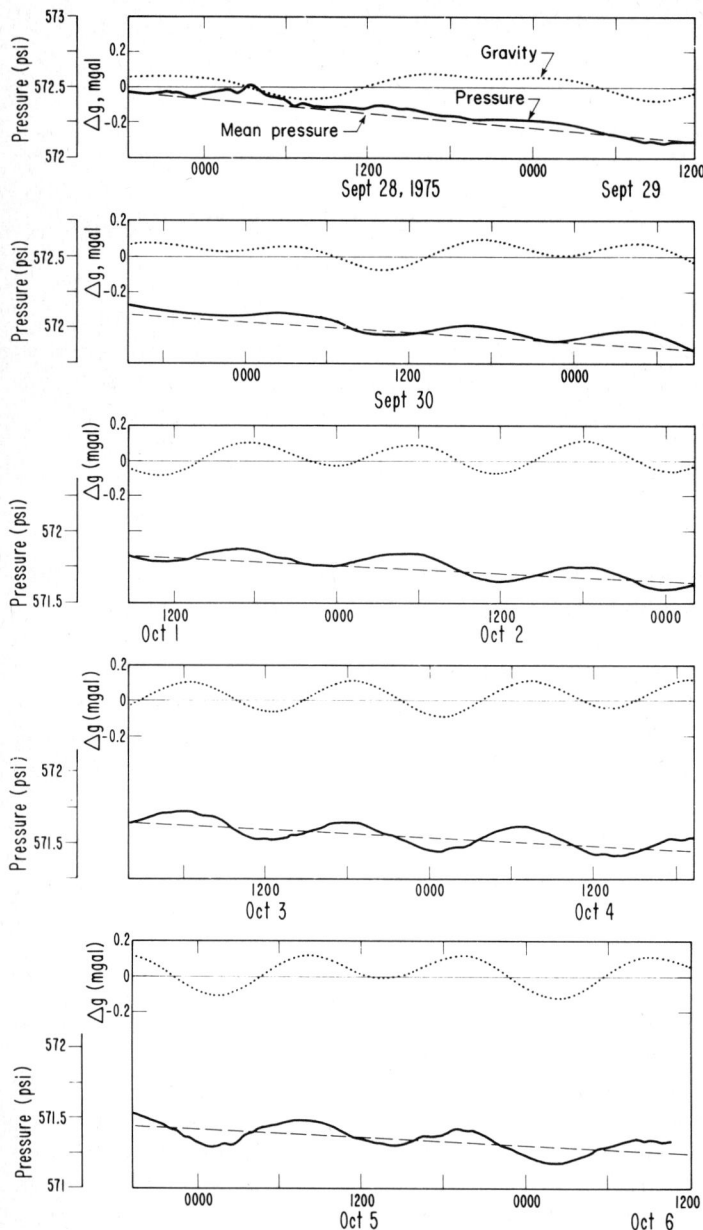

Fig. 4. Correlation between reservoir pressure and computed changes in the Earth's gravity field in a geothermal well in the Raft River Valley of Idaho.

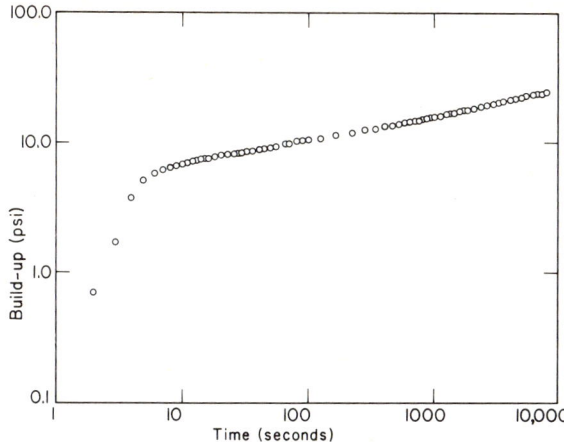

Fig. 5. Buildup data from a geothermal well in Raft River Valley, Idaho, showing resolution of early-time data.

In order to give an idea of the frequency with which drawdown or buildup can be measured with the electronic devices, the buildup data from a producing geothermal well in the Raft River Valley of Idaho is given in Fig. 5. Note from this figure that within the first 60 s more than 40 data points are available for analysis. In this particular case the instrument was located at a depth of 1,500 m where the reservoir temperature was about 148°C. Although in this case the early-time data failed to reveal the presence of significant well-bore storage effects (unit slope) or the presence of fractures (half slope), the value of such early-time data in well-test interpretation cannot overlooked.

SOME FIELD SAMPLES

The first example relates to an interference test conducted at the Raft River Valley geothermal field by Witherspoon et al. (1978). The interference data from this test is given in Fig. 6. The primary problem in analyzing this data was to eliminate the perturbation on the fluid pressure data caused by earth tides. The fact that the data showed an excellent correlation between computed earth tides (dotted line in Fig. 4) and the fluid pressure in the well, enabled a simple method of elimination of earth-tide effects. Thus, one had only to choose the pressure values corresponding to those instants when the computed earth-tide effect was zero and join those points by a smooth line (Fig. 4). This smooth line was then used for interpretation using the conventional Theis type-curve analysis, as in Fig. 6. Note from Fig. 6 that a barrier boundary is clearly suggested by the drawdown data.

The next illustration relates to a step-drawdown test conducted on a producing geothermal well at East Mesa in the Imperial Valley of southern

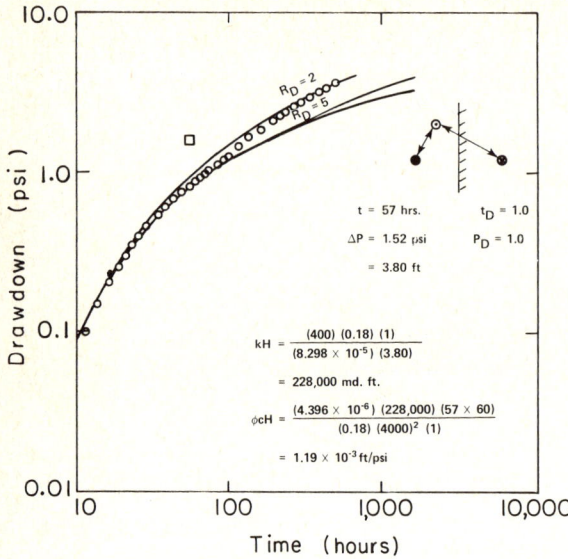

Fig. 6. Interpretation of interference test between two geothermal wells in the Raft River Valley, Idaho.

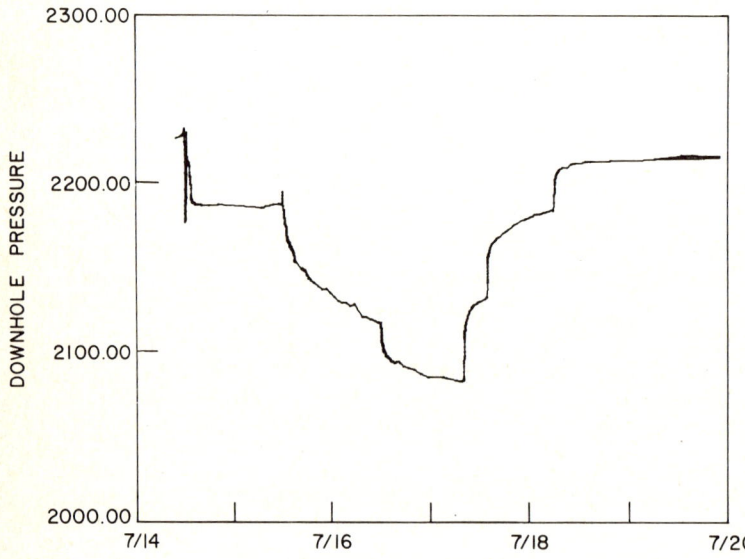

Fig. 7. Step-drawdown data from a geothermal well at East Mesa in California.

California (Narashimhan et al., 1978). In this well, about 2200 m deep, down-hole pressure data were monitored during production with a Sperry-Sun® nitrogen gas system. The pressure data so collected is shown in Fig. 7. Note from Fig. 7 that at the commencement of production and at each time the flow rate was raised, the measured pressures show a perceptible increase. This is due to the fact that every time the flow rate is increased, the temperature regime in the well also increased, causing heating of the nitrogen gas.

As can be inferred from Fig. 7, the flow rate during each step was variable and as such the data cannot be reliably analyzed using constant flow rate concepts. Therefore, the data were analyzed using the variable flow rate analysis of Tsang et al. (1977). This computer-assisted analysis was used to treat all the data in Fig. 7 (drawdown as well as buildup). The analysis indicated a permeability—thickness product (kH) of 730 mD m (millidarcy meters).

Note that in the case of geothermal reservoirs, the conventional concept of transmissivity (which is defined for water at a temperature of 16°C) is of little use since different reservoirs have markedly different temperatures and hence markedly different fluid viscosities. Therefore, a more meaningful parameter to use is the absolute permeability, k (millidarcies) or the product, kH.

The next example serves to illustrate the fact if detailed interference data are available from a single-phase geothermal reservoir, one could use the hydrogeological techniques of well-test interpretation to decipher reservoir

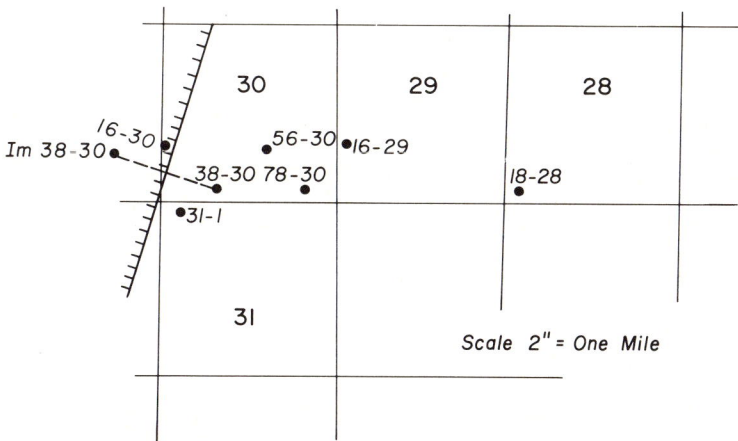

Fig. 8. Well-test analysis leading to inference of barrier boundary, East Mesa geothermal reservoir, California.

geometry. Several interference tests were conducted in the East Geothermal Field at East Mesa (Narasimhan et al., 1978) using multiple observation wells equipped with quartz crystal pressure devices. The layout of the well field can be seen in Fig. 8.

During the interference tests, wells *38-30* and *16-29* served alternately as production wells while wells *31-1*, *56-30* and *16-30* were used as observation wells. The data from *31-1* and *56-30* clearly picked up the effect of a barrier boundary and arcs drawn from these wells showed two possible locations for the image well. However, observations made on well *16-30* showed that this well did not show any communication with *38-30* or *16-29*. Based on this third piece of information the image well location was uniquely chosen as shown in Fig. 8 and the location of the barrier boundary was fixed

The next illustration pertains to a non-artesian geothermal well in the former French territory of Afars and Issas in Africa (Gringarten, 1978). This well was characterized by two-phase flow in the well bore although the reservoir itself remained water-dominated. The 1130 m deep well pierces a reservoir with a temperature of about 250°C and a pressure of about 8.3 MPa (1,200 lb. in.$^{-2}$). When shut-in and cold, the well is non-flowing with fluid level at about 200 m below the surface. On stimulation, the well could self flow at rates varying from 6.3 to 23 kg/s with the flash point (boiling point) within the well varying between 700 and 870 m below ground level. The well was subjected to seven different flow periods ranging from a few hours to several days with intervening shut-in durations. The flow rates were variable as already indicated. During the tests, a bomb-type pressure device was used to measure down-hole pressure at a depth of 1050 m. A careful study of the pressure data showed that well-bore storage effects were clearly discernable during early times of drawdown and buildup. Moreover, two of the tests, involving low flow rates, indicated a change in the nature of the well-bore storage effect, from one in which it was controlled by the liquid level change to one in which it was controlled by the compressibility of the steam—water mixture. Gringarten showed that if sufficiently long-duration data are available so that the semi-log approximation to the Theis equation is valid, then one can compute the reservoir permeability (in this case kH = 16 D m), and evaluate the extent of well-bore damage. To take into account the variable flow rate prevalent during the test, Gringarten was able to use a technique proposed by Odeh and Jones (1974), which essentially consists in superposing the semi-log solutions corresponding to each flow segment. It was found during this test that well-bore effects lasted for a longer period during production than during buildup. Moreover, after ~ 6 h of shutting down production, the buildup data was so markedly perturbed by effects of steam condensation that it was of no more practical use in interpretation.

A log-log plot of the pressure data collected during the different tests is presented in Fig. 9.

The final illustration pertains to a geothermal well test from a steam well at The Geyser's geothermal field in northern California. It is very well known

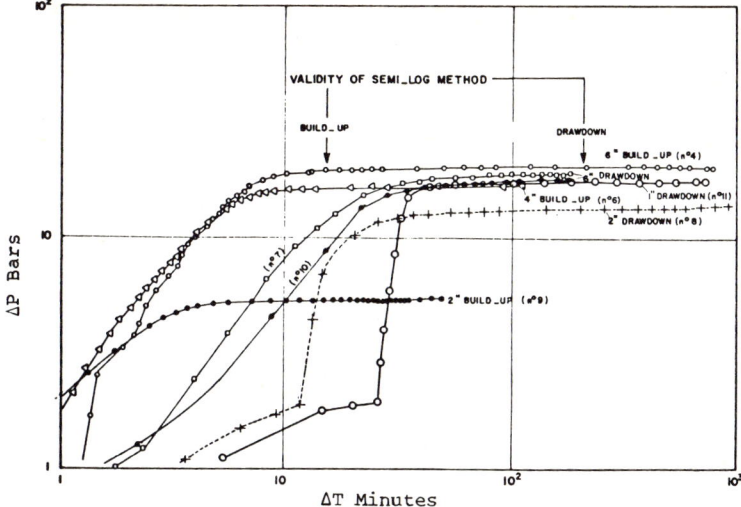

Fig. 9. Log-log plot of pressure transient data from several tests on a two-phase geothermal well in Africa (after Gringarten, 1978).

in the petroleum literature that the radial flow solutions valid for liquid-filled systems (aquifers, oil reservoirs) can be applied to gas-filled systems if one were to replace drawdown, p by the quantity $(p^2 - p_i^2)$ where p is the pressure at a given time and p_i is the initial static pressure. Using this approach, Ramey and Gringarten (1975) analyzed data from a steam producing well at The Geyser, as illustrated in Fig. 10. Note that the data could be analyzed to compute such information as reservoir permeability,

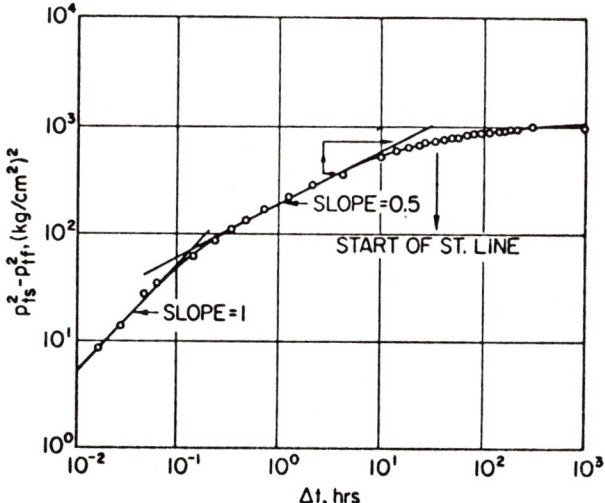

Fig. 10. Well-test interpretation of a steam well at The Geyser (after Ramey and Gringarten, 1975).

storativity, well-bore storage (unit slope section, and presence of fractures (half-slope section).

CONCLUDING REMARKS

Geothermal well testing is, in many respects, similar to well testing as practiced in hydrogeology and petroleum engineering. Experience so far gained in testing geothermal wells shows that under many field situations it is still possible to directly apply conventional well-test theory to evaluate geothermal reservoir parameters, geometry and well-bore damage when the reservoir is either liquid-dominated or vapor-dominated. The greatest challenge to geothermal well testing, then, consists in developing instrumentation to provide down-hole pressure data from producing geothermal wells. This implies that we need instruments with ability to measure down-hole pressures accurately over prolonged periods of time at temperatures in excess of $150°C$.

On the theoretical side, existing well-testing techniques are inadequate to handle those conditions in which the well pierces a reservoir under two-phase conditions. Considering the fact that geothermal systems are usually several hundred meters thick, it is likely that a single well may be exposed to pure water at the bottom and steam at the top of the reservoir. The testing procedure becomes complicated in this case and it may be essential to make pressure measurements at more than one location within the well for proper interpretation. Only very recently have workers turned their attention to a systematic study of evaluating two-phase geothermal reservoirs. As a first step in the study, Garg (1978) and Pruess et al. (1978) have considered a simple system in which the flash front is vertical and moves laterally outward from the well. Their study shows that under certain assumptions, the pressure transient behavior in such a system is similar to the semi-log solution frequently encountered in hydrogeology. Nevertheless, these studies have not yet been applied to actual field situations. Further research work in this area should prove to be of great practical importance.

ACKNOWLEDGEMENTS

We thank R.C. Schroeder, D.G. McEdwards and B.Y. Kanehiro of Lawrence Berkeley Laboratory for constructive criticisms and suggestions, This work was supported by the Division of Geothermal Energy of the U.S. Department of Energy, under contract number W-7405-Eng-48.

REFERENCES

Anderson, K.E., 1977. Water Well Handbook. Missouri Water Well and Pump Contractors Association, Inc., Rolla, Mo.

Frick, T.C. and Taylor, R.W., 1962. Mathematics and production equipment. In: T.C. Frick (Editor-in-chief) Petroleum Production Handbook, Society of Petroleum Engineers, A.I.M.E., Dallas, Texas.

Garg, S.K., 1978. Pressure transient analysis for two-phase (liquid water/steam) geothermal reservoirs. Systems, Science and Software, Inc., La Jolla, Calif., Rep. No. SS-IR-78-3568(R2).

Gringarten, A.C., 1978. Well testing in two-phase geothermal wells. A.I.M.E., Annu. Meet., Houston, Texas, Pap. No. SPE 7480.

James, R., 1963—1964. Maximum steam flow through pipes to the atmosphere. Proc., Inst. Mech. Eng., 178 (18): 473—483.

Lippmann, M.J., Bodvarsson, G.S., Witherspoon, P.A. and Rivera, J.R., 1978. Preliminary simulation studies related to the Cerro Prieto field. Proc. 1st Symp. on Cerro Prieto Geothermal Field, Baja California, Mexico, Lawrence Berkeley Laboratory, Berkeley, Calif., (in press).

Miller, C.W. and Haney, J., 1978. Response of pressure changes in a fluid-filled capillary tube. Proc. 2nd Invitational Well-Testing Symposium, Lawrence Berkeley Laboratory, Berekeley, Calif., Oct. 1978.

Narasimhan, T.N., Schroeder, R.C., Goranson, C.B. and Benson, S.M., 1978. Results of reservoir engineering tests, 1977, East Mesa, California. Soc. Pet. Eng., A.I.M.E., 53rd Annu. Meet., Oct. 2—5, 1979, Houston, Texas, Pap. No. SPE 7482.

Odeh, A.S. and Jones, L.G., 1974. Two-rate flow test, variable rate case — application to gas lift and pumping wells. J. Pet. Technol., 26: 93—99.

Pruess, K., Schroeder, R.C. and Zerzan J.M., 1978. Studies of flow problems with SHAFT 78. Pap. presented at 4th Annu. Geothermal Workshop, Stanford, Calif., Dec. 1978.

Ramey, H.J., 1978. Hand computer program for James' lip pressure steam flow rate. Trans., Geothermal Resour. Counc., 2: 555—557.

Ramey, H.J. and Gringarten, A.C., 1975. Effect of high volume vertical fracture on geothermal steam well behavior. Proc. 2nd U.N. Symp. on Development and Use of Geothermal Resources, San Francisco, Calif., 3: 1759—1762.

Tsang, C.F., McEdwards, D.G., Narasimhan, T.N. and Witherspoon, P.A., 1977. Variable flow well test analysis by a computer assisted matching procedure. Soc. Pet. Eng., A.I.M.E., Bakersfield, Calif., Pap. No. SPE 6547.

Veneruso, A.F., 1977. High temperature instrumentation. Proc. 1st Invitational Well-Testing Symposium, Lawrence Berkeley Laboratory, Berekeley, Calif., Rep. No. LBL-7027, pp. 45—51.

Witherspoon, P.A., Narasimhan, T.N. and McEdwards, D.C., 1978. Results of interference tests from two geothermal reservoirs. J. Pet. Technol., 30: 10—16.

[2]

GROUNDWATER: NEW DIRECTIONS
Where we've been and where we're going

RAPHAEL G. KAZMANN

231 du'Plantier Boulevard, Baton Rouge, LA 70808 (U.S.A.)

(Accepted for publication March 6, 1979)

ABSTRACT

Kazmann, R.G., 1979. Groundwater: new directions — Where we've been and where we're going. In: W. Back and D.A. Stephenson (Guest-Editors), Contemporary Hydrogeology — The George Burke Maxey Memorial Volume. J. Hydrol., 43: 555—569.

During Burke Maxey's lifetime the potentiometric analysis of groundwater problems, pioneered by Theis, was completely worked out by theoreticians. Such "macro-methods" are now taught in universities and are widely used. In the future they will be supplemented by "micro-methods" that are site specific and best described as involving miscible displacement: a well known application is the leach mining of copper or uranium.

Saline aquifers are already used for the storage of natural gas and as permanent containments for liquid wastes. Future uses of such aquifers will include storage of freshwater and, possibly more important, storage of hot water (water that has been heated by waste heat from generating plants or industrial plants) for later retrieval and use in space heating and other applications of low-grade heat.

THE PAST IS PROLOGUE

It seems very long ago that I first met Burke Maxey. We were very young in late 1942 and early 1943 — and so was the practice of groundwater engineering: there was not even a table of well function generally available so that you could read $W(u)$ to the second decimal place for any value of u that you might compute. L.K. Wenzel's *U.S. Geological Survey Water-Supply Paper* 887 was still in page proof. The concept of river infiltration was understood, qualitatively, but there were no accepted methods of determining whether or not it would occur. An engineer was considered to be well-qualified if he could perform a Theis analysis of an aquifer test. C.E. Jacob was just starting to work on the problem of drawdown corrections needed when tests were performed on water-table aquifers.

During Burke's professional lifetime the mother-lode of potentiometric analysis, applied to a multitude of field conditions, was pretty well worked out. Glover (1974) and before him the Symposium on Transient Ground Water Hydraulics (1963) pretty well summed up the theoretical background available for field application. The theory of groundwater movement to wells, ditches, recharge galleries, in short to "sinks" of all sorts, has been refined and re-refined. Some papers contain solutions to problems that have not, and may never, occur in the field.

During Burke's career he encountered a variety of hydrogeologic problems. It is not possible to summarize them here: his bibliography will have to suffice. One can obtain a feeling for the milieu by reading the work of McGuiness (1963) who, with the cooperation of the U.S. Geological Survey Ground Water Branch, published an overall look at the role of ground water in the national water situation. The work has not, as yet, been updated and it may be that it never will be. And even that survey did not take fully into account all of the privately prepared reports by consultants, reports by various state agencies, and reports prepared specifically for other federal agencies. Since that time a multitude of new studies on groundwater conditions in different areas has appeared — and the field of interest has broadened from potentiometric problems to problems involving more than one fluid, to mining problems, leaching problems, geochemical (osmotic) hydrodynamics, ion exchange by clays, etc. In the past five or ten years new directions in the use of what is now termed "inner space" are being manifested — but more of this later.

One of the areas of Burke's interest was in the conjunctive development of surface and groundwater with his principal focus on the geohydrology. Such developments impinge on the general field of "systems", and while the "system" concept is superficially attractive, and much work has been done on the mathematical problems involved, the practical results, as far as the writer can determine, have been trivial. The earliest attempts in this direction involved electrical analog models. Later, with the high-speed digital computers, digital models were developed. Then hybrid models, using the analog to get approximate answers and the digital to get more exact answers, came into use.

These models were rather successful in predicting water levels under various conditions of groundwater pumpage; in showing how shifts in location or pumping rate would affect conditions throughout the aquifer in question; and as a basis for predicting costs of operation.

However, when such models, including those using so-called "dynamic programming", were applied to real-world situations, the result became progressively less useful as the scale increased. What might work for a single company on a small tract, produced what might be termed "disinformation" when applied to the large-scale problems of a water district. The reason for this is not hard to understand: in the real-world operation of wells, the occurrence and availability of surface water for recharge and conjunctive use cannot be properly expressed by mathematical or analog models that are simple enough to program. The goal of such models (computer programs) is usually to "optimize" the conjunctive operation, which usually means "lowest cost operation". As everyone knows, operating costs plus capital costs constitute the annual cost of the project — target for optimization. But the target is a moving one: optimization involves economics and not only are discount rates (which translate into the annual cost of capital) changing, but so is the demography and the profitability of various commodities that use water as one factor of production. There are also wild cards like the oil cost squeeze by the OPEC monopolists, a reduction in the anchovy catch, crop failures in the U.S.S.R. (some would say this is to be expected), and new technologic developments that can either reduce or increase the need for water. As a result of these unpredictable factors most, if not all, of the dynamic programming for purposes of optimization is obsolete even before it is published or presented. By 1977 the optimization fad of the previous decade had pretty well run its course. From my discussions with Burke I am sure he bade them farewell with little regret.

During the same period aquifer testing was perfected and developed into a routine, to be learned partly in college and partly in the field. Not only were boundaries on one sort or another detected and their positions determined, but a vast effort was expended on so-called "leaky aquifers". A whole series of type-curves, similar to the well-function used by C.V. Theis, was produced and occasionally utilized. The practice of artificial recharge, both by flooding (induced river infiltration) and through the operation of injection wells, became widespread. Los Angeles County and Orange County in California provide outstanding examples of some of these practices.

The side effects of groundwater withdrawals became more important toward the end of Burke's professional life. Subsidence was recognized in a number of areas: the Central Valley of California; Houston and Galveston, Texas; and New Orleans, Louisiana — to mention a few areas where a declining water level produced subsequent compaction of the confining clays. Other causes of subsidence were epitomized by the local collapse of areas overlaying cavernous limestone, such as those in Alabama and Florida, where water levels in the aquifers were lowered and the pressure of water no longer served to support the overlying formations. In such instances subsidence was sudden and spectacular, as compared to subsidence due to the campaction of clays, which is slow and relatively unobtrusive.

A relatively minor subject, at least in the U.S.A., was the prevention of seawater intrusion. At both shores of the continent: in Long Island, New York, and at numerous localities on the Pacific Coast of California, seawater intrusion was nature's response to the withdrawal of groundwater just inland of the coast. And in much the same way that men fight their common enemy, flood, efforts were made to combat the degradation of groundwater supplies. These efforts resulted in the creation of on-shore pressure ridges produced by injecting freshwater into the invaded aquifer. And these efforts also resulted in the construction and operation of off-shore pressure troughs, created by pumping wells that drew on the saline water in the aquifer and prevented movement of salt water inland. Such projects, involving two fluids in the same aquifer, might be considered forerunners of the new uses for underground space.

Another use of underground space is, of course, the disposal of difficult industrial wastes in saline aquifers. The earliest efforts in this direction started in the 1950's with mixed success. The wells often were plugged by the wastes; pressure rises occurred and when the owners tried to overcome the pressure by the use of brute force, the wastes sometimes spewed forth on the biosphere, causing trouble to the perpetrators and bystanders. Nonetheless, deep-well disposal is currently an acceptable industrial practice which is governed by regulations that were set forth long before the critical parameters had been identified and evaluated.

The enlargement of our understanding of the geochemistry of groundwater has been long in coming. However, Jones (1969) pointed out in some detail what might be termed "geochemical hydrodynamics", the process of natural osmosis, an osmosis that utilizes clays as the semi-pervious membranes and differences in water quality (TDS) as the driving force. His work has helped us understand why reversing the original seaward flow in aquifers has not resulted in the intrusion of seawater in Gulf-fronting Louisiana or even the New Jersey coast.

The role of clay and dispersed silts and clays in the geochemistry of aquifers received its first serious attention in the late 1960's and early 1970's. The role of montmorillonite and illite clays, and their abilities to replace the calcium in water with sodium, thus softening it, have given hydrogeologists and groundwater engineers the idea of injecting hard water into an aquifer and pumping out soft water. Although, to be objective, if the association of the occurrence of heart disease with the use of soft water (lacking calcium) is ever proven, perhaps such ideas will die aborning.

Probably enough has been said to indicate that the subject of fluids in the natural pore spaces of earth materials is still in its infancy and that the past is indeed prologue to a challenging and productive future.

THE FUTURE IS A-BIRTHING

It must not be thought that future developments in groundwater hydrology will make obsolete all of our current practices and uses. What will probably happen is that the new developments will be added to methods currently in use. On an educational level, instead of a few courses in groundwater geology and groundwater engineering, a suite of specialized courses will ultimately be created, bringing the now-esoteric ideas into common practice. This will require at least another generation.

A characteristic of present-day groundwater evaluation is the averaging of hydraulic and geologic properties. We tend to use averages and act, for most part, as though the mathematical model based on the average properties of an aquifer properly portrays the widely differing properties of the erratically stratified, heterogenous formation that actually exists. And, surprisingly enough, for most of the studies made thus far, both the quantities of water produced and potentiometric changes accompanying such productions have been predicted with sufficient reliability for useful applications, thus forming the basis for extensive investments of capital and talent. Such predictions, for example, can indicate to a city, or industry, the drop in water level to be expected at any specified rate of water production (thus enabling pump selection and estimates of power cost). Or the existence of recharge boundaries can be determined, such recharge boundaries can be utilizes for the induced infiltration of river water to wells (as in Miami, Ohio; Canton, Ohio; East St. Louis, Illinois; the Los Angeles County water spreading grounds, etc.). Or the lack of recharge from a stream, or even rainfall, can be documented and the proper conclusions drawn as to permanence of supply. Examples of this include the High Plains area of Texas and most of the State of New Mexico.

Even the slow drainage of aquicludes is now included in the evaluation of water supplies, using the potentiometric methods pioneered by Theis (1935). One result of such slow drainage is the gradual subsidence of the land surface as evidenced in the San Joaquin Valley of California, the Houston—Galveston area of Texas, and the New Orleans area. Not so well recognized are the changes in water quality in the aquifer that result from the compaction of the clays. The water expelled by the pressure of the overlying materials is invariably different in composition from the native groundwater. Moreover, as Jones (1969) has shown, the water derived from the clays and shales is of varying quality: the most mineralized water is expelled from the interstices first and as compaction progresses the TDS content of the expelled water slowly decreases.

In short, potentiometric methods and analysis, useful as they have been and will continue to be, can be considered coarse-grain, or "macro" methods. As the needs of society change, as human intervention shapes the environment that human beings must inhabit, the coarse-grain methods must be supplemented by more precise, more complex methods of analysis — such approaches might be termed "fine-grain" or "micro" approaches.

It is always professionally hazardous to predict the future course of research and application in any field and the field of groundwater hydrology, now undergoing a great expansion, is no exception. At the boundary there is an overlap with petroleum engineering, in another dimension there is geochemistry, in still another dimension the field of clay mineralogy impinges, in yet another problems of heat flow and disposal of heat are important, in still another area there is an overlap with the discipline formerly known as soil mechanics (geotechnology), and so it goes. It might be said that groundwater is a multi-dimensional science adding to the four dimensions of height, width and length and time, at least the additional parameters of chemical reaction, clay mineaology and multi-liquid flow (miscible displacement of waters of differing densities, not just freshwater and seawater).

Let us first expand our viewpoint to include, as groundwater, all water found in the interstices of the geologic matrix. This would exclude oil and gas from our purview but would include all sorts of mineralized, i.e. non-potable, water, water which, with few exceptions, during most of Maxey's professional life was not included as "groundwater" at all.

Not all of subsurface fluids can be considered as industrial minerals stocked underground by nature. We are well acquainted with such natural "stocks": freshwater, slightly saline water, and bromine-laden brines come to mind. But the pore spaces of geologic strata area, for most part, full of unusable liquids. In such instances it is not the naturally occurring fluid, but the pore space that it occupies, that must be regarded as the resource of interest.

Brine disposal

The first use of such pore space was by the petroleum industry: vast quantities of brine were produced in association with the desired oil, and the operators had to get rid of it. What better place than some overlying formation that already was full of salty water? By one estimate some 400 million barrels ($6.4 \cdot 10^7$ m^3) of brine are currently injected into the ground each day, some of which was to maintain the field pressure of the oil reservoir, but a good fraction of which was to keep the brine from contaminating the biosphere. Such operations have been in progress for more than thirty years, and have received relatively little attention from the majority of groundwater engineers or hydrogeologists.

Waste disposal

In the early 1950's the petrochemical and chemical industries found themselves with volumes of intractible liquid wastes, large volumes by their standards. With the example of the reservoir engineers of the petroleum industry before them, they searched for, and found, saline aquifers that could be used as receptacles for these wastes. Judging from the number of failures due to plugging of the injection wells and deterioration of screens, the writer has the impression that most of these wells were built using oil-field rather than water-well technology — and despite certain overlaps, there is a great difference of design and approach. For the vast majority of groundwater hydrologists (both engineers and geologists), waste disposal was, and still (1978) is, terra incognita. Understandably this is so since the number of industrial and commercial water wells drilled each year approaches 150,000, while the total number of wells for the disposal of industrial waste drilled since the first one was installed some thirty years ago, including those abandoned due to plugging is probably less than 1000. Yet there is an economic demand for a method of removing such hard-to-treat, possibly poisonous, liquids from the biosphere. And use of underground pore space, space that is now filled with a useless, mildly poisonous liquid (if various types of salty water can be considered mildly poisonous), has a number of quantifiable advantages: on-site disposal, reduction of hazard to the general population because there is no need to transport containers of waste over public roads or railroad roadbeds where accidents can and do occur, and preserving the land surface for other uses than those of waste storage. The practitioners of ground-water hydrology must apply their knowledge and experience to the tasks of making the disposal of wastes underground not only economic, but safe. Parameters such as geochemical reactions, density difference, viscosity difference, dispersivity, porosity, dip, location of faults, pressure rises, osmotic flows and porosity, must all evaluated in addition to the parameters familiar to groundwater hydrologists, transmissivity and storativity. Of course, testing for the competence of the confining beds cannot be neglected — and this takes care of such a parameter as "leakance".

Aquifer storage of natural gas

However, waste disposal is not the only current use of underground space: there is also the storage of natural gas. It came about in this way: in the early 1950's when plentiful gas supplies were available, pipeline companies found themselves with surplus pipeline capacity during periods of hot weather or periods of industrial slow-down. Yet the pipelines were unable to transport gas fast enough during severely cold weather or other periods when demand peaked. There was a need for storage of gas near the point of use, so that temporary peaks could be satisfied by withdrawals from storage. The logical storage reservoirs were, of course, depleted gas reservoirs. Gas could be pumped in during the off-peak periods, and withdrawn during peak periods.

Unfortunately there was one small problem: near the major points of gas utilization, abandoned gas reservoirs were in short supply. However, there were salaquifers nearby and it occurred to someone that it would be a good idea to pump gas into the aquifer, force the water away from the injection well, increase the pressure of the aquifer water, and let the gas sit there until needed. Inasmuch as the bottom hole pressure was usually measured in the hundreds of pounds per square inch (depending on depth of aquifer), the gas would flow under pressure to the surface where it could be compressed to main-line pressures and injected into the pipeline. The geologic search for water-filled, porous structures was successful and about 20% of all gas provided for peaking in the U.S.A. is from aquifer storage (American Gas Association, 1974).

The story would not be complete without mention of the major economic problem: the problem of "cushion gas". There is an analog here with a familiar phenomenon: salt water is denser than freshwater so when you pump freshwater from an aquifer that also contains underlying salty water, water that is in hydraulic contact with the freshwater, if you pump at too great a rate a phenomenon known as "coning" occurs. Thus all of the freshwater is unrecoverable. Similarly with gas, only more so. Experience has shown that about two-thirds of the volume in gas originally stored must remain in storage until the project is abandoned at which time some fraction of the "cushion storage" may be ultimately recovered. As long as natural gas was selling at ten cents per thousand cubic feet, this was not too expensive: most of the retail cost of natural gas originates in the cost of transportation (pipelines, compressors, etc.) and the well-head price was only a small fraction of the price to the consumer. But with the OPEC caused rise in the price of oil and natural gas, the cost of cushion gas became a major capital expenditure amounting, it is estimated, to more than half the cost of the storage project (rather than about 30%).

Many suggestions have been made to reduce the cost of cushion gas, one of which was to inject a large volume of cheap flue gas to act as cushion gas, and to have the zone of active gas consist of methane. There is still much work to be done in the area of miscible displacement so that the mixing of the flue gas and natural gas can be predicted and minimized (this affects the economics of any project). So here we have an instance of the utilization of geologic pore volume combined with the displacement of saline water by gas and, in addition, the displacement of one gas by another.

Leach mining

Leach mining, or in-situ leaching, is accomplished by injecting a solvent (usually an acid) that selectively dissolves the substance of interest, normally uranium or copper. In practice several wells are drilled around a central extraction well. The peripheral wells are used as injection wells for the acid solution that will dissolve the uranium or copper from the country rock. The enriched solution is pumped out of the central well and the metal is extracted by electrolysis. The technique has the great advantage of not disturbing the land surface. Moreover, it makes available, as ore, concentrations of material that would be uneconomic to extract by normal material-handling mining methods be they-open pit mining or underground mining.

The hydraulics of the system is analogous to the secondary recovery operations practiced by the oil industry except that it is a miscible displacement process rather than an immiscible displacement process. Thus the recovery, theoretically, is far superior to that of oil from secondary recovery operations. In designing such a system the sweep efficiency is of utmost importance (sweep efficiency is crudely defined as the percent of area where the original fluid is replaced by the injected fluid, divided by the total area between the injection wells and the production well). Acid is expensive and the concentration of metal in the spent solution must be great enough to justify the operation. The leaching of uranium from its sand, as well as the leach-mining of copper from waste piles, can be included as novel applications of ground-water hydraulics.

Water storage

Yet, I believe, the main thrust in the utilization of underground pore space will be in the field of water. One of the principal problems of water supply is provision of adequate storage, both for temporary peaking and for long-term year-to-year carry over. Until now this requirement has invariably been met by the damming of streams, building artifical lakes, or, in extreme instances, construction of steel or concrete tankage.

There are two rather somber viewpoints on the storage of water in reservoirs: that from a long-term viewpoint surface reservoirs are a wasting asset due to the deposition of silt and the consequent, inevitable reduction in storage capacity and second, even if siltation were not a problem, the availability of reservoir sites where such reservoirs are needed, in urbanized areas where the population is concentrated, is sharply limited and in the future another method has to be found to store water. It is not easy to produce refutations of either of these arguments. And, if the arguments are valid, then hydrogeology has much to contribute to the continued production of potable water, for beneath our feet we can usually find aquifers filled with saline water. And since the volume of freshwater to be stored is small relative to the volume of saline water already contained in an aquifer, the pressure change to be expected from the storage of freshwater in a salaquifer should not be large. Thus the response of the groundwater hydrologist and engineer must be development of a method to store freshwater in a salaquifer and to retrieve the water as needed after some storage period.

The first requirement is, of course, that the freshwater be free from particulates — and this requirement can be satisfied by use of filtered water, as from a city filtration plant. And, in fact, Cederstrom (1947) injected freshwater into a brackish aquifer in Virginia, a decade later, Moulder and Frazor (1957) also made such a test and, more recently, Brown and Silvey (1973) reported on a similar test in Norfolk, Virginia. As this is being written the writer and his colleagues at Louisiana State University, are supervising the preliminary drilling and testing prior to undertaking a similar test in Houma, Louisiana.

What then are the problems? Why, in the thirty-odd years since Cederstrom's first test has there been no wide-scale application of the concept of freshwater storage in salaquifers? The shortest answer is that there is a long distance between a successful experiment and a general procedure that can be used by engineers and geologists under a variety of conditions. For one thing, design principles have been lacking. More importantly, a theoretical generalized approach to the problem that includes all the parameters involved has yet to be developed, although much progress has been made.

To the normal parameters involved in groundwater studies (including transmissivity, storativity and determination of boundaries) we must add porosity, dispersivity, density difference, geopressure, and such geochemical attributes as ion-exchange capacity of silts and clays, interaction between injected freshwater and the water and solids of the salaquifer, the dissolved oxygen content of the freshwater and its chlorine content, and the pH of the fresh and saline waters — to mention only a few. There are also several interesting problems involved in well screen design, in the construction and operation of the essential monitoring wells, in the operation of bounding wells to create an isopotential plateau and thus counteract the effect of pre-existing groundwater flow, and in the effect of viscosity differences between the salaquifer water and the injected water. Some of these parameters were discussed by Kimbler et al. (1975), still others are just now being worked on.

We can clearly discern the outlines of a new field of hydrogeologic practice, one that will become ever more important as surface reservoir sites become scarce and less accessible. A new era in groundwater development is evidently beginning. Its principal characteristic is a complexity beyond the reach of classical mathematical techniques — without the step-wise computational approach made available by the digital computer, the new era would never become a reality. Another characteristic is that the problems involved are only slightly generic — the problems are primarily site-specific. A third distinguishing characteristic is that geochemical reactions, of one sort or another are critical. All of these parameters are to be added to those that we are already familiar with in hydrogeology.

In designing such a project we can use the average characteristics of the aquifer to solve the potentiometric problems associated with injection, storage and pumpage: we are all familiar with the superposition of effects produced by the pumping of wells. Instead of a cone of depression, one obtains a cone of impression, meaning that the direction of potentiometric change is up rather than down. Thus the cost of power can be predetermined based on the results of field tests. And the rates of injection and withdrawal can be specified with an accuracy and reliability that is adequate for engineering purposes.

Then we are faced with some less tractable problems: the freshwater can disperse the clays and silts in the aquifer matrix and thus lead to the plugging of the injection well. If the freshwater is oxygen-laden and the aquifer contains ferrous iron, the iron will precipitate and plug the formation — a free chlorine residual will accomplish the same result. Ion-exchange reactions will occur, the extent and importance of which will vary as the characteristics of the clay involved vary. For example, injection of a calcium-bearing water into an aquifer that contains a small amount of montmorillonite clay will cause an ion-exchange reaction between calcium and sodium (softening of the water), so the water that is pumped out will be different in character from the water that was stored. Moreover obscure sorptions of the minor constituents of the freshwater will occur and minor releases of trace elements from the aquifer matrix will also occur — sometimes this will be beneficial, sometimes deleterious. The saving grade may be that with each succeeding I—S—P cycle (called an I—S—P cycle because it involves injection, storage and production) such geochemical problems may decrease. If this is verified by experience then ultimately the quality of water produced will be very close to the quality of water injected — but the duration of a cycle may be a year or more, so the hoped-for improvement may be a long time in arriving. Only the future will reveal whether this analysis is correct — and it probably will not apply everywhere.

At the present time we are not yet certain as to the field-test procedures and it may well be that after we have tested all of the critical parameters that we now know about, a small-scale injection—production test, using the water that is to be stored, will be needed before the final design of the I—S—P well can proceed. In any event an entirely new and productive area of research including hydrogeology, geochemistry and clay minerology is now opening up.

Heat storage wells

The increases in the cost of fuel, the problems with exotic new sources of energy, be they nuclear, solar, or geothermal, have forcibly brought to the attention of engineers and geologists the need for energy conservation. Examination of the present uses of energy shows that about 40% of our energy requirements are "low grade". This means that for all practical purposes we do not need source temperatures of 150°C or higher: we need heat mainly for space heating, preheating of industrial materials, and water heating. Only minor quantities of fuel are used for cooking in the home and even in industry the need for very high temperature requires only a small percentage of the total fuel used.

The industrial use of fuel for the generation of electric power, for example, requires the rejection of about 60% of all of the energy in the fuel — it appears in the form of waste heat from power plants and constitutes what is termed "thermal pollution" of our surface waters. So large a percentage of our energy consumption is presently "wasted" that the possibility of salvaging even as little as 10% would have a discernable impact on our fuel requirements. (The word "wasted" is in quotation marks because as long as fuel was cheap and plentiful there was no economic reason to try to salvage it and from the standpoint of economics there was no waste involved.)

If we could store even a part of the energy that now goes to the atmosphere or surface water and later retrieve if for use, it would be worthwhile. Such a process would require that the temperature of the water be raised by the waste heat, and the water then be injected into a heat storage well or well field, and later retrieved for use. Essentially this would amount to the artificial establishment of geothermal sources of heat near our industrial centers and using them for purposes where low-temperature heat is most appropriate. Couple such a source with heat pumps as needed to produce even hotter water, or with an absorption cooling refrigeration system, and such an artificial geothermal source would satisfy the domestic requirements for heat in every respect except heat for cooking.

This, then, is the prospect before us: the prize is worthwhile and success depends on our ability to solve the problems of the subsurface environment.

The task, however, is a formidable one. If the storage of freshwater in saline aquifers seems to be challenging, the storage of hot water in saline aquifers seems to be challenging, the storage of hot water in saline aquifers is a challenge whose complexity is an order of magnitude higher. First, why saline aquifers? Well, for one thing, they are not being used and the utilization of the water in a salaquifer is expected to be trivial as compared to the use of water in a pumped aquifer. Thus there will be no conflict between current users of a potable aquifer and the organization planning to store water in an underlying saline aquifer. Moreover, the cost of maintaining an isopotential surface and thus the position of the "bubble" of hot water in the storage aquifer, will be small. Second, in many, if not most, sedimentary basins, saline aquifers underlie the shallower aquifers that contain freshwater. So there are storage aquifers available in the areas that are industrialized and produce waste heat. The local populations can utilize the new, artificial thermal source.

There are a number of interesting technical questions associated with the storage of heated water in a saline aquifer. The first is the state of the hot water: is it sub-boiling in temperature or is it superheated? To reduce the scope of this discussion, let us assume that the water is to be stored at a sub-boiling temperature, for example, 90°C.

The next question is: where is the water that is to be heated to come from: will it be freshwater from a shallow aquifer, water from a surface source, or water from a saline aquifer (possibly the same one in which the heated water is to be stored)? Next question: what is to be done with the spent water, now at a temperature of 50° or 60°C?

Hot water closely resembles the universal solvent. So when it is injected into the salaquifer it will tend to dissolve a portion of the surrounding matrix. As the water particles travel further and further from the injection well they will exchange heat with the solid material and will mix with the native water, and their temperature will decrease. Some of the material that was dissolved will reprecipitate. As injection continues the heated zone will increase, there will be a continuous process of solution, deposition, resolution, redeposition, and so on. Simultaneously there will be heat losses to the roof and floor of the salaquifer. Since the hot water is less viscous by a factor of three or four than the native water, small differences in aquifer permeability will result in the birth of "fingers" of injected water. These fingers will, in turn, give birth to additional fingers as the hotter water moves more rapidly in the aquifer's most permeable zones. The normal mixed zone of injected and native water will increase dramatically and the apparent dispersivitity of the salaquifer will increase.

About half of the heat, however, will be stored in the solid material in the vicinity of the well or well field. So around the point or points of injection there will be a zone where the temperature will closely approach the temperature of the injected fluid. Most of this heat will be recaptured when the production phase of the cycle is reached.

The higher temperature of the injected water will undoubtedly alter the characteristics of the clays and silts included in the aquifer and may well do the same for the proximate portions of the confining layers. Just what these reactions will do to the feasibility of the project cannot be predicted now. It all depends on the type, position and volume of the clay and silts fraction. Of course many chemical reactions proceed more rapidly under conditions of elevated temperature and pressure than they do under the original temperature of the aquifer.

The monitoring of the potentiometric and geochemical processes in such an I—S—P project will turn out to be quite a technical achievement. Instead of water-level measurements we will have to measure pressures by means of transducers in monitoring wells. It is quite possible that some sort of non-reactive material, such as PVC, will prove suitable for use in the construction of monitoring wells. But the problem of obtaining water samples for chemical analyses will be complicated by the need to maintain both temperature and pressure of the sample, so that the fluid does not alter in character between the moment of sampling and the time of analysis. We may recognize this problem as an intensification of one that we have already encountered: determination of the in situ iron, pH, and free carbon dioxide in water from a deep artesian aquifer. Much ingenuity and experimentation will be needed before standard, effective methods of sampling are devised.

There will be a need to prepare mathematical descriptions of the coupled processes involved. The effect of density difference (bouyancy) must be included, the effect of a varying viscosity (that varies with both time and space) must be, somehow, expressed mathematically, probably by some sort of a step-wise approximation of varying transmissivity. The flow of heat through the upper and lower boundaries of the aquifer, which not only affects the thermal recovery ratio, but also has a direct effect on flow patterns, must somehow be brought into the computational process. There may also be some artificially generated osmotic flows through the confining layers and it is not clear to me how these are to be detected, predicted or taken into account. The flow of heat from the aquifer while the hot water is in storage must also be taken into account.

There is also a small matter of material balance, in this instance, water. Since there is a lower temperature limit to the utility of the stored water, production can be expected to stop when the produced water reaches a temperature of about $60°C$. This means that cycle after cycle more water will be injected into the well field than will be removed. Sooner or later some of this excess water must be bled off. What do you do with it and from from what part of the aquifer do you remove it? If you do not do something the pressure in the aquifer will slowly rise and the entire operation will become more and more expensive. It is true, however, that a vast volume of water will be needed before the rise in the potentiometric surface will cause operational problems — and the thicker the aquifer, the longer it will take for these problems to manifest themselves. There are some important economic parameters that stem from this material balance: some heat will inevitably be lost together with the surplus water injected into the aquifer and this phenomenon will have to be quantified on a case-by-case basis in order to accomplish an economic evaluation.

SUMMARY AND OUTLOOK

The groundwater technology based primarily on potentiometric relationships, that was developed during Maxey's professional lifetime, seems to have matured and its practice codified. Included in this rubric are such undertakings as the prediction of well-field yields, recharge, detection of no-flow boundaries, drains, dewatering of underground mines and surface excavations, and the conjunctive use of surface and groundwater. Prevention of seawater intrusion through the creation of pressure ridges or pumping throughs is a relatively minor area of practice. The evaluation of freshwater movement under the influence of osmotic forces has not yet been undertaken, although the phenomenon is recognized.

The upcoming area of groundwater research and technology is in the evaluation and application of the process of miscible displacement. Included here are the movements of leachate from landfills, deep-well disposal of liquid wastes, storage of gas in saline aquifers, leach-mining of non-ferrous metals, the storage of freshwater in saline aquifers, and the possible storage of waste heat in saline aquifers. All of these possibilities, if achieved, offer a rich reward to society. All of these possibilities offer unexplored fields of research to the imaginative and innovative investigator. All of these possibilities offer an unexcelled opportunity for failure. The era of the use of miscible displacement in porous media is just beginning. Burke Maxey would have responded to the challenge and contributed much, had he lived. My generation is counting on Burke's former students and protegees, among others, to undertake and successfully complete the needed research. Good luck.

REFERENCES

American Gas Association, 1974. Underground storage of gas in the United States and Canada. American Gas Association Committee on Underground Storage, 23rd Annu. Rep. Stat., Rep. XV 0274, 22 pp.
Brown, D.C. and Silvey, W.D., 1973. Underground storage and retrieval of fresh water from a brackish aquifer. Am. Assoc. Pet. Geol. 2nd Int. Symp. on Underground Waste Management and Artificial Recharge, New Orleans, La., 1: 379—419.
Cederstrom, D.J., 1947. Artificial recharge of a brackish water well. The Commonwealth, Va., Dec. 1947, pp. 31, 71—73.

Glover, R.E., 1974. Transient ground water hydraulics. Dep. Civ. Eng., Colo. State Univ. Fort Collins, Colo., 413 pp.

Jones, P.H., 1969. Hydrology of Neogene deposits in the Gulf of Mexico Basin. Louisiana State University, Baton Rouge, La., La. Water Resour. Res. Inst. Bull. GT-2, 105 pp.

Kimbler, O.K., Kazmann, R.G. and Whitehead, W.R., 1975. Cyclic storage of fresh water in saline aquifers. La. Water Resour. Res. Inst., Bull. 10, 78 pp. (5 appendices).

Maasland, D.E.L. and Bittinger, M.W., (Editors), 1963. Proc. Symp. on Transient Ground Water Hydraulics. Colo. State Univ., Fort Collins, Colo., 223 pp.

McGuinness, C.L., 1963. Ground water in the national water situation. U.S. Geol. Surv., Water-Supply Pap., 1800.

Moulder, E.A. and Frazor, D.R., 1957. Artificial recharge experiments at McDonald well field, Amarrillo, Texas. Texas Board Water Eng. Bull., 5701.

Theis, C.V., 1935. The relation between the lowering of the piezometric surface and the rate and duration of discharge of a well using ground-water storage. Am. Geophys. Union, Trans., 16: 519—524.